Phycoremediation of Wastewater

Phycoremediation is an alternative method of water and wastewater remediation, which includes the use of algae for treatment, and is an environmentally friendly and sustainable technology. More conventional methods of wastewater treatment have been successful in the removal of conventional contaminants from the water; however, these techniques typically require more time and energy than phycoremediation. *Phycoremediation of Wastewater: Practical Applications for Sustainability* focuses on the latest developments in water remediation as well as the major challenges faced by municipalities implementing large-scale phycoremediation operations. It addresses the latest advancements in the field as well as the future applications and techniques to make water remediation processes more environmentally sustainable.

- It focuses on the latest developments in phycoremediation and outlines the major challenges in large-scale operation and implementation.
- It explores the future scope of the remediation techniques to make processes more sustainable going forward.

Dr. Maulin P. Shah has been an active researcher and scientific writer in his field for over 20 years. He received a B.Sc. (1999) in Microbiology from Gujarat University, Godhra (Gujarat), India. He also earned his Ph.D. (2005) in Environmental Microbiology from Sardar Patel University, Vallabh Vidyanagar (Gujarat), India. His research interests include biological wastewater treatment, environmental microbiology, biodegradation, bioremediation, and phytoremediation of environmental pollutants from industrial wastewaters. He has published more than 350 research papers in national and international journals of repute on various aspects of microbial biodegradation and bioremediation of environmental pollutants. He is the editor of 150 books of international repute, and he has edited 25 special issues, specifically in industrial wastewater research, microbial remediation, and biorefinery of wastewater treatment area. He is associated as an editorial board member in 25 highly reputed journals.

Dr. Günay Yıldız Töre has a Ph.D. in Environmental Science. She completed her postdoctoral position at Istanbul Technical University, and is now a professor at Tekirdağ Namık Kemal University. Her research interests include the biological treatment of industrial wastewater, industrial waste management, industrial reclamation and reuse with emerging treatment technologies. She has published papers in international peer-reviewed journals as well as more than 20 national projects about industrial wastewater treatment and waste management. She has also published one book titled *Advanced Oxidation Processes for Wastewater Treatment* and several book chapters. Currently, she is studying antibiotic and microplastic removal with emerging treatment technologies from treated urban and industrial wastewater for agricultural reuse and its effect on activated sludge.

Phycoremediation of Wastewater

Practical Applications for Sustainability

Edited by

Maulin P. Shah and Günay Yıldız Töre

CRC Press
Taylor & Francis Group
Boca Raton London New York

CRC Press is an imprint of the
Taylor & Francis Group, an **Informa** business

Designed cover image: CRC Press

First edition published 2025
by CRC Press
6000 Broken Sound Parkway NW, Suite 300, Boca Raton, FL 33487-2742

and by CRC Press
4 Park Square, Milton Park, Abingdon, Oxon, OX14 4RN

CRC Press is an imprint of Taylor & Francis Group, LLC

© 2025 selection and editorial matter, Maulin P. Shah and Günay Yıldız Töre; individual chapters, the contributors

ISBN: 978-1-032-48675-8 (hbk)
ISBN: 978-1-032-48676-5 (pbk)
ISBN: 978-1-003-39021-3 (ebk)

DOI: 10.1201/9781003390213

Typeset in Times
by codeMantra

Contents

List of Contributors

Pınar Akdoğan Şirin
Fatsa Faculty of Marine Science
Department of Fisheries Technology
  Engineering
Ordu University, 52400 Fatsa, Ordu,
  Turkey

Vincent Braganza
Loyola Centre for Research and
  Development
Xavier Research Foundation
St. Xavier's College Campus
Navrangpura, Ahmedabad, Gujarat,
  India

Meltem Çelen
Gebze Technical University
Institute of Earth and Marine Sciences
Gebze Kocaeli, Turkey

Ankita Chatterjee
Department of Biotechnology
School of Applied Sciences
REVA University, Bangalore,
  Karnataka, India

Shailesh R. Dave
Loyola Centre for Research and
  Development
Xavier Research Foundation
St. Xavier's College Campus
Navrangpura, Ahmedabad, Gujarat,
  India

Yousra A. El-Maradny
Biotechnology department
City of Scientific Research and
  Technological Applications
Borg El Arab, Alexandrina, Egypt

Mohamed A. Etman
Professor of Pharmaceutics
Educational Department Head
Arab academy for Science, Technology
  and Maritime Transport (AASTMT),
  Alamein, Egypt

Kaniye Güneş
Izmir Institute of Technology University
Faculty of Engineering
Department of Environmental
  Engineering, İzmir, Turkey

Mehmet Ali Gürbüz
TAGEM
Atatürk Soil, Water and Agricultural
  Meteorology Research Institute
  Directorate
Kırklareli Turkey

Muhammad Uzair Javed
Institute of Industrial Biotechnology
Government College University
Lahore, Pakistan

B. Batuhan Kaplangı
Izmir Institute of Technology University
Faculty of Engineering
Department of Environmental
  Engineering, İzmir, Turkey

Meltem Kizilca Coruh
Department of Chemistry
Faculty of Science
Ataturk University, Erzurum, Turkey

Mehmet Ali Kucuker
Izmir Institute of Technology University
Faculty of Engineering
Department of Environmental
  Engineering, İzmir, Turkey

Afeefa Khalid
Professor of Biotechnology
Institute of Industrial Biotechnology
Government College University, Lahore,
    Pakistan

Dina M. Mahdy
Pharmaceutical science park unit
Faculty of Pharmacy, Alexandria
    University, Egypt

Tehreem Mahmood
Department of Biotechnology
Quaid-i-Azam University, Islamabad,
    Pakistan

Deepika Malik
Department of Food Science
    (Microbial & Food Technology)
Mehr Chand Mahajan DAV College for
    Women
Sector 36A, Chandigarh

Hamid Mukhtar
Professor of Biotechnology
Institute of Industrial Biotechnology,
Government College University, Lahore,
    Pakistan

Enes Özgenç
Trakya University
Health Vocational School

Altan Özkan
Izmir Institute of Technology University
Faculty of Engineering
Department of Environmental
    Engineering, İzmir, Turkey

Anjali V. Prajapati
Department of Microbiology and
    Biotechnology
School of Sciences
Gujarat University, Ahmedabad,
    Gujarat, India

Sudha Sahay
Loyola Centre for Research and
    Development
Xavier Research Foundation
St. Xavier's College Campus
Navrangpura, Ahmedabad, Gujarat,
    India

Gargi Sarkar
Medical Laboratory Technology
Institute of Genetic Engineering
Badu, Kolkata, West Bengal, India

Devayani R. Tipre
Department of Microbiology and
    Biotechnology
School of Sciences
Gujarat University, Ahmedabad, India

1 Mechanisms and Major Influencing Factors for Phycoremediation of Metallic Pollutants from Industrial Effluents

Anjali V. Prajapati, Shailesh R. Dave, and Devayani R. Tipre

1 INTRODUCTION

Worldwide, over 80% of wastewater is discharged into the ecosystem without proper treatment, which accounts for nearly 28% of industrial wastewater containing many metals as pollutants (Mao et al., 2022). The use of water from such heavy metal-contaminated sources leads to several health issues for living beings, including humans. Heavy metals contain an atomic weight between 63.5 and 200.6 amu and a specific density >5 g/cm^3 (Briffa et al., 2020; Al-Dhabi & Arasu, 2022; Rajoria et al., 2022). They are categorized into two parts: essential (copper, iron, nickel, magnesium, and zinc) and nonessential metals/metalloids (arsenic, cadmium, lead, mercury, and tin) (Shamim, 2018). Heavy metals are introduced into environments through various anthropogenic activities, but industrial effluents are major contributors to the heavy metal contamination of soil and water, which not only contaminates the groundwater but also enters the food chain. Many plants can accumulate nonessential metals like lead, cadmium, and mercury, which are toxic to humans, through ingestion of metal-contaminated water and plants (Bauddh & Korstad, 2022). Thus, remediation of metallic pollutants from the contaminated aquatic system has become essential to provide pollutant-free water that is appropriate for human health and the environment.

Conventional techniques have been employed for effluent treatment and metal removal for many years. However, these strategies are not only costly and huge amounts of chemical sludge, which is a major concern to treat. Removal of metal contaminants can be done by conventional techniques such as coagulation/flocculation, ion exchange, adsorption, chemical precipitation, membrane filtration, polymer micro-encapsulations, and current techniques such as electrochemical techniques, electrocoagulation, electrodeposition, electrodialysis, electroflotation, electrooxidation, photocatalysis, and the electron-Fenton method. Figure 1.1 illustrates an overview of the major conventional techniques used for metal remediation.

DOI: 10.1201/9781003390213-1

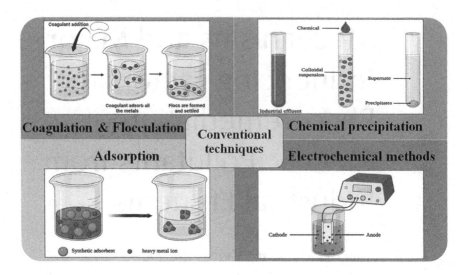

FIGURE 1.1   General overview and mechanisms of conventional techniques.

However, all these techniques produce high chemical waste and require higher operating costs and current efficiency (Barquilha et al., 2019; Jaafari & Yaghmaeian, 2019; Plöhn et al., 2021; Rajoria et al., 2022; Razzak et al., 2022). Moreover, many of these techniques are inadequate to meet the required regulatory standards. So there is an urgent need for innovative, sustainable technology for metal removal from polluted water.

Biosorption is an economical and eco-friendly concept for removing heavy metals from an aqueous system up to a very minute concentration. Many natural adsorbents, like plants, fungi, algae, and bacteria, have been utilized for the biosorption of heavy metals in the last 25 years. Biological adsorbents are the most promising material for metal sorption, leading to more green practices (Shah et al., 2017; Khan et al., 2022a; Al-Dhabi & Arasu, 2022). Among biological sorbents, algal biomass has several essential features to be considered an ideal biosorbent.

This chapter reviews the recent developments in the field of microalgae-based biosorption of various heavy and nonessential metals. In this context, the chapter first focuses on identifying different potential metal contaminants generated by various industries and their effects on humans and the environment. Thereafter, different microalgae were reported for metal removal and their detailed mechanisms for the biosorption of heavy metals will be discussed. The chapter also discusses various factors that affect the biosorption process, pre-treatments for improving sorption efficiency, major pros and cons of using live and non-living biomass, future directions, and the possibility of commercialization.

1.2   SOURCES OF METAL POLLUTION

Heavy metals are generally found in the earth's crust in various forms. Natural heavy metal pollution may be caused by volcanic activities, metal corrosion, soil erosion, and geological weathering (Briffa et al., 2020). Rapid urbanization and industrialization are causing an imminent surge of heavy metal pollution. In developing countries, major contributors to heavy metal pollution are electroplating, metallurgical, textile and dying industries, tannery industries, and many chemical industries as well as mining operations (Rajoria et al., 2022). The discharge of effluent from textile industries is highly contaminated with various types of organic pollutants and heavy metals. Approximately 35 billion tons of wastewater are generated by textile industries (Mubashar et al., 2020). Another major contributor to heavy metal pollution is the electroplating industry. These industries discharge wastewater containing around 29% of toxic metals and waste. Electroplating industrial effluent contains heavy metals/metalloids such as arsenic (As), cadmium (Cd), chromium (Cr), cobalt (Co), copper (Cu), mercury (Hg), and nickel (Ni) (Rajoria et al., 2022). The paint, paper, leather tanning, and steel fabrication industries are causing chromium pollution. Lead pollution is primarily caused by waste effluents from the petroleum and battery industries. Cadmium pollution is mainly caused by the cement, ceramic, and metallurgical industries, as well as by fuel combustion. It is highly toxic to the environment and living beings (Soliman & Moustafa, 2020). Dye manufacturing industrial effluent contains copper as a major toxic compound; chromium, zinc, and lead are the toxic pollutants found in paint manufacturing industries. Partially treated industrial effluents containing metal when thrown into an ecosystem can be carcinogenic and teratogenic; they may cause oxidative stress, organ damage, nervous system failure, a reduction in the body weight of neonates, delayed eye opening, retarded growth and development (Ahmed et al., 2021; Alengebawy et al., 2021).

1.3 ALGAE INVOLVED IN METAL REMEDIATION FROM INDUSTRIAL EFFLUENT

Plentiful microalgae are naturally occurring unicellular organisms in various freshwater resources like lakes, ponds, and rivers, as well as being found in waste or contaminated water. Various types of microalgae have been studied based on their application point of view worldwide. They are photosynthetic organisms. Microalgae have been used many times for the removal of various organic and inorganic metal pollutants. According to the literature, more than 50,000 algal species are found as biosorbent material for remediation purposes (Spain et al., 2021; Khan et al., 2022a). Microalgae are widely used for domestic wastewater treatments as well as industrial effluents; however, industrial effluents have more toxic pollutants compared to fresh water, which retard the growth of microalgae, so the use of live biomass is tedious and has several limitations. Inactive or non-living biomass can be cheaper, more economical, and reusable for many cycles with metal recovery comparable to the use of live biomass (Zhou et al., 2021; Premaratne et al., 2021; Priya et al., 2022). The application of microalgae as biosorbent materials has various advantages: (1) a variety of functional groups present on the cellular surface; (2) a few steps are required in the preparation of biosorbent; (3) there are no requirements for hazardous chemicals; (4) easy production and reusable biosorbent; and (5) a low amount of chemical sludge production (Bilal et al., 2018). Species of *Chlorella, Chlamydomonas, Desmodesmus, Spirulina, Oscillatoria, Microspora, Scenedesmus, Spirogyra, Gelidium, Chlorophyceae, Cystoseira, Nannochloropsi*etc, *Aphanotheca*, etc. have been reported for their biosorption of heavy metals. Even several *Cyanobacterial* species are reported for metal removal from wastewater, including *Spirulina, Oscillatoria,* and *Microspora* sp. (Urrutia et al., 2019; Khan et al., 2022a). Some details about the pH, temperature, and sorption magnitude of microalgae recently reported for metal removal from various industrial and domestic wastewater are shown in Table 1.1. Species of *Chlorella, Scenedesmus, Chlamydomonas, Spirogyra,* and *Spirulina* showed the capacity to remove Fe (all at most 100%), Zn (98%), Cu (76%), and Pb (78%). Even a consortium of microalgae efficiently removes various metal pollutants (Samal et al., 2020). In the case of dead algal biomass, the cell surface and functional groups present on it are responsible for the metal sorption. Different metals bind to a specific functional group via various interactions. In the case of live biomass, intracellular uptake of metal takes place along with sorption on the cell surface (Shamim, 2018).

1.4 MECHANISMS OF METAL REMOVAL

Microalgae have potential applicability in removing metal ions from various types of metal-contaminated wastewater and industrial effluents. Many studies have investigated the efficient removal of metals from the aqueous system as potential sorbent materials. Microalgae possess various mechanisms for heavy metal removal. These mechanisms could be divided into two major types: (1) metabolism-independent mechanisms and (2) metabolism-dependent mechanisms. The general overview of mechanistic approaches carried out by microalgae is shown in Figure 1.2a and b (Costa & Tavares, 2018; Barquilha et al., 2019; Tripathi & Poluri, 2021; Ahmad et al., 2022; Razzak et al., 2022).

1.4.1 METABOLISM-INDEPENDENT MECHANISMS

Metabolism-independent processes are generally physicochemical. Microalgae can potentially remove heavy metals through biosorption by microprecipitation; apart from microprecipitation complexation, adsorption and ion exchange play a significant role in metal sorption. These mechanisms are discussed in brief below.

1.4.1.1 Microprecipitation

Microalgae generally contain sulfate and phosphate groups on their cellular surfaces. The metal ions present in an aqueous solution may precipitate. These metal precipitates can easily be absorbed through the biosorption process by microalga. Due to metal toxicity, many microalgae produce certain compounds that enhance the formation of metal precipitation to overcome the toxic effect of metal ions. However, microprecipitation involves both active and passive metal uptake (Javanbakht et al., 2014; Tripathi & Poluri, 2021; Danouche et al., 2021).

1.4.1.2 Complexation

Microalgae show the ability to form a metal complex on their cellular surface with the help of functional groups present on it. The functional groups that play a vital role in the complexation process are hydroxyl, sulfhydryl, phosphate, carbonyl,

TABLE 1.1

Recent Discoveries in the Field of Biosorption of Heavy Metals by Various Microalgae

Algal Strain	Metal	pH	Temperature (°C)	Biosorbent (g/L)	Adsorption Capacity (mg/g)	Metal Removal (%)	Remarks	References
Chlorella vulgaris	Cd	-	25	-	-	95.2 (L) 96.8 (D)	Dead biomass and live biomass	Cheng et al. (2017)
Chlorella vulgaris	Cu Ni	-	22	-	-	39 32	Incubation time - 12 days; live biomass	Rugnini et al. (2017)
Gelidium amansii	Pb	4.5	-	-	-	100	Contact time – 60 min; electrostatic attraction of metals (dried biomass); immobilization: 100% in 3 h	El-Naggar et al. (2018)
Chlorella vulgaris *Scenedesmus almeriensis* *Chlorophyceae* spp.	Mn As B Zn Cu	7.0 9.5 5.5 5.5 7.0				99.4 (3h) 40.7 (3h) 38.6 (10min) 91.9 (3h) 88 (10min)	Monometallic and multi-metallic studies; *Chlorophyceae* sp. shows potential metal removal in multi-metallic solution	Saavedra et al. (2018)
Chlorella vulgaris	Fe Mn Zn	6.0	25	0.4	129.83 115.90 105.29		Ca-alginate beads with immobilized biomass; contact time 300 min; removal from Palm oil mill effluent	Ahmad et al. (2018)
Cystoseira barbata	Cr	4.5	60	2.0	-	70.7	Dried biomass at 60°C for 24 h; contact time - 120 min	Yalçın and Özyürek (2018)
Sargassum dentifolium	Cr	7.0	50	-	-	99.6	Washing and dried at 50°C for 24 h	Husien et al. (2019)
Nannochloropsis oculate	Cu	8.5		-	-	99.9	Incubation time - 21 days	Martínez-Macias et al. (2019)

(Continued)

TABLE 1.1 (Continued)
Recent Discoveries in the Field of Biosorption of Heavy Metals by Various Microalgae

Algal Strain	Metal	pH	Temperature (°C)	Biosorbent (g/L)	Adsorption Capacity (mg/g)	Metal Removal (%)	Remarks	References
Scenedesmus quadricada based biochar	Cr	2.0	35	–	–	100	Contact time - 4 h	Daneshvar et al. (2019)
Desmodesmus sp. MAS1 *Heterochlorella* sp. MAS3	Cu	3.5	-	–	–	27	Biodiesel production at acidic 3.5 pH (30%–40% yield)	Abinandan et al. (2019a)
	Fe					43		
	Mn					Total metals		
	Zn							
Desmodesmus sp. MAS1 *Heterochlorella* sp. MAS3	Cd	3.5	–	–	0.77	58	Initial concentration 5 and 20 mg/L; biodiesel rich biomass	Abinandan et al. (2019b)
					0.36			
Chlorella coloniales	Cr	–	–	2.4–2.9	–	97.8	Scale-up process; incubation time 95–111 h; living biomass	Jaafari and Yaghmaeian (2019)
	Co					97.05		
	Cd					95.15		
	Fe					98.6		
	As					96.5		
Cytoseira indica	Cu	4	25	0.6	72.1	–	Sulfonate, carboxyl and amine groups as a binding site	Roozegar and Behnam (2019)
Consortium of Chlorella vulgaris Enterobacter sp. MN17	Cr	7.0	25	2.18	–	79	Decolourize the dyes from textile wastewater; 74%	Mubashar et al. (2020)
	Cd					93		
	Cu					72		
	Pb					79		
Chlorella kessleri	Pb	6.3	27	1.5	–	99.5	Multi-metal removal Pb > Co > Cu > Cd > Cr	Sultana et al. (2020)
Chlorella vulgaris and *Coelastrella* sp.	Cd	5.5	20	0.4	49	72	Live biomass; adsorption on cell wall through various functional groups	Plöhn et al. (2021)
					65	82		

(Continued)

TABLE 1.1 (Continued)
Recent Discoveries in the Field of Biosorption of Heavy Metals by Various Microalgae

Algal Strain	Metal	pH	Temperature (°C)	Biosorbent (g/L)	Adsorption Capacity (mg/g)	Metal Removal (%)	Remarks	References
Aphanotheca sp.	Pb	–	–	–	185.4	99.9	Contact time 30 min; initial concentration 18.6 mg/L; dried biomass; cultivation time 14 days	Keryanti and Mulyono (2021)
Spirulina platensis	Al	5.5	25	2.5	–	95	Contact time 5–100 min; Higher electronegative groups found on surface acid (sulfuric acid) treated biomass	Almomani and Bhosale (2021)
	Ni	6.0				87		
	Cu	7.0				63		
Chlorella vulgaris	Al	5.0	25	4.8	–	87		
	Ni	6.0				79		
	Cu	7.0				80		
Nordic microalgae	Cd						Intimal concentration	
Chlorella vulgaris					49	72	2.5 mg/L; contact time 24h	Plöhn et al. (2021)
Coelastrella sp.					65	82		
Scenedesmus obliquus					25			
Chlorella vulgaris and Spirulina platensis	Cr	6	25	0.4		100	Sulfuric acid pretreated biomass; intimal concertation 25 mg/L	Musah et al. (2022)
	Fe	3				100		
Turbinaria ornate	Cu	4.5	–	16.2	–	99.8	Biomass dried at 80°C for 6h	Al-Dhabi and Arasu (2022)
Chlorella sorokiniana	U	2.5	25	0.02	–	66.6	Contact time 90 min; rpm 250 non-living biomass; carboxyl, amino, hydroxyl groups involved in adsorption	Embaby et al. (2022)
Chlamydomonas sp.	As	4.0	25	0.6	53.8	95.2	Dried biomass; rpm 300 contact time 60 min	Mohamed et al. (2022)

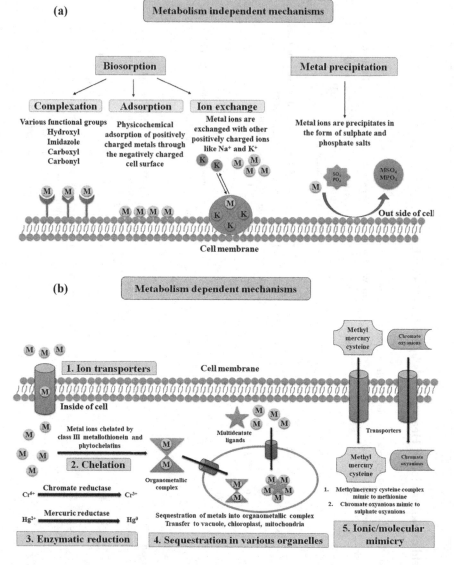

FIGURE 1.2 (a) Metabolism-independent metal-uptake mechanisms in microalgae. (b) Metabolism-dependent metal-uptake mechanisms in microalgae. (Costa & Tavares, 2018; Barquilha et al., 2019; Tripathi & Poluri, 2021; Ahmad et al., 2022).

imidazole, amide, and thiol (Chu & Phang, 2019; Abinandan et al., 2019a; Tripathi & Poluri, 2021). Metal sequestration is also a type of complexation of heavy metals through the multidentate ligand present on the cell surface of algae.

Certain metal ions can coordinate with the phosphate and hydroxy-carbonyl to form a metal complex. Complex compounds consist of one or more central atoms as metal ions bound with various ligands, which are useful for the sorption of such complex compounds into microalgae. Many species of *Cyanobacteria* carry out complexation for antimony removal (Javanbakht et al., 2014). In the case of *Chlamydomonas reinhardtii,* the complexation takes place through the carboxylic functional group (Chu & Phang, 2019). El-Naggar et al. (2018) reported that carboxyl and amino functional groups are substantial binding sites for lead biosorption.

1.4.1.3 Adsorption

The adsorption process depends on the nature of the adsorbate and adsorbent material. In the case of dead algal biomass, heavy metal removal can be carried out through the adsorption process. The efficiency of the adsorption process may be increased by increasing the surface area of the biosorbent material and the temperature. The adsorption process can be of two types: (1) Physisorption, in which the functional groups present on the cell wall of algae and metal ions are bound by weak interactions such as Van der Waals forces, electrostatic attraction, and Coulombic attraction. This process is rapid, unspecific, reversible, and forms a multi-ion layer on the surface of the biosorbent. (2) Chemisorption is very specific; only specific metal ions can bind to specific functional groups on the cellular surface through strong bonds such as ionic or covalent interactions. In chemisorption, the sorption process can be in the mono-ionic layer (Javanbakht et al., 2014; Soliman & Moustafa, 2020). Many studies have shown that extracellular polymeric substances also play a role in the adsorption of metal ions; hence, they significantly pose metal tolerance and self-defense to algae (Danouche et al., 2021; Hu et al., 2022). In adsorption process, the carboxyl group present on the algal cell wall plays a key role. Majorly the carboxyl group adsorbed heavy metals like Cd, Cu, and Pb (Spain et al., 2021).

1.4.1.4 Ion Exchange

Ion exchange is also a potential mechanism in biosorption by microalgae. The algal cell wall contains a variety of polysaccharides, which play a significant role in the exchange of bivalent or monovalent ions with counter ions present in polysaccharides, proteins, and lipids such as H^+, Na^+, Ca^{2+}, etc. Nevertheless, the ion exchange process can be reversible (Javanbakht et al., 2014; Danouche et al., 2021; Ahmad et al., 2022). The ion exchange mechanism depends on the pH; a simultaneous increase of H^+ in an aqueous solution indicates the uptake of metal ions (Javanbakht et al., 2014).

1.4.2 Metabolism-Dependent Mechanisms

Metabolism-dependent mechanisms involve the intracellular transport of metal ions by living cells. The intracellular transport of metal by living algal cells is known as bioaccumulation. Bioaccumulation can be carried out in two major ways: (1) active transport and (2) sequestration. In the active transport of metal carrier proteins, ion channels, enzymatic reduction, chelation, and molecular mimicry play a crucial role, while sequestration can be done through the accumulation of metals in vacuoles (Tripathi & Poluri, 2021; Ahmad et al., 2022).

1.4.2.1 Metal Sequestration by Bioaccumulation

The metal sequestration process is carried out in the cytosol by chelation mechanisms. A group of ions chelated by multi-dentate ligands with class III metallothionein forms organometallic complexes that detoxify the metal ions for the survival of the cell. These complexes are transported to vacuoles. The vacuolar transporter carried out the transport of these complexes. This vacuolar compartmentalization is a vital part of heavy metal detoxification. Sometimes these complexes are also transported to the chloroplast and mitochondria. In *Euglena*, cadmium metal sequestration is carried out in the cytosol, and the Cd-MTs (III) complex is transported into the cytosol by various types of ATP-dependent transporters (Danouche et al., 2021; Ahmad et al., 2022; Pradhan et al., 2022).

1.4.2.2   Enzymatic Reduction

Heavy metals are non-degradable, but to avoid their hazardous effects, certain algae possess various enzymes that convert them from one oxidation state to another or low-toxicity forms. Scanty research has been carried out on the enzymatic mechanism for metal remediation in algae. The most widely reported redox enzymes in microalgae are chromate reductase, mercuric reductase, and arsenic reductase. These enzymes have been reported in *Chlorella vulgaris*, which converts Cr^{6+} to Cr^{3+}; *C. reinhardtii*, which transforms As^{3+} into a less toxic form of As^{5+}; and *Selenastrum minutum*, which converts Hg^{2+} to elemental Hg^0, for the detoxification of heavy metals. Some studies are also carried out on enzymatic reduction as a biosorption mechanism in microalgae (Leong & Chang, 2020; Danouche et al., 2021).

1.4.2.3   Ion Channels/Carrier Proteins/Transporters

Many algae possess various types of transporters and ion channels, which are responsible for the influx and efflux of metal ions. There are mainly two types of transporters that play a role in the influx of metals. Group A transporters influx the metals in the cytosol from the extracellular environment, like the Fe Cu transporters. They are also found in the vacuole membrane. Group B transporters play a role in the efflux of metal from the cytosol to the extracellular environment by FerroPortiN and P1B-type ATPases (Danouche et al., 2021; Pradhan et al., 2022). The ion exchange mechanism was associated with copper removal in *Cytoseira indica*; Cu^{2+} ions were exchanged by Ca^{2+} at pH 4.0 (Roozegar & Behnam, 2019).

1.4.2.4   Chelation

Many algae have chelating molecules for their survival in heavy metal-contaminated habitats. These chelator molecules are generally metal-binding proteins that play a vital role in maintaining intracellular metal concentration. *Chlorella, Symbiodinium*, and *Nannochlorpsis* contain various chelating molecules, like class III metallothionines and cysteines containing small peptides with low molecular weight found in the cytosol. Moreover, some algae also synthesize phytochelatins, which are thiol-containing peptides of three amino acids: glutamate, cysteine, and glycine. Phytochelatins are produced by *Chlamydomonas* sp. and *Dunaliella* sp. Cu, Pb, Cd, and As are chelated by phytochelatin and converted into organometallic complexes with the key ligand glutathione. These complexes are stored in vacuoles (Balzano et al., 2020; Danouche et al., 2021; Ahmad et al., 2022; Pradhan et al., 2022; Bauddh & Korstad, 2022).

1.4.2.5 Molecular/Ionic Mimicry

In molecular mimicry, organic forms of metals bind to some cellular molecules. The methylmercury cysteine complex mimics methionine molecules and is carried in cells through various transporters. Ionic mimicry can be defined by the ability of cationic species of metals behave or mimic the structural or functional similarity of another element at the site of carrier proteins. Toxic heavy metals are transported to the cytosol through such mechanisms. There is a deceptive similarity that can be seen in the chromate and sulfate oxyanions, so these chromate ions transport to the cell through this ionic mimicry (Mantzorou et al., 2018). Toxic metals follow molecular mimicry with ions like Mn, Ca, and Zn (Ahmad et al., 2022).

1.5 FACTORS AFFECTING PHYCOREMEDIATION

Phycoremediation is a physicochemical as well as complex mechanism, so there are various factors that affect the removal process in different ways. Due to that, several factors should be considered in metal removal. The process is mainly affected by pH, initial metal concentration, contact time, biomass dosage, temperature, various pre-treatments, and multi-metals for competitive sorption. All these factors are discussed below briefly.

1.5.1 EFFECT OF pH

In aqueous systems, pH is the major factor that affects the speciation of metal ions (Chu & Phang, 2019). The sorption capacity of any biosorbent material can be affected by the solubility of metal ions and charges on the binding sites (Shamim, 2018). At an acidic pH, the surface of any biomass becomes positively charged due to a higher concentration of the H_3O^+ ions present in the solution, whereas at a pH above 6.0, biomass surface becomes negatively charged due to the OH^- ions present in the solution (Sultana et al., 2020). Additionally, above the maximum pH range, there will be non-availability of soluble metal ions due to the precipitation of metals. For instance, precipitation occurs at pH > 6 for copper and nickel, pH > 5 for lead, and pH > 4 for chromium. These precipitated metals can hinder metal uptake and drastically reduce biosorption (Salam, 2019). However, in the case of chromium biosorption, maximum sorption can be achieved at a lower pH because, above pH 4.5, there is the possibility of starting the formation of hydroxide ions, so at this pH, Cr^{3+} ion retention is higher as compared to Cr^{6+}. It is also reported that at pH 2.0, the Cr^{6+} ion is found in the forms of $Cr_2O_7^{2-}$ and $HCrO_4^-$, so positively charged biomass will favor the sorption of such forms (Yalçın & Özyürek, 2018; Sutkowy & Kłosowski, 2018; Sibi, 2019). Moreover, according to Al-Dhabi and Arasu (2022), copper biosorption by marine algae can also reach a maximum at pH 4.5 due to the higher solubility of copper ions in the range of four to five, and above pH 6, copper ions are converted into precipitate forms of copper hydroxides, which overall deduct the sorption by algal biomass (Sibi, 2019). However, Saavedra et al. (2018) reported that maximum copper biosorption of 88% was achieved at 7.0 pH by *Chlorophyceae* spp.

1.5.2 EFFECT OF INITIAL METAL CONCENTRATION

The biosorption of heavy metals by algal biomass is significantly influenced by the initial metal concentration. The initial metal concentration is an important factor in overcoming the mass transfer resistance of the metals between the aqueous and solid phases (Costa & Tavares, 2018). The sorption rate increases with the increase in initial metal concentration up to the saturation point at a stable metal concentration (Sibi, 2019). *Chlorella coloniales* showed an increased biosorption capacity of Cd as the initial concentration was increased from 5 to 12 mg/L; thereafter, no significant effects were observed with further increases in Cd concentration (Jaafari & Yaghmaeian, 2019). Moreover, Embaby et al. (2022) reported that the uranium uptake by biomass increased significantly with the increase in the initial concentration from 400 to 1,200 mg/L, and this increase could be due to collision between the metal and biosorbent. A further increase in the metal concentration decreased the metal uptake, which may be due to the saturation of all the active sites on the cell surface of *Chlorella sorokiniana* biomass.

1.5.3 EFFECT OF CONTACT TIME

The contact time between biomass and metal significantly affects the biosorption process. In the initial stage, rapid metal sorption can be seen, typically attributed to the large number of binding sites available. However, after saturation of all the binding sites, increasing the contact time does not affect the biosorption efficiency significantly (Saavedra et al., 2018). The biosorption process takes place in two stages: in the first stage, rapid and passive metal sorption takes place, while in the second stage, active metal accumulation takes place for living cells, which become stable after a suitable contact time (Sibi, 2019). The maximum biosorption of metal was observed in the contact time of the first 60 min; after that, biosorption capacity of biomass decreased due to an unstable and weak interaction between metal ions and biosorbent materials. Lead biosorption as high as 98%–99% was observed in 30 min by *Aphanothece* sp., and after 120 min, biosorption efficiency of decreased due to the saturation of binding sites (Keryanti & Mulyono, 2021). Moreover, the uranium biosorption efficiency by *C. sorokiniana* increased with a rise in contact time, which reached a maximum of 82.2 mg/g at 90 min. After that, it became constant. Optimum contact time is mostly considered the equilibrium of the process, which is further utilized for adsorption studies (Embaby et al., 2022).

1.5.4 EFFECT OF BIOMASS DOSAGE

The effect of biomass concentration is also a key influencing factor for metal sorption in an aqueous system. An increase in biomass dosage rapidly increases the biosorption rate of any biosorbent material due to the presence of binding sites in abundance, though after critical concentration there is a decrease in the sorption capacity due to the high aggregation of biomass formed in solution as well as a decrease in the average distance between the sorption sites (Sibi, 2019). Lead biosorption efficiency was increased by increasing the biomass concentration from 0.5 to 1.5 g/L due to more binding sites available for the sorption of Pb on the biomass surface (Sultana et al., 2020). Uranium removal was rapidly augmented by increasing the biomass dosage from 0.005 to 0.02 g, and further amplification in the biomass dosage did not affect the sorption process because of the lower concentration of metal present in the solution (Embaby et al., 2022).

1.5.5 EFFECT OF TEMPERATURE

Temperature variations can cause changes in the thermodynamic properties, which affect the sorption capacity of the sorbents. The effect of temperature can be endothermic or exothermic. In the case of an endothermic reaction, an elevation in temperature may increase the sorption efficiency, while an exothermic reaction deduces the overall sorption by increasing the temperature (Bilal et al., 2018). According to Sulaymon et al. (2013), maximum biosorption of Pb, Cd, and As has been obtained in a temperature range of 15°C–30°C. However, some studies showed that an increase in temperature negatively influences the biosorption capacity of algae due to increasing the pore size of biomass, so metal ions can easily escape from the biomass, resulting in an overall reduction of heavy metal sorption. Biosorption of Fe^{2+}, Mn^{2+}, and Zn^{2+} by free and immobilized *C. vulgaris* decreased with increasing temperature due to an exothermic process. Extremely high temperatures can cause the denaturation of proteins and the deactivation of the functional groups on the cell surface (Sulaymon et al., 2013; Ahmad et al., 2018). According to Alothman et al. (2020), the rate of biosorption increased with raising the temperature from 10°C to 60°C due to the increased affinity of the biomass for metal ions (Zamora-Ledezma et al., 2021). Moreover, major investigations have proven that increasing the temperature may reduce the biosorption efficiency of biomass, which can be accredited to the deactivation of proteins in the case of live biomass and the destruction of active sites for dead biomass (Embaby et al., 2022).

1.5.6 EFFECT OF VARIOUS PRE-TREATMENTS

The biosorption capacities of dead or live algal biomass can be enhanced by modification of their cell surface with the help of various pre-treatments. It can be physical and/or chemical pre-treatments that improve or modify the functional groups of the cell surface to increase the metal affinity of the biomass. It has been investigated that many physical processes, such as drying, autoclaving, and pyrolysis, have been used for better sorption of metals. In the chemical treatments, various acids, alkalis, solvents, and different chemicals have been applied to biomass (Legorreta-Castañeda et al., 2020). It has been reported that chemical pre-treatments more efficiently increase the sorption capacity of biomass in the case of *C. vulgaris* and *Spirulina platensis* biomass when treated with 1% HNO_3, 1% H_2SO_4, 1% H_3PO_4, and 1% citric acid. The efficiency of Cr and Fe sorption marginally increased for both pretreated algae (Musah et al., 2022). Many studies also revealed that alkali and salt pre-treatments were the best strategies to increase the sorption capacity for Cu ions, while acid treatments decreased the sorption efficiency (Barquilha et al., 2019). The alkali treatment may have exposed the functional groups present in the biomass, which were previously masked by lipids and proteins. However, acid treatment may cause the protonation of biomass, which changes its electronegativity, resulting in over-all reduction in biosorption capacity. However, pre-treatment with $CaCl_2$ and NaCl enhanced the biosorption capacity up to 54–62 mg/g (Roozegar & Behnam, 2019). Nevertheless, the effect of acidic treatment increased the sorption capacity of *S. platensis* and *C. vulgaris* for Al, Ni, and Cu. The acid treatment may chemically alter the functional groups to enhance and improve the biosorption rate (Almomani & Bhosale, 2021).

1.5.7 EFFECT OF THE PRESENCE OF MULTI-METAL (COMPETITIVE) ON SORPTION

The biosorption of heavy metals by various biosorbent materials has been widely studied for single-metal sorption and non-competitive biosorption. However, practically, actual wastewater is a cocktail of multiple metals, dyes, and organic as well as inorganic contaminants. In the case of non-competitive biosorption, there are plenty of binding sites present for the single metal, but in competitive biosorption, different metals have to compete for binding on the same site, so the presence of other metals affects the overall sorption process negatively (Costa & Tavares, 2018). For these reasons, it is obligatory to study the effect of multi-metal solutions on an industrial scale. Moreover, it is necessary to understand the chemical nature of every metal, which makes them suitable for binding to the sorbent over the other metal. The selectivity of metal ions is due to differences in ionic properties like electronegativity, ionic radius, hydration energy, and the redox potential of the metal ions. The net sorption is lower in the competitive case as compared to single-metal sorption. In the multi-metal system, the preferential uptake of metals can be different from the non-competitive sorption (Legorreta-Castañeda et al., 2020). Furthermore, Sultana et al. (2020) reported that in the competitive sorption by *Chlorella kessleri*, the removal efficiency for each heavy metal was found in the following order: Pb > Co > Cu > Cd > Cr. The explanation for such a trend is that the different electronegativity and ionic radius of each of the metals influence their affinity to the biomass. Lead sorption was preferred over the other metals due to its higher electronegativity.

1.6 METAL RECOVERY AND REGENERATION OF ALGAL BIOMASS

Biosorption is an eco-friendly concept, so it is necessary to carry out the desorption of metal from algal biomass for the recovery and reusability of biomass. Moreover, desorption is important for the recovery of valuable metals and to avoid the generation of secondary solid waste during the effluent treatment processes. As per the literature, various chemical agents like chelating agents (thiosulfate, EDTA), mineral acids (HNO_3, HCl, H_2SO_4), and organic acids (acetic acid, citric acid) have been employed, which are depicted in Table 1.2 (Shamim, 2018). Moreover, after the desorption experiments, the disposal of desorbing eluents into the environment is another major issue. So, the recovery of metal from these desorbing eluents becomes mandatory for an environment-friendly approach (Daneshvar et al., 2019). Desorbing agents can be selected based on factors such as their efficient desorption rate, being harmless to biomass, and being non-pollutant. It is necessary to optimize the contact between biomass and desorbing agents; otherwise, it will cause a detrimental effect on biomass and can change the structural characteristics of biomass (Chatterjee & Abraham, 2019).

TABLE 1.2
Regeneration of Biomass and Recovery of Metals

Algae	Metal	Desorbing Agent	Desorption (%)	Recovering Agent	Recovery (%)	References
Microalgal biomass + biochar	Cr	0.1 M NaOH + shaking/ sonication	51	0.5 mM BaCl$_2$	97	Daneshvar et al. (2019)
Acid-based modified *Chlorella vulgaris* and *Spirulina platensis*	Cr Fe	0.1 M HNO$_3$	75.5	–	–	Musah et al. (2022)
Naostoc sp. and *Turbinaria vulgaris*	Cr^{6+}	0.1 M EDTA 0.1 M HNO$_3$	90 80	–	–	Khan et al. (2022b)
Chlamydomonas sp.	As^{3+}	EDTA HNO$_3$	80.59 76.79	–	–	Mohamed et al. (2022)
C. sorokiniana	U	1 M HCl 1 M NaCl	92.3 60	–	–	Embaby et al. (2022)

1.7 KNOWLEDGE GAPS AND FUTURE PROSPECTS

Many fresh and marine water algae have been reported for phycoremediation. Non-living biomass is more economical and reusable, but industrial effluent is rich in organic nutrients as well, under these conditions, it is beneficial to use such water to cultivate microalgae as well as for metal removal as a sustainable bio-based circular economy. Various pre-treatments need to be developed that will not affect the functionality of biomass and augment the sorption capacities of multi-metals in actual effluent. Moreover, the phycoremediation concept is tested for laboratory-made or synthetic wastewater, but very few studies have been carried out using actual industrial effluents that are required for more sustainable approaches. Reduction of biochemical oxygen demand and chemical oxygen demand from wastewater through microalgae is widely established, though there are very few studies reported for multi-metal(s) removal and its mechanisms from wastewater by microalgae, and there will be more opportunities for researchers to explore green technology in the field of microalgal-based remediation of various industrial effluent treatments on a large scale. Industrial wastewater does not have the balance of nutrients required for the optimum growth of algae; thus, analysis of the wastewater is necessary, and deficient nutrients may be supplemented. Algae biomass can be modified with the application of extracellular substances to increase sorption capacity and ease desorption. The study may also be needed to achieve the metal detection and remediation of specific metals simultaneously even at low concentrations (Chugh et al., 2022). Algae can also be used for biofuel production, so the remediating tool can also be applied for fuel production. The process should be developed to avoid the generation of secondary solid waste and better eluting chemicals for the desorption of heavy metals from microalgae-based biomass, which can improve the reusability of biomass for more cycles. It is also necessary to optimize the process for selective recovery of desired metals from the eluting solution, as scanty data are available for the recovery of metals from the eluting solution and the applicability of those metals in reuse.

1.8 CONCLUSION

Heavy metals and their compounds pose various threats to humans and all living beings, as well as polluting the soil, water, and air. Day by day, due to rapid industrialization, huge amounts of metals are added to the ecosystem, which has a detrimental effect on the environment. Conventional and chemical techniques have been utilized for the last few decades, but certain effluents contain a very low level of metal concentration, so the application of chemical treatment becomes costly and not effective for the removal of such a low amount. Microalgae have the potential to remove metal from the lowest metal concentration effectively. Moreover, microalgal-based biosorbent materials are easy to reuse and recover valuable metals. Non-living or dead biomass of algae will be a cheaper and more eco-friendly concept for metal removal from industrial effluent treatments. The use of algae-based metal remediation may result in the development of circular technology with zero or minimal waste.

REFERENCES

Abinandan S, Subashchandrabose SR, Panneerselvan L, Venkateswarlu K, Megharaj M. (2019a) Potential of acid-tolerant microalgae, *Desmodesmus* sp. MAS1 and *Heterochlorella* sp. MAS3, in heavy metal removal and biodiesel production at acidic pH. *Bioresour Technol* 278:9–16. https://doi.org/10.1016/j.biortech.2019.01.053.

Abinandan S, Subashchandrabose SR, Venkateswarlu K, Perera IA, Megharaj M. (2019b) Acid-tolerant microalgae can withstand higher concentrations of invasive cadmium and produce sustainable biomass and biodiesel at pH 3.5. *Bioresour Technol* 281:469–473. https://doi.org/10.1016/j.biortech.2019.03.001.

Ahmad A, Banat F, Alsafar H, Hasan SW. (2022) Algae biotechnology for industrial wastewater treatment, bioenergy production, and high-value bioproducts. *Sci Total Environ* 806:150585. https://doi.org/10.1016/j.scitotenv.2021.150585.

Ahmad A, Bhat AH, Buang A. (2018) Biosorption of transition metals by freely suspended and Ca-alginate immobilised with *Chlorella vulgaris*: Kinetic and equilibrium modeling. *J Clean Prod* 171:1361–1375. https://doi.org/10.1016/j.jclepro.2017.09.252.

Ahmed J, Thakur A, Goyal A. (2021). Industrial wastewater and its toxic effects. In: *Biological Treatment of Industrial Wastewater 2021 E-Book Collection*, edited by Shah MP. London, UK: Royal Society of Chemistry. https://doi.org/10.1039/9781839165399-00001.

Al-Dhabi NA, Arasu MV. (2022) Biosorption of hazardous waste from the municipal wastewater by marine algal biomass. *Environ Res* 204:112115. https://doi.org/10.1016/j.envres.2021.112115.

Alengebawy A, Abdelkhalek ST, Qureshi SR, Wang MQ. (2021) Heavy metals and pesticide toxicity in agricultural soil and plants: Ecological risks and human health implications. *Toxics* 9(3):42. https://doi.org/10.3390/toxics9030042.

Almomani F, Bhosale RR. (2021). Bio-sorption of toxic metals from industrial wastewater by algae strains *Spirulina platensis* and *Chlorella vulgaris*: Application of isotherm, kinetic models and process optimization. *Sci Total Environ* 755:142654. https://doi.org/10.1016/j.scitotenv.2020.142654.

Alothman ZA, Bahkali AH, Khiyami MA, Alfadul SM, Wabaidur SM, Alam M, Alfarhan BZ. (2020) Low cost biosorbents from fungi for heavy metals removal from wastewater. *Sep Sci Technol* 55(10):1766–1775. https://doi.org/10.1080/01496395.2019.1608242.

Balzano S, Sardo A, Blasio M, Chahine TB, Dell'Anno F, Sansone C, Brunet C. (2020) Microalgal metallothioneins and phytochelatins and their potential use in bioremediation. *Front Microbiol* 11:517. https://doi.org/10.3389/fmicb.2020.00517.

Barquilha CE, Cossich ES, Tavares CR, da Silva EA. (2019) Biosorption of nickel and copper ions from synthetic solution and electroplating effluent using fixed bed column of immobilized brown algae. *J Water Process Eng* 32:100904. https://doi.org/10.1016/j.jwpe.2019.100904.

Bauddh K, Korstad J. (2022) Phycoremediation: Use of algae to sequester heavy metals. *Hydrobiology* 1(3):288–303. https://doi.org/10.3390/hydrobiology1030021.

Bilal M, Rasheed T, Sosa-Hernández JE, Raza A, Nabeel F, Iqbal HM. (2018) Biosorption: An interplay between marine algae and potentially toxic elements-a review. *Marine Drugs* 16(2):65. https://doi.org/10.3390/md16020065.

Briffa J, Sinagra E, Blundell R. (2020) Heavy metal pollution in the environment and their toxicological effects on humans. *Heliyon* 6(9):e04691. https://doi.org/10.1016/j.heliyon.2020.e04691.

Chatterjee A, Abraham J. (2019) Desorption of heavy metals from metal loaded sorbents and e-wastes: A review. *Biotechnol Lett* 41(3):319–333. https://doi.org/10.1007/s10529-019-02650-0.

Cheng J, Yin W, Chang Z, Lundholm N, Jiang Z. (2017) Biosorption capacity and kinetics of cadmium (II) on live and dead *Chlorella vulgaris*. *J Appl Phycol* 29:211–221. https://doi.org/10.1007/s10811-016-0916-2.

Chu WL, Phang SM. (2019). Biosorption of heavy metals and dyes from industrial effluents by microalgae. In: *Microalgae Biotechnology for Development of Biofuel and Wastewater Treatment*, edited by Alam M, Wang Z. Singapore: Springer, pp. 599–634. https://doi.org/10.1007/978-981-13-2264-8_23.

Chugh M, Kuma L, Shah MP, Bharadvaja N. (2022). Algal Bioremediation of heavy metals: An insight into removal mechanisms, recovery of by-products, challenges, and future opportunities. *Energy Nexus* 7:100129.

Costa F, Tavares T. (2018) Biosorption of multicomponent solutions: A state of the art of the understudy case. In: Biosorption, edited by Derco J, Vrana B. United Kingdom: Intech Open. https://doi.org/10.5772/intechopen.72179.

Daneshvar E, Zarrinmehr MJ, Kousha M, Hashtjin AM, Saratale GD, Maiti A, Bhatnagar A. (2019) Hexavalent chromium removal from water by microalgal-based materials: Adsorption, desorption and recovery studies. *Bioresour Technol* 293:122064. https://doi.org/10.1016/j.biortech.2019.122064.

Danouche M, El Ghachtouli N, El Aroussi, H. (2021) Phycoremediation mechanisms of heavy metals using living green microalgae: Physicochemical and molecular approaches for enhancing selectivity and removal capacity. *Heliyon* 7(7):e07609.

El-Naggar NEA, Hamouda RA, Mousa IE, Abdel-Hamid MS, Rabei NH. (2018) Biosorption optimization, characterization, immobilization and application of *Gelidium amansii* biomass for complete Pb2+ removal from aqueous solutions. *Sci Rep* 8:13456. DOI: 10.1038/s41598-018-31660-7.

Embaby MA, Haggag ESA, El-Sheikh AS, Marrez DA. (2022) Biosorption of uranium from aqueous solution by green microalga Chlorella sorokiniana. *Environ Sci Pollut Res* 1–17. https://doi.org/10.1007/s11356-022-19827-2.

Hu R, Cao Y, Chen X, Zhan J, Luo G, Ngo HH, Zhang S. (2022) Progress on microalgae biomass production from wastewater phycoremediation: Metabolic mechanism, response behavior, improvement strategy and principle. *J Chem Eng* 137187. https://doi.org/10.1016/j.cej.2022.137187.

Husien S, Labena A, El-Belely EF, Mahmoud HM, Hamouda AS. (2019) Adsorption studies of hexavalent chromium [Cr (VI)] on micro-scale biomass of *Sargassum dentifolium*, Seaweed. *J Environ Chem Eng* 7(6):103444. https://doi.org/10.1016/j.jece.2019.103444.

Jaafari J, Yaghmaeian K. (2019) Optimization of heavy metal biosorption onto freshwater algae (*Chlorella coloniales*) using response surface methodology (RSM). *Chemosphere* 217:447–455. https://doi.org/10.1016/j.chemosphere.2018.10.205.

Javanbakht V, Alavi SA, Zilouei H. (2014) Mechanisms of heavy metal removal using microorganisms as biosorbent. *Water Sci Technol* 69(9):1775–1787. https://doi.org/10.2166/wst.2013.718.

Keryanti K, Mulyono EWS. (2021) Determination of optimum condition of lead (Pb) biosorption using dried biomass microalgae *Aphanothece* sp. *Period Polytech Chem Eng* 65(1):116–123. https://doi.org/10.3311/PPch.15773.

Khan AA, Gul J, Naqvi SR, Ali I, Farooq W, Liaqat R, Juchelková D. (2022a) Recent progress in microalgae-derived biochar for the treatment of textile industry wastewater. *Chemosphere* 135565. https://doi.org/10.1016/j.chemosphere.2022.135565.

Khan AA, Mukherjee S, Mondal M, Boddu S, Subbaiah T, Halder G. (2022b) Assessment of algal biomass towards removal of Cr (VI) from tannery effluent: A sustainable approach. *Environ Sci Pollut Res* 29(41):61856–61869. https://doi.org/10.1007/s11356-021-16102-8.

Legorreta-Castañeda AJ, Lucho-Constantino CA, Beltrán-Hernández RI, Coronel-Olivares C, Vázquez-Rodríguez GA. (2020) Biosorption of water pollutants by fungal pellets. *Water* 12(4):1155. https://doi.org/10.3390/w12041155.

Leong YK, Chang JS. (2020) Bioremediation of heavy metals using microalgae: Recent advances and mechanisms. *Bioresour Technol* 303:122886. https://doi.org/10.1016/j.biortech.2020.122886.

Mantzorou A, Navakoudis E, Paschalidis K, Ververidis F. (2018) Microalgae: A potential tool for remediating aquatic environments from toxic metals. *Int J Environ Sci Technol* 15(8):1815–1830. https://doi.org/10.1007/s13762-018-1783-y.

Mao G, Han Y, Liu X, Crittenden J, Huang N, Ahmad UM. (2022) Technology status and trends of industrial wastewater treatment: A patent analysis. *Chemosphere* 288:132483.

Martínez-Macias MDR, Correa-Murrieta MA, Villegas-Peralta Y, Dévora-Isiordia GE, Álvarez-Sánchez J, Saldivar-Cabrales J, Sánchez-Duarte RG. (2019) Uptake of copper from acid mine drainage by the microalgae *Nannochloropsis oculata*. *Environ Sci Pollut Res* 26:6311–6318. https://doi.org/10.1007/s11356-018-3963-1.

Mohamed MS, Hozayen WG, Alharbi RM, Ibraheem IBM. (2022) Adsorptive recovery of arsenic (III) ions from aqueous solutions using dried *Chlamydomonas* sp. *Heliyon* 8(12):e12398. https://doi.org/10.1016/j.heliyon.2022.e12398.

Mubashar M, Naveed M, Mustafa A, Ashraf S, Shehzad Baig K, Alamri S, Kalaji HM. (2020) Experimental investigation of *Chlorella vulgaris* and *Enterobacter* sp. MN17 for decolorization and removal of heavy metals from textile wastewater. *Water* 12(11):3034. https://doi.org/10.3390/w12113034.

Musah BI, Wan P, Xu Y, Liang C, Peng L. (2022) Biosorption of chromium (VI) and iron (II) by acid-based modified Chlorella vulgaris and Spirulina platensis: Isotherms and thermodynamics. *Int J Environ Sci Technol* 1–16. https://doi.org/10.1007/s13762-021-03873-3.

Plöhn M, Escudero-Onate C, Funk C. (2021) Biosorption of Cd (II) by Nordic microalgae: Tolerance, kinetics and equilibrium studies. *Algal Res* 59:102471. https://doi.org/10.1016/j.algal.2021.102471.

Pradhan B, Bhuyan PP, Nayak R, Patra S, Behera C, Ki JS, Jena M. (2022) Microalgal phycoremediation: A glimpse into a sustainable environment. *Toxics* 10(9):525. https://doi.org/10.3390/toxics10090525.

Premaratne M, Nishshanka GKSH, Liyanaarachchi VC, Nimarshana PHV, Ariyadasa TU. (2021) Bioremediation of textile dye wastewater using microalgae: Current trends and future perspectives. *J Chem Technol Biotech* 96(12):3249–3258. https://doi.org/10.1002/jctb.6845.

Priya AK, Jalil AA, Vadivel S, Dutta K, Rajendran S, Fujii M, Soto-Moscoso M. (2022) Heavy metal remediation from wastewater using microalgae: Recent advances and future trends. *Chemosphere* 305:135375. https://doi.org/10.1016/j.chemosphere.2022.135375.

Rajoria S, Vashishtha M, Sangal VK. (2022) Treatment of electroplating industry wastewater: A review on the various techniques. *Environ Sci Pollut Res* 1–51. https://doi.org/10.1007/s11356-022-18643-y.

Razzak SA, Farooque MO, Alsheikh Z, Alsheikhmohamad L, Alkuroud D, Alfayez A, Hossain MM. (2022) A comprehensive review on conventional and biological-driven heavy metals removal from industrial wastewater. *Environ Adv* 100168. https://doi.org/10.1016/j.envadv.2022.100168.

Roozegar M, Behnam S. (2019) An eco-friendly approach for copper (II) biosorption on alga *Cystoseira indica* and its characterization. *Environ Prog Sustain* 38(s1):S323–S330. https://doi.org/10.1002/ep.13044.

Rugnini L, Costa G, Congestri R, Bruno L. (2017) Testing of two different strains of green microalgae for Cu and Ni removal from aqueous media. *Sci Total Environ* 601:959–967. https://doi.org/10.1016/j.scitotenv.2017.05.222.

Saavedra R, Muñoz R, Taboada ME, Vega M, Bolado S. (2018) Comparative uptake study of arsenic, boron, copper, manganese and zinc from water by different green microalgae. *Bioresour Technol* 263:49–57. https://doi.org/10.1016/j.biortech.2018.04.101.

Salam KA. (2019) Towards sustainable development of microalgal biosorption for treating effluents containing heavy metals. *Biofuel Res J* 6(2):948–961. DOI: 10.18331/BRJ2019.6.2.2.

Samal DK, Sukla LB, Pattanaik A, Pradhan D. (2020) Role of microalgae in treatment of acid mine drainage and recovery of valuable metals. *Mater Today Proc* 30:346–350. https://doi.org/10.1016/j.matpr.2020.02.165.

Shah KR, Duggirala SM, Tipre DR, Dave SR. (2017) Mechanistic aspects of Au (III) sorption by *Aspergillus terreus* SRD49. *J Taiwan Inst Chem Eng* 80:46–51. https://doi.org/10.1016/j.jtice.2017.08.001.

Shamim S. (2018) Biosorption of heavy metals. In: *Biosorption*, edited by Derco J, Vrana B. United Kingdom: Intech Open. https://doi.org/10.5772/intechopen.72099.

Sibi G. (2019) Factors influencing heavy metal removal by microalgae: A review. *J Crit Rev* 6:29–32. https://dx.doi.org/10.22159/jcr.2019v6i6.35600.

Soliman NK, Moustafa AF. (2020) Industrial solid waste for heavy metals adsorption features and challenges: A review. *J Mater Res Technol* 9(5):10235–10253. https://doi.org/10.1016/j.jmrt.2020.07.045.

Spain O, Plöhn M, Funk C. (2021) The cell wall of green microalgae and its role in heavy metal removal. *Physiol Plant* 173(2):526–535. https://doi.org/10.1111/ppl.13405.

Sulaymon AH, Mohammed AA, Al-Musawi TJ. (2013) Competitive biosorption of lead, cadmium, copper, and arsenic ions using algae. *Environ Sci Pollut Res* 20(5):3011–3023. https://doi.org/10.1007/s11356-012-1208-2.

Sultana N, Hossain SM, Mohammed ME, Irfan MF, Haq B, Faruque MO, Hossain MM. (2020) Experimental study and parameters optimization of microalgae based heavy metals removal process using a hybrid response surface methodology-crow search algorithm. *Sci Rep* 10(1):1–15. https://doi.org/10.1038/s41598-020-72236-8.

Sutkowy M, Kłosowski G. (2018) Use of the coenobial green algae *Pseudopediastrum boryanum* (Chlorophyceae) to remove hexavalent chromium from contaminated aquatic ecosystems and industrial wastewaters. *Water* 10(6):712. https://doi.org/10.3390/w10060712.

Tripathi S, Poluri KM. (2021) Heavy metal detoxification mechanisms by microalgae: Insights from transcriptomics analysis. *Environ Pollut* 285:117443. https://doi.org/10.1016/j.envpol.2021.117443.

Urrutia C, Yañez-Mansilla E, Jeison D. (2019) Bioremoval of heavy metals from metal mine tailings water using microalgae biomass. *Algal Res* 43:101659. https://doi.org/10.1016/j.algal.2019.101659.

Yalçın S, Özyürek M. (2018) Biosorption potential of two brown seaweeds in the removal of chromium. *Water Sci Technol* 78(12):2564–2576. https://doi.org/10.2166/wst.2019.007.

Zamora-Ledezma C, Negrete-Bolagay D, Figueroa F, Zamora-Ledezma E, Ni M, Alexis F, Guerrero VH. (2021) Heavy metal water pollution: A fresh look about hazards, novel and conventional remediation methods. *Environ Technol Innov* 22:101504. https://doi.org/10.1016/j.eti.2021.101504.

Zhou H, Zhao X, Kumar K, Kunetz T, Zhang Y, Gross M, Wen Z. (2021) Removing high concentration of nickel (II) ions from synthetic wastewater by an indigenous microalgae consortium with a revolving algal biofilm (RAB) system. *Algal Res* 59:102464. https://doi.org/10.1016/j.algal.2021.102464.

2 Applications of Pyhcoremediation Technologies for Wastewater Treatment

Removal, Mechanisms and Perspectives of Micropollutants

Muhammad Uzair Javed, Hamid Mukhtar, and Tehreem Mahmood

2.1 INTRODUCTION

A great number of challenges are being faced by our planet in current times. Our economy and environment are significantly affected by challenges such as global warming, conventional depletion of energy, and water contamination. Industrialization, expanding population, and prompt urbanization contribute to several facets of environmental contamination in modern eras (Koul et al., 2022).

Massive amounts of both solid and liquid waste are created annually around the globe. Only a minute amount of waste is recycled, while the remaining one is either discarded or left unprocessed, resulting in several problems like water and air pollution as well as thermal contamination. Moreover, there is a concern regarding the treatment of wastewater (WW) that will otherwise get discharged into the surroundings of developing nations. Insufficient treatment of WW and fecal sludge can result in the expansion of antimicrobial resistance and can also spread a number of diseases. It is anticipated that by the year 2050, the quantity of waste will rise from 2.01 to 3.40 billion metric tons worldwide. As a result, ecologically sound and cost-efficient WW treatment methods with limited infrastructure and inputs are required (López-Serrano et al., 2020).

WW mainly consists artificial elements and organic and inorganic compounds. Amino acids, proteins, fats, and volatile acids constitute the majority of organic carbon in WW. Calcium, ammonium, phosphate, magnesium, sulfur, sodium, and heavy metals are abundant in inorganic quantities. If, by any chance, these contaminants

DOI: 10.1201/9781003390213-2

get transmitted to living creatures, they can trigger bioaccumulation and infections (Akhil et al., 2021).

Methods that are conventionally employed for the treatment of WW include anaerobic digestion, the electro-Fenton process, the activated sludge method, the advanced oxidation process, and the method of membrane filtration (Krishnan et al., 2021). Several problems exist with these conventional methods, such as aeration-associated costs, enormous energy expenses, and sludge control. In addition, due to the intensification of human activities, effluent compositions are becoming more complex. Thus, there is a dire need to develop suitable WW treatment methods that are easy to employ, cost-effective, and ecologically responsible for reducing water pollution (Hena et al., 2021).

Phycoremediation is a technique that employs either micro or macroalgae for removing or biotransforming pollutants like heavy metals, nutrients, and xenobiotics from WW. Phycoremediation seems to be a sustainable method in contrast to conventional ones (Koul & Taak, 2018; Upadhyay et al., 2019a). The word algae has been derived from "phyco" (Greek for "alga") and is considered to be a group comprised heterogeneous, predominantly eukaryotic marine creatures that diverge from individual cells to extremely differentiated plants. Algae can fix carbon and release oxygen into the atmosphere at the expense of sunlight. More than 50% of the overall photosynthetic activity on this planet is performed by algae, and thus the foundation of the food chain is formed by them (Day et al., 2017). Algal species can be efficiently employed for treating WW as they have an innate characteristic of removing metals, nutrients, and organic compounds from WW (Laurens et al., 2017). It can also be used for the mitigation of carbon dioxide in the environment owing to its capacity to sequester CO_2. The employment of algae for the management of WW was first reported in 1960. The capability of the scheme for the generation of energy from the algal biomass was also described by Oswald and Golueke (1960). In multiple regions, oxidation ponds have been used where algae and photosynthetic bacteria are managed to grow, as a result of which WW is oxidized with the fixation of carbon dioxide and the release of oxygen (Milano et al., 2016).

Microalgae offer a cost-effective and significant scheme for removing an excessive amount of toxic compounds from WW while growing potentially beneficial biomass since they can uptake inorganic nutrients. It would be advantageous to employ microalgae in the treatment methods because the manufacturers demand a cost-efficient and constant process of treatment. Algal cells can be grown and manipulated easily in laboratory conditions, making them an ideal system for remediation studies. Moreover, phycoremediation offers more benefits in contrast to physicochemical processes like dialysis, ion exchange, activated carbon adsorption, reverse osmosis, membrane separation, electro-dialysis, and chemical oxidation or reduction because of its small costs of operation and the acclimatization of nitrogen (N) and phosphorus (P) into algal biomass. Thus, phycoremediation prohibits the requirement of sludge handling and oxygenation of seepage prior to its emission into the aquatic body. This method is considered to be environmentally friendly because once the nutrients are removed, algal biomass can be again utilized as fertilizer without growing any subsequent pollutant (Liu et al., 2016). Microalgae demonstrate efficient removal of both inorganic and organic pollutants despite the high nutrient content found in agro-industrial WW (Das et al., 2019).

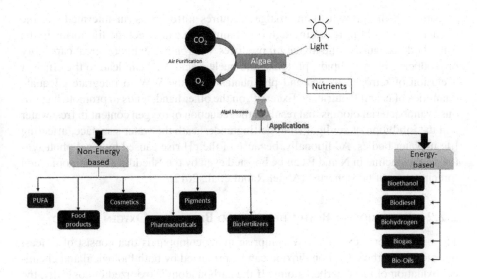

FIGURE 2.1   Energy and non-energy-based applications of algae.

Researchers have successfully engineered such strains of algae that can be used for WW treatment by detoxifying the pollutants. By employing molecular techniques and the approach of functional genomics, their photosynthetic effectiveness has been enhanced. Moreover, the techniques were used to increase their resistance to climate change (Lutzu et al., 2021). A diverse range of non-pathogenic algae is being used for treating WW, such as *Chlamydomonas* sp., *Nostoc* sp., *Oscillatoria* sp., *Scenedesmus* sp., *Chlorella* sp., and *Spirulina* sp. When *Chlorella reinhardtii* and *Chlorella vulgaris* were grown in swine WW, the amounts of total nitrogen (TN) and chemical oxygen demand (COD) increased significantly. *Chlorella reinhardtii* decreased COD by 46% and TN by 90%, and *Chlorella vulgaris* decreased COD and TN by 59% and 93%, respectively (Emparan et al., 2019).

Algal biomass has a broad array of uses via numerous biological conversion routes (Figure 2.1). The production of algae on a huge scale for WW treatment can also lead to a wider range of uses, such as the production of biogas and biochar, which can be employed as a replacement for fuel or coal. Algae-based fertilizers also offer many advantages for circular bioeconomy and soil fertilization (Sharma et al., 2021).

2.2   WASTEWATER PHYCOREMEDIATION

2.2.1   NUTRIENTS REMOVAL

Different types of nutrients, like phosphorus and nitrogen, are taken up by algae. However, the amount of nutrients within and around the cell affects the process of treatment. How nutrients are taken up by algae is also affected by the rate of diffusion. Both the algal growth and the WW treatment procedure are affected by the ratios and concentrations of metabolites and nutrients. Nitrogen is present in WW primarily in the form of nitrate, nitrogen dioxide, or ammonium, while phosphorus is present as a phosphate or orthophosphate ion (Sen et al., 2013).

Although the growth of microalgae requires nitrogen as an intermediate, the greater amounts of nitrogen in the form of ammonia can affect the life forms in the water body because of the toxicity it produces. Microalgae exhibit a great capability of producing phospholipids, proteins, and nucleic acids and can lead to the efficient exclusion of nitrogen oxides and phosphates from the WW to integrate adequate quantities of N and P nutrients. Excess P, on the other hand, tends to promote hazardous cyanobacterial blooms that result in the reduction of oxygen content in freshwater and the inhibition of sunlight from getting underneath the water's surface, affecting life in water bodies. Additionally, because of the pH rise caused by algal photosynthesis, the decline in N and P can be boosted even by the 'shedding of ammonia' and 'precipitation of phosphorus' (Abdel-Raouf et al., 2012).

2.2.2 REDUCTION OF BOTH CHEMICAL AND BIOLOGICAL OXYGEN DEMAND

The organic complexes of WW comprise many compounds that consist of at least one atom of carbon. Carbon dioxide can be produced by both biological and chemical oxidation of these carbon atoms. If the carbon atom is oxidized biologically, the test is termed Biochemical Oxygen Demand (BOD), while for chemical oxidation, the test is termed COD. BOD takes advantage of the microbial ability to oxidize the organic matter into water and CO_2 by utilizing the oxidizing agent, which is molecular oxygen. Thus, BOD refers to the degree of the respiratory need of the bacteria, which is involved in metabolizing the organic material existing in a body of water. Surplus BOD can diminish the dissolved oxygen and consequently result in the death of aquatic life, so its elimination is a principal objective of WW treatment (Colak & Kaya, 1988).

As microalgae create oxygen through photosynthesis, they can alleviate BOD in WW. The elimination of phenolic compounds also decreases the waterbody's BOD. It has been reported by several experimental studies that some species of microalgae, such as *Spirulina* sp., *Chlorella kessleri*, and *Chlorella pyrenoidosa*, exhibit the potential to remove phenolic components of water bodies (Zhang et al., 2020a). However, in the process of WW treatment, removing phenols can be considered a problem by algae, as it can biodegrade phenolic compounds only under restricted carbon sources because, in such circumstances, phenols can be used by algae as a substitute for carbon sources. While WWs are often high in carbon sources for algae to absorb, the prospective usage of phenolic chemicals as an alternative means of energy is reduced (Table 2.1).

2.2.3 HEAVY METAL REMOVAL

Heavy metals that are present in household WW mainly constitute Cr, Hg, Cu, Zn, Pb, Cd, and As (Gupta & Bux, 2019). It is known that non-natural elements and heavy metals can be volatilized or detoxified by the metabolism of algae. The application of algal biomass for absorbing heavy metals has been reported as an alternative route. Several benefits are provided by algal species, which makes them a promising applicant for WW treatment. Such benefits include their capacity to grow in WW, their low nutrient demand, and their rapid rate of growth. Moreover, algae can grow

TABLE 2.1
Wastewater Treatment Efficacy of Several Algal Species

S.no.	Algal Species	Habitat	Contaminant Removal	Type of WW Treatment	References
1	Scenedesmus dimorphus	Freshwater	Ammonia and phosphorus	Industrial wastewater treatment; phosphate elimination 20%–55%	González et al. (1997)
2	Choleralla vulgaris	Freshwater	Ammonia and phosphorus	Domestic sewage treatment	González et al. (1997) and Sydney et al. (2011)
3	Botryococcus braunii	Freshwater	Hydrocarbons	Secondary treated sewage wastewater treatment	Mohamed et al. (2017, 2018)
4	Arthrospira platensis	Freshwater and brackish water	Zinc Yttrium (III)	Industrial wastewater treatment; nitrogen removal 96%–100%; phosphate removal 87%–99%	Phang et al., (2000), Yushin et al. (2022), and Zinicovscaia et al. (2018)
5	Ettlia oleoabundans	Freshwater	Nitrogen and phosphorus	Agricultural anaerobic trash runoff treatment	Baldisserotto et al. (2020) and Yang et al. (2011)
6	Chlorella kessleri	Freshwater	Heavy metals	Artificial medium treatment by algae advances to nitrogen exclusion 8%–19%; phosphate elimination 8%–20%	Lee and Lee (2001)
7	Spirulina maxima	Marine and freshwater	Ammonia and phenol	Sugarcane vinasse	Hamouda and El-Naggar (2021) and Lee et al. (2015)
8	Chlamydomonas debaryana	Freshwater	Chromium (VI) Chromium (III)	Swine wastewater treatment	Hasan et al. (2014)

robustly in severe climatic conditions without any growth requirement for land space. The formation of biofuel-like products from the developed biomass adds more to their potential for treating WW (Salama et al., 2019b).

It has been described that the employment of *Spirogyra communis* and *Chlorella pyrenoidosa* for treating WW with a large number of metal ions resulted in the reduction of Cr(IV), Pb(II), and Cu(II) from a 20% effluent quantity in 20 days (Sati et al., 2016). At the temperature range of 10°C–28°C, the green microalga *Chlorella minutissima* successfully eliminates zinc, cadmium, and copper due to its enhanced biosorption efficiency (Yang et al., 2021). At pH 2, *Codium tomentosum,* a sea green macroalgae, has an adsorption ability of 5.032 ± 0.644 for the Cr(VI) compound (Anandaraj et al., 2018). Different algal species, such as *Anabaena* sp., *Nostoc* sp., and *Ankistrodesmus* sp., were employed in an association for the removal of iron, copper, chromium, zinc, and lead from the dumpsite leachate. The findings described the order of elimination as lead> copper> zinc> iron> chromium (Iqbal et al., 2022). Algal species including *Chlorella* sp., *Scenedesmus* sp., and *Spirulina* sp. were evaluated for the sorption of cell phone shades with the 200 mg/kg concentration of indium and were found to have an effectiveness of ~70% (Chugainova, 2021).

The heavy metal removal process can be successful or unsuccessful depending on the species, the sort of WW, the kind of ion metals available, and the number of dead algae cells in the water body. Since live microalgal cells use heavy metals through biosorption and bioaccumulation, dead algal biomass may also remove heavy metals from WW by way of biosorption, though with considerably less efficiency than alive ones. The ability of biosorption not only depends upon the species of algae used but is also affected by the fact that either the algae are present as a free cell or an immobilized one. Another factor that needs to be considered while designing the WW treatment scheme is the type of contaminant. In addition to the identification of a suitable algal species, the approach of molecular genetics has also been employed for engineering a strain with enhanced qualities and capacities for eliminating heavy metals from water bodies (Salama et al., 2019b).

2.2.4 Bioremediation of PAHs and PCBs by Algae

Polycyclic aromatic hydrocarbons (PAH) and polychlorinated biphenyls (PCB) are the two major classes of contaminants. They are organic in nature and are closely controlled and screened because of their pliability, toxic nature, and extensive dispersal in the environment (Baghour, 2019). PCBs are generally employed in industries for different uses. These complexes are thought to be the main contaminants that can cause serious effects on the health of the public. These contaminants are also carcinogenic to both humans and animals. Because of their negative environmental impact, the United States Environmental Protection Agency assessed and designated these chemicals as the main contaminants (Lu et al., 2021).

Diverse strains of algae are utilized for degrading PAHs and PCBs. *Ulva lactuca* was found to be a promising candidate for accumulating PCBs detected within the range of 7–13 µg per 1 kg of algae-derived dry biomass (Net et al., 2015). Some other efficient species that can be used for removing PCBs are *Fucus vesiculosus, Caepidium antarcticum, Selenastrum capricornutum, Gracilaria gracilis, Cystoseira barbata, Desmarestia* sp., *Nitzschia* sp., *Skeletonema costatum*, and *Fucus virsoides*. Benzo[a] pyrene can also be effectively removed by *Scenedesmus acutus* and *Selenastrum capricornutum*, with the highest elimination rates of 99% and 95%, respectively. This maximal removal was achieved when cells of *Selenastrum capricornutum* and *Scenedesmus acutus* were exposed to benzo(a)pyrene for 15 and 72 h, respectively (García de Llasera et al., 2016). *Desmodesmus* sp., which has been utilized in the preparation of BioMnOx, showed the degradation of Bisphenol A with the highest reduction rate of 78% (Table 2.2).

TABLE 2.2
Examples of Algal Species Capable of Degrading PHAs and PCBs

Pollutants	Algal Species	References
Naproxen	*Cymbella* *Aspergillus niger* *Scenedesmus quadricauda*	Aracagök et al. (2017) and Ding et al. (2017)
Diazinon	*Selenastrum capricornutum* *Chlorella vulgaris*	EPA (2003) and Kurade et al. (2016)
Benzo(a)pyrene	*Selenastrum capricornutum* *Oscillatoria* (Cyanophyta) *Chlorella* (chlorophyta)	Aldaby and Mawad (2019) and García de Llasera et al. (2016)
Bisphenol A	*Chlorella vulgaris* *Aeromonas hydrophilia* *Chlamydomonas Mexicana* *Chlorella fusca* *Desmodesmus*	Gulnaz and Dincer (2009), Ji et al. (2014), and Wang et al. (2017)
Chlorobenzenes	*Chlorella pyrenoidosa*	Zhang et al. (2016)
Chlorpyrifos	*Cladosporium cladosporioides* *Chlorella vulgaris* *Spirulina platensis* *Synechocystis* *Merismopedia*	Chen et al. (2012, 2016) and Yadav et al. (2016)

2.2.5 CONTRIBUTION OF ALGAE TO WW DISINFECTION

Some pathogens, such as viruses, protozoa, and a few bacterial species, that exist in WW are of great concern. The grade of total coliform elimination in the aquatic body is used to assess the efficacy of WW treatment. Substances that promote algal development are detrimental to the existence of coliforms. The use of *Scenedesmus obliquus* in sewage effluent in high-rate algal ponds schemes resulted in overall *E. coli* exclusion in 4 days because of the high pH in the pond system being greater than pH 9.4, whereas it was stated that 2 days at pH 11 were appropriate for overall *E. coli* removal from the high-rate algal ponds. Waste stabilization pond systems outperform traditional sewage treatment methods in the phycoremediation process. It has been claimed that in stabilization ponds, up to 99.6% of coliform bacteria may be eliminated from WW. The high-rate algal ponds have reported a comparable percentage of coliform clearance of 99% (Abdel-Raouf et al., 2012).

2.3 APPLICATIONS OF HARVESTED BIOMASS

Algae have achieved phenomenal prominence worldwide as the most efficient source of green fuels. There is an expanding demand for algae because of their antioxidant, antiviral, anticancer, and cholesterol-reducing potential. It shows a wide range of uses in the medicinal as well as nutraceutical, pharmaceutical, cosmeceutical, and paper industries. Proteins (6%–52%), carbohydrates (5%–23%), and lipids (7%–23%)

constitute the biomass of algae, making it a significant feedstock for the commercial production of food, medicines, and cosmetics (Javed et al., 2022). Algal bioproducts can impact global change since they are sustainable, recyclable, ecologically sound, and give a path to a more sustainable society. The following section discusses some of the applications of algal biomass.

2.3.1 ALGAE AS A FOOD INGREDIENT

Algae offer a significant likelihood of fulfilling the expanding demand for proteins and can be used as a resource of bioactive substances like fats, amino acids, minerals, and vitamins for the improvement of conventional algal food products. Algae and its byproducts can be effectively utilized for manufacturing innovative food, supplying valuable biomass as an alternative to plant-based meat products (Onwezen et al., 2021).

Consumers often pay attention to several ingredients before following a plant-based diet. Foods that are rich in vitamins, minerals, proteins, and fats are usually consumed at higher rates. There is also a great focus on essential omega-3 fatty acids as a constituents of plant-based foods. Moreover, some of the additives or adjuvants that are employed in the food industry are usually polysaccharides such as alginate, carrageen, and agar-agar isolated from macroalgae and seaweed. These agents are generally used for thickening food products such as ice cream, drinks, and jellies. For example, soybean flour and sodium alginate can be assorted together to make an analog of fish or meat (Zhang et al., 2020b).

Recent investigations have described the utilization of algae or its extract for the production of innovative foods. Research has been made on the accessibility and digestibility of algal biomass in various food matrixes. For instance, *Arthrospira platensis* was investigated to evaluate its sensory, physical, and nutritional properties so that it can be used as a snack enrichment agent (Lucas et al., 2018). The digestibility of algal biomass was also checked in vitro to determine the possibility of using algae as a substitutive component in cookies (Batista et al., 2017). *Ascophyllum nodosum*, a brown alga, was also investigated for enhancing the physical and antioxidant characteristics of gluten-free bread (Różyło et al., 2017). The bread wheat pasta also indicated an enhanced nutritional quality due to the impact of *Spirulina* biomass (De Marco et al., 2014). Under *in vitro* research associated with the bioavailability of nutrients, all the above-mentioned investigations provide insight into the promising effects of utilizing algae-based food products.

However, some studies also report the negative effects of added algal biomass on the flavor and appearance of the resulting product, thus causing a reduction in overall consumers' acceptance rates. To avoid this problem, a maximum of 5% (w/w) of algal biomass is usually used and included in algae-based foods (Batista et al., 2017).

2.3.2 BIOFUELS

Third-generation biofuels are created by using microalgae-derived biomass. An additional benefit provided by this scheme of biofuel production is the significant production of lipids. Microalgae may operate as microbial factories, producing a range of compounds as well as lipids for biodiesel generation. Some algae species, such as *Chlorella* sp. or *Nannochloropsis* sp., may maintain up to 60%–70% of their storage components as lipids or polysaccharides when nutrients are scarce (John et al., 2020). By functioning as a biofuel source, microalgae have several benefits over angiospermic plant biomass.

i. More lipid compounds per unit area can be produced by using microalgae as compared to the use of traditional crops.
ii. Areas that are not suitable for traditional farming are favorable for the development of microalgae.
iii. They can flourish in restricted conditions.
iv. Marine environments and brackish water bodies also support the growth of microalgae but do not favor conventional farming.
v. TN or total phosphorus can be restored by microalgae.

However, conventional cultivation of crops is cost-effective as compared to microalgal cultivation, which necessitates the resources for blending, crushing, and processing (Liu et al., 2021).

2.3.3 BIOLOGICALLY ACTIVE COMPOUNDS

In addition to the energy sector, algae offers numerous other applications, some of which are yet to be fully discovered because of the broad and unexploited potential of this valuable resource. Even though a variety of algae-derived products are commercially accessible in the marketplace in the form of medicines, food, and cosmetics, there is still a demand for broad investigations in this regard. There are ~30,000 species of algae documented, but only a few among them have been thought to harness prospective value-added chemicals. They are remarkable lipid compounds made by marine microalgae, specifically *Dunaliella* sp., *Chlorella* sp., and *Spirulina* sp., which are the producers of polyunsaturated fatty acid, *Phaeodactylum tricornutum* and *Odontella aurita* as makers of eicosapentaenoic acid, and *Schizochytrium* sp. as manufacturers of docosahexaeno. These lipid molecules are being researched for the prevention of a range of diseases like asthma, cancer, cardiovascular disorders, skin and renal problems, and neurological illnesses such as depression and schizophrenia (Sathasivam et al., 2019; Souza et al., 2019).

Because of their renewing and antioxidant effects on human skin, cosmetics derived from lipids, such as lotions or creams, and microalgal biomass extracts produced by using ethanol or supercritical carbon dioxide, are currently growing industries (Kumar et al., 2020).

Another family of compounds with multiple therapeutic properties is microalgal polysaccharides. For example, certain heavily sulfated polysaccharides activate the immune response on a cellular or humoral level. Researchers have successfully identified polysaccharides that exhibit effectiveness, especially in *Cyanobacteria*, and also in green and red microalgae. In the domain of nanomedicine, the use of algae to produce bio-nanoparticles is gaining prominence. For instance, Ar-Ag nanoparticles generated from the marine red algae *Amphiroa rigida* demonstrated great efficacy by acting as reducing agents. Additionally, *Staphylococcus aureus* and *Pseudomonas aeruginosa* are also susceptible to the larvicidal and antibacterial actions of Ar-Ag nanoparticles. Ar-Ag nanoparticles were also found to be toxic against MCF-7 human breast cancer cells (Gopu et al., 2021).

2.4 MECHANISMS OF PHYCOREMEDIATION

Sedimentation, flocculation, and rhizofiltration are the elementary methods by which algae eliminate pollutants (Stauch-White et al., 2017). Microalgae may absorb and ingest heavy metals, plant nutrients, organic and inorganic pollutants, pesticides, and radioactive materials in their unicellular bodies. Its potential for lowering N, P, and HM has resulted in several benefits, notably in improving water properties, and offers an easier, more suitable, and cost-effective alternative to traditional environmental clean-up efforts. Biochemical, physical, and biological approaches are used to decrease contaminants like inorganic and organic pollutants. Absorption, oxidation/reduction, cation, anion exchange, and precipitation are all part of the biochemical method, while the open pond system/waste stabilization pond is a biological method (Upadhyay et al., 2019b). The schematic illustration in Figure 2.2 provides the basic outline of phycoremediation.

2.4.1 CATION/ANION EXCHANGE

The presence of various functional groups on ($-$COOH, $-$OH, PO$_3^{2-}$, $-$P2O3, $-$NH2, $-$SH, aromatic, carboxyl, alkyl, and amide) results in a negative charge on the cell wall and enables the adsorption and absorption of metals (cations) on it. Consequently, they serve as sturdy binding sites for metal cations and engage in metal exchange using the ion exchange technique. This approach to biochemically removing HM from aquatic systems (due to cation/anion exchange) is particularly successful since it provides an opportunity for remediation and metal extraction from WW (Upadhyay et al., 2019b).

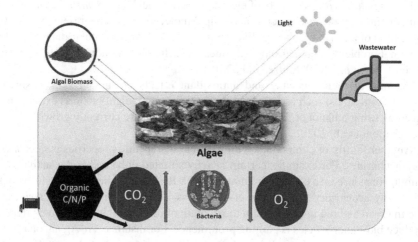

FIGURE 2.2 Illustration of the schematic phycoremediation process.

2.4.2 ABSORPTION

WW contains a lot of inorganic ions, such as phosphates, nitrates, and heavy metals. Microalgae employ the assimilation process to transform inorganic N into organic N. Inorganic N is translocated into the cytoplasm during this process. When cytoplasm is comprised of nitrate reductases, a series of oxidation and reduction processes take place, turning inorganic N into ammonium, which is subsequently absorbed. All the lipids, proteins, and nucleic acids are primarily composed of phosphorus, which is utilized in algal metabolism as dihydrogen phosphate and hydrogen phosphate. The phosphorylation step further converts orthophosphate into organic compounds in algae. With enhanced electronegativity and reduced ionic radii, algal biomass rapidly absorbs metal ions (Upadhyay et al., 2019a).

2.4.3 PRECIPITATION

In WW, algae release a variety of compounds, such as organic acids and secondary metabolites, causing a reduction in the nearby pH and thus facilitating the precipitation of numerous toxic pollutants. Precipitation facilitates P drop in WW by soluble Fe, Al, or Ca. At low pH, the cell wall's active sites are coupled with protons, leaving no free sites for metal cations to attach. As a consequence, as pH starts to rise, the number of negatively charged sites increases, causing the adherence of metal cations to the surface of cells, reducing their bioavailability even further (Leong & Chang, 2020).

2.5 PERSPECTIVES OF MICROPLASTICS

Owing to the economic viability and accessibility of plastic products in various parts of daily life, the manufacture of plastic items has grown dramatically. Plastics are classified as non-biodegradable pollutants since they cannot be decomposed naturally and are therefore very persistent pollution for the environment. According to Plastics Europe research, worldwide plastics output has surpassed 350 million tons, and experts warn that it may exceed 500 million tons by 2025 if we do not take fast efforts to mitigate it (Geyer et al., 2017). Plastic products disintegrate into small particle sizes as a result of numerous physicochemical and biological processes, resulting in nanoplastics (5–50 cm), megaplastics (>50 cm), or microplastics (Lebreton et al., 2018).

Microplastics are plastic particles that are <5 mm in size (Wang et al., 2019). Because of their pervasiveness and endurance in the atmosphere, Microplastics have become a source of research and societal concern (Galgani et al., 2013). Because of their polymeric nature and ease of transit between various environments, microplastics are of great concern to biologists and environmentalists.

2.5.1 DIVERSITY AND FATE OF MPs

Plastic pollution primarily involves microplastics that can induce hazardous conditions in our environment. Microplastics are among the top ten pollutants in the world that need to be addressed immediately (UNEP, 2011). Microplastics offer a diverse nature and can be classified as primary MPs and secondary MPs based on their derivation. Primary MPs are little plastic bits produced as nurdles (containing microbeads and glitter) so that they can be utilized in cosmeceutical or skin care products (Europe, 2019). Plastic particles (like road wear, tire abrasion, and films) generated from big polymers by mechanical fragmentation and photooxidative reactions are referred to as secondary MPs (Horton et al., 2017). Primary MPs account for 21% of all MPs released into aquatic ecosystems each year, while secondary MPs include the remaining 79% (Menéndez-Pedriza & Jaumot, 2020). One of the critical challenges relating to food contamination and social well-being is the interaction of MPs and the soil-plant system (Allouzi et al., 2021).

Usually, there are two major aspects regarding microplastics and their toxic nature:

i. Microplastics often show self-toxic properties, as they can be accumulated in living beings and disrupt their usual metabolic functions. Several reproductive, intestinal, and neurological activities can be disturbed due to the toxicity of microplastics (Zhu et al., 2020). The natural biota of our environment can also be damaged due to the release of additives from the surface of microplastics. Under the circumstances of air, UV, or water stress, microplastics can release various additives like biocides, flame retardants, UV stabilizers, etc. However, such additives are not considered a natural composition of MPs (Lambert et al., 2014).

ii. MPs exhibit an extremely large surface area due to their minute size. They show great potential for adsorbing different materials, like heavy metals or persistent organic pollutants (POPs). The adsorbed matter can move from one habitat to another through the food chain and cause numerous toxic effects on our environment, leading to the disruption of the whole natural food web (Du et al., 2021).

It is important to eradicate MPs and their toxic effects if the fate of the environment needs to be determined. MPs elimination is difficult since they cannot be recycled or separated once reaching freshwater or marine habitats (Thompson & De Falco, 2020). As a result, the invention of methods to aid in plastic breakdown has enticed increasing research.

2.5.2 POTENTIAL RISKS OF AVAILABLE STRATEGIES FOR MP REMOVAL

Strategies that are commonly employed for the removal of microplastics include photocatalytic degradation, electrocoagulation, dynamic membranes, membrane bioreactors, conventional activated sludge processes, and the installation of a WW treatment plant. Large-sized MPs are removed by water-surface cleaning methods such as nets, but minute plastic particles are not removed. As a result, a different strategy is necessary to remove the smaller pieces of plastic trash. As compared to traditional water treatment methods, the technique using the membrane bioreactor can eliminate more MPs (>90%) from water-based solutions (Talvitie et al., 2017). Nevertheless, these filtering modules are costly and readily broken, resulting in greater operational expenses. Also, the used membranes need extensive washing to avoid severe membrane fouling. Moreover, depending on the shape of the membrane, the rate of removal varies, and owing to the flat structure, sludge can accumulate in them (Dyachenko et al., 2019).

Zinc oxide photocatalytic decomposition is considered a useful technique to be employed in WW treatment plants (Razali et al., 2020). In this process, strong oxidizing radicals act on the surface of microplastics, as a result of which, polymer chains in microplastics get oxidized and ruptured. The major drawback of this method is associated with the management cost, which restricts its functional use. Furthermore, certain photocatalytic cracking end products can be harmful to both animals and humans and therefore turn out to be a hindrance at the wastewater treatment plants level. Some other disadvantages of employing this process include an absence of selectivity, minor effectiveness in contrast to other approaches, the complexity of scaling up, the production of secondary-type organic contaminants, enormous photoreactor arrangements, extreme energy consumption, and difficulty in photocatalyst recovery (Wang et al., 2018).

Traditional methods (like dissolved air flotation, coagulation/flocculation, and filtering) can retain some of the big and higher-density MPs (Wang et al., 2020). Smaller, low-density MPs, on the other hand, are not entrapped and are instead let out into their surroundings (Perren et al., 2018). Chemical addition to the medium is ineffective for large microplastics (Talvitie et al., 2017). Long retention durations in the tank, a huge sedimentation area, high energy costs, and sludge processing and disposal are all disadvantages of conventional activated sludge (Wang et al., 2018). As a result, an appropriate approach to microplastic remediation is necessary to avoid these issues and accomplish considerable removal and degradation of microplastics.

2.5.3 ALGAE AS A POTENTIAL CANDIDATE FOR MP REMOVAL

Microalgae and their enzymes can be employed to successfully degrade polymeric materials (Chia et al., 2020). Since algae can disintegrate complex polymers and do not include endotoxins, bioremediation is the best method for eliminating MPs (Yang et al., 2014). The key benefit is that, unlike other biological systems, they do not need a rich carbon supply for development and are adaptable to a wide range of settings where most microplastics exist (Yan et al., 2016). Also, the algae-based technique requires less maintenance than the traditional photocatalytic breakdown method (Wang et al., 2018). Even if MPs are contaminant vectors in an ecosystem, algae can repair pollutants linked to MPs because algae lower a variety of heavy metals and pollutants in aquatic environments (Oyebamiji et al., 2021; Hena et al., 2020). Furthermore, algae can lower turbidity, total soluble solids, alkalinity, sulfates, and nitrates in the water while raising the pH from 7.40 to 8.20 (Ugya et al., 2021). To lessen the danger to the environmental integrity and working of natural aquatic food webs, it is vital to investigate the mechanism of contact among MPs and algae (Figure 2.3).

Many algae species are effective at biodegrading microplastics. Since low-density polyethylene is utilized as a main source of energy and carbon for algae, *Phormidium lucidum* and *Oscillatoria subbrevis* have shown a significant potential for microplastic biodegradation (Sarmah & Rout, 2018b). Certain microalgae, like the green photosynthetic alga *Scenedesmus dimorphus*, the diatom *Navicula pupula*, and notably the blue-green photosynthetic alga *Anabaena spiroides*, may digest low- and

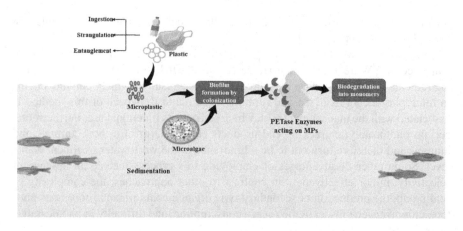

FIGURE 2.3 Fate and degradation of MPs by algae.

high-density polyethylene, with low-density polyethylene degrading more efficiently (Kumar et al., 2017). Microplastics, on the other hand, severely hinder the growth of *Spirulina* sp. as well as other photosynthetic algae because they restrict the quantity of light that moves into the system and hence halt the photosynthetic potential of algae (Campanale et al., 2020). As a result, more study into the algal biodegradation of microplastics is required.

2.5.4 VIABLE REMOVAL MECHANISM OF MPS BY ALGAE

Settlement, aggregation, and adsorption are the main processes involved in MP removal by algae. Since algae's extracellular polymeric substance has a great propensity to cling to and capture MPs, it can aggregate with them to produce hetero-aggregations (Li et al., 2020). Moreover, microalgae colonize the surface of microplastics and cause chemical variations in the MPs, which might be considered for removal (Kumar et al., 2017). Because various polymers promote their synthesis in different ways, their amounts and compositions vary. This controls the variety in biofilm cohesiveness and, eventually, the biomass of organisms that might colonize them (Lagarde et al., 2016). Moreover, several parameters, such as plastic aging, biogeography (Amaral-Zettler et al., 2020), temperature (Chen et al., 2019), salinity, available nutrients, and material exposure time in the environment, play an essential influence on the algal colonization of microplastic surfaces (Fu et al., 2019; Li et al., 2019).

Algae interactions with plastic polymer debris can result in synergetic alterations, which have consequences for the fate of MPs in aquatic settings. Biofilms alter MP stickiness, polymer density, and porosity (Nava & Leoni, 2021). Certain microbes, including algae, survive MP toxicity, with an initial period of weakening followed by adaptive effects that aid in regeneration; consequently, the effects observed in algae seem to be temporary (Dehghani et al., 2017).

When microplastics contact algae, they are detoxified. Various detoxifying processes, such as membrane thickening, surface exposure reduction via homo-aggregation, and hetero-aggregation, have been proposed to impact microalgal activity recovery. The adhesion of algae to the surface initiates biodegradation, and the synthesis of ligninolytic and exopolysaccharide enzymes is a critical stage in the breakdown process (Sarmah & Rout, 2018a). The algae enzymes in the liquid medium subsequently engage with molecules on the plastic surface, causing biodegradation (Chinaglia et al., 2018). The degrading mechanisms of MPs differ according to their chemical makeup and must be explored further to protect the ecosystem (Yoshida et al., 2016).

In comparison to higher plants, eukaryotic cells, and other sophisticated creatures, algae have a simpler genetic structure. Omics technology paired with synthetic biology techniques can give a detailed insight into the methods and genetic features of algae that may help to combat MP contamination (Salama et al., 2019a). Factory cells that are accountable for the degradation and disposal of microplastics can be created by genetically modifying algae with plastic-degrading enzymes. C. reinhardtii was recently engineered to produce polyethylene terephthalate (PET)-degrading enzymes (PETase), and the modified cell was co-incubated with PET, resulting in scuffs and fractures on the PET film surface as well as the creation of terephthalic acid, which is a PET breakdown product (Kim et al., 2020). By employing the microalgae *Phaeodactylum tricornutum* as a host, PETase was successfully produced that can act against PET and the copolymer PET glycol (Moog et al., 2019).

These investigations show that algae provide a fundamental podium for plastic exclusion, which might be a potential and viable substitute for using bacteria for biological plastic breakdown. Therefore, the use in algomics might assist to recognizing the metabolism, evolution, and response of algae species, and the hazards related to suspected MP toxicity.

2.6   CONCLUSION

Owing to the unique characteristics of algae, great progress has been made in the techniques of phycoremediation. The use of algae offers better remediation of pollutants as compared to conventional techniques. This is because algae grow at a rapid rate and can adapt to a variety of habitats where it is difficult to employ other remediation methods. Algae also exhibit a great potential for degrading microplastics that may otherwise lead to the destruction of the natural food web. However, there is a dire need to focus on the concern regarding the contracted growth of algae in the existence of MPs. The production and applications of genetically engineered algal strains explore new possibilities to investigate their potential against pollutants such as heavy metals and microplastics. Studies have shown that algae provide a fundamental podium for pollutant exclusion and, therefore, must be investigated more properly to identify the metabolic reactions, evolution, activities of algal species, and threats related to the alleged toxicity of environmental pollutants.

REFERENCES

Abdel-Raouf, N., Al-Homaidan, A. A., & Ibraheem, I. B. M. (2012). Microalgae and wastewater treatment. *Saudi Journal of Biological Sciences*, *19*(3), 257–275. https://doi.org/10.1016/j.sjbs.2012.04.005.

Akhil, D., Lakshmi, D., Senthil Kumar, P., Vo, D. V. N., & Kartik, A. (2021). Occurrence and removal of antibiotics from industrial wastewater. *Environmental Chemistry Letters, 19*, 1477–1507.

Aldaby, E. S. E., & Mawad, A. M. M. (2019). Pyrene biodegradation capability of two different microalgal strains. *Global Nest Journal, 21*(3), 290–295. https://doi.org/10.30955/gnj.002767.

Allouzi, M. M. A., Tang, D. Y. Y., Chew, K. W., Rinklebe, J., Bolan, N., Allouzi, S. M. A., & Show, P. L. (2021). Micro (nano) plastic pollution: The ecological influence on soil-plant system and human health. *Science of the Total Environment, 788*, 147815.

Amaral-Zettler, L. A., Zettler, E. R., & Mincer, T. J. (2020). Ecology of the plastisphere. *Nature Reviews Microbiology, 18*(3), 139–151.

Anandaraj, B., Eswaramoorthi, S., Rajesh, T. P., Aravind, J., & Suresh Babu, P. (2018). Chromium (VI) adsorption by Codium tomentosum: Evidence for adsorption by porous media from sigmoidal dose-response curve. *International Journal of Environmental Science and Technology, 15*, 2595–2606.

Aracagök, Y. D., Göker, H., & Cihangir, N. (2017). Biodegradation of micropollutant naproxen with a selected fungal strain and identification of metabolites. *Zeitschrift Fur Naturforschung - Section C Journal of Biosciences, 72*(5–6), 173–179. https://doi.org/10.1515/znc-2016-0162.

Baghour, M. (2019). Algal degradation of organic pollutants. *Handbook of Ecomaterials, 1*, 565–586.

Baldisserotto, C., Demaria, S., Accoto, O., Marchesini, R., Zanella, M., Benetti, L., Avolio, F., Maglie, M., Ferroni, L., & Pancaldi, S. (2020). Removal of nitrogen and phosphorus from thickening effluent of an urban wastewater treatment plant by an isolated green microalga. *Plants, 9*(12), 1–23. https://doi.org/10.3390/plants9121802.

Batista, A. P., Niccolai, A., Fradinho, P., Fragoso, S., Bursic, I., Rodolfi, L., Biondi, N., Tredici, M. R., Sousa, I., & Raymundo, A. (2017). Microalgae biomass as an alternative ingredient in cookies: Sensory, physical and chemical properties, antioxidant activity and in vitro digestibility. *Algal Research, 26*, 161–171.

Campanale, C., Massarelli, C., Savino, I., Locaputo, V., & Uricchio, V. F. (2020). A detailed review study on potential effects of microplastics and additives of concern on human health. *International Journal of Environmental Research and Public Health, 17*(4), 1212.

Chen, S., Chen, M., Wang, Z., Qiu, W., Wang, J., Shen, Y., Wang, Y., & Ge, S. (2016). Toxicological effects of chlorpyrifos on growth, enzyme activity and chlorophyll a synthesis of freshwater microalgae. *Environmental Toxicology and Pharmacology, 45*, 179–186.

Chen, S., Liu, C., Peng, C., Liu, H., Hu, M., & Zhong, G. (2012). Biodegradation of chlorpyrifos and its hydrolysis product 3,5,6-trichloro-2-pyridinol by a new fungal strain Cladosporium cladosporioides Hu-01. *PLoS One, 7*(10), e47205. https://doi.org/10.1371/journal.pone.0047205.

Chen, X., Xiong, X., Jiang, X., Shi, H., & Wu, C. (2019). Sinking of floating plastic debris caused by biofilm development in a freshwater lake. *Chemosphere, 222*, 856–864.

Chia, W. Y., Tang, D. Y. Y., Khoo, K. S., Lup, A. N. K., & Chew, K. W. (2020). Nature's fight against plastic pollution: Algae for plastic biodegradation and bioplastics production. *Environmental Science and Ecotechnology, 4*, 100065.

Chinaglia, S., Tosin, M., & Degli-Innocenti, F. (2018). Biodegradation rate of biodegradable plastics at molecular level. *Polymer Degradation and Stability, 147*, 237–244.

Chugainova, A. A. (2021). Efficiency of sorption of metals from electronic waste by microscopic algae. *IOP Conference Series: Earth and Environmental Science, 723*(4), 42055.

Colak, O., & Kaya, Z. (1988). A study on the possibilities of biological wastewater treatment using algae. *Doga Biyolji Serisi, 12*(1), 18–29.

Das, A., Adhikari, S., & Kundu, P. (2019). Bioremediation of wastewater using microalgae. In: R. Kundu, R. Narula, R. Paul, & S. Mukherjee (eds), *Environmental Biotechnology for Soil and Wastewater Implications on Ecosystems*, pp. 55–60. Springer: Singapore.

Day, J. G., Gong, Y., & Hu, Q. (2017). Microzooplanktonic grazers: A potentially devastating threat to the commercial success of microalgal mass culture. *Algal Research*, *27*, 356–365.

De Marco, E. R., Steffolani, M. E., Martínez, C. S., & León, A. E. (2014). Effects of spirulina biomass on the technological and nutritional quality of bread wheat pasta. *LWT-Food Science and Technology*, *58*(1), 102–108.

Dehghani, S., Moore, F., & Akhbarizadeh, R. (2017). Microplastic pollution in deposited urban dust, Tehran metropolis, Iran. *Environmental Science and Pollution Research*, *24*, 20360–20371.

Ding, T., Lin, K., Yang, B., Yang, M., Li, J., Li, W., & Gan, J. (2017). Biodegradation of naproxen by freshwater algae *Cymbella* sp. and *Scenedesmus quadricauda* and the comparative toxicity. *Bioresource Technology*, *238*, 164–173. https://doi.org/10.1016/j.biortech.2017.04.018.

Du, H., Xie, Y., & Wang, J. (2021). Microplastic degradation methods and corresponding degradation mechanism: Research status and future perspectives. *Journal of Hazardous Materials*, *418*, 126377.

Dyachenko, A., Lash, M., & Arsem, N. (2019). Method development for microplastic analysis in wastewater. In: H. K. Karapanagioti, & I. K. Kalavrouziotis (eds.), *Microplastics in Water and Wastewater*, pp. 63–83. IWA Publishing: London, UK.

Emparan, Q., Harun, R., & Danquah, M. K. (2019). Role of phycoremediation for nutrient removal from wastewaters: A review. *Applied Ecology and Environmental Research*, *17*(1), 889–915.

Epa, U. (2003). Aquatic life ambient water quality criteria: Diazinon, dinal. December. https://www.epa.gov/wqc/final-recommended-aquatic-life-ambient-water-quality-criteria-diazinon.

Europe, P. (2019). Plastics-the facts: An analysis of European plastics production, demand and waste data. Plastics Europe, Brussels. Https://Www.Plasticseurope.Org/Download_file/Force/2367/181, Accessed, 11.

Fu, D., Zhang, Q., Fan, Z., Qi, H., Wang, Z., & Peng, L. (2019). Aged microplastics polyvinyl chloride interact with copper and cause oxidative stress towards microalgae Chlorella vulgaris. *Aquatic Toxicology*, *216*, 105319.

Galgani, F., Hanke, G., Werner, S., & De Vrees, L. (2013). Marine litter within the European marine strategy framework directive. *ICES Journal of Marine Science*, *70*(6), 1055–1064.

García de Llasera, M. P., de Jesús Olmos-Espejel, J., Díaz-Flores, G., & Montaño-Montiel, A. (2016). Biodegradation of benzo (a) pyrene by two freshwater microalgae *Selenastrum capricornutum* and *Scenedesmus acutus*: A comparative study useful for bioremediation. *Environmental Science and Pollution Research*, *23*, 3365–3375.

Geyer, R., Jambeck, J. R., & Law, K. L. (2017). Production, use, and fate of all plastics ever made. *Science Advances*, *3*(7), e1700782.

González, L. E., Cañizares, R. O., & Baena, S. (1997). Efficiency of ammonia and phosphorus removal from a *Colombian agroindustrial* wastewater by the microalgae *Chlorella vulgaris* and *Scenedesmus dimorphus*. *Bioresource Technology*, *60*(3), 259–262.

Gopu, M., Kumar, P., Selvankumar, T., Senthilkumar, B., Sudhakar, C., Govarthanan, M., Selva Kumar, R., & Selvam, K. (2021). Green biomimetic silver nanoparticles utilizing the red algae *Amphiroa rigida* and its potent antibacterial, cytotoxicity and larvicidal efficiency. *Bioprocess and Biosystems Engineering*, *44*, 217–223.

Gulnaz, O., & Dincer, S. (2009). Biodegradation of bisphenol a by *Chlorella vulgaris* and *Aeromonas Hydrophilia*. *Journal of Applied Biological Sciences*, *3*(2), 7. www.nobel.gen.tr.

Gupta, S., & Bux, F. (2019). *Application of Microalgae in Wastewater Treatment*. Springer: Cham.

Hamouda, R. A., & El-Naggar, N. E. A. (2021). Chapter 14 - Cyanobacteria-based microbial cell factories for production of industrial products. In: V. Singh (ed.), *Microbial Cell Factories Engineering for Production of Biomolecules*, pp. 277–302. Academic Press. https://doi.org/https://doi.org/10.1016/B978-0-12-821477-0.00007-6.

Hasan, R., Zhang, B., Wang, L., & Shahbazi, A. (2014). Bioremediation of swine wastewater and biofuel potential by using *Chlorella vulgaris, Chlamydomonas reinhardtii*, and *Chlamydomonas debaryana*. *Journal of Petroleum & Environmental Biotechnology*, *5*(3), 175–180.

Hena, S., Gutierrez, L., & Croué, J.-P. (2020). Removal of metronidazole from aqueous media by *C. vulgaris*. *Journal of Hazardous Materials*, *384*, 121400.

Hena, S., Gutierrez, L., & Croué, J.-P. (2021). Removal of pharmaceutical and personal care products (PPCPs) from wastewater using microalgae: A review. *Journal of Hazardous Materials*, *403*, 124041.

Horton, A. A., Svendsen, C., Williams, R. J., Spurgeon, D. J., & Lahive, E. (2017). Large microplastic particles in sediments of tributaries of the River Thames, UK-Abundance, sources and methods for effective quantification. *Marine Pollution Bulletin*, *114*(1), 218–226.

Iqbal, J., Javed, A., & Baig, M. A. (2022). Heavy metals removal from dumpsite leachate by algae and cyanobacteria. *Bioremediation Journal*, *26*(1), 31–40.

Javed, M. U., Mukhtar, H., Hayat, M. T., Rashid, U., Mumtaz, M. W., & Ngamcharussrivichai, C. (2022). Sustainable processing of algal biomass for a comprehensive biorefinery. *Journal of Biotechnology*, *352*, 47–58. https://doi.org/10.1016/j.jbiotec.2022.05.009.

Ji, M.-K., Kabra, A., Choi, J., Hwang, J.-H., Kim, J. R., Abou-shanab, R., Oh, Y.-K., & Jeon, B.-H. (2014). Biodegradation of bisphenol A by the freshwater microalgae *Chlamydomonas mexicana* and *Chlorella vulgaris*. *Ecological Engineering*, *73*, 260–269. https://doi.org/10.1016/j.ecoleng.2014.09.070.

John, E. M., Sureshkumar, S., Sankar, T. V, & Divya, K. R. (2020). Phycoremediation in aquaculture; a win-win paradigm. *Environmental Technology Reviews*, *9*(1), 67–84.

Kim, J. W., Park, S.-B., Tran, Q.-G., Cho, D.-H., Choi, D.-Y., Lee, Y. J., & Kim, H.-S. (2020). Functional expression of polyethylene terephthalate-degrading enzyme (PETase) in green microalgae. *Microbial Cell Factories*, *19*(1), 1–9.

Koul, B., Poonia, A. K., Singh, R., & Kajla, S. (2022). Strategies to cope with the emerging waste water contaminants through adsorption regimes. In: S. Rodriguez-Couto, M. P. Shah, & J. K. Biswas (eds.), *Development in Wastewater Treatment Research and Processes*, pp. 61–106. Elsevier: Amsterdam, Netherlands.

Koul, B., & Taak, P. (2018). *Biotechnological Strategies for Effective Remediation of Polluted Soils*. Springer: Berlin, Germany.

Krishnan, R. Y., Manikandan, S., Subbaiya, R., Biruntha, M., Govarthanan, M., & Karmegam, N. (2021). Removal of emerging micropollutants originating from pharmaceuticals and personal care products (PPCPs) in water and wastewater by advanced oxidation processes: A review. *Environmental Technology & Innovation*, *23*, 101757.

Kumar, R., Ghosh, A. K., & Pal, P. (2020). Synergy of biofuel production with waste remediation along with value-added co-products recovery through microalgae cultivation: A review of membrane-integrated green approach. *Science of the Total Environment*, *698*, 134169.

Kumar, R. V., Kanna, G. R., & Elumalai, S. (2017). Biodegradation of polyethylene by green photosynthetic microalgae. *Journal of Bioremediation & Biodegradation*, *8*(381), 2.

Kurade, M. B., Kim, J. R., Govindwar, S. P., & Jeon, B.-H. (2016). Insights into microalgae mediated biodegradation of diazinon by *Chlorella vulgaris*: Microalgal tolerance to xenobiotic pollutants and metabolism. *Algal Research*, *20*, 126–134. https://doi.org/https://doi.org/10.1016/j.algal.2016.10.003.

Lagarde, F., Olivier, O., Zanella, M., Daniel, P., Hiard, S., & Caruso, A. (2016). Microplastic interactions with freshwater microalgae: Hetero-aggregation and changes in plastic density appear strongly dependent on polymer type. *Environmental Pollution, 215*, 331–339.

Lambert, S., Sinclair, C., & Boxall, A. (2014). Occurrence, degradation, and effect of polymer-based materials in the environment. *Reviews of Environmental Contamination and Toxicology, 227*, 1–53.

Laurens, L. M. L., Chen-Glasser, M., & McMillan, J. D. (2017). A perspective on renewable bioenergy from photosynthetic algae as feedstock for biofuels and bioproducts. *Algal Research, 24*, 261–264.

Lebreton, L., Slat, B., Ferrari, F., Sainte-Rose, B., Aitken, J., Marthouse, R., Hajbane, S., Cunsolo, S., Schwarz, A., & Levivier, A. (2018). Evidence that the great pacific garbage patch is rapidly accumulating plastic. *Scientific Reports, 8*(1), 1–15.

Lee, H. C., Lee, M., & Den, W. (2015). Spirulina maxima for phenol removal: Study on its tolerance, biodegradability and phenol-carbon assimilability. *Water, Air, and Soil Pollution, 226*(12). https://doi.org/10.1007/s11270-015-2664-3.

Lee, K., & Lee, C.-G. (2001). Effect of light/dark cycles on wastewater treatments by microalgae. *Biotechnology and Bioprocess Engineering, 6*, 194–199.

Leong, Y. K., & Chang, J.-S. (2020). Bioremediation of heavy metals using microalgae: Recent advances and mechanisms. *Bioresource Technology, 303*, 122886.

Li, S., Wang, P., Zhang, C., Zhou, X., Yin, Z., Hu, T., Hu, D., Liu, C., & Zhu, L. (2020). Influence of polystyrene microplastics on the growth, photosynthetic efficiency and aggregation of freshwater microalgae *Chlamydomonas reinhardtii*. *Science of the Total Environment, 714*, 136767.

Li, W., Zhang, Y., Wu, N., Zhao, Z., Xu, W., Ma, Y., & Niu, Z. (2019). Colonization characteristics of bacterial communities on plastic debris influenced by environmental factors and polymer types in the Haihe Estuary of Bohai Bay, China. *Environmental Science & Technology, 53*(18), 10763–10773.

Liu, C., Subashchandrabose, S., Ming, H., Xiao, B., Naidu, R., & Megharaj, M. (2016). Phycoremediation of dairy and winery wastewater using *Diplosphaera* sp. MM1. *Journal of Applied Phycology, 28*, 3331–3341.

Liu, Z.-Q., Huang, C., Li, J.-Y., Yang, J., Qu, B., Yang, S.-Q., Cui, Y.-H., Yan, Y., Sun, S., & Wu, X. (2021). Activated carbon catalytic ozonation of reverse osmosis concentrate after coagulation pretreatment from coal gasification wastewater reclamation for zero liquid discharge. *Journal of Cleaner Production, 286*, 124951.

López-Serrano, M. J., Velasco-Muñoz, J. F., Aznar-Sánchez, J. A., & Román-Sánchez, I. M. (2020). Sustainable use of wastewater in agriculture: A bibliometric analysis of worldwide research. *Sustainability, 12*(21), 8948.

Lu, T., Zhang, Q., Zhang, Z., Hu, B., Chen, J., Chen, J., & Qian, H. (2021). Pollutant toxicology with respect to microalgae and cyanobacteria. *Journal of Environmental Sciences, 99*, 175–186.

Lucas, B. F., de Morais, M. G., Santos, T. D., & Costa, J. A. V. (2018). Spirulina for snack enrichment: Nutritional, physical and sensory evaluations. *LWT, 90*, 270–276.

Lutzu, G. A., Ciurli, A., Chiellini, C., Di Caprio, F., Concas, A., & Dunford, N. T. (2021). Latest developments in wastewater treatment and biopolymer production by microalgae. *Journal of Environmental Chemical Engineering, 9*(1), 104926.

Menéndez-Pedriza, A., & Jaumot, J. (2020). Interaction of environmental pollutants with microplastics: A critical review of sorption factors, bioaccumulation and ecotoxicological effects. *Toxics, 8*(2), 40.

Milano, J., Ong, H. C., Masjuki, H. H., Chong, W. T., Lam, M. K., Loh, P. K., & Vellayan, V. (2016). Microalgae biofuels as an alternative to fossil fuel for power generation. *Renewable and Sustainable Energy Reviews, 58*, 180–197.

Mohamed, R. M., Al-Gheethi, A. A., Aznin, S. S., Hasila, A. H., Wurochekke, A. A., & Kassim, A. H. (2017). Removal of nutrients and organic pollutants from household greywater by phycoremediation for safe disposal. *International Journal of Energy and Environmental Engineering, 8*(3), 259–272. https://doi.org/10.1007/s40095-017-0236-6.

Mohamed, R. M. S. R., Al-Gheethi, A. A., Wurochekke, A. A., Maizatul, A. Y., Matias-Peralta, H. M., & Mohd Kassim, A. H. (2018). Nutrients removal from artificial bathroom greywater using *Botryococcus* sp. Strain. *IOP Conference Series: Earth and Environmental Science, 140*(1). https://doi.org/10.1088/1755-1315/140/1/012026.

Moog, D., Schmitt, J., Senger, J., Zarzycki, J., Rexer, K.-H., Linne, U., Erb, T., & Maier, U. G. (2019). Using a marine microalga as a chassis for polyethylene terephthalate (PET) degradation. *Microbial Cell Factories, 18*(1), 1–15.

Nava, V., & Leoni, B. (2021). A critical review of interactions between microplastics, microalgae and aquatic ecosystem function. *Water Research, 188*, 116476.

Net, S., Henry, F., Rabodonirina, S., Diop, M., Merhaby, D., Mahfouz, C., Amara, R., & Ouddane, B. (2015). Accumulation of PAHs, Me-PAHs, PCBs and total mercury in sediments and marine species in coastal areas of Dakar, Senegal: Contamination level and impact. *International Journal of Environmental Research, 9*(2), 419–432.

Onwezen, M. C., Bouwman, E. P., Reinders, M. J., & Dagevos, H. (2021). A systematic review on consumer acceptance of alternative proteins: Pulses, algae, insects, plant-based meat alternatives, and cultured meat. *Appetite, 159*, 105058.

Oswald, W. J., & Golueke, C. G. (1960). Biological transformation of solar energy. In W.W Umbreit (Ed.), *Advances in Applied Microbiology*, vol. 2, pp. 223–262. Elsevier: Amsterdam, Netherlands.

Oyebamiji, O. O., Corcoran, A. A., Pérez, E. N., Ilori, M. O., Amund, O. O., Holguin, F. O., & Boeing, W. J. (2021). Lead tolerance and bioremoval by four strains of green algae from Nigerian fish ponds. *Algal Research, 58*, 102403.

Perren, W., Wojtasik, A., & Cai, Q. (2018). Removal of microbeads from wastewater using electrocoagulation. *ACS Omega, 3*(3), 3357–3364.

Phang, S. M., Miah, M. S., Yeoh, B. G., & Hashim, M. A. (2000). Spirulina cultivation in digested sago starch factory wastewater. *Journal of Applied Phycology, 12*, 395–400.

Razali, N., Abdullah, W. R. W., & Zikir, N. M. (2020). Effect of thermo-photocatalytic process using zinc oxide on degradation of macro/micro-plastic in aqueous environment. *Journal of Sustainability Science and Management, 15*, 1–14.

Różyło, R., Hameed Hassoon, W., Gawlik-Dziki, U., Siastała, M., & Dziki, D. (2017). Study on the physical and antioxidant properties of gluten-free bread with brown algae. *CyTA-Journal of Food, 15*(2), 196–203.

Salama, E.-S., Govindwar, S. P., Khandare, R. V, Roh, H.-S., Jeon, B.-H., & Li, X. (2019a). Can omics approaches improve microalgal biofuels under abiotic stress? *Trends in Plant Science, 24*(7), 611–624.

Salama, E.-S., Roh, H.-S., Dev, S., Khan, M. A., Abou-Shanab, R. A. I., Chang, S. W., & Jeon, B.-H. (2019b). Algae as a green technology for heavy metals removal from various wastewater. *World Journal of Microbiology and Biotechnology, 35*, 1–19.

Sarmah, P., & Rout, J. (2018a). Algal colonization on polythene carry bags in a domestic solid waste dumping site of Silchar town in Assam. *Phykos, 48*(67), e77.

Sarmah, P., & Rout, J. (2018b). Efficient biodegradation of low-density polyethylene by cyanobacteria isolated from submerged polyethylene surface in domestic sewage water. *Environmental Science and Pollution Research, 25*, 33508–33520.

Sathasivam, R., Radhakrishnan, R., Hashem, A., & Abd_Allah, E. F. (2019). Microalgae metabolites: A rich source for food and medicine. *Saudi Journal of Biological Sciences, 26*(4), 709–722.

Sati, M., Verma, M., & Rai, J. P. N. (2016). Phycoremediation of heavy metals by *Chlorella pyrenoidosa* and *Spirogyra communis*. *International Journal of Current Microbiology and Applied Sciences*, 5(10), 920–930.

Sen, B., Alp, M. T., Sonmez, F., Kocer, M. A. T., & Canpolat, O. (2013). Relationship of algae to water pollution and waste water treatment. *Water Treatment*, 14, 335–354.

Sharma, G. K., Khan, S. A., Shrivastava, M., Bhattacharyya, R., Sharma, A., Gupta, D. K., Kishore, P., & Gupta, N. (2021). Circular economy fertilization: Phycoremediated algal biomass as biofertilizers for sustainable crop production. *Journal of Environmental Management*, 287, 112295.

Souza, C. M. M., de Lima, D. C., Bastos, T. S., de Oliveira, S. G., Beirão, B. C. B., & Félix, A. P. (2019). Microalgae *Schizochytrium* sp. as a source of docosahexaenoic acid (DHA): Effects on diet digestibility, oxidation and palatability and on immunity and inflammatory indices in dogs. *Animal Science Journal*, 90(12), 1567–1574.

Stauch-White, K., Srinivasan, V. N., Camilla Kuo-Dahab, W., Park, C., & Butler, C. S. (2017). The role of inorganic nitrogen in successful formation of granular biofilms for wastewater treatment that support cyanobacteria and bacteria. *AMB Express*, 7, 1–10.

Sydney, E. B. D, Da Silva, T. E., Tokarski, A., Novak, A. C. D, De Carvalho, J. C., Woiciecohwski, A. L., Larroche, C., & Soccol, C. R. (2011). Screening of microalgae with potential for biodiesel production and nutrient removal from treated domestic sewage. *Applied Energy*, 88(10), 3291–3294.

Talvitie, J., Mikola, A., Koistinen, A., & Setälä, O. (2017). Solutions to microplastic pollution-removal of microplastics from wastewater effluent with advanced wastewater treatment technologies. *Water Research*, 123, 401–407.

Thompson, R. C., & De Falco, F. (2020). Marine litter: Are there solutions to this environmental challenge? In: M. Cocca, E. Di Pace, M. E. Errico, G. Gentile, A. Montarsolo, R. Mossotti, & M. Avella (eds.), *Proceedings of the 2nd International Conference on Microplastic Pollution in the Mediterranean Sea*, pp. 39–44. Springer Nature: Berlin.

Ugya, A. Y., Ajibade, F. O., & Hua, X. (2021). The efficiency of microalgae biofilm in the phycoremediation of water from River Kaduna. *Journal of Environmental Management*, 295, 113109.

UNEP, U. (2011). *Year Book 2011: Emerging Issues in Our Global Environment*. United Nations Environment Programme: Nairobi.

Upadhyay, A. K., Singh, R., & Singh, D. P. (2019a). Phycotechnological approaches toward wastewater management. In: R. N. Bharagava & P. Chowdhary (eds.), *Emerging and Eco-Friendly Approaches for Waste Management*, pp. 423–435. Springer: Berlin, Heidelberg.

Upadhyay, A. K., Singh, R., Singh, J. S., & Singh, D. P. (2019b). Microalgae-assisted phycoremediation and energy crisis solution: challenges and opportunity. In: V. G. Gupta (ed.), *New and Future Developments in Microbial Biotechnology and Bioengineering*, pp. 295–307. Elsevier: Amsterdam, Netherlands.

Wang, R., Wang, S., Tai, Y., Tao, R., Dai, Y., Guo, J., Yang, Y., & Duan, S. (2017). Biogenic manganese oxides generated by green algae *Desmodesmus* sp. WR1 to improve bisphenol: A removal. *Journal of Hazardous Materials*, 339, 310–319.

Wang, S., Lydon, K. A., White, E. M., Grubbs III, J. B., Lipp, E. K., Locklin, J., & Jambeck, J. R. (2018). Biodegradation of poly (3-hydroxybutyrate-co-3-hydroxyhexanoate) plastic under anaerobic sludge and aerobic seawater conditions: Gas evolution and microbial diversity. *Environmental Science & Technology*, 52(10), 5700–5709.

Wang, X., Liu, L., Zheng, H., Wang, M., Fu, Y., Luo, X., Li, F., & Wang, Z. (2020). Polystyrene microplastics impaired the feeding and swimming behavior of mysid shrimp *Neomysis japonica*. *Marine Pollution Bulletin*, 150, 110660.

Wang, Z., Qin, Y., Li, W., Yang, W., Meng, Q., & Yang, J. (2019). Microplastic contamination in freshwater: First observation in lake ulansuhai, yellow river basin, China. *Environmental Chemistry Letters*, 17, 1821–1830.

Yadav, M., Shukla, A. K., Srivastva, N., Upadhyay, S. N., & Dubey, S. K. (2016). Utilization of microbial community potential for removal of chlorpyrifos: A review. *Critical Reviews in Biotechnology*, *36*(4), 727–742. https://doi.org/10.3109/07388551.2015.1015958.

Yan, N., Fan, C., Chen, Y., & Hu, Z. (2016). The potential for microalgae as bioreactors to produce pharmaceuticals. *International Journal of Molecular Sciences*, *17*(6), 962.

Yang, J., Yang, Y., Wu, W.-M., Zhao, J., & Jiang, L. (2014). Evidence of polyethylene biodegradation by bacterial strains from the guts of plastic-eating waxworms. *Environmental Science & Technology*, *48*(23), 13776–13784.

Yang, X., Wang, H., Zhang, L., Kong, L., Chen, Y., He, Q., Li, L., Grossart, H.-P., & Ju, F. (2021). Marine algae facilitate transfer of microplastics and associated pollutants into food webs. *The Science of the Total Environment*, *787*, 147535.

Yang, Y., Xu, J., Vail, D., & Weathers, P. (2011). Ettlia oleoabundans growth and oil production on agricultural anaerobic waste effluents. *Bioresource Technology*, *102*(8), 5076–5082.

Yoshida, S., Hiraga, K., Takehana, T., Taniguchi, I., Yamaji, H., Maeda, Y., Toyohara, K., Miyamoto, K., Kimura, Y., & Oda, K. (2016). A bacterium that degrades and assimilates poly (ethylene terephthalate). *Science*, *351*(6278), 1196–1199.

Yushin, N., Zinicovscaia, I., Cepoi, L., Chiriac, T., Rudi, L., & Grozdov, D. (2022). Biosorption and bioaccumulation capacity of arthrospira platensis toward yttrium ions. *Metals*, *12*(9). https://doi.org/10.3390/met12091465.

Zhang, C., Wang, X., Ma, Z., Luan, Z., Wang, Y., Wang, Z., & Wang, L. (2020a). Removal of phenolic substances from wastewater by algae: A review. *Environmental Chemistry Letters*, *18*, 377–392.

Zhang, J., Liu, L., Jiang, Y., Shah, F., Xu, Y., & Wang, Q. (2020b). High-moisture extrusion of peanut protein-/carrageenan/sodium alginate/wheat starch mixtures: Effect of different exogenous polysaccharides on the process forming a fibrous structure. *Food Hydrocolloids*, *99*, 105311.

Zhang, S., Lin, D., & Wu, F. (2016). The effect of natural organic matter on bioaccumulation and toxicity of chlorobenzenes to green algae. *Journal of Hazardous Materials*, *311*, 186–193. https://doi.org/https://doi.org/10.1016/j.jhazmat.2016.03.017.

Zhu, L., Zhao, S., Bittar, T. B., Stubbins, A., & Li, D. (2020). Photochemical dissolution of buoyant microplastics to dissolved organic carbon: Rates and microbial impacts. *Journal of Hazardous Materials*, *383*, 121065.

Zinicovscaia, I., Yushin, N., Shvetsova, M., & Frontasyeva, M. (2018). Zinc removal from model solution and wastewater by Arthrospira (Spirulina) Platensis biomass. *International Journal of Phytoremediation*, *20*(9), 901–908. https://doi.org/10.1080/15226514.2018.1448358.

3 Phycoremediation of Sewage Wastewater

Afeefa Khalid and Hamid Mukhtar

3.1 INTRODUCTION

Wastewater treatment is a crucial process for protecting public health and the environment. Unprocessed industrial and commercial sewage consists of a combination of organic and inorganic elements, some of which are non-biodegradable. Failure to appropriately treat such sewage can create unfavorable results for both aquatic ecosystems and human health. Organic matter can consume oxygen in water, leading to low oxygen levels that can harm aquatic life, and inorganic substances such as heavy metals can accumulate in aquatic organisms, leading to toxic effects.

Conventional techniques for treating wastewater, such as activated sludge schemes, trickling filters, and rotating biological contactors, necessitate substantial resources to ensure optimum performance. They often need large land areas, consume high amounts of energy, and generate significant amounts of sludge, which requires further treatment or disposal. Furthermore, these processes may not be effective in removing some of the pollutants found in industrial wastewater, such as heavy metals and non-biodegradable chemicals.

Phycoremediation, a promising alternative technology, offers a sustainable solution for treating wastewater effluent. This process utilizes microalgae to remove pollutants from wastewater effluent, including nutrients, heavy metals, organic pollutants, and biological oxygen demand. The microalgae absorb these pollutants during their growth, converting them into biomass, which can be harvested and used for various applications such as biofuels, fertilizers, and animal feed. Algae, originating from the Greek term 'phyco', represents a diverse set of primarily eukaryotic species varying in size from standalone cells to established plants. Their photosynthetic abilities allow them to utilize CO_2 and produce oxygen (O_2) under sunlight, thus playing a crucial role in upholding ecologically balanced environments. Algal cells account for over 50% of the total photosynthetic activity and form an essential component of the food chain.

Algae are proficient in generating considerable amounts of biomass and energy, with some species containing as much as 60% lipid content for increased combustion heat and energy values. Recent research demonstrates that a combination of *Chlorella sorokiniana*, *Coelastrella* sp., and *Acutodesmus nygaardii* can significantly diminish phosphorus, chemical oxygen demand (COD), and ammonia after just two days of nutrient starvation, further proving the notable prospects of phycoremediation in industrial settings.

DOI: 10.1201/9781003390213-3

Waterways contamination remains a prevalent and enduring issue, sourced from various origins and contributing to the problem in diverse ways. Wastewater is particularly challenging to manage, as it presents a complex blend of both organic and inorganic compounds, reflecting the current lifestyle and technology. The components of wastewater can be broadly classified into two groups: organic and inorganic compounds. Organic components consist of carbohydrates, fats, sugars, and amino acids. Among these, amino acids constitute ~75% of the organic carbon content. In contrast, the inorganic compounds present in wastewater are diverse, including elements such as sodium, calcium, potassium, magnesium, arsenic, sulfur, phosphorus, bicarbonate, and ammonium salts, as well as heavy metals. Furthermore, wastewater also contains significant amounts of persistent organic pollutants, including aromatic and chlorinated pollutants such as polycyclic aromatic hydrocarbons (PAHs), polychlorinated biphenyls (PCBs), and organochlorinated pesticides. Among these, PAHs stand out as a particularly concerning category of persistent organic pollutants commonly found in wastewater. These pollutants are byproducts of the incomplete burning of fossil fuels such as coal and petroleum, as well as other sources like biomass combustion, industrial manufacturing, household heating, greenhouse gas emissions, landfills, and wildfires. Notably, pyrene, a specific type of PAH, is particularly abundant in wastewater generated by petroleum-related industries. The toxic effects of PAHs on living organisms have garnered increasing attention, making the removal of these compounds from wastewater an essential task (Dayana Priyadharshini et al., 2021).

Waterways contamination can originate from various sources, such as untreated or treated water discharge from rural and urban settings, industrial emissions, agricultural runoffs, and leachates from waste disposal sites, as indicated in Table 3.1. The attainment of optimal wastewater treatment necessitates effective monitoring of several vital parameters, including but not limited to pH, color, odor, total nitrogen (TN) and total phosphorus (TP) levels, COD and biochemical oxygen demand (BOD),

TABLE 3.1
Pollutant Sources and Their Effects on the Environment

Pollutant	Source	Effects
Nutrients (nitrogen and phosphorus)	Human and animal waste, agricultural runoff, industrial discharge	Eutrophication, harmful algal blooms, oxygen depletion, water quality degradation
Organic matter	Human and animal waste, food processing, paper mills, breweries	Depletion of dissolved oxygen, unpleasant odor, potential health risks
Heavy metals (e.g., lead, mercury, cadmium)	Industrial discharge, mining activities, stormwater runoff	Toxic to aquatic organisms, accumulation in food chains, potential human health risks
Pathogens (e.g., bacteria, viruses, parasites)	Human and animal waste, untreated sewage discharge	Spread of waterborne diseases, potential human health risks
Chemical substances (e.g., pesticidal products, pharmaceuticals, personal care products)	Agricultural runoff, industrial discharge, domestic use	Ecological damage, potential human health risks, persistence in the environment

total suspended solids, total dissolved solids, and metal ion concentrations. To combat the negative impact of human activities on water pollution, wastewater treatment procedures must adapt and address the increasingly complex effluent compositions. The efficacy of well-established methodologies such as chemical treatment, activated sludge processes, anaerobic digestion, membrane filtration, advanced oxidation processes, and electro-Fenton processes has been substantiated through their successful implementation in the past. Yet, given the new challenges, innovative, environmentally friendly, and direct techniques must be developed to ensure the best possible outcomes.

For every 1 kg of COD removal, the traditional wastewater treatment process generates a secondary solid waste that ranges between 0.3 and 0.5 kg of dry biomass, resulting in sludge treatment costs ranging from $150 to $300. The activated sludge technique is a widely employed approach to handling sewage in wastewater treatment facilities aimed at eliminating organic compounds as well as other impurities. Despite its proven efficiency, this process significantly impacts the environment through the release of carbon dioxide equivalent to about $0.78 \, kg/m^3$ of treated water. This value is representative of other greenhouse gases, including methane and nitrous oxide, generated during the process.

In the field of membrane reactors, the anaerobic process has gained significant popularity due to its numerous benefits, including quicker start-up times, reduced treatment periods, enhanced COD removal, and the potential for nutrient recovery. While it has its favorable features, there are downsides to this process, such as methane disintegration in the treated water and fouling. Alternatively, microbial fuel cells are dependent on microorganisms for the conversion of organic substances present in wastewater into electrical energy. These cells typically consist of two compartments, namely the anode and cathode, with an ionic membrane separating them. The oxidation process of organic compounds at the anode generates electricity and CO_2, providing a COD reduction of over 50%. Nevertheless, microbial fuel cells demonstrate a lower COD removal efficiency as compared to the anaerobic process (Figure 3.1).

Extensive research has been devoted to the investigation of phycoremediation as an ecologically sound method of reducing environmental pollutants. Recent research studies have unveiled the promising potential of specific non-pathogenic algae species, such as *Spirulina* sp., *Chlorella* sp., *Chlamydomonas* sp., *Oscillatoria* sp., *Nostoc* sp., and *Scenedesmus* sp., to serve as highly efficient agents for treating wastewater. These selected algae species possess distinctive properties that enable them to flourish in contaminated water and proficiently eradicate harmful pollutants. Studies have shown that *Chlorella vulgaris* and *Chlorella reinhardtii* are capable of significantly reducing total nitrogen and COD concentrations in swine wastewater.

Algal biomass holds potential for a plethora of applications, including the creation of biochar, biogas, and diverse biofuels. Phycoremediation is a multifaceted concept that encompasses a diverse range of uses, including the elimination of nutrient and xenobiotic substances, the processing of heavy metal effluents, CO_2 mitigation, and the use of algae as biosensors in the monitoring of potentially toxic substances.

Microalgae possess a crucial ability to mitigate the pollution caused by industrial effluents. The chief aim of this study is to evaluate the efficacy of different types of algae in treating diverse wastewater types and pollutants. These pollutants comprise metal ions, PCBs, PAHs, dyes, and microplastics. In addition, this review examines the

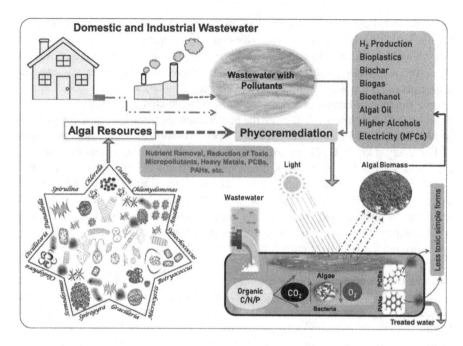

FIGURE 3.1 A fundamental overview of phycoremediation is shown in the schematic illustration.

drivers that influence phycoremediation, such as nutrients and innovative cultivation strategies, and their application in different wastewater contexts. Finally, this analysis delves into recent trends in phycoremediation and the future prospects within the field.

Several types of microalgae, such as *Chlorella* sp., *Scenedesmus* sp., and filamentous algae like *Oedogonium* sp. and Rhizoclonium, have been found to be efficient in wastewater degradation. Studies have indicated that *Chlorella vulgaris*, *Chlorella pyrenoidosa*, *Chlorella sorokiniana*, *Chlorella variabilis*, and *Scenedesmus acutus* and quadricauda from the *Chlorella* species possess significant potential for treating wastewater. Filamentous algae, specifically *Klebsormidium* sp., *Cladophora* sp., and *Stigeoclonium*, have been shown to have high potential for effectively treating wastewater. A study found that co-digestion of varying proportions of cow dung, chicken waste, and *Chlorella pyrenoidosa* grown in digestate water resulted in high-efficiency biogas production. The optimal substrate ratio for maximum methane production (68%) was found to be 2:1:2. The synthesis of magnetic nanocomposite particles (Fe_3O_4@EPS) by co-precipitating iron (III) chloride and iron (II) sulfate with exopolysaccharides (EPS) derived from *Chlorella vulgaris* proved to be extremely effective in reducing PO_4^{3-} wastewater.

3.2 TREATMENT OF SEWAGE WASTEWATER USING PHYCOREMEDIATION

3.2.1 NUTRIENT REMOVAL

Phycoremediation presents a sustainable and economical approach to wastewater treatment, utilizing the capacity of algae and other aquatic flora to eliminate pollutants and surplus nutrients from sewage and industrial effluent. This procedural approach has a clear edge over traditional methods of wastewater treatment, owing to its ability to effectively eliminate nutrients and heavy metals in one go while also displaying remarkable treatment efficiency.

Microalgae are highly efficient at removing nitrogen and phosphorus from wastewater. Various species of microalgae, including *Chlorella sp., Scenedesmus sp.,* and *Chlorella vulgaris,* have exhibited exceptional efficacy in processing nutrients present in wastewater. With their remarkable growth rate and photosynthetic efficiency, these microalgae can rapidly and efficiently remove nutrients from wastewater. Through extensive research and investigation, it has been revealed that the use of microalgae can result in nutrient removal rates of up to 90%.

Apart from their ability to efficiently remove nutrients, the incorporation of microalgae in wastewater treatment techniques can aid in mitigating the issue of eutrophication. Eutrophication is a harmful phenomenon where an overabundance of nutrients in water bodies leads to the growth of toxic algal blooms and a decline in oxygen levels, posing a threat to aquatic organisms. Moreover, microalgae may be employed in the production of biofuels and various value-added goods, rendering the phycoremediation approach for wastewater management a sustainable and practical solution.

3.2.1.1 Mechanisms of Nutrient Uptake by Algae

Algae utilize two primary mechanisms, namely nitrogen assimilation and phosphorylation, to extract nutrients from their surroundings. The assimilation of nitrogen is a crucial process that involves the conversion of inorganic forms, namely nitrate, nitrite, ammonium, and ammonia, into organic compounds. Organic compounds serve as fundamental constituents for a diverse range of biological assemblages, encompassing enzymes, proteins, chlorophylls, ADP and ATP, and nucleic acids such as RNA and DNA. Vital to this mechanism is the conversion of nitrate and nitrite to ammonium, which is facilitated by nitrate and nitrite reductases. The amino acid glutamine, supported by glutamate and ATP, provides aid in the process of nitrogen assimilation. However, unlike bacteria, algae display unique nitrogen uptake strategies. Algae achieve nitrogen removal through nitrification, followed by denitrification, or by the process of anaerobic ammonium oxidation, also referred to as ANAMMOX (Nguyen et al., 2022) (Figure 3.2).

Microalgae rely on phosphorus as well as nitrogen for their metabolism and growth. They use inorganic phosphorus ($H_2PO_4^-$ and HPO_4^{2-}) to create nucleic acids, lipids, and proteins through the process of phosphorylation, which involves several phosphate transporters on their plasma membrane. This process also produces ATP from ADP and polyphosphate, such as acid-soluble and acid-insoluble polyphosphate, through the action of polyphosphate kinase during photosynthesis. The uptake patterns of nutrients in algae are influenced by nutrient availability. Macroalgae with slow growth rates are capable of accumulating significant nutrient pools during periods of nutrient abundance in preparation for growth during periods of scarcity (luxury uptake).

Opportunistic algae, such as *Cladophora glomerata, Enteromorpha ahlneriana,* and *Scytosiphon lomentaria,* have adopted nutrient assimilation as a viable strategy to sustain high growth rates and synthesize nutrient-rich organic compounds, such as amino acids and proteins. These adaptive traits enable these algae to flourish in

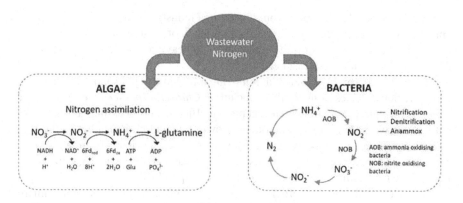

FIGURE 3.2   Two primary approaches can be employed for the removal of nitrogen from wastewater. The algae-based and bacteria-based mechanisms and pathways employed to eliminate nitrogen from wastewater vary considerably, as depicted in the illustrated process.

nutrient-rich environments, such as wastewater. Recent studies have indicated that macroalgae belonging to the green, brown, and red species cultivated in wastewater environments exhibit enhanced growth rates and protein-rich biomass. Of these macroalgae, green macroalgae have demonstrated remarkable tolerance toward environmental fluctuations, thus emphasizing their potential as an ideal species for wastewater cultivation. The capacity of macroalgae to absorb nutrients and develop biomass rich in protein within wastewater settings provides a resolution for wastewater treatment and an avenue to create valuable resources, such as biofuels and fertilizers, while simultaneously promoting sustainable practices for managing wastewater.

The availability of nitrogen is a crucial factor in regulating the growth rate of both microalgae and macroalgae in the context of cultivating algae. Ammonium is considered an ideal nitrogen source for algae due to its low energy consumption for assimilation compared to other forms of inorganic nitrogen. It has been suggested that high nitrogen concentrations in the culture medium can positively impact the assimilation of nitrogen into essential amino acids and proteins, thereby enhancing overall biomass productivity. This affirms the potential of nitrogen-rich wastewater as a viable source for algal cultivation, given its typically high ammonium content. Recent research has demonstrated that algae grown in nitrogen-rich wastewater can possess up to four times the crude protein content of those cultivated in seawater.

Elevated concentrations of ammonia and ammonium may have adverse impacts on algal growth by causing intracellular oxidative stress and disrupting cellular metabolism. Microalgae can endure up to $100\,\text{mg}\,\text{NH}_4^+\text{-N/L}$ of ammonium, while macroalgae can withstand higher levels of up to $250\,\text{mg}\,\text{NH}_4^+\text{-N/L}$. However, some algal species have displayed a higher tolerance for ammonium levels. Moreover, cultures rich in nitrogen can lead to the accumulation of nitrogen-containing photosynthetic pigments in algae, resulting in a more intense and rich green coloration toward the conclusion of the growth cycle. Whilst the negative effects of ammonium on algal growth are recognized, it is paramount to acknowledge the noteworthy capacity of algae to endure and integrate nitrogen. This characteristic is imperative for numerous applications, such as wastewater treatment, biofuel production, and carbon sequestration.

3.2.1.2 Unit-Time Nutrient Uptake

Various species of algae exhibit differing degrees of effectiveness and efficiency in eliminating nitrogen and phosphorus from wastewater, contingent on the N/P ratio present in the wastewater. Macroalgae achieve optimal results when the N/P ratio ranges from 10 N:1P to 80 N:1P, with the highest efficacy observed at a ratio of 30 N:1P. In comparison, microalgae demonstrate the highest removal rates under N/P ratios of 5–30 N:1P. Chlorella sp. and Scenedesmus sp. are among the algae species that can be successfully cultivated in wastewater, given their adaptability to varying environmental conditions and a high tolerance for nitrogen and phosphorus. The utilization of these algae species in wastewater treatment is considered a promising approach to reducing environmental pollution and creating a renewable source of biomass with numerous applications. The increasing demand for sustainable and eco-friendly wastewater treatment methods further highlights the potential of these algae species to address the challenges associated with nutrient removal. The nutrient uptake rate of macroalgae is closely linked to their morphological structure. Certain species, such as opportunistic, filamentous, delicately branched, or monochromatic macroalgae, show consistently higher nutrient uptake rates. It is crucial to note that increasing water flow within the culture system can enhance nutrient flux to the surface of the thalli, which in turn can stimulate nutrient uptake rates (Nguyen et al., 2022).

3.2.2 Heavy Metals Removal (HMR)

Phycoremediation is an effective approach that not only eliminates surplus nutrients from wastewater but also offers a viable solution for heavy metal ion removal. Heavy metal ions such as cadmium, lead, and copper are known to be lethal to aquatic life and may cause extensive ecological harm if not treated promptly.

The utilization of microalgae for phycoremediation of heavy metals involves a two-part procedure that employs bio-binding or bio-removal techniques. The initial phase encompasses biosorption, where heavy metals are quickly absorbed onto the surface of microalgae cells. The passive extracellular process that enables the binding of heavy metals to the functional groups present on the algal cell surface may or may not involve cell metabolism. Due to their remarkable affinity for heavy metals, microalgal cells with large surface areas can bind up to 10% of their total biomass with these metals. Factors influencing this process include metal bioavailability, accessibility of metal binding groups on the algal cell surface, and the efficiency of algal cells in metal uptake and storage (Chugh et al., 2022).

Biosorption presents itself as a valuable technique for the removal of pollutants from wastewater, which can be accomplished using either living or non-living biomass. Essentially, two distinct mechanisms are responsible for biosorption: ion exchange and complex formation with functional groups situated on the cell's surface. Algal cell walls possess diverse functional groups such as hydroxyl, sulfate, amino, and carboxyl, which function as binding sites for the pollutants. Biosorption is a cost-effective and efficient alternative to live biomass since it doesn't require substantial amounts of energy or metabolic activity. Based on various studies, a plethora of biosorbents have been exhaustively researched, showcasing their improved potential for effectively reducing pollutants. A common outcome of biosorption is the formation of flocs, which diminishes water pollutants (Figure 3.3).

In the second phase of phycoremediation, microalgae actively transport heavy metals found in wastewater across their cell membrane, subsequently accumulating them in their cells or organelles via a bioaccumulation process that requires requisite metabolic activity and expended energy. In contrast to biosorption, where heavy metals bind passively to the cell walls, bioaccumulation solely occurs in living biomass. The accumulation of heavy metals can be subjected to detoxification, compartmentalization, or complexation to form non-toxic entities, while the resulting algal biomass can be harnessed to recover valuable products such as biofuels, pigments, and pharmaceuticals. The use of microalgae for heavy metal removal presents an attractive option for wastewater treatment due to its effectiveness and environmentally friendly approach, in addition to its potential for generating value-added products.

3.2.2.1   Self-Defense Mechanisms to Counter HMT

Algae possess unique strategies to counteract heavy metal toxicity, including gene regulation, complexation, ion exchange, chelation, and the production of reducing agents and antioxidants. These mechanisms are vital for preventing the buildup of heavy metals and play a crucial role in enabling microalgae to combat heavy metal toxicity. Untreated heavy metal toxicity can hinder the metabolism of cells, interfere with enzyme function, and disrupt both normal physiological and morphological development. It may also lead to protein denaturation, defects in nucleic acid, and cellular membrane impairment. Algae have developed a natural defense system that allows them to thrive in wastewater contaminated with high levels of heavy metals. These self-protection mechanisms help to mitigate the detrimental effects of

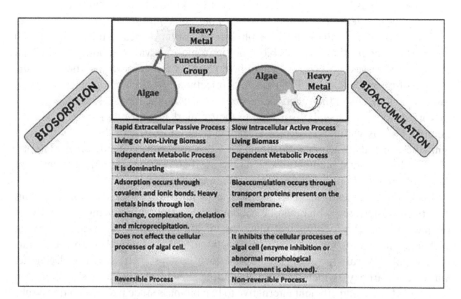

FIGURE 3.3   An illustration showing a dual-step approach toward the restoration of heavy metals.

toxic heavy metals. The self-defense mechanisms of algae can be categorized into biosorption, bioaccumulation and compartmentalization, biotransformation, metal detoxification, and metal exclusion, as depicted in Figure 3.4. In the subsequent section, we will examine the significance of these defense mechanisms in the removal of heavy metals from wastewater.

3.2.2.2 Biosorption: Mechanisms of Extracellular Uptake against Hmt

Microalgae possess a highly effective self-defense mechanism that prevents the uptake of heavy metals into their cellular structure. This mechanism operates by adsorbing heavy metals onto the surface of cells, which is facilitated by functional groups of polysaccharides, lipids, proteins, and monomeric alcohols that are rich in hydroxyl, carboxyl, and phosphate functional groups. These functional groups employ various bonding mechanisms, such as van der Waals, ion exchange, covalent, or combined mechanisms, to bind positively charged metal ions. The extent of metal sorption by different strains of algae varies depending on their unique cellular compositions. Freshwater and marine algae have demonstrated the ability to remove heavy metals, including copper, lead, zinc, and cadmium, at concentrations of up to 40 ppm, by efficiently utilizing the functional groups on their cell walls.

Brown algae possess a remarkable capacity for heavy metal remediation owing to the existence of various functional groups such as cellulose, alginic acid, and other polymers. This is evidenced by the successful removal of metals such as zinc, cadmium, nickel, and lead through Fucus vesiculosus, and the reduction of nickel and cadmium through *Cystoseria indica*. Moreover, *Anbaena spharica* has been found to

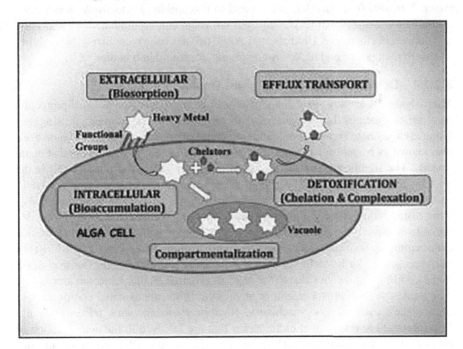

FIGURE 3.4 Self-defense mechanisms found in algae to counter the toxicity of heavy metals.

remediate cadmium and lead by means of amino, hydroxyl, and carboxyl functional groups, with pH levels playing a crucial role in metal uptake. The desorption of heavy metals is observed at lower pH values, while alkaline pH values promote their uptake. This inherent self-protection mechanism is efficient, cost-effective, and produces no toxic metabolites, thereby enabling the preservation of cellular metabolism and co-product accumulation.

3.2.2.3 Bioaccumulation and Compartmentalization: Intracellular Uptake Mechanisms for HMT

Microalgae employ a second self-protection mechanism against heavy metals, called bioaccumulation or compartmentalization, to remediate heavy metals. Although heavy metals are initially prevented from entering the cell, they can enter if they are present at high concentrations in the surroundings. Microalgae undergo bioaccumulation, wherein heavy metal ions are accumulated inside the cell, while compartmentalization manifests when these ions are stored in organelles like vacuoles or thylakoids. The aforementioned procedure encompasses the absorption of metal ions and their subsequent transportation into the cell or organelles, which is accomplished through ion-selective transport proteins located on the cell membrane.

Microalgae display a substantial surface area-to-volume ratio, enabling them to take up metal ions in an effortless manner. The presence of sulfur (S), iron (Fe), and nitrogen (N) in the environment has been recognized as crucial to this process. For instance, *Chlorococcum sp.* has demonstrated remarkable potential as a bioindicator and bioremediator, being able to accumulate up to 239.09 µg/g of arsenic from wastewater. *Tetraselmis suecica* has been found to be capable of effectively storing cadmium in its vacuoles, while *Skeletonema costatum* can accumulate both cadmium and copper. Furthermore, Ulothrix sp. LAFIC 010 has been recognized for its ability to absorb manganese and nickel by employing electron-dense bodies.

3.2.2.4 Detoxifying HMs: Mechanisms Employed by Algae

Microalgal cells employ metal detoxification or biotransformation as a self-protection mechanism, whereby they convert heavy metals into non-toxic compounds using chelating agents. These organic compounds combine with metal ions to create complexes, also known as complexation. Chelating agents are released by algae to combat metal toxicity. The significant contribution of sulfur (S), iron (Fe), and nitrogen (N) in the relevant process is widely acknowledged. *Chlorococcum* sp., with its exceptional ability to accumulate up to 239.09 µg/g of arsenic from wastewater, has been identified as a promising bioindicator and bioremediator. Similarly, *Tetraselmis suecica* has been found to be effective in storing cadmium in its vacuoles, while *Skeletonema costatum* can accumulate both cadmium and copper. Furthermore, *Ulothrix* sp. LAFIC 010 has been recognized for its ability to absorb manganese and nickel using electron-dense bodies, adding significant value to the ongoing research on this subject matter.

Peptides, such as phytochelatin or metallothioneins, are essential components of microalgae's detoxification machinery. Notably, *Phaeodactylum tricornutum* has been observed to release metallothioneins that are involved in the detoxification of copper ions. This alga can effectively bind cadmium with metallothioneins and sulfide ions to form a cadmium-metallothioneins complex. According to the findings reported by Nowicka et al., *Chlamydomonas reinhardtii* employed a combination of detoxifying enzymes, hydrophilic antioxidants, and reactive oxygen species to mitigate the toxic impact of cadmium and chromium.

3.2.2.5 Metal Exclusion Mechanisms Employed for HMT

Efflux transport is a self-defense mechanism utilized by microalgal cells to rid themselves of heavy metals. By doing so, the cells can maintain healthy levels of these metals inside their cytoplasm. Furthermore, microalgae employ an active transport system to expel heavy metals that could pose a threat to the cell by either eliminating them altogether or converting them into a non-toxic form. Ion-selective transporters also aid in this exclusion process. In addition, algae possess the ability to modify their cell membranes to reduce the influx of heavy metals. Although comparable self-defense mechanisms can be detected in other microorganisms like bacteria and fungi, additional investigation is imperative to comprehensively grasp the efflux mechanism functioning in microalgae.

3.2.2.6 Algal Phycoremediation of Cadmium (Cd), Arsenic (Ar), Chromium (Cr), and Mercury (Hg): Mechanisms and Applications

Traditional methods for eliminating heavy metals from wastewater, such as ion exchange, evaporation, electrolysis, osmosis, and precipitation, have been in use for an extended period. Nonetheless, these methods are comparatively less effective than modern techniques and also tend to be quite expensive. Conversely, bioremediation is a powerful and eco-friendly technology that is capable of efficiently removing pollutants, even when present in low concentrations that other physicochemical approaches struggle to eliminate. Additionally, bioremediation generates biomass, making it a valuable tool for purifying the environment. The biomass generated from this methodology can also be leveraged to develop products with greater economic value.

The process of utilizing living organisms such as bacteria, fungi, and algae to eliminate dangerous heavy metals from the environment is called bioremediation. Through scientific research, it has been established that algae possess exceptional abilities to eliminate heavy metals, exhibiting remediation rates of up to 84.6% when compared to fungi and bacteria. Utilizing algae for bioremediation purposes presents several advantages over other available options, owing to its cost-effective nature, adeptness for high heavy metal absorption, ability to renew, and consistent availability of algal biomass year-round. Moreover, the implementation of this approach does not result in the generation of any hazardous byproducts.

The reduction of heavy metals using algae is a two-stage process. Firstly, functional groups such as thioether, carboxylic, hydroxyl, amide, sulfhydryl, and imidazole on the cell wall attract positively charged metal ions toward negatively charged functional groups, leading to their removal. This process is known as biosorption, whereby heavy metal ions are absorbed by the cell surface of the algae. In the second stage, the metal ions undergo bioaccumulation and are gradually transported into the cell. Following this, various mechanisms such as detoxification, efflux, or compartmentalization come into play to ensure the safe removal of these metal ions (Chugh et al., 2022).

3.2.3 REMOVAL OF ORGANIC POLLUTANTS BY ALGAE

Several studies have shown that microalgae species, such as Scenedesmus acutus and Chlorella pyrenoidosa, can effectively remove PAHs and PCBs from wastewater. For instance, Pang et al. (2015) found that *Scenedesmus acutus* could remove up to 85% of PAHs from wastewater. Tan et al. (2018) reported that *Chlorella pyrenoidosa* could effectively remove PCBs from wastewater. Moreover, Pandey et al. (2021) conducted a study that revealed that *Chlorella vulgaris* is capable of eliminating as much as 85% of dyes found in wastewater within a time span of only 72 h. The removal of these organic pollutants occurs through several mechanisms. Microalgae can metabolize these compounds, convert them into less harmful compounds, or mineralize them completely.

Persistent organic pollutants have the potential to travel significant distances and accumulate in water systems and the environment, leading to toxicity. These pollutants arise from a variety of sources, including agricultural and household waste, industrial discharge, and urban runoff. In addition to degrading water quality, they can also have negative impacts on aquatic organisms' health. Exposure to such pollutants has been linked to the development of a range of conditions, including cancer, birth defects, diabetes, cardiovascular disease, reproductive and immune system dysfunction, and endocrine disruption. Figure 3.5 provides a depiction of these adverse health effects (Dayana Priyadharshini et al., 2021).

The accumulation and prolonged exposure of organic pollutants in the human body is a significant global concern, particularly in developing countries, as highlighted in the study by Guo et al. (2019). Contaminated animal-based food, especially fish,

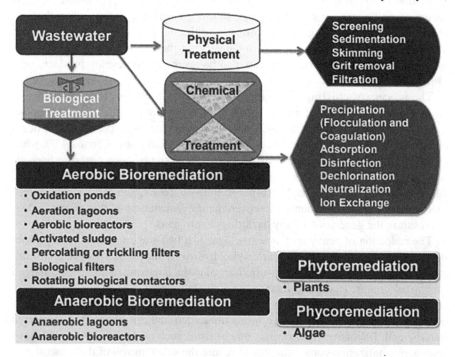

FIGURE 3.5   A comprehensive outline showcasing diverse wastewater treatment facilities.

is the leading source of human exposure to these pollutants containing pesticide residues. This group of organic pollutants consists of organic hydrocarbons, including polychlorinated dibenzofurans and dibenzo-para-dioxins (PCDF/PCDD), PAHs, di-(2-Ethylhexyl) phthalate (DEHP), surfactants, and PCBs. The hydrophobic nature of these chemicals allows them to remain attached to aquatic debris for extended periods, posing a significant threat to individuals' health and the environment.

3.2.4 BOD REDUCTION

BOD is a critical parameter for assessing the quality of wastewater. High levels of BOD indicate a high concentration of organic material in the water, which can lead to oxygen depletion and negatively impact aquatic life. Phycoremediation has shown promise in reducing BOD in industrial effluent, making it a viable method for wastewater treatment. Several studies have investigated the effectiveness of microalgae in reducing BOD in wastewater. *Chlorella vulgaris* and *Chlorella sorokiniana* have been identified as potential microalgae species for this purpose. Their study found that phycoremediation with *Chlorella vulgaris* and *Scenedesmus* sp. resulted in significant reductions in BOD levels, with reductions of up to 73.9% and 79.8%, respectively. These results indicate the potential for microalgae to serve as an effective treatment method for reducing BOD in various types of wastewater.

BOD reduction is critical for improving water quality and protecting aquatic ecosystems, as high levels of BOD can lead to reduced oxygen levels and the death of aquatic organisms. The ability of microalgae to reduce BOD through phycoremediation offers a promising and sustainable solution for addressing water pollution and improving overall water quality. The researchers found that microalgae were able to reduce BOD by up to 94%, indicating the potential of phycoremediation in treating wastewater from industries with high organic content (Dayana Priyadharshini et al., 2021).

3.3 POTENTIAL STRATEGIES FOR PHYCOREMEDIATION

Microalgae have shown great promise in addressing the issue of contaminated soil and water due to their ability to accumulate liquid waste and metals. By discovering species of algae with high accumulation capacities, several techniques have been developed to exploit microalgae's potential in removing hazardous compounds, notably heavy metals, and organic and inorganic waste. These techniques have been extensively researched and improved to enhance their effectiveness and efficiency.

There are a variety of techniques that are frequently employed in environmental remediation efforts, including popular methods such as phytoremediation, biosorption, bioaccumulation, and phytovolatilization. In the case of phytoremediation, microalgae are utilized to absorb and accumulate pollutants from water or soil. Alternatively, in biosorption, contaminants are bound to the exterior of microalgae, simplifying their removal through harvesting. Bioaccumulation refers to the process by which microalgae take up and accumulate toxins within their cellular structures. Lastly, phytovolatilization describes the mechanism whereby microalgae convert pollutants into gaseous forms, making it easier to eliminate them from the ecosystem (Upadhyay et al., 2019) (Table 3.2).

TABLE 3.2

Type of Waste Treated through Phycoremediation by Microalgal Species

Algae	Type of Waste	References
Spirulina	Anaerobic effluents of pig waste	Lincoln et al. (1996)
Phormidium bohneri	Municipal wastewater	Talbot and De la Noüe (1993)
Chlorella sp.	Municipal wastewater/domestic sewage	Choi and Lee (2015)
Euglena	Domestic wastewater	Mahapatra et al. (2013)
Desmodesmus sp. TAI-1 and Chlamydomonas	Industrial wastewater	Wu et al. (2012)
Scenedesmus quadricauda	Campus sewage	Han et al. (2015)

The application of these methods has exhibited optimistic outcomes in eliminating perilous substances from polluted soil and water. The various remediation approaches developed for microalgae utilization in purifying wastewater and removing metals from contaminated soil and water are summed up in Figure 3.6. These techniques present an efficient and enduring resolution to the mounting issue of environmental contamination attributed to hazardous compounds (Upadhyay et al., 2019).

3.3.1 Phycovolatilization: Transformation of Pollutants into Volatile Substances

In order to gain a thorough understanding of the concept of phytovolatilization, it is essential to have knowledge of the different types involved. The first type, direct phyto-volatilization, entails the uptake of contaminants by plants, which are then transported and released through the leaves and stem/trunk. This process is akin to the transpiration

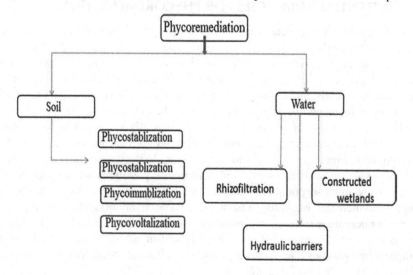

FIGURE 3.6   Schematic representation of phycoremediation mechanisms in microalgae.

of water through the plant's vascular network and has been extensively studied. On the other hand, indirect phytovolatilization is a secondary process that occurs as a result of the microbial degradation of contaminants in the rhizosphere, with the subsequent release of volatile compounds. These compounds can subsequently be taken up by the plant and released as well. Although indirect phytovolatilization is less well understood than direct phytovolatilization, it still plays a significant part in the overall process.

The process of direct phytovolatilization is multifaceted, requiring the absorption, movement, and release of contaminants by plants. It occurs when plants take in substances that are moderately hydrophobic and transport them across hydrophobic barriers present in either the plant's epidermis or woody dermal tissues. For the compound to be classified as directly phytovolatilized, it must undergo uptake, translocation, and volatilization by the plant. Compounds that are produced or modified by the plant are not considered direct phytovolatilization. It is essential to differentiate between direct phytovolatilization and various volatile organic compounds (VOCs) produced and emitted by plants, and the phytotransformation of selenite into dimethyl selenide, as they do not fall under the direct phytovolatilization category (Limmer and Burken, 2016) (Figure 3.7).

The process of indirect phytovolatilization, whereby plant roots heighten the flow of volatile contaminants from below the surface, is an intriguing phenomenon. Given the substantial quantity of water that plants circulate globally (~62,000 km^3/year) and their thorough probing of soil, the consequences for the transportation and destiny of substances within the subsurface are noteworthy. Several mechanisms may contribute to raising the flux of volatile contaminants in roots, such as water table reduction, advection associated with gas fluxes due to diel fluctuations in the water table, improved soil permeability, hydraulic redistribution in chemical transportation, advective movement of water toward the surface, and containment of rainfall that would otherwise infiltrate and dilute, leading to the advection of VOCs beyond the surface. Indirect phytovolatilization has the capability of assisting volatile contaminants in transportation, thereby increasing the likelihood of pollution spread and possible exposure of humans to hazardous chemicals. Consequently, a comprehensive grasp of indirect phytovolatilization mechanisms is crucial for devising approaches toward managing associated risks (Limmer and Burken, 2016).

The scope of operation of phycovolatilization is wide-ranging and effective against various contaminants, such as inorganic, organic, and metal substances. For example, several inorganic compounds like Se, As, and Hg can be volatilized from algae. It is also noteworthy to mention that this technique holds promise in the treatment of radioactive water (T_2O) (Upadhyay et al., 2019).

Algae act as powerful agents that can enhance the rates of volatile contaminant flux by reducing water tables, increasing soil permeability, transferring chemicals through surface and vacuole mechanisms, driving advection with water toward the surface, and modifying rainfall patterns that would normally displace and dilute the VOCs from the surface. The process of direct phycovolatilization is a notable contributor to the phenomenon of volatilization. This mechanism ensues when volatile compounds are emitted directly from algae via metabolic processes such as decomposition or respiration. The discharge of VOCs from algae can be influenced by different factors, including the species of algae, the concentration and type of pollutants, and the prevailing environmental conditions.

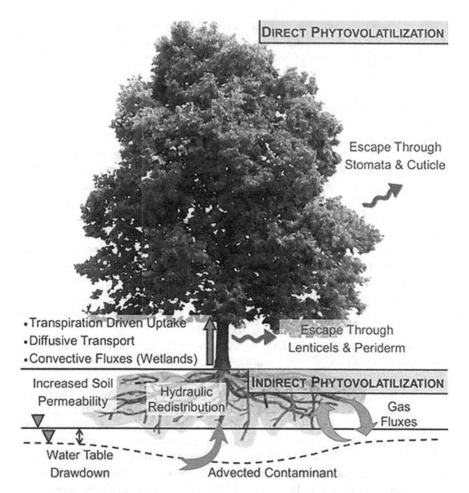

FIGURE 3.7 Phytovolatilization: understanding the working mechanisms of direct and indirect processes.

Carrasco Gil et al. (2013) detailed a methodology wherein varied concentrations of algae are combined with pollutants in a laboratory environment and grown hydroponically. After a specified period of growth, the resulting biomass is analyzed to gauge the extent of pollutant elimination from the treatment solution. To ascertain the quantity of contaminant removed through phytovolatilization, one approach is to measure the amount of contaminant mass present in the open system, as per Harper's (2000) instructions, by examining the air that passes through a sorbent trap (Upadhyay et al., 2019).

3.3.2 RHIZOFILTRATION: INVOLVEMENT OF MICROALGAE AND AQUATIC PLANTS

Rhizofiltration refers to a remediation strategy that entails the utilization of aquatic plants and microalgae for treating wastewater and soil that is contaminated with poisonous metals and organic pollutants. Several research studies have substantiated the efficiency of rhizofiltration in the elimination of heavy metals such as Pb, Zn, Cd, Cu, Ni, and Cr. A range of plants, including microalgae and blue-green algae, are recognized for their effectiveness in rhizofiltration. These include *Chlorella vulgaris, Scenedesmus obliquus, Spirulina platensis, Nannochloropsis* sp., *Dunaliella tertiolecta, Tetraselmis suecica, Chlamydomonas reinhardtii, Ankistrodesmus falcatus, Euglena gracilis, Botryococcus braunii,* spinach, tobacco, and Indian mustard. For this exercise, marine algae and terrestrial plants have been selected due to their capability to swiftly establish root systems.

Rhizofiltration systems provide an opportunity to build floating rafts on ponds or still water and utilize the growth of plants and algae for remediation purposes. However, a drawback of this method lies in the requirement of cultivating plants and algae in a greenhouse before relocating them to the remediation site. It is also essential to uphold optimal pH, temperature, and humidity levels for effective remediation. Despite these challenges, rhizofiltration stands as an eco-friendly solution that does not entail the production of secondary pollutants during the treatment process. It is essential to regularly monitor the sites to ensure the maintenance of optimum conditions (Upadhyay et al., 2019) (Table 3.3).

3.3.3 Phytostabilization: Removal from the Top of Phreatic Zones

Phytostabilization is a phycoremediation strategy that involves the use of algae to stabilize contaminated soils or sediments by reducing the mobility of pollutants. The implementation of this technique proves to be effective in obstructing the movement of pollutants from the surface of phreatic zones. Microalgae have certain benefits for phytostabilization, including their rapid growth rate, increased efficiency in metal uptake, and adaptability to diverse environmental conditions. Based on these studies, it can be inferred that utilizing microalgae for phytostabilization can be a constructive method for restoring heavy metal-contaminated soils and sediments (Bolan et al., 2011).

The implementation of phytostabilization is a promising method for the restoration of metalloid-polluted soils. It entails the application of plant coverage to hinder the movement of contaminants within the soil by confining them in the root zone or rhizosphere (Bolan et al., 2011).

TABLE 3.3
The Metal Accumulation Capacity of Various Microalgae Species

Microalgae Species	Metals Accumulated
Chlorella vulgaris	Copper, zinc, cadmium, lead, chromium
Scenedesmus obliquus	Copper, zinc, cadmium, lead, chromium, nickel, cobalt
Spirulina platensis	Copper, zinc, lead
Nannochloropsis sp.	Copper, zinc, cadmium, lead
Dunaliella tertiolecta	Copper, zinc, lead
Tetraselmis suecica	Copper, zinc, cadmium, lead
Chlamydomonas reinhardtii	Copper, zinc, cadmium, lead, nickel
Ankistrodesmus falcatus	Copper, zinc, cadmium, lead, chromium
Euglena gracilis	Copper, zinc, cadmium, lead, nickel
Botryococcus braunii	Copper, zinc, cadmium, lead

3.3.3.1   Reduction of Contaminant Mobility

Utilizing plants to immobilize soil contaminants and prevent their migration to other ecosystems is known as phytostabilization. The roots of the plants create a shield, and the transpiration process curtails soil moisture, which impedes the movement of pollutants. The organic matter present in the soil boosts the plant's binding capability and decreases the availability of contaminants. Phytostabilization is an efficacious, economical, and sustainable soil remediation strategy with proven results.

3.3.3.2   Microbial Activity

The microbial processes occurring in conjunction with the roots of plants have the potential to expedite the conversion of harmful organic substances, including hydrocarbons and pesticides, into less hazardous forms. Implementing this technique can contribute to mitigating overall soil pollution levels.

3.3.3.3   Soil Amendments

The efficacy of phytostabilization techniques can be optimized through the application of soil amendments that can immobilize metal (loid)s. These amendments can be strategically used in conjunction with plant species that are well-equipped to thrive in low-fertility soil and high-contaminant deposits. It should be noted that periodic reapplication of soil amendments may be necessary to sustain their potency.

3.3.3.4   Regular Monitoring

Regular monitoring is crucial in phytostabilization to ensure that the stabilization conditions remain intact for consistent effectiveness. It is imperative to acknowledge that the potency of soil amendments may diminish over time.

3.3.3.5   Applicability

The efficiency of phytostabilization in confining metalloids depends on the site-specific conditions. To ensure effectiveness, the selection of appropriate plant species, soil enhancements, and management strategies ought to be customized according to the site's unique properties.

3.3.4   PHYCOEXTRACTION: A PROMISING APPROACH FOR HMR

Phycoextraction involves the use of microalgae to eradicate hazardous anthropogenic pollutants from soil and water by facilitating uptake, accumulation, and sequestration. Such pollutants may be assimilated by the algae for nourishment and growth or may be transported into the cell through active or passive mechanisms alongside the water. Upon entry into the cell, metals are sequestered within vacuoles, assisted by phytochelatins present in the cell. To cope with pollutants, such as wastewater and heavy metal ions, algae can withstand them via an avoidance mechanism whereby contaminants are immobilized on the surface or cell walls. The sequestration process by vacuoles and the binding of contaminants with various ligands, such as proteins, peptides, organic acids, and enzymes, including high phytoextracted ions, are the fundamental basis for achieving tolerance to wastewater contaminants.

For algae strains to be deemed fit for this method, they must display the ability to withstand elevated metal concentrations, exhibit rapid growth rates, have a robust surface system, possess the potential to amass heavy metals in their harvestable parts, and show promising results in generating high biomass yields under field conditions. Furthermore, studies have documented hyperaccumulators that accumulate pollutants such as Ni, Cu, and potentially Zn in their dry weight, reaching levels between 1% and 5%.

Unfortunately, the growth rates of the vast majority of hyperaccumulators are currently slow, and at present, we do not possess the necessary technology to cultivate them on a large scale. Several studies have evidenced that the natural pollutant hyperaccumulator phenotype is of greater importance than the higher plant ability when using plants as a means of remediating polluted soil and water. Upadhyay et al. (2016) described two methods for phycoextraction, namely chelate-assisted surface bonding and continuous phycoextraction. Chelating agents have been utilized as soil extractants and as a source of macro- and micronutrient fertilizers to maintain the solubility and bonding of micronutrients in hydroponic forms (Upadhyay et al., 2019).

3.3.5 Constructed Wetland: Cost-Effective Green Technology

The green algae and cyanobacteria that are present in wetlands have demonstrated a superior ability to convert carbon dioxide (CO_2) into beneficial substrates when compared to other land-based vegetation. The implementation of constructed wetlands (CWs) that incorporate algae as a treatment component has been established as an effective method for addressing a range of pollutants, such as nitrogen, phosphorus, heavy metals, and organic pollutants. Algae are integral to the removal of nutrients, particularly nitrogen and phosphorus, from wastewater via biological nutrient removal.

The efficiency of CW systems is influenced by multiple factors, such as the assortment of vegetation and microbial species in the area, the duration of hydraulic retention time, and the quality of the incoming wastewater. When algae are present in CWs, they can bolster the system's overall effectiveness due to their unique ability to eliminate various pollutants and generate biomass that can serve as a sustainable energy source. Nonetheless, it is crucial to exercise caution when cultivating algae in CWs, as excessive growth could impede proper pollutant removal and undermine system efficacy.

Algae are increasingly being recognized as a viable option for wastewater treatment due to their remarkable ability to remove excess dissolved nutrients from effluents. This eco-friendly technology not only ensures cost-effectiveness but has also been proven to be efficient in reducing pollutants such as nitrogen, phosphorous, and organic pollutants. Moreover, algae possess the impressive capacity to eliminate heavy metals from contaminated water via two critical mechanisms: metabolism-dependent uptake in cells and non-active adsorption. This metal-binding capacity of algae has also made them widely used as biomonitoring agents in ecosystems (Upadhyay et al., 2019) (Figure 3.8).

CWs are acknowledged as a viable and sustainable alternative to conventional wastewater treatment methods, owing to their low energy requirements, cost-effectiveness, and high-contaminant elimination efficacy.

Vertical flow planted filters, 1st treatment stage, scheme

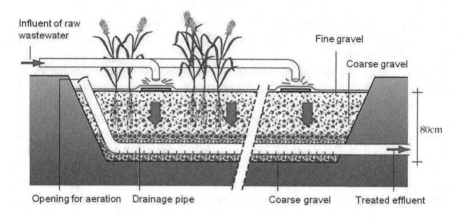

Influent of raw wastewater

Fine gravel

Coarse gravel

80cm

Opening for aeration   Drainage pipe   Coarse gravel   Treated effluent

FIGURE 3.8   Constructed wetlands in their initial stages.

The detrimental effects of CWs can potentially result in the release of greenhouse gases, namely methane and nitrous oxide, and the pollution of effluent, negatively impacting public health and the environment. The widely held perception that CWs technology is a foolproof, environmentally friendly method for treating wastewater has left the environment susceptible to undesirable consequences over time. The reasons for this include a lack of a comprehensive and strategic sustainability assessment protocol, breakdowns in operational processes, inadequate upkeep, and the impact of interactions between CWs and ecosystems. Although CWs are frequently viewed as a reliable method for commercial irrigation agriculture, the failure to implement standard operating procedures and to adopt critical thresholds for contaminants means that most CWs systems have a low concentration of dissolved oxygen in their influent, which leads to incomplete N-oxidation and an elevated total N concentration in the resulting effluent.

3.3.6   HYDRAULIC BARRIER: REMOVAL FROM LARGER WATER BODIES

A collective of filamentous, green, and blue-green algae forms these algal mats, which develop into a compact layer that drifts through the water, effectively removing dissolved pollutants through filtration. This mat works akin to a trickling filter, where the water's flow across the mat is impeded, thus increasing retention time and reducing the rate of water flow. Over the course of time, algal mats accumulate a substantial quantity of impurities that can be effortlessly removed for disposal, rendering the technology simple to maintain and operate. Studies have revealed that algal mats are highly efficient at eliminating a diverse range of pollutants, including organic contaminants, nitrogen, phosphorus, and heavy metals.

Apart from their proficiency as hydraulic barriers, research shows that algal mats have the potential to be an invaluable source of renewable energy. The cultivated mats can be utilized to generate a variety of biofuels, such as bioethanol and biodiesel, which can function as a substitute for fossil fuels. Furthermore, the biomass produced by algal mats exhibits the potential to be utilized as a fertilizer, animal feed, or even as a source of high-value chemicals (Upadhyay et al., 2019).

A novel method has been created to capitalize on the waste heat and superfluous nutrients available in the cooling waters integrated with nuclear reactors and fossil-fueled power plants. This procedure involves the cultivation of precise categories of thermotolerant microalgae in the discharged, heated waters. These microorganisms can then be utilized to extract energy, nutrients, and valuable products. The process has been designed for implementation at large cooling reservoirs, receiving high volumes of secondary cooling water containing available phosphorus and nitrogen. Based on the nutrient load, an estimated 100 ha of land would be required for the effective implementation of the process. Assuming a 1% P content in the biomass, the productivity of the microalgae has been evaluated at 10 g/m^{2}day. This technology presents a promising solution for effectively utilizing waste heat and excess nutrients while simultaneously providing a sustainable source of energy and valuable commodities.

For the effective cultivation of filamentous, thermotolerant, nitrogen-fixing blue-green algae, it is recommended to establish an environment that favorably promotes selective growth. One approach to achieving this goal involves the use of hydraulic barriers made of submerged plastic curtains to isolate the algal production area, or "cultivation zone," at the inlet end of the reservoir. The algal culture can then be introduced into the thermal plume and collected near the far-end barrier of the cultivation zone using rotating, backwashed, fine mesh screens commonly referred to as "microstrainers." The harvested biomass is then utilized to sustain a dense culture at the inoculation site. By reducing thermal and nutrient loadings on water bodies that receive such runoffs, this process holds great potential for significant positive outcomes (Wilde et al., 1991).

The implementation of reactive porous barriers for heterotrophic denitrification has proven to be a viable, environmentally conscious, and economical solution for treating nitrate-contaminated groundwater. By utilizing waste cellulose solids as a carbon source, the nitrate concentration was significantly lowered by an average of 80%–91% across three different sites. The consistent and efficient performance of these reactors indicates that they could be used as a practical alternative to the traditional treatment methods that require substantial financial investments. The results of the field tests suggest that the reactive barriers have the potential to treat nitrate for at least 10 years without the need for carbon replenishments, making them an affordable and low-maintenance option for nitrate remediation.

Similarly, hydraulic barriers have several advantages for phycoremediation, including cost-effectiveness, ease of installation, and minimal maintenance requirements. The use of algae in hydraulic barriers provides an ecological solution to pollution, converting pollutants into biomass. However, successful implementation of hydraulic barriers for phycoremediation depends on factors such as appropriate algae species selection, barrier design, and water flow rate. Therefore, further research is needed to optimize the use of hydraulic barriers for phycoremediation (Robertson et al., 2000).

3.4 MECHANISMS OF PHYCOREMEDIATION

3.4.1 BIOLOGICAL MECHANISM OF REMEDIATION: EFFICIENCY OF OPEN POND SYSTEMS/WASTE STABILIZATION PONDS

One of the key approaches to phycoremediation is implementing open pond systems, which are commonly known as waste stabilization ponds (WSPs). Such systems employ a range of physical, chemical, and biological methods to purify wastewater. The WSPs consist of interconnected basins, each designated for specific functions such as sedimentation, biological degradation, and clarification. Within these ponds, microalgae are cultivated to remove organic and inorganic contaminants from the wastewater.

3.4.2 BIOCHEMICAL MECHANISMS OF ALGAL MAT

The biochemical mechanisms involved in the removal of pollutants by algal mats include cationic exchange, anionic exchange, absorption/assimilation, and precipitation through pH alteration.

3.4.2.1 Cationic Exchange/Anionic Exchange

The exchange of charged ions, known as a cationic and anionic exchange, is utilized by algae to eliminate pollutants from wastewater. This process involves exchanging charged ions present in the wastewater with similarly charged ions found on the surface of the algal cells. The effectiveness of this method is determined by the charges of the contaminants and the surface charge of the algal cells. According to the research conducted by Jiang and colleagues in 2021, microalgae exhibit great potential for effectively removing heavy metals from wastewater. In particular, the utilization of two strains of microalgae, namely *Chlorella* sp. and *Scenedesmus* sp., was observed to facilitate the removal of heavy metals via cationic and anionic exchange mechanisms. The former involves the binding of positively charged heavy metal ions to the negatively charged functional groups on the microalgae's surface, while the latter pertains to the binding of negatively charged heavy metal ions to the positively charged functional groups. The results of this study indicate that the application of microalgae for wastewater treatment provides an eco-friendly and cost-effective approach compared to traditional methods. Additionally, utilizing microalgae for this purpose offers the possibility of generating biomass, which can be utilized as a source of biofuels or other valuable commodities.

Algae employ the ion exchange mechanism to effectively and economically eliminate metals from aquatic habitats, thereby enabling soil remediation and the retrieval of metals from wastewater. The functional groups present on the outer surface of algae cells assist in the adsorption and absorption of metals by forming ionic exchange reactions. The outcomes of this study reveal that the technique can efficiently extract metal ions from the solution and confine them within the confines of algae cells.

The utilization of the ion exchange technique by algae to eradicate metals from aquatic environments is a viable and cost-effective method for soil remediation and metal retrieval from wastewater. The high specificity and selectivity of this approach, due to the distinctive functional groups on the algae cell walls, facilitate the selective removal of targeted metals from wastewater. This attribute is particularly beneficial for the recovery of valuable precious metals. Furthermore, the algae-based ion exchange method is relatively simple and budget-friendly, thus presenting significant potential in the field of environmental remediation and resource recovery (Upadhyay et al., 2019).

3.4.2.2 Absorption/Assimilation

The processes of absorption and assimilation are recognized as the mechanisms by which algae consume nutrients and organic compounds present in wastewater. Algae can absorb pollutants, including nitrogen and phosphorus, subsequently adding them to their biomass. This mechanism effectively decreases the concentration of pollutants in the wastewater while facilitating algae growth. A recent investigation by Singh et al. (2021) revealed that microalgae, specifically *Chlorella* sp., could eliminate up to 95% of nitrogen from wastewater through absorption and assimilation.

The identification of significant quantities of inorganic nutrients like nitrate, phosphate, and heavy metals in wastewater has been associated with the development of algae. The gathering of these nutrients is highly significant in the growth and survival of algae, and uncontrolled accumulation may lead to damaging environmental outcomes. Nevertheless, the assimilation process, as presented by Cai et al. (2013), enables microalgae to transform inorganic nitrogen into organic nitrogen, thus presenting a unique benefit to the ecosystem.

During the assimilation process, algae uptake inorganic nitrogen into their cytoplasm. In this specialized cellular environment, nitrate and nitrite reductase enzymes catalyze a series of oxidation and reduction reactions leading to the conversion of inorganic nitrogen into the essential nutrient NH_4. The transformed ammonia is then incorporated into the intracellular fluid, where it plays a crucial role in various cellular processes such as protein synthesis and cell division. The significance of this assimilation process to the survival and growth of algae cannot be overstated, as it confers a source of organic nitrogen that can serve as fuel for energy production and biosynthetic pathways.

Phosphorus, a vital nutrient, plays an essential role in the growth and metabolic regulation of algae. Its significance lies in the fact that it is an indispensable constituent of nucleic acids, lipids, and proteins. Algae acquire phosphorus in the forms of H_2PO_4 and H_2PO_4 morphs. Through phosphorylation mechanisms, these forms are metabolized to assimilate phosphate compounds into organic compounds. It is further explained that this process allows the algae to maintain their metabolic processes and thrive in their environment. Specifically, phosphorylation entails a process whereby a phosphate group is transferred from ATP to a substrate molecule, forming a phosphorylated compound that serves as a crucial constituent for a range of cellular processes (Upadhyay et al., 2019).

3.4.2.3 Precipitation: pH Alteration

Precipitation, brought about by pH modification of wastewater to encourage the development of insoluble precipitates, is another method by which algae can extract contaminants. When dealing with wastewater treatment, residual matter refers to the solid substances that remain insoluble after a reaction occurs between soluble reactants. Such materials typically consist of a variety of minerals, organic matter, and solid particles that are not easily dissolved or broken down in water. Algae, found in wastewater, play a key role in the formation and elimination of residual matter by producing various chemical compounds, organic acids, and secondary metabolites.

Algae play a significant role in eliminating residual matter by generating organic acids, which, upon release into the surrounding water, bring down the pH level of the immediate environment. Consequently, the acidic conditions promote the precipitation of toxic chemicals and phosphorus, as revealed by De-Bashan and Bashan's study in 2004. The efficacy of this process is enhanced in the event of the existence of soluble iron, aluminum, or calcium in wastewater, owing to their capability to facilitate phosphorus reduction through precipitation.

Removing phosphorus from wastewater is crucial to prevent eutrophication and other environmental issues caused by its excessive levels. Precipitation, as a mechanism, is considered significant in this regard. This process leads to the formation of insoluble phosphorus compounds that help sequester the nutrient, preventing it from contributing to harmful algal blooms or causing any other adverse environmental impacts. Consequently, the precipitation of phosphorus is recognized as an effective approach to addressing the environmental challenges associated with high levels of this nutrient (Upadhyay et al., 2019).

3.5 THE POTENTIAL RESOURCE FOR BIOFUEL PRODUCTION: ALGAE

3.5.1 CULTIVATION AND AGRONOMIC DATA FOR MICROALGAE PRODUCTION

In general, microalgae obtain their energy and carbon sources through photosynthesis, a process called photoautotrophic culture, utilizing light and carbon dioxide. However, certain species can also be grown in the absence of light by using organic carbons like glucose or acetate, in a process known as heterotrophic culture. Although heterotrophic culture offers various benefits, such as lower costs, it is not widely employed in biodiesel production owing to the significant capital and operational expenses involved.

The prevailing approach for the production of algal biofuels is still photoautotrophic culture. This method effectively utilizes sunlight as a plentiful and affordable light source. As such, it continues to be the preferred method for algae-based biofuel production. For phototrophic microalgae to thrive, specific growth conditions must be met, including the presence of water, carbon dioxide, inorganic salts, and essential elements such as nitrogen, phosphorus, iron, and occasionally silicon. The ideal temperature range for growth is between 15°C and 30°C. Constant agitation is crucial to prevent settling, and nutrient supplementation during the day is necessary. Successfully cultivating and producing algae calls for conscientious management of ecological variables and nutrient availability. Multiple culture systems exist, such as suspension-based and solid surface-attached cultures, open ponds, and enclosed photobioreactors. In the subsequent section, the benefits and drawbacks of each system will be thoroughly examined (Chisti, 2007; Fernández et al., 1999; Metting, 1996).

The traditional and widely utilized method for the mass cultivation of microalgae is through open ponds. These ponds are typically shallow and roughly one foot deep, and are designed in a raceway form to promote optimal circulation and mixing of nutrients and algal cells, as demonstrated in Figure 3.9. The channel of the pond is

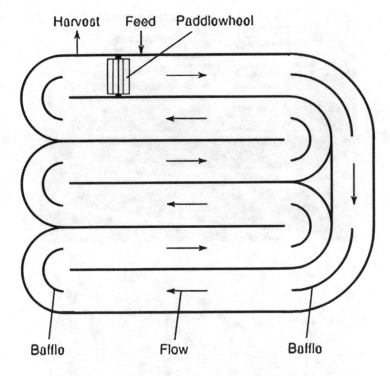

FIGURE 3.9   Algae-based open pond system schematic flowchart.

typically composed of concrete or dug into the ground and is lined with plastic to minimize water loss. A nutritious feed is continuously added to the culture using a paddle wheel. Open ponds have numerous advantages, including their low operational and capital expenses and ease of scaling. Furthermore, they are capable of supporting the cultivation of multiple species of microalgae, which is crucial for producing a wide range of products. Nevertheless, open ponds have certain drawbacks, such as water evaporation, contamination, and vulnerability to weather variations. The paddle wheel facilitates the circulation of the algal broth throughout the pond, and the collected biomass is harvested at the rear end (as depicted in Figure 3.9) (Figure 3.10).

The utilization of open ponds for microalgae cultivation is highly adaptable and can adapt to various sources of wastewater. Despite their drawbacks, such as water loss, contamination, challenges in maintaining optimal culture conditions, and limited biomass yield due to inadequate carbon dioxide uptake, open ponds still hold an advantage over enclosed photobioreactors in terms of cost-effectiveness.

One of the advantages of closed photobioreactors is that they offer a solution to the issues that can arise in open ponds, such as contamination and evaporation. By using a tubular configuration with transparent tubes, more light can be absorbed to stimulate the growth of healthy algae. Maintaining a consistent culture of algae requires the use of uninterrupted pumping and light exposure before returning the algal broth to the reservoir. To prevent the settling of algal biomass, a highly turbulent flow must be maintained within the reactor. This enables continuous cultivation and the ability

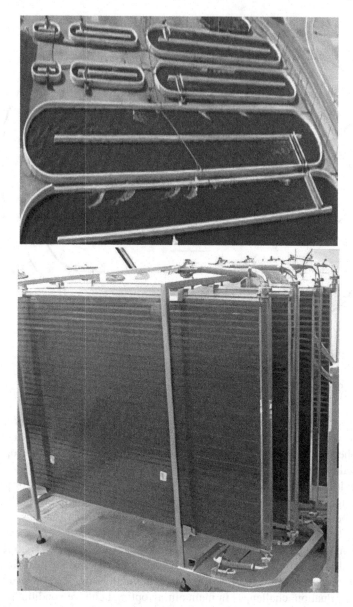

FIGURE 3.10 An open pond system and an algae photobioreactor are demonstrated here (Metting, 1996; Fernández et al., 1999; Chisti, 2007).

to harvest a portion of the algae after it passes through the solar collection tubes. Although a helical tubular photobioreactor may be used for high-value products, it requires additional illumination and is not commonly utilized for regular biodiesel production (Chisti, 2007; Fernández et al., 1999; Metting, 1996) (Figure 3.11).

The process of photosynthesis generates oxygen as a secondary outcome, which can prove detrimental to the health of algae in a closed photobioreactor. To counter this, the culture must be transferred periodically to a degassing zone. Furthermore,

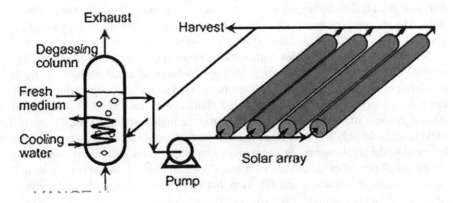

FIGURE 3.11 Schematic flowchart illustration of a tubular photobioreactor.

microalgae require carbon dioxide to flourish, but excessive amounts may cause carbon starvation and an increase in pH levels. Sustaining an ideal balance in carbon dioxide levels is crucial for healthy growth, and a steady stream of carbon dioxide supply is necessary for large-scale cultivation that can be sustained over time.

Photobioreactors offer several benefits in comparison to open ponds, including the ability to prevent contamination and evaporation while achieving greater concentrations and productivity of biomass. Nevertheless, photobioreactors are subject to higher costs and scalability issues, with temperature and light intensity fluctuations affecting microalgae growth. Additionally, collecting microalgae biomass for processing usually requires gravity settlement or centrifugation methods. Finally, the extraction of oil from microalgae biomass entails employing solvent extraction techniques, followed by further processes to produce biodiesel (Metting, 1996; Fernández et al., 1999; Chisti, 2007).

3.5.2 Yield Potential of Microalgae

The quantity of microalgae produced is contingent upon the type of culture system employed, namely open ponds or enclosed photobioreactors. The typical biomass yield per square meter of surface area per day for open ponds is between 5 and 10 g, with certain research indicating potential yields of up to 50 g/m²/day. Enclosed photobioreactors measure yield in terms of biomass per unit of reactor volume, with an average output of 2–3 g/L/day. The oil content of dry biomass varies from strain to strain, with some strains exhibiting up to 80% oil content. Gravity settlement or centrifugation methods can be employed for harvesting, resulting in solvent extraction for the production of biodiesel (Chisti, 2007; Fernández et al., 1999; Metting, 1996).

3.5.3 Challenges Associated with Microalgae Production

During the oil crisis of the 1970s, the United States Department of Energy initiated the ASP program with a particular emphasis on the exploration of alternative energy sources. The program recognized the potential of algal biofuels as a viable energy source, given their high productivity and ability to thrive in non-arable areas.

Furthermore, algal biofuels were viewed as a means to minimize greenhouse gas emissions. The program involved a significant amount of research on the biology of microalgae, as well as the development of cultivation methods and harvesting techniques.

Maintaining optimal species within the cultivation infrastructure presents a significant challenge in the production of algal biofuels. Microalgae have different requirements for light, nutrients, and temperature, and maintaining optimal conditions for growth can be difficult. Another challenge is the low yield of algal oil. Although some strains of microalgae can produce high amounts of oil, the overall yield is still relatively low compared to other biofuel crops such as corn or soybeans. Harvesting the algal biomass is also a major challenge, as the cells are typically very small and dispersed in the culture medium. Common methods of harvesting include centrifugation, flocculation, and filtration, but all of these methods have drawbacks in terms of energy requirements, cost, and efficiency.

The ultimate objective is to commercialize the production of algal biofuel. This has resulted in extensive research and development efforts in numerous areas, including elevating the oil content of existing strains and identifying new high-oil content strains. There are various methods being undertaken to enhance the cultivation of algae, including genetic modification and growth technologies aimed at improving growth rates. Other approaches include developing systems that are suited for open-air and closed environments, exploring options for co-product development, utilizing algae for bioremediation, and creating more efficient techniques for oil extraction (Metting, 1996; Fernández et al., 1999; Chisti, 2007).

3.5.4 Estimating the Cost of Microalgae Production

Various factors impact the expenses associated with algal oil production. These include biomass yield, oil content, production capacity, and expenditures related to oil recovery. Despite ongoing efforts to improve the efficacy of algal biofuel production, the cost of producing algal oil is still noticeably higher than that of petroleum diesel fuels. It was estimated that the cost of producing algal oil from a photobioreactor with a yearly production volume of 10,000 tons and an assumed oil composition of 30% would be $2.80/L ($10.50/gallon). It is important to note that this estimate does not encompass the expenses associated with converting algal oil to biodiesel, distribution and marketing, or taxes. On the other hand, the cost of petroleum diesel was found to range between $2.00 and $3.00 per gallon. The potential of algal oil as a biofuel source largely depends on the cost of petroleum oil. Formula to assess the cost of algal oil and determine whether it could serve as a competitive substitute for petroleum diesel. According to the formula, algal oil contains ~80% of the caloric energy value of crude petroleum. To remain competitive with petroleum diesel, the price of algal oil should not exceed $2.59/gallon when petroleum is priced at $100/barrel.

3.5.5 The Production of Microalgae: Environmental and Sustainability Concerns

In addition to serving as a promising source of biofuel, algae exhibits great potential for various other applications, such as serving as fertilizer and aiding in pollution control efforts. Certain varieties of algae possess organic fertilizing properties and may be employed in either a raw or semi-decomposed state. Through the cultivation of algae in ponds, the collection of fertilizer runoff from farms can be optimized. Consequently, the nutrient-rich algae can be used as a means of fertilizer, rendering the process potentially cost-effective and efficient for crop production (Chisti, 2007; Fernández et al., 1999; Metting, 1996).

3.6 GENETICALLY ENGINEERED ALGAE IN WASTEWATER TREATMENT

The application of genetic modification techniques can enhance the efficacy of PW bioremediation initiatives through the augmentation of algae's ability to efficiently eliminate water pollutants. For example, overexpression of genes involved in nutrient uptake, carbon fixation, and photosynthesis can enhance the growth and biomass yield of algae. Knocking down genes responsible for the synthesis of unwanted metabolites, such as pigments or toxins, can improve the suitability of algae for PW treatment. Gene editing can be used to develop algal strains that are more tolerant to high salinity or other environmental stressors commonly found in polluted water (Hassanien et al., 2023).

An acclimatization process for industrial wastewater can be effectively utilized to boost the biomass production and lipid content of *Chlorella vulgaris* and *Chlorella sorokiniana* UKM3. Detailed information on recent studies that focused on algal-mediated produced water treatment, including the experimental and cultivation parameters used in addition to the strains employed, can be referenced in Table 3.4 (Hassanien et al., 2023).

3.6.1 INDUCING GENETIC VARIATION: RANDOM MUTAGENESIS TECHNIQUES

Random mutagenesis is a valuable technique that can effectively induce trait modification in microalgae. Extensive research has evidenced its ability to consistently generate strains that exhibit enhanced resistance to contaminants and stress, heightened productivities, and greater metabolite production rates. Random mutagenesis experiments commonly involve the employment of chemical, nuclear irradiation, plasmas, and ultraviolet mutagenesis to achieve the desired results with efficiency.

Qi et al. (2018) have demonstrated the efficacy of ultraviolet mutagenesis and selective pressure in their development of a mutant variant of *Scenedesmus obliquus* that displays enhanced genetic stability and an increased capability to tolerate CO_2. The mutant strain exhibited marked improvements in both biomass productivity and light conversion efficiencies when subjected to higher CO_2 concentrations, as compared to its parental strains, in the course of the experiments conducted. The carbohydrate and lipid contents of the mutant strain were observed to be higher by 37% and 25%, respectively.

In 2018, a study was conducted by Ammar and colleagues to investigate the viability of utilizing radon mutagenesis on algae strains, combined with exposure to high loads of polluted wastewater. The objective was to determine whether this approach could be effective for bioremediation purposes without the need for a dilution step. This approach has shown great promise in increasing the industrial and economic feasibility of algae-based PW treatment options (Hassanien et al., 2023).

TABLE 3.4

Recent Scientific Publications Regarding the Utilization of Algae to Remediate Polluted Water Sources Have Been Carefully Reviewed (Hassanien et al., 2023)

Strain	Wastewater Type	Cultivation Conditions	Highest Biomass Yield (g/L)	Pollutants Removed	Highest Removal Efficiency (%)	References
Chlorella sp.	Produced water from a qatari local petroleum company	In a temperature-controlled room, a glass bottle was agitated with compressed air and illuminated with a light intensity of 600 mot photons-m^2/s	1.72	TOC Total nitrogen	73 92	Das et al. (2019)
Chlorella vulgaris	Produced water from an oil and gas facility in the United States	Tissue culture roller drum apparatus inside an incubator with a constant level of CO_2 of 296% to 3% (v/v), a temperature of 28°C with 16:8h light: dark cycle and an illumination of ~4,000 lux	3.1–0.5	Total nitrogen Phosphorus	100 2–74.2	Rahman et al. (2021)
Chlorella vulgaris	PW from dumping site generated by oil wells in Colombia	Fluorescent light at an irradiance of 36.8 ± 4.2 vino' photons m' r' at the surface of the culture medium, temperature at 20°C and permanent aeration supplied by a blower	—	Total hydrocarbons		Calderen-delgado et al. (2019)

(Continued)

TABLE 3.4 (Continued)

Recent Scientific Publications Regarding the Utilization of Algae to Remediate Polluted Water Sources Have Been Carefully Reviewed (Hassanien et al., 2023)

Strain	Wastewater Type	Cultivation Conditions	Highest Biomass Yield (g/L)	Pollutants Removed	Highest Removal Efficiency (%)	References
Chlorella pyrenoidosa	PW from oilfield in Algeria	Outdoor, under sunlight radiation, using an open system sited in the desert area in the winter season. The temperatures fluctuated from 26°C to 31°C during the day	1.15	COD Ammonium nitrogen Total nitrogen Total phosphorus Copper Lead cadmium	89.67% 100% 57.14%, 75.51% 73 39 72.80 48.42	Rahmani et al. (2022)
Nannochloropsis oculata	Produced water from oil field in Iraq	Florescence light (2,000 lux) at and a light photoperiod of 18:6h light dark, 25°C±1°C, continuous filtered air at a constant flow rate via two aquarium air pumps	1.13	Oil COD	66.5 54	Ammar et al. (2018)
Nannochloropsis oculata	Produced water from a TOTAL operating site in France	14/10h light/dark periods, by LED lamps, temperature at (21°C±1°C) autotrophic conditions with air. CO$_2$ was added in pulse, 5s each 20–40min and pH between 7.5 and 9	—	Ammonium nitrogen COD Iron	~100 70 100	Parsy et al. (2020)

(Continued)

TABLE 3.4 (Continued)

Recent Scientific Publications Regarding the Utilization of Algae to Remediate Polluted Water Sources Have Been Carefully Reviewed (Hassanien et al., 2023)

Strain	Wastewater Type	Cultivation Conditions	Highest Biomass Yield (g/L)	Pollutants Removed	Highest Removal Efficiency (%)	References
Nannochloropsis oculata	Produced water from an oil field in Brazil	Aerated photobioreactors (3 L min⁻¹), cold white LED lamps with light intensity of 57 ttmol m²/s'', photoperiod of 12:12 h. Temperature controlled at 21°C±0.9°C. The pH was fixed at 7	—	PAHs NAP APT FLU PHE BbF DA BaP Iron	94 96 95 91 83 95 90 95 96.80	Marques et al. (2021)
Galdieria sulphuraria	Produced water from an oil and gas facility in United States	Tissue culture roller drum apparatus inside an incubator CO$_2$ level was kept constant at 2%–3% (v/v), temperature 42°C with 24 h of continuous illumination (~4,000 lux)	5.12 ± 0.28	Total nitrogen	8 N	Rahman et al. (2021)
Isochrysis galbana	Produced water from oil field in Iraq	Florescence light (2,000 lux), and a photoperiod of 18:6 h light: dark, 256°C ± 1°C, continuous filtered air at a constant flow rate via two aquarium air pumps	1.01	Oil COD	68 56	Ammar et al. (2018)
Dunaliella tertiolecta	Produced water from an oil production facility in the Permian Basin of southeast New	Temperature controlled at 24°C in a growth chamber with fluorescent illumination of 100 pmol photons nc.2 s''. agitation was set at 140 rpm with	4.3	Nitrate Phosphate	2 = 99.6 0.49.6	Hopkins et al. (2019)

3.6.2 PRECISION ENGINEERING OF GENOMES: TARGETED GENETIC MODIFICATION METHODS

Targeted genetic modification methods can offer even more precise control over the modification of specific traits in algae compared to random mutagenesis. In particular, the heterologous expression of foreign genes has shown great potential for enhancing specific traits of microalgae. This process involves introducing DNA sequences from another organism into the microalgae, allowing them to produce new enzymes or proteins that can improve their performance under certain conditions.

In recent years, the field of microalgae has witnessed a surge in genetic engineering, thanks to the emergence of advanced genetic tools that enable researchers to reshape and enhance metabolic pathways, thereby uncovering novel prospects for the industrial growth of microalgae. In recent years, gene-editing tools such as TALENs, ZFNs, and CRISPR/Cas9 have gained popularity among recombinant DNA technologies. These tools have been effectively used in various microorganisms, including microalgae, over the past decade. These tools allow researchers to make precise modifications to specific genes or DNA sequences within the algae genome (Hassanien et al., 2023).

3.6.2.1 Zinc-Finger Nucleases for Gene Editing Technology

Zinc-finger nucleases (ZFNs) are modular components renowned for their proficiency in binding specifically to particular DNA sequences and creating DNA double-strand breaks. These mechanisms allow a vast spectrum of genetic modifications through error corrections. In the context of microalgae, ZFNs have found utility in genomic DNA modification, such as targeting the COP3 gene in *Chlamydomonas reinhardtii*. As a result, stable transformed colonies were established. The use of electroporation rather than glass beads in the ZFN protocol has exhibited positive outcomes, particularly in the reliable editing of genes by homologous recombination in the wild type and various strains of *Chlamydomonas*.

While ZFN technology offers multiple advantages, such as its precision in DNA editing, it is crucial to consider its limitations. There exists a constraint concerning the restricted quantity of sites amenable for nuclease selection, which results in a heightened probability of inadvertently causing dual-strand breaks at untargeted sites. Figure 3.12 provides a concise depiction of the ZFN procedure (Hassanien et al., 2023).

3.6.2.2 Exploring the Talens for Genome Editing

TALENs (Transcription Activator-Like Effector Nucleases) constitute a potent technique to modify genes with precision. They are created by combining TALE proteins with the FokI nuclease, which can create double-strand breaks in DNA. TALE proteins are naturally found in Xanthomonas bacteria and are known for their ability to bind to specific DNA sequences.

TALENs possess the advantageous ability to be tailored to target precise locations within the genome, offering an array of benefits. They are also easier to design than ZFNs and are commercially available. Additionally, TALENs can bind to longer DNA sequences than other gene editing tools. However, the larger size of TALENs can make them less specific, and they may be harder to deliver to cells.

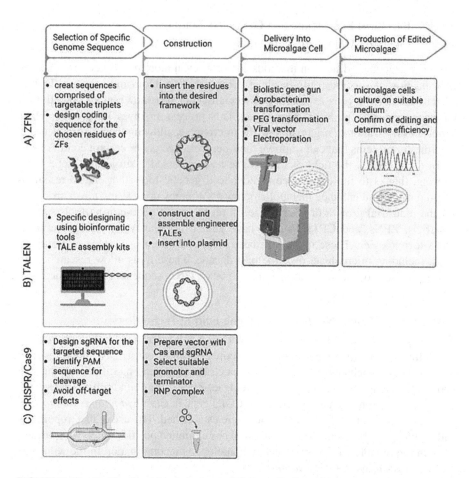

FIGURE 3.12   Outlined succinctly is the workflow applied to utilizing three genome editing tools, namely ZFN, TALEN, and CRISPR/Cas9, in microalgae. This systematic approach is made up of four vital stages: (1) identification and development of the target sequence with the aid of bioinformatics, (2) selection of a suitable vector to create a delivery system, (3) introduction of the plasmid into the microalgae cell using an appropriate method of delivery, and (4) evaluation of the gene editing efficiency by sub culturing the microalgae on a selective medium (Hassanien et al., 2023).

Despite obstacles encountered, researchers have achieved successful genome modification of microalgae through the use of TALENs. This powerful tool was employed to amplify lipid metabolic reactions in the diatom Phaeodactylum tricornutum and the green microalga *Coccomyxa* sp. Furthermore, *Nannochloropsis oceanica* underwent a transformative process as TALENs were utilized for the manipulation of both the nitrate reductase and acyltransferase genes. A diagrammatic illustration of the TALEN-based genome modification process for microalgae can be found in Figure 3.12 (Hassanien et al., 2023).

3.6.2.3 The Crispr/Cas9 System: Consequential Effects on Various Fields

The genome editing technique of CRISPR/Cas9 is highly promising given its simplicity and versatility. It has been successfully demonstrated across various organisms, offering a streamlined and efficient means of working. Further bolstering its reliability, several mitigation strategies have been implemented to significantly reduce off-target effects. Furthermore, the implementation of a CRISPR-Cas9 system that is devoid of plasmids has resulted in the formation of dependable RNPs inside the cells under investigation.

Achieving stable gene expression over an extended period is crucial for the industrial-scale application of genetically modified algae. Hence, it is imperative to undertake measures that enable the integration of gene cassettes into the transformed strain's genome for large-scale algal cultivation. Several examples of CRISPR/Cas9-engineered microalgae have demonstrated consistent and enduring mutations, thereby resolving the issue of uncertain mutations. Figure 3.13 illustrates the CRISPR/Cas9 system's approach to genome manipulation of microalgae (Hassanien et al., 2023).

The CRISPR/Cas9 technology has shown great success in microalgae, particularly in *Chlamydomonas reinhardtii*. However, further adjustments are required to minimize Cas9's negative impact on the algae strain. As a result, CRISPR/Cas9 has been frequently employed to enhance various aspects of Chlamydomonas, such as increasing lipid and pigment content, improving triacylglycerol yield and lipid deposition, and understanding CO_2 sequestration mechanisms. Research efforts have primarily focused on the enhancement of algal cell lipid content through genetic modifications utilizing CRISPR/Cas9 techniques.

Moreover, it is feasible to improve the lipid content of *Chlorella vulgaris* by precisely editing the fad3 gene. The resulting modified strain yielded up to a 46% increase in lipid content, with biomass concentrations reaching 20% higher than the unmodified wild-type strain. Several other studies have documented the potential of enhancing lipid yields in varied species of microalgae by disabling genes associated with the degradation of fatty acids (Hassanien et al., 2023).

3.6.3 IMPROVING THE EFFICACY OF PHYTOREMEDIATION THROUGH CRISPR/CAS9-MEDIATED TARGET GENE IDENTIFICATION

The identification of multiple candidate genes that are capable of modifying metabolic pathways to enhance biomass production and bioremediation has been established. To achieve the most efficient microalgal wastewater treatment standards using genetic engineering, two primary strategies can be adopted. The first approach entails enhancing the strains' ability to withstand pollutants like hydrocarbons and heavy metals, while the second approach involves the expression and production of degradation-promoting molecules such as antifouling agents and surfactants. In either scenario, the identification of the specific target genes that require transformation is a crucial preliminary step toward achieving desirable outcomes.

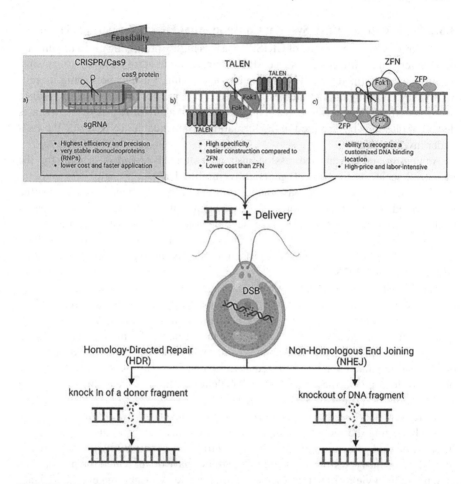

FIGURE 3.13 Specialized molecular tools such as CRISPR/Cas, TALEN, and ZFN complexes offer a viable method for modifying the genetic makeup of microalgae. These tools work through the creation of double-stranded breaks (DSBs) at predetermined locations on the genome, initiating mechanisms such as non-homologous end joining (NHEJ) or homology-directed repair (HDR) if a donor template is present. The NHEJ approach frequently results in the insertion and/or deletion (indels) of various sizes at the DSB site, which may interfere with the targeted gene's reading frame. In contrast, HDR offers a precise insertion or deletion achieved via homologous recombination. Among the available options, CRISPR/Cas9 is a preferred tool due to its fast implementation, accuracy, and user-friendly interface, as emphasized (Hassanien et al., 2023).

3.6.3.1 Plants' Capacity to Endure and Store HMs: Mechanisms and Genetic Factors

Gene overexpression and the construction of transgenic algal strains are the primary techniques utilized for genetic manipulation in algal-based metal recovery. Notably, the application of genetic manipulation for heavy metal tolerance in microalgae has not been extensively studied, unlike plants and bacteria. The incorporation of

ACC deaminase and iaaM genes in Petunia hybrida Vilm significantly enhanced heavy metal tolerance, resulting in healthier and larger plants in contaminated soil. Similarly, the introduction of phytochelating synthase genes in *Mesorhizobium huakuii* produced a transgenic strain that accumulated Cd^{+2} up to 19-fold more than the phytochelating synthase-free strain.

A further study indicates the crucial role of KoCBF1 and KoCBF3 genes in promoting growth and heavy metal accumulation in *Kandelia obovata* when subjected to lead (Pb $(NO_3)_2$) exposure, as they demonstrated significant expression levels. Moreover, an experiment on *C. reinhardtii* species demonstrated elevated resistance to cadmium stress in a transgenic strain with elevated expression of the CrMTP4 gene, in contrast with the wild type that could not endure the same conditions.

3.6.3.2 Biodegradation of Hydrocarbons: Genetics and Applications in Bioremediation

The composition of wastewater, including petroleum wastewater (PW), is subject to considerable variation based on its source, which directly impacts the levels of total organic carbon (TOC) and hydrocarbons present. For instance, empirical data demonstrates that PW collected from an industrial site in Qatar showed a TOC benchmark of 720.33 mg/L, with reported values soaring as high as 2,430 mg/L, implying significant fluctuations.

A vector was developed to express alkane hydroxylase in *Escherichia coli* (*E. coli*) DH5α, which was subsequently utilized in biodegradation assays within diesel-containing media. The combination of Acinetobacter and a transgenic *E. coli* strain was found to result in a significant enhancement of up to 49% in diesel biodegradation. A flavin-diffusible monooxygenase gene (cph) from Arthrobacter chlorophenolicus, comprising two components, was cloned and subsequently overexpressed to achieve efficient degradation and removal of 4-chlorophenol. The modified strains showed impressive removal rates, with up to 82.7% of 4-chlorophenol removed from the media.

PW has a well-established reputation for exhibiting significant variability in its composition, particularly with regard to its TOC and hydrocarbon levels. The TOC values of PW collected from a petroleum industrial site in Qatar have been reported at both moderate levels (such as 720.33 mg/L) and exceptionally high concentrations (up to 2,430 mg/L), stressing the need for advanced bioremediation practices to be employed.

Limited bioremediation techniques have been developed to address this issue. In one study, a pCom8 vector was implemented to express alkane hydroxylase in *E. coli* DH5α, which was introduced to diesel-containing media for biodegradation assays. The utilization of Acinetobacter in combination with the transgenic *E. coli* strain resulted in a significant improvement of up to 49% in diesel biodegradation. In another examination, a two-component flavin-diffusible monooxygenase gene (cph) from Arthrobacter chlorophenolicus was cloned for enzyme overexpression. The newly engineered enzyme could successfully remove up to 82.7% of 4-chlorophenol present in the media. A range of microalgae and cyanobacteria species, such as *Dunaliella salina* and *Porphyridium cruentum*, have undergone extensive testing due to their ability to generate biosurfactants. These organisms create extracellular polymeric substances, acting as emulsifiers to support the metabolic deterioration of oil hydrocarbons (Table 3.5).

TABLE 3.5

Genes That Are Potential Candidates for Heavy Metal Tolerance and Accumulation, as well as Hydrocarbon Degradation

Gene	Plasmid	Promotor	Application	References
PCSAt	PBBRMCS-2 PMP220	nifH	Accumulation of Cd	Sriprang et al. (2003)
KoCBF3	–	–	Accumulation of Pb	Peng et al. (2020)
ACC deaminase	PBI-iaaM/ACC	CaMV 35S	Accumulation of copper and cobalt	Zhang et al. (2008)
iaaM	PBI-iaaM	GPR	Accumulation of copper and cobalt	Zhang et al. (2008)
CrMPT4	CrMTP4gDNA-pH2GW7	–	Increase in Mn and Cd content in the cell	Ibuot et al. (2017)
alkane hydroxylase (alkB)	pCom8	–	Degrading diesel oil	Luo et al. (2015)
cph	pET-24a	–	Removal of 4-chlorophenol	Kang et al. (2017)
mat1	pET15b	–	Biosurfactant	Hewald et al. (2003)

Correlation between the manifestation of mannosylerythritol lipids, a type of biosurfactant found in diverse microalgae, and the *Ustilago maydis* gene strains Emt1, Mmc1, Mac1, and Mac2. The utilization of wastewater as a substrate demonstrated a significant augmentation in mannosylerythritol lipids manufacturing in *Pseudozyma tsukubaensis*. A comprehensive summary of these and other genes that impact the degradation of hydrocarbons and heavy metals is furnished in Table 3.4. Microalgae have significant potential for bioremediation of water that has been heavily contaminated with organic pollutants and heavy metals. Their remarkable ability to adapt and utilize these contaminants to promote their growth is a pivotal factor.

3.6.4 The Obstacles Encountered in Utilizing Genetic Engineering Techniques to Improve the Efficacy

Microbial degradation is a widely-used method for treating water contaminated with pollutants, and there has been increasing interest in genetically modifying these microorganisms to enhance bioremediation efficiency. Achieving successful genetic transformation relies heavily on the selection of an appropriate species. *Thalassiosira weissflogii*, *Ulva lactuca*, and *Gracilaria changii* are examples of species that have exhibited low stability after undergoing nuclear transformation. When utilizing CRISPR/Cas9, it is critical to consider the possibility of off-target effects, in addition to the technical limitations associated with genome editing techniques such as TALEN.

For the optimization of engineered cells designed for bioremediation, a comprehensive understanding of their metabolic and structural responses under metal stress is crucial, particularly in the removal of heavy metals. A genetic modification of microalgae offers a promising opportunity to enhance bioremediation capabilities, though questions remain over the sustainability and feasibility of large-scale usage. Hence, exploring strategies to optimize the sustainable application of engineered microalgae is vital (Hassanien et al., 2023).

3.7 POSTHARVEST UTILIZATION OF MICROALGAE

Microalgae are increasingly being recognized as a promising and sustainable option to harness a diverse array of biologically active compounds. Their ability to accumulate high concentrations of valuable molecules such as antioxidants, polyunsaturated fatty acids, lipids, antiviral, and anticytostatic agents makes them an attractive alternative to traditional sources. The use of microalgae biomass has found a range of applications across various industries, including dietary supplements, feed, cosmetics, and the manufacturing of chemicals and pharmaceuticals.

Microalgae have a wide range of applications, with their primary usage being in the form of a nutrient source for human consumption. These tiny organisms are rich in essential amino acids, vitamins, and minerals, which makes them ideal for use as supplements in functional foods, nutraceuticals, and dietary supplements. Moreover, microalgae are particularly valued for their high content of polyunsaturated fatty acid, including docosahexaenoic acid and eicosapentaenoic acid, which are known for their various health benefits, such as reducing inflammation and enhancing cardiovascular health. Additionally, microalgae also possess high levels of antioxidants, which have been shown to mitigate the risk of chronic diseases, including cancer and heart disease.

The utilization of microalgae for various industries is becoming increasingly popular owing to its numerous advantages, such as sustainability, high productivity, and versatility. Typically, the total lipid content of microalgae accounts for 20%–50% of its cell dry weight; however, under certain circumstances, it can increase up to 80%. Fatty acids with carbon atoms falling within the range of 14–20 can be employed in the production of biodiesel, whereas those with more than 20 carbon atoms, such as docosahexaenoic acid and eicosapentaenoic acid, serve as essential components of dietary supplements. The biochemical composition of microalgae, encompassing lipids, proteins, and carbohydrates, varies among different species and can be regulated by tweaking the cultivation conditions.

Microalgae are highly advantageous not only for their exceptional nutritional properties but also for their tremendous potential to manufacture biologically active compounds for various sectors. The polysaccharides present in microalgae possess noteworthy antiviral and anticytostatic characteristics, effectively rendering them a potent ingredient for use in the pharmaceutical, chemical, and cosmetic industries. In addition, the pigments naturally present in microalgae, such as chlorophylls, carotenoids, and phycobiliproteins, are widely utilized as organic coloring agents for cosmetics and food products. Moreover, microalgae also comprise bioactive peptides recognized for their wide-ranging health benefits, including antioxidant and antihypertensive effects.

The lipid concentration in microalgae is influenced by several factors, including the species, cultivation conditions, and approach to harvesting. Nevertheless, microalgae have the potential to accumulate up to 80% of dry cell weight as lipids, rendering them a viable source of biofuels such as biodiesel. Biodiesel production requires fatty acids with carbon chain lengths ranging from 14 to 20 atoms, whereas PUFA with over 20 carbon atoms serve as dietary supplements. The capacity of microalgae to generate significant quantities of lipids, coupled with their other advantageous compounds, makes them a compelling prospect for renewable energy and bioproducts. Given their broad applicability in various industries, researchers and entrepreneurs worldwide are increasingly directing investments to explore and optimize the potential of microalgae. Exploring and fully realizing the potential benefits of microalgae could make a considerable contribution to a sustainable and ecologically sound future (Dayana Priyadharshini et al., 2021).

3.7.1 MICROALGAE AS A SOURCE OF NUTRIENT SUPPLEMENTS

The biomass produced by microalgae is a precious commodity that presents a multitude of health benefits. It is a wealth of unsaturated fatty acids, vital vitamins, and carotenoids, which contribute to mitigating the risk of different ailments and enhancing immune system functioning. Additionally, this biomass serves as a supplementary feed for animals, and it offers a considerable input of nutrients to enhance their product traits and overall well-being.

Microalgal biomass is a precious commodity that offers a wealth of health benefits, attributed to its high content of unsaturated fatty acids and crucial vitamins. These unsaturated fatty acids, such as omega-3 and omega-6, are instrumental in providing essential nutrients that aid in mitigating a wide range of conditions, such as hyperlipidemia, cardiovascular disease, hypertension, arteriosclerosis, and rheumatoid arthritis. Moreover, these fatty acids have been shown to fortify the immune system, thus offering protection against diverse infections and ailments.

Alongside fatty acids, microalgal biomass is a rich source of all essential vitamins, including B1, B2, B12, K, E, C, and nicotinic acid. These vitamins play a crucial role in promoting optimal health and enabling various metabolic processes in the body. Moreover, microalgae exhibit a higher concentration of carotenoids than higher plants, which includes β-carotene, astaxanthin, canthaxanthin, lutein, violaxanthin, zeaxanthin, and neoxanthin. Of these carotenoids, astaxanthin and canthaxanthin are widely used as coloring agents and antioxidants in aqua-farming systems (Dayana Priyadharshini et al., 2021) (Table 3.6).

The advantages of microalgal biomass extend beyond human consumption, as they serve as a nutritional supplement for goats, ornamental fish, ruminants, and poultry. Incorporating microalgae into their diets can confer benefits such as improved product traits, disease resistance, and enhancements to hair and/or skin color. In addition to providing valuable nutrients, microalgae can also contribute to the overall health and well-being of these animals (Dayana Priyadharshini et al., 2021).

3.7.2 POSTHARVEST EXTRACTION OF BIOLOGICALLY ACTIVE COMPOUNDS FROM MICROALGAE

The use of microalgal biomass and lipids in cosmetic products is increasingly gaining popularity owing to their rejuvenating and antioxidant properties for human skin. Lipid-based creams, lotions, and microalgal biomasses obtained through supercritical CO_2 or ethanol extraction have demonstrated remarkable benefits for the skin.

Further research indicates that microalgal polysaccharides have remarkable pharmacological properties that can stimulate the human immune system on a

TABLE 3.6
Biochemical Compositions of Various Algal Species (Dayana Priyadharshini et al., 2021)

Name of the Algae	Biomass (mg/L day)	Protein (%)	Carbohydrate (%)	Lipid (%)	References
Spirulina platensis	60–4,300	31.4–68.2	8–24.9	4–16.6	Laurens et al. (2017)
Spirulina maxima	210–250	60–71	13–16	4–9	Sun et al. (2018)
Chlorella vulgaris	20–200	51–58	9–17	5–58	Deshmukh et al. (2019) and Ungureanu et al. (2021)
Chlorella pyrenoidosa ·	290–760	57–60.4	26	2–37	Yadavalli and Heggers (2013)
Chlorella sp.	20–2,500	51–58	12–17	10–48	Ganeshkumar et al. (2018)
Scenedesmus quadricauda	190	47	NA	1.9–18.4	Sirakov et al. (2013)
Scenedesmus obliquus	4–74	11.8–56	10–17	11–55	Mata et al. (2012)
Scenedesmus dimorphus	1.523	8–43	16–52	16–40	Xu et al. (2015)
Synechococcus sp.	–	63	15	11	Agawin et al. (2003)
Chlamydomonas reinhardtii	820–2,000	48	17	16.6–25.3	Toyama et al. (2018)
Chlamydomonas debaryana	8–15	15	18	16–20	Hasan (2014)
Botryococcus braunii	600–1,800	48	17	25–75	Gouveia et al. (2017) and Sirakov et al. (2013)
Dunaliella salina	220–340	25.7–57	16.3–32	6–25.3	Passos et al. (2015)
Chlorella sorokiniana	8.08 g/L	0.272 g/L/d	–	–	Lee et al. (2021)

cellular or humoral level. Notably, highly sulfated polysaccharides, found explicitly in Cyanobacteria and green and red microalgae, have shown promising results. These polysaccharide fragments can be leveraged for the development of newer and more potent drugs and therapies in the future.

The application of microalgae in the synthesis of bio-nanoparticles is currently drawing considerable attention in the field of nanomedicine. One example is the AR-Ag nanoparticles derived from *Amphiroa rigida*, a species of marine red algae, which have shown remarkable efficiency as reducing agents. These nanoparticles exhibit potent antibacterial properties, effectively combating *Staphylococcus aureus* and *Pseudomonas aeruginosa*, while also demonstrating high levels of larvicidal activity. Moreover, the AR-Ag nanoparticles have demonstrated remarkable cytotoxicity in MCF-7 human breast cancer cells. This discovery highlights the significant potential of microalgal biomass and lipids in various fields, such as cosmetics, drug development, and nanomedicine, offering more sustainable and environmentally friendly alternatives to conventional manufacturing methods (Dayana Priyadharshini et al., 2021).

3.7.3 Microalgae as a Feedstock for Postharvest Biofuel Production

Microalgae are an emerging and promising source of renewable energy that has captured the attention of many seeking alternative solutions to traditional fossil fuels. Unlike terrestrial plants, microalgae can produce a greater biomass per unit area, making them a more efficient source of biofuels. Furthermore, microalgae can be cultivated in areas that are unsuitable for traditional crop growth, such as deserts, rooftops, and other non-agricultural lands, increasing the potential for scalability.

Due to their quick and efficient growing cycles, microalgae are an ideal candidate for biofuel production. When grown in nutrient-limited environments, certain algal species have been found to accumulate substantial amounts of lipids or polysaccharides, which can be converted into biofuels. For example, algal strains such as *Chlorella* sp. and *Nannochloropsis* sp. can retain up to 60%–70% of their storage compounds as lipids or polysaccharides, even under nutrient-limiting conditions. This indicates that microalgae have the potential to be a major contributor to the global push toward sustainable and renewable energy sources.

The utilization of microalgae in biofuel production yields environmental advantages, as they consume carbon dioxide during growth and produce oxygen as a by-product, thereby contributing to the reduction of greenhouse gas emissions. Moreover, microalgae cultivation aids in nutrient remediation by eliminating surplus nutrients, such as nitrogen and phosphorus, from wastewater and agricultural runoff and effectively curtails eutrophication in aquatic ecosystems.

Microalgae exhibit several advantages over angiospermic plant biomass as a biofuel resource. Firstly, microalgae demonstrate a significantly higher capacity for lipid compound production per unit area in comparison to traditional crops. Indeed, numerous studies have shown that microalgae productivity can be up to 100 times greater than that of conventional crops. This heightened productivity renders microalgae a particularly appealing choice for biofuel generation. Secondly, microalgae can be cultivated in areas that are typically unsuitable for conventional farming, such as arid regions or locales with poor soil quality. The reason for this is that microalgae cultivation can occur in a regulated environment, such as a photobioreactor, enabling optimal growth conditions to be sustained. Accordingly, microalgae cultivation is not restricted to areas with specific soil or climatic prerequisites. Thirdly, microalgae cultivation presents a viable solution for areas with limited freshwater resources, as the organisms can be grown in marine or brackish water that is not suitable for traditional farming practices. This provides new prospects for biofuel generation in regions facing water scarcity.

Moreover, microalgae can aid in the remediation of water bodies contaminated with excess nutrients, including TN and TP, often resulting from agricultural or industrial runoff. By utilizing these excess nutrients, microalgae can potentially improve the water quality of such bodies. However, cultivating microalgae can entail higher costs than conventional crop cultivation due to the need for resources such as mixing, crushing, and processing. Additionally, difficulties in scaling up production to meet commercial demands pose significant challenges. Nonetheless, ongoing research aims to address these concerns and explore effective methods to optimize microalgae usage for biofuel production (Dayana Priyadharshini et al., 2021).

3.8 CONCLUSION

The treatment of sewage wastewater is a critical issue facing modern society, with conventional treatment processes often necessitating significant land, energy resources, and finances to operate effectively. Fortunately, phycoremediation offers a practical substitute treatment option for addressing pollutants found in industrial and commercial sewage wastewater. This approach utilizes microalgae to clean the wastewater effluent and can effectively remove harmful chemicals, heavy metals, and nutrients, among other contaminants, while also reducing the biological oxygen demand (BOD) of the wastewater. Within this chapter, a range of potential strategies have been examined for the purpose of phycoremediation. These methodologies encompass phycovolatilization, rhizofiltration, phytostabilization, CWs, and hydraulic barriers.

Each approach provides distinctive benefits and mechanisms for effectively eliminating pollutants from wastewater. For instance, phycovolatilization involves the transformation of pollutants into volatile substances that can be easily removed from wastewater. Rhizofiltration involves the involvement of microalgae and aquatic plants in the treatment process, while phytostabilization involves the removal of pollutants from the top of phreatic zones. CWs are a cost-effective green technology that can be utilized to remove pollutants from wastewater, and hydraulic barriers can be used to remove pollutants from larger water bodies via algae. Moreover, genetic engineering has emerged as a promising tool to enhance bioremediation efficiencies, with techniques such as random mutagenesis and targeted genetic engineering showing great potential for improving the effectiveness of phycoremediation.

One potential application of genetic engineering involves specifically targeting genes related to pollutant degradation or uptake in microalgae to enhance their efficacy for environmental remediation purposes. Furthermore, CRISPR/Cas9 technology can be applied to target specific genes and improve the efficiency of phycoremediation. Lastly, postharvest utilization of microalgae provides an added benefit to phycoremediation by generating value-added products such as biofuels, fertilizers, and biologically active compounds. This can further contribute to the sustainability of the process by reducing waste and providing alternative sources of energy and nutrients.

REFERENCES

Agawin, N.S.R., Duarte, C.M., Agustí, S., McManus, L., 2003. Abundance, biomass and growth rates of Synechococcus sp. in a tropical coastal ecosystem (Philippines, South China Sea). Estuar. Coast Shelf Sci. 56 https://doi.org/10.1016/S0272-7714(02) 00200-7.

Ahmad, S., Pandey, A., Pathak, V.V., Tyagi, V.V., Kothari, R., 2020. Phycoremediation: algae as eco-friendly tools for the removal of heavy metals from wastewaters. In: Bioremediation of Industrial Waste for Environmental Safety. Springer Singapore, pp. 53–76. https://doi.org/10.1007/978-981-13-3426-9_3.

Ammar, S. H., Khadim, H. J., and Isam, A. (2018). Cultivation of Nannochloropsis oculata and Isochrysis galbana microalgae in produced water for bioremediation and biomass production. Environ. Technol. Innovation 10, 132–142. doi:10.1016/j.eti.2018.02.002

Bolan, N. S., Park, J. H., Robinson, B., Naidu, R. & Huh, K. Y. 2011. Chapter four - phytostabilization: A green approach to contaminant containment. In: Sparks, D. L. (ed.), *Advances in Agronomy*. Cambridge, MA: Academic Press.

Calderón-delgado, I. C., Mora-solarte, D. A., and Velasco-Santamaría, Y. M. (2019). Physiological and enzymatic responses of Chlorella vulgaris exposed to produced water and its potential for bioremediation. Environ. Monit. Assess. 191, 399. doi:10.1007/s10661-019-7519-8.

Cai, W., Zhao, Z., Li, D., Lei, Z., Zhang, Z., Lee, D.J., 2019. Algae granulation for nutrients uptake and algae harvesting during wastewater treatment. Chemosphere 214, 55–59. https://doi.org/10.1016/j.chemosphere.2018.09.107.

Carrasco Gil S, Siebner H, Le Duc DL, Webb SM, Millán R, Andrews JC, Hernández LE (2013) Mercury localization and speciation in plants grown hydroponically or in a natural environment. Environ Sci Technol 47:3082–3090

Chisti, Y. 2007. Biodiesel from microalgae. *Biotechnology Advances,* 25, 294–306.

Choi HJ, Lee SM (2015) Effect of the N/P ratio on biomass productivity and nutrient removal from municipal wastewater. Bioprocess and Biosystems Engineering 38(4):761–766

Chugh, M., Kumar, L., Shah, M. P. & Bharadvaja, N. 2022. Algal Bioremediation of heavy metals: An insight into removal mechanisms, recovery of by-products, challenges, and future opportunities. *Energy Nexus,* 7, 100129.

Das, P.K., Rani, J., Rawat, S., Kumar, S., 2021. Microalgal co-cultivation for biofuel production and bioremediation: current status and benefits. Bioenergy Res. https:// HYPERLINK "http://doi.org/10.1007/s12155-021-10254-8" doi.org/10.1007/s12155-021-10254-8.

Dayana Priyadharshini, S., Suresh Babu, P., Manikandan, S., Subbaiya, R., Govarthanan, M. & Karmegam, N. 2021. Phycoremediation of wastewater for pollutant removal: A green approach to environmental protection and long-term remediation. *Environmental Pollution,* 290, 117989.

Deshmukh, S., Bala, K., Kumar, R., 2019. Selection of microalgae species based on their lipid content, fatty acid profile and apparent fuel properties for biodiesel production. Environ. Sci. Pollut. Res. 26, 24462–24473. https://doi.org/10.1007/s11356-019- 05692-z.

Fernández, F. A., Camacho, F. G. & Chisti, Y. 1999. Photobioreactors: Light regime, mass transfer, and scaleup. *Progress in Industrial Microbiology*, 35, 231–247.

Ganeshkumar, V., Subashchandrabose, S.R., Dharmarajan, R., Venkateswarlu, K., Naidu, R., Megharaj, M., 2018. Use of mixed wastewaters from piggery and winery for nutrient removal and lipid production by Chlorella sp. MM3. Bioresour. Technol. 256 https://doi. org/10.1016/j.biortech.2018.02.025.

Gouveia, J.D., Ruiz, J., van den Broek, L.A.M., Hesselink, T., Peters, S., Kleinegris, D.M. M., Smith, A.G., van der Veen, D., Barbosa, M.J., Wijffels, R.H., 2017. Botryococcus braunii strains compared for biomass productivity, hydrocarbon and carbohydrate content. J. Biotechnol. 248, 77–86. https://doi.org/10.1016/j.jbiotec.2017.03.008.

Hassanien, A., Saadaoui, I., Schipper, K., Al-Marri, S., Dalgamouni, T., Aouida, M., Saeed, S. & Al-Jabri, H. M. 2023. Genetic engineering to enhance microalgal-based produced water treatment with emphasis on CRISPR/Cas9: A review. *Frontiers in Bioengineering and Biotechnology,* 10, 1104914.

Han L, Pei H, Hu W, Jiang L, Ma G, Zhang S, Han F (2015) Integrated campus sewage treatment and biomass production by Scenedesmus quadricauda SDEC-13. Bioresour Technol 175:262–268

Harper M (2000) Sorbent trapping of volatile organic compounds from air. J Chromatogr A 885:129–151

Hasan, R., 2014. Bioremediation of swine wastewater and biofuel potential by using Chlorella vulgaris, Chlamydomonas reinhardtii, and Chlamydomonas debaryana. J. Petrol Environ. Biotechnol. 5, 1–20. https://doi.org/10.4172/2157-7463.1000175, 0.

Hewald, S., Linne, U., Scherer, M., Marahiel, M. A., Bo, M., and Bölker, M. (2006). Identification of a gene cluster for biosynthesis of mannosylerythritol lipids in the basidiomycetous fungus Ustilago maydis. Appl. Environ. Microbiol. 72, 5469–5477. doi:10.1128/AEM.00506-06

Hopkins, T. C., Graham, E. J. S., and Schuler, A. J. (2019). Biomass and lipid productivity of Dunaliella tertiolecta in a produced water-based medium over a range of salinities. J. Appl. Phycol. 31, 3349–3358. doi:10.1007/s10811-019-01836-3

Ibuot, A., Dean, A. P., McIntosh, O. A., and Pittman, J. K. (2017). Metal bioremediation by CrMTP4 over-expressing Chlamydomonas reinhardtii in comparison to natural wastewater-tolerant microalgae strains. Algal Res. 24, 89–96. doi:10.1016/j.algal.2017.03.002

Kang, C., Yang, J. W., Cho, W., Kwak, S., Park, S., Lim, Y., et al. (2017). Oxidative biodegradation of 4-chlorophenol by using recombinant monooxygenase cloned and overexpressed from Arthrobacter chlorophenolicus A6. Bioresour. Technol. 240, 123–129. doi:10.1016/j.biortech.2017.03.078

Laurens, L.M.L., Chen-Glasser, M., McMillan, J.D., 2017. A perspective on renewable bioenergy from photosynthetic algae as feedstock for biofuels and bioproducts. Algal Res. 24, 261–264. https://doi.org/10.1016/j.algal.2017.04.002.

Lee, S.A., Lee, N., Oh, H.M., Ahn, C.Y., 2021. Stepwise treatment of undiluted raw piggery wastewater, using three microalgal species adapted to high ammonia. Chemosphere 263, 127934. https://doi.org/10.1016/j.chemosphere.2020.127934.

Limmer, M. & Burken, J. 2016. Phytovolatilization of organic contaminants. *Environmental Science & Technology,* 50, 6632–6643.

Lincoln EP, Wilkie AC, French BT (1996) Cyanobacterial process for renovating dairy wastewater. Biomass Bioenergy 10:63–68

Luo, Q., He, Y., Hou, D. Y., Zhang, J. G., and Shen, X. R. (2015). GPo1 <italic>alkB</ italic> gene expression for improvement of the degradation of diesel oil by a bacterial consortium. Braz. J. Microbiol. 46, 649–657. doi:10.1590/S1517-838246320120226

Mahapatra DM, Chanakya HN, Ramachandra TV (2013) Euglena sp. as a suitable source of lipids for potential use as biofuel and sustainable wastewater treatment. J Appl Phycol 25:855–865

Marques, I. M., Oliveira, A. C. V., de Oliveira, O. M. C., Sales, E. A., and Moreira, Í. T. A. (2021). A photobioreactor using Nannochloropsis oculata marine microalgae for removal of polycyclic aromatic hydrocarbons and sorption of metals in produced water. Chemosphere 281, 130775. doi:10.1016/j.chemosphere.2021.130775

Mata, T.M., Melo, A.C., Sim˜oes, M., Caetano, N.S., 2012. Parametric study of a brewery effluent treatment by microalgae Scenedesmus obliquus. Bioresour. Technol. 107, 151–158. https://doi.org/10.1016/j.biortech.2011.12.109.

Metting, F. B. 1996. Biodiversity and application of microalgae. *Journal of Industrial Microbiology,* 17, 477–489.

Nguyen, L. N., Aditya, L., Vu, H. P., Johir, A. H., Bennar, L., Ralph, P., Hoang, N. B., Zdarta, J. & Nghiem, L. D. 2022. Nutrient removal by algae-based wastewater treatment. *Current Pollution Reports,* 8, 369–383.

B. Nowicka, T. Fesenko, J. Walczak, J. Kruk, The inhibitor-evoked shortage of tocopherol and plastoquinol is compensated by other antioxidant mechanisms in Chlamydomonas reinhardtii exposed to toxic concentrations of cadmium and chromium ions Ecotoxicol. Environ. Saf., 191 (October 2019) (2020), pp. 1–12, 10.1016/j.ecoenv.2020.110241.

Robertson, W. D., Blowes, D. W., Ptacek, C. J. & Cherry, J. A. 2000. Long-term performance of in situ reactive barriers for nitrate remediation. *Groundwater,* 38, 689–695.

Parsy, A., Sambusiti, C., Baldoni-andrey, P., Elan, T., and Périé, F. (2020). Cultivation of Nannochloropsis oculata in saline oil & gas wastewater supplemented with anaerobic digestion effluent as nutrient source. Algal Res. 50, 101966. doi:10.1016/j.algal.2020.101966.

Phang, S.-M., Chu, W.-L., Rabiei, R., 2015. Phycoremediation. In: Sahoo, D., Seckbach, J. (Eds.), The Algae World. Cellular Origin, Life in Extreme Habitats and Astrobiology, vol. 26. Springer, Dordrecht, pp. 357–389. https://doi.org/10.1007/978-94-017-7321-8_13.

Peng, Y.Y., Gao, F., Yang, H.L., Wu, H.W.J., Li, C., Lu, M.M., Yang, Z.Y., 2020. Simultaneous removal of nutrient and sulfonamides from marine aquaculture wastewater by concentrated and attached cultivation of Chlorella vulgaris in an algal biofilm membrane photobioreactor (BF-MPBR). Sci. Total Environ. 725, 138524.

Qi, F., Wu, D., Mu, R., Zhang, S., and Xu, X. (2018). Characterization of a microalgal UV mutant for CO_2 biofixation and biomass production. BioMed Res. Int. 2018, 1–8. doi:10.1155/2018/4375170.

Rahman, A., Pan, S., Houston, C., and Selvaratnam, T. (2021). Evaluation of galdieria sulphuraria and chlorella vulgaris for the bioremediation of produced water. WaterSwitzerl. 13 (9), 1183. doi:10.3390/w13091183.

Rahmani, A., Zerrouki, D., Tabchouche, A., and Djafer, L. (2022). Oilfield - produced water as a medium for the growth of Chlorella pyrenoidosa outdoor in an arid region. Environ. Sci. Pollut. Res. 29, 87509–87518. doi:10.1007/s11356-022-21916-1.

Sirakov, I., Velichkova, K., Beev, G., Staykov, Y., 2013. The influence of organic carbon on bioremediation process of wastewater originate from aquaculture with use of microalgae from genera Botryococcus and Scenedesmus. Agric. Sci. Technol. 5, 443–447.

Singh, A., Ummalyma, S. B., and Sahoo, D. (2020). Bioremediation and biomass production of microalgae cultivation in river water contaminated with pharmaceutical effluent. Bioresour. Technol. 307, 123233. doi:10.1016/j.biortech.2020.123233.

Sriprang, R., Hayashi, M., Ono, H., Takagi, M., Hirata, K., and Murooka, Y. (2003). Enhanced accumulation of Cd2+ by a Mesorhizobium sp. transformed with a gene from Arabidopsis thaliana coding for phytochelatin synthase. Appl. Environ. Microbiol. 69, 1791–1796. doi:10.1128/AEM.69.3.1791-1796.2003.

Sun, X.M., Ren, L.J., Zhao, Q.Y., Ji, X.J., Huang, H., 2018. Microalgae for the production of lipid and carotenoids: a review with focus on stress regulation and adaptation. Biotechnol. Biofuels 11, 272. https://doi.org/10.1186/s13068-018-1275-9.

Talbot P, De la Noüe J (1993) Tertiary treatment of wastewater with Phormidium bohneri (Schmidle) under various light and temperature conditions. Water Res 27:153–159.

Tan, X.B., Zhao, X.C., Zhang, Y.L., Zhou, Y.Y., Yang, L. Bin, Zhang, W.W., 2018. Enhanced lipid and biomass production using alcohol wastewater as carbon source for Chlorella pyrenoidosa cultivation in anaerobically digested starch wastewater in outdoors. Bioresour. Technol. 247, 784–793. https://doi.org/10.1016/j. biortech.2017.09.152.

Toyama, T., Kasuya, M., Hanaoka, T., Kobayashi, N., Tanaka, Y., Inoue, D., Sei, K., Morikawa, M., Mori, K., 2018. Growth promotion of three microalgae, Chlamydomonas reinhardtii, Chlorella vulgaris and Euglena gracilis, by in situ indigenous bacteria in wastewater effluent. Biotechnol. Biofuels 11, 1–12. https://doi.org/10.1186/s13068-018-1174-0.

Ungureanu, N., Vladut, V., Popa Ivanciu, M., 2021. Nutrient removal from wastewater by microalgae Chlorella vulgaris. XIV. Acta Tech Corvin. Bull. Eng. 331–340

Upadhyay AK, Singh NK, Singh R, Rai UN (2016) Amelioration of arsenic toxicity in rice: comparative effect of inoculation of Chlorella vulgaris and Nannochloropsis sp. on growth, biochemical changes and arsenic uptake. Eco Environ Saf 124:68–73.

Upadhyay, A. K., Singh, R. & Singh, D. 2019. Phycotechnological approaches toward wastewater management. In: Bharagava, R. N. & Chowdhary, P. (eds.), Emerging and Eco-Friendly Approaches for Waste Management, pp. 423–435. Singapore: Springer.

Wang, R., Wang, S., Tai, Y., Tao, R., Dai, Y., Guo, J., Yang, Y., Duan, S., 2017. Biogenic manganese oxides generated by green algae Desmodesmus sp. WR1 to improve bisphenol A removal. J. Hazard Mater. 339, 310–319. https://doi.org/10.1016/j.jhazmat.2017.06.026.

Wilde, E. W., Benemann, J. R., Weissman, J. C. & Tillett, D. M. 1991. Cultivation of algae and nutrient removal in a waste heat utilization process. Journal of Applied Phycology, 3, 159–167.

Xu, X., Shen, Y., Chen, J., 2015. Cultivation of Scenedesmus dimorphus for C/N/P removal and lipid production. Electron. J. Biotechnol. 18, 46–50. https://doi.org/10.1016/j. ejbt.2014.12.003.

Yadavalli, R., Heggers, G.R.V.N., 2013. Two stage treatment of dairy effluent using immobilized Chlorella pyrenoidosa. J. Environ. Heal. Sci. Eng. 11, 36. https://doi. org/10.1186/2052-336x-11-36.

Zhang, Y., Zhao, L., Wang, Y., Yang, B., and Chen, S. (2008). Enhancement of heavy metal accumulation by tissue specific co-expression of iaaM and ACC deaminase genes in plants. Chemosphere 72, 564–571. doi:10.1016/j.chemosphere.2008.03.043.

4 Applications of Algae for Industrial Wastewater Treatment

Deepika Malik

4.1 INTRODUCTION

Industrial activities generate large volumes of wastewater, which contain a wide range of pollutants, such as organic compounds, heavy metals, nutrients, and pathogens. These pollutants can have adverse effects on the environment, public health, and economic activities if not properly treated and disposed of. Some of the main challenges of industrial wastewater treatment include the complexity and variability of wastewater composition, the toxicity and recalcitrance of pollutants, the energy and resource demands of treatment systems, and compliance with regulations and standards. Industrial wastewater can vary widely in terms of composition, pH, temperature, and flow rate, depending on the type of industry, production processes, and raw materials used. This variability can make it difficult to design and operate treatment systems that are effective and efficient for different types of wastewater (Mara and Pearson, 1998). Many of the pollutants found in industrial wastewater are toxic or difficult to biodegrade, such as heavy metals, chlorinated compounds, and persistent organic pollutants. These pollutants can inhibit microbial activity and reduce the efficiency of biological treatment systems. The wastewater treatment systems can require large amounts of energy, chemicals, and water for operation, maintenance, and monitoring. These demands can have economic and environmental costs, such as greenhouse gas emissions, water scarcity, and chemical waste. In addition, industrial activities are subject to various local, national, and international regulations and standards regarding wastewater discharge, quality, and safety. Meeting these requirements can be challenging, especially for small and medium-sized enterprises that lack resources and expertise.

Phycoremediation, also known as algae-based wastewater treatment or algal remediation, is a sustainable and environmentally acceptable method of controlling and remediating water pollution. It entails using algae, primarily microalgae and macroalgae, to remove contaminants and excess nutrients from polluted bodies of water. Algae-based systems have shown great potential as a sustainable and effective solution for wastewater treatment. Algae are photosynthetic microorganisms that can remove pollutants from wastewater by utilizing them as nutrients for growth, producing oxygen, and generating biomass that can be harvested and used for various applications. Algae-based systems have several advantages over

DOI: 10.1201/9781003390213-4

traditional wastewater treatment methods, such as lower energy requirements, higher removal rates, and the potential for value-added products. Algae-based systems can be designed and operated in different configurations, such as open ponds, raceways, closed photobioreactors (PBRs), and hybrid systems. The selection of the system type and design depends on several factors, such as the wastewater characteristics, the treatment objectives, the site conditions, and the economic feasibility. Several studies have shown the effectiveness of algae-based systems for treating different types of wastewater, including dairy, textile, petroleum, and municipal wastewater. The objectives of the review are to provide information on the use of algae-based systems for treating industrial wastewater from various sources, such as food and beverage processing, textile manufacturing, pulp and paper mills, oil and gas refineries, and pharmaceutical production, including their performance and limitations, to highlight the potential value-added products that can be generated from algae biomass, such as biofuels, bioplastics, and fertilizers, and to discuss the economic feasibility and scalability of algae-based systems for industrial wastewater treatment.

4.2 ALGAE AS A NATURAL SOLUTION FOR WASTEWATER TREATMENT

Algae play a vital role in the ecosystem and can have both positive and negative impacts on water quality. As photosynthetic microorganisms, algae contribute significantly to the global carbon and oxygen cycles by producing oxygen and fixing carbon dioxide. However, excessive growth of algae, known as algal blooms, can cause significant water quality problems such as decreased light penetration, oxygen depletion, and toxic metabolite production. Algae can also act as indicators of water quality, as their growth and composition are influenced by various factors such as nutrient concentrations, temperature, light intensity, and pH. Changes in algae populations and species composition can provide valuable information about the health and quality of aquatic ecosystems. Furthermore, algae-based systems have been used for water quality improvement through the removal of nutrients and pollutants from wastewater, stormwater, and other sources of contaminated water. Algae can also be used for bioremediation at contaminated sites and as a tool for monitoring water quality. There are several types of algae that are commonly used for wastewater treatment, as mentioned in Table 4.1.

4.3 ALGAE-BASED TECHNOLOGIES FOR INDUSTRIAL WASTEWATER TREATMENT

Algae-based technologies have been shown to be effective for industrial wastewater treatment.

4.3.1 ALGAL PONDS

Algal ponds are shallow, open-air ponds used for the cultivation of algae. They are a cost-effective and sustainable method for treating wastewater and have been shown

TABLE 4.1

Different Types of Algae Used for Wastewater Treatment

S.No.	Algae	Uses	References
1	*Chlorella vulgaris* (green microalgae) *Chlorella sorokiniana*	• Effectively remove nitrogen and phosphorus from wastewater, and has high biomass productivity.	Kwon et al. (2020)
2	*Spirulina* (blue-green microalgae)	• Has been shown to effectively remove heavy metals and organic pollutants from wastewater. • Has high biomass productivity and can be easily harvested.	Zhang et al. (2020a)
3	*Scenedesmus* (green microalgae)	• Is commonly used for nutrient removal in wastewater treatment. • It has been shown to effectively remove nitrogen and phosphorus from wastewater, and has high biomass productivity.	Pham and Bui (2020)
4	*Nannochloropsis* (green microalgae)	• Has been shown to effectively remove nitrogen and phosphorus from wastewater. • It also has high lipid content, making it a potential source for biofuel production.	Ledda et al. (2015)
5	*Dunaliella* (green microalgae)	• Effectively remove nitrogen and phosphorus from wastewater. • It also has high salt tolerance, making it a potential candidate for wastewater treatment in saline environments.	Wu et al. (2015)

to effectively remove nutrients and pollutants from water. Algal ponds use sunlight and nutrients in the wastewater to promote the growth of algae, which can then be harvested for use as biofuels, animal feed, or fertilizer. One of the key advantages of algal ponds is their low energy requirements. Unlike traditional wastewater treatment methods, such as activated sludge, algal ponds do not require significant energy inputs for aeration or mixing. Algal ponds can also be operated at a lower cost than other treatment methods, making them a viable option for resource-limited communities. Studies have shown that algal ponds can effectively remove nutrients from wastewater. For example, a study by Liu et al. (2022) found that algal ponds were effective at removing nitrogen and phosphorus from swine wastewater. Another study by Dwivedi (2012) demonstrated that algal ponds could effectively remove pollutants, such as heavy metals, from wastewater. Overall, algal ponds have the potential to remove nutrients and pollutants from wastewater while also providing a source of renewable energy or other value-added products.

4.3.2 RACEWAY SYSTEMS

A raceway system is another type of algal cultivation system commonly used for wastewater treatment. It consists of a long, narrow channel that circulates water and promotes algal growth. The system, just like algal ponds, is low in energy requirements and relatively easy to operate and maintain. Studies have shown the effectiveness of raceway systems in treating wastewater. For example, a study by Cho et al. (2015) found that a raceway system was effective at removing nutrients from municipal wastewater. Another study by Fazal et al. (2018) demonstrated the feasibility of using a raceway system for treating textile wastewater. Also, Pankratz et al. (2020) reported the feasibility of using an open pond system for treating dairy wastewater.

4.3.3 CLOSED PHOTOBIOREACTORS

PBRs are closed systems designed to cultivate photosynthetic microorganisms, such as algae and cyanobacteria. These systems are engineered to provide optimal conditions for the growth of these organisms, including the regulation of light, temperature, nutrients, and gas exchange. PBRs have become increasingly popular in recent years due to their potential to produce biomass for various applications, including biofuels, pharmaceuticals, and food supplements (Wang et al., 2012). PBRs come in various designs, but the most common type is the tubular PBR, which consists of a long, transparent tube wrapped in a helix shape around a central column. The tube is usually made of glass or plastic, and the inner surface is coated with a material that enhances light penetration while minimizing reflection. The microorganisms are suspended in a nutrient-rich medium that flows through the tube, and light is provided by an external source, such as LED lights. The temperature and gas exchange are regulated by controlling the flow rate of the medium and the gas mixture, respectively.

The advantages of PBRs over open ponds include higher productivity, better control of culture conditions, and reduced water usage. PBRs can achieve higher biomass productivity due to their higher cell density, which results from the efficient use of light and nutrients. Furthermore, PBRs can maintain a more stable culture environment, which reduces the risk of contamination and improves the quality of the biomass. Finally, PBRs can reduce water usage by recycling the nutrient-rich medium, which results in lower overall water consumption compared to open ponds. PBRs have many potential applications, but the most promising one is the production of biofuels. Algae can produce various types of biofuels, such as biodiesel, bioethanol, and biogas, which can replace fossil fuels and reduce greenhouse gas emissions. PBRs can also be used to produce high-value products, such as pharmaceuticals and food supplements, as algae can synthesize a wide range of bioactive compounds, including antioxidants, pigments, and polysaccharides. With further development, optimization, and scale-up, PBRs have the potential to become a major contributor to the transition to a more sustainable and environmentally friendly economy (Benner et al., 2022).

4.3.4 HYBRID SYSTEMS

Hybrid systems for algae-based wastewater treatment offer a promising alternative to conventional methods. By combining algae-based systems with other treatment methods, these systems can achieve improved efficiency and lower costs while producing high-quality effluent that can be discharged or reused. As research in this area continues, it is likely that hybrid systems will become increasingly popular for wastewater treatment, contributing to a more sustainable and environmentally friendly future (Gao et al., 2021).

One example of a hybrid system is the use of an algae-based wastewater treatment system coupled with a membrane filtration system. In this system, the wastewater is first treated using an algae-based system to remove nutrients and organic matter. The biomass produced during this process is then separated from the treated wastewater using a membrane filtration system, resulting in a high-quality effluent that can be discharged or reused (Sun et al., 2018). Another example of a hybrid system is the use of an algae-based wastewater treatment system in combination with a bioelectrochemical system. In this system, the algae-based system is used to remove nutrients and organic matter from the wastewater while generating electricity through the photosynthesis process. The bioelectrochemical system, in turn, uses this electricity to remove additional nutrients from the wastewater, resulting in a highly purified effluent (Li et al., 2022).

4.4 SELECTION CRITERIA FOR ALGAE-BASED SYSTEMS

The selection of appropriate algae-based systems depends on several factors, including wastewater characteristics, treatment objectives, and the system design (Figure 4.1). For example, high concentrations of nutrients like nitrogen and phosphorus are often

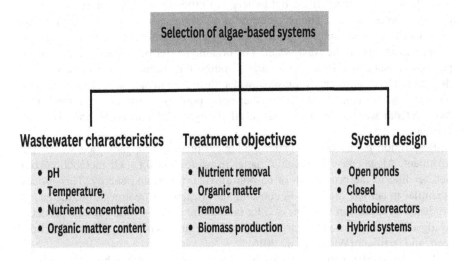

FIGURE 4.1 Factors on which the selection of appropriate algae-based systems depends.

found in municipal wastewater, and algae species that can efficiently remove these nutrients, such as *Chlorella vulgaris* and *Scenedesmus* sp., can be selected for treatment (Pham and Bui, 2020). Similarly, wastewater with high organic matter content, such as industrial effluents, requires algae species with high organic matter uptake rates, such as *Nannochloropsis* sp. (Udaiyappan et al., 2017). For biomass production, algae species that can produce high biomass yields, such as *Spirulina* sp., can be selected (Ramanan et al., 2010). Open ponds are suitable for large-scale wastewater treatment but require large land areas and are sensitive to environmental conditions (Boruff et al., 2015). Closed PBRs are suitable for small-scale wastewater treatment but require high capital costs and are sensitive to operational conditions (Acién et al., 2017). Hybrid systems, which combine open ponds or closed PBRs with other treatment methods, offer improved efficiency and lower costs (Narala et al., 2016).

4.5 MECHANISMS OF ALGAE-BASED WASTEWATER TREATMENT

The mechanisms underlying algae-based treatment involve the unique capabilities of algae to remove pollutants and nutrients from the water through various biological, physicochemical, microbial consortia, and ecosystem-based approaches.

4.5.1 BIOLOGICAL PROCESSES (PHOTOSYNTHESIS, RESPIRATION, AND NUTRIENT UPTAKE)

Biological processes such as photosynthesis, respiration, and nutrient uptake are crucial mechanisms in algae-based wastewater treatment. These processes can efficiently remove organic matter, nutrients, and pollutants from wastewater while producing valuable biomass and oxygen (Goli et al., 2016).

Photosynthesis is one of the primary biological processes in algae-based wastewater treatment. Algae utilize light energy to produce organic matter and oxygen through photosynthesis. The oxygen produced during photosynthesis can promote the growth of aerobic microorganisms, which subsequently enhance the removal of organic matter from the wastewater. The organic matter produced by algae during photosynthesis can also act as a carbon source for microorganisms and enhance the removal of nutrients from wastewater through biological processes such as nitrification and denitrification. Furthermore, photosynthesis can remove carbon dioxide from wastewater and produce alkalinity, which can buffer the pH of the wastewater.

Respiration is another important biological process in algae-based wastewater treatment. Algae consume organic matter in wastewater for their metabolic activities, leading to the production of carbon dioxide, water, and energy. This process is similar to aerobic digestion in conventional wastewater treatment, but the energy produced by algae can be harnessed for biomass production. The biomass produced by algae can then be used for various applications, such as biofuel production, animal feed, and fertilizer (Wuang et al., 2016).

Nutrient uptake is also a crucial biological process in algae-based wastewater treatment. Nitrogen and phosphorus are major pollutants in wastewater that can cause eutrophication and algal blooms in water bodies. Algae-based wastewater treatment can effectively remove these nutrients and prevent their release into the environment. Furthermore, algae can accumulate heavy metals and other contaminants in their cells, leading to the removal of these pollutants from wastewater (Singh et al., 2021).

The efficiency of biological processes in algae-based wastewater treatment can be enhanced through various strategies. For example, optimizing the light intensity and wavelength can increase the photosynthetic activity of algae, leading to higher biomass production and organic matter removal. Controlling the dissolved oxygen concentration can promote the growth of specific microorganisms and enhance the removal of specific pollutants. Additionally, nutrient supplementation can increase the growth rate and nutrient uptake of algae, leading to faster and more efficient wastewater treatment (Yadav et al., 2021).

4.5.2 PHYSICOCHEMICAL PROCESSES (ADSORPTION, COAGULATION, AND FLOCCULATION)

Algae-based wastewater treatment by physicochemical processes such as adsorption, coagulation, and flocculation has been shown to be effective in removing pollutants from industrial wastewater. Adsorption is a process in which pollutants are removed from wastewater by adhering to the surface of a solid material, such as algae biomass. Algae biomass has a high surface area and contains functional groups such as carboxyl, hydroxyl, and amino groups, which can effectively adsorb pollutants such as heavy metals, dyes, and organic compounds (Elgarahy et al., 2021). The adsorption process is influenced by factors such as pH, temperature, initial pollutant concentration, and contact time. Algae-based adsorption has been shown to be effective in removing pollutants from wastewater, with removal efficiencies of up to 90% reported in various studies (Singh et al., 2022).

Coagulation is a process in which colloidal particles in wastewater are destabilized and aggregated into larger particles through the addition of a coagulant. Algae-based coagulation involves the use of algae biomass as a coagulant agent. Algae biomass contains polysaccharides and proteins, which can act as natural coagulants and effectively remove suspended particles and turbidity from wastewater. This method is effective in removing pollutants such as heavy metals and suspended solids, with removal efficiencies of up to 99% reported in various studies (Ali et al., 2023).

Flocculation is a process in which small particles in wastewater are aggregated into larger particles, or flocs, through the addition of a flocculant. Algae-based flocculation involves the use of algae biomass as a flocculant agent. Algae biomass contains polysaccharides and proteins, which can act as natural flocculants and effectively remove suspended particles and turbidity from wastewater. This approach has been shown to be effective in removing pollutants such as heavy metals, dyes, and suspended solids, with removal efficiencies of up to 99% reported in various studies.

The effectiveness of algae-based wastewater treatment by physicochemical processes depends on various factors, such as the type and concentration of pollutants, the characteristics of algae biomass, and the operating conditions. In addition, the combination of different physicochemical processes can enhance the effectiveness of algae-based wastewater treatment. For example, the combination of adsorption and flocculation has been shown to be effective in removing heavy metals and suspended solids from wastewater.

4.5.3 MICROBIAL CONSORTIA AND ECOSYSTEM-BASED APPROACHES

Microbial consortia are groups of microorganisms that work together to perform specific functions. In algae-based wastewater treatment, microbial consortia play a vital role in the treatment process by facilitating the growth of algae, enhancing nutrient removal, and promoting the breakdown of pollutants. These consortia consist of different types of bacteria, fungi, and other microorganisms that interact with algae to create a dynamic ecosystem. The interactions between these microorganisms are complex, and their exact mechanisms are still being studied. However, some researchers have found that microbial consortia can enhance the growth of algae by providing them with essential nutrients and producing enzymes and other compounds that break down pollutants in wastewater.

In addition to microbial consortia, ecosystem-based approaches can also be used in algae-based wastewater treatment. These approaches focus on creating a balanced and diverse ecosystem in which algae and other microorganisms can thrive. This involves providing the appropriate environmental conditions, such as light, temperature, and pH, and adding different types of microorganisms to the system. By creating a diverse ecosystem, the treatment process can be optimized for maximum nutrient removal and pollutant degradation.

4.6 APPLICATIONS OF ALGAE-BASED WASTEWATER TREATMENT

Algae have shown great potential for industrial wastewater treatment due to their unique properties and capabilities (Figure 4.2). The following are some key applications of algae for industrial wastewater treatment:

4.6.1 NUTRIENT REMOVAL

Industrial wastewater often contains high concentrations of nutrients, which can have detrimental effects on aquatic ecosystems if discharged without proper treatment. Algae-based wastewater treatment utilizes the natural ability of algae to assimilate and metabolize these nutrients. Algae, including microalgae and macroalgae, can uptake and store significant amounts of nitrogen and phosphorus through processes such as adsorption, absorption, and bioaccumulation. The efficiency of nutrient removal by algae depends on various factors, including algal species, wastewater characteristics, and operating conditions. Various species of algae have demonstrated high nutrient uptake rates, reducing the risk of eutrophication and its associated environmental consequences.

4.6.2 ORGANIC POLLUTANT DEGRADATION

Industrial wastewater often contains a variety of organic pollutants that can have adverse effects on the environment and human health if discharged without proper

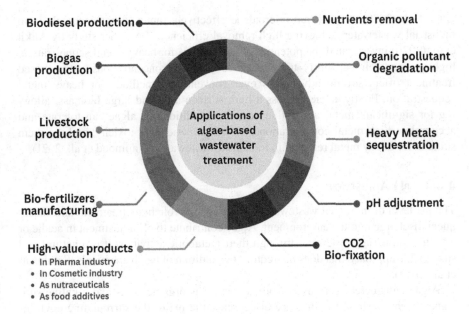

FIGURE 4.2 Various uses of algae-based wastewater treatment.

treatment. Algae possess unique capabilities to degrade and remove organic pollutants such as hydrocarbons, phenols, pesticides, and pharmaceutical compounds through various mechanisms. They can produce a wide range of enzymes and metabolites that facilitate the breakdown and conversion of organic compounds into simpler forms. Numerous studies have demonstrated the efficacy of algae-based wastewater treatment for organic pollutant degradation. For instance, research conducted by Ummalyma et al. (2018) showed that microalgae effectively removed organic pollutants, including pharmaceuticals, from industrial wastewater. Another study by Tripathi et al. (2019) investigated the degradation of phenolic compounds by macroalgae, highlighting their potential for organic pollutant removal. The unique capabilities of algae to degrade organic compounds, coupled with their environmental sustainability and potential for biomass valorization, make them an attractive option for organic pollutant removal.

4.6.3 HEAVY METAL SEQUESTRATION

Industrial wastewater often contains elevated levels of heavy metals. Algae exhibit a remarkable ability to sequester heavy metals through a process known as biosorption. Certain species of algae have metal-binding capabilities, making them effective in removing toxic heavy metals from wastewater. The sequestered heavy metals can be recovered and recycled for potential reuse or properly disposed of, minimizing the environmental impact (Nagarajan et al., 2020). They can bind heavy metal ions through various mechanisms, such as adsorption, bioaccumulation, and precipitation. Algae-based systems can effectively remove heavy metals such as cadmium, lead, copper, zinc, and mercury from industrial wastewater (Znad et al., 2022).

Guo et al. (2020) showed that microalgae effectively removed heavy metals from industrial wastewater, achieving high removal efficiencies. Another study by Ankit et al. (2020) investigated the potential of macroalgae for heavy metal sequestration, highlighting their capability to reduce heavy metal concentrations. Algae-based treatment offers several advantages over conventional methods for heavy metal sequestration. Firstly, algae possess a high surface area and large biomass, allowing for significant metal adsorption capacity. Additionally, algae can tolerate and accumulate high metal concentrations without experiencing toxicity, making them suitable for heavy metal removal in industrial wastewater (Mahmoud et al., 2021).

4.6.4 pH Adjustment

The pH level of industrial wastewater plays a crucial role in its treatment and subsequent discharge into the environment. Algae contribute to pH adjustment in acidic or alkaline industrial wastewater through their metabolic activities. It is a natural and sustainable approach that does not require the addition of harsh chemicals (Sambusiti et al., 2015).

Algae can secrete various compounds, such as organic acids and alkaline substances, which can either increase or decrease the pH of the surrounding environment. This pH adjustment capacity of algae makes them suitable for managing the pH levels of industrial wastewater. Research conducted by Zhao and Chen (2019) showed that microalgae effectively adjusted the pH of industrial wastewater, bringing it within the desired range for further treatment processes. Another study by Alami et al. (2021) investigated the use of macroalgae for pH adjustment, highlighting their potential for maintaining optimal pH conditions. Also, during photosynthesis, algae convert carbon dioxide (CO_2) into oxygen and biomass, helping to neutralize the pH of wastewater and reduce CO_2 emissions.

4.6.5 CO_2 Biofixation

Coupling wastewater treatment with CO_2 biofixation is an innovative approach that aims to address two environmental challenges concurrently – wastewater treatment and the reduction of carbon dioxide (CO_2) emissions. The production and release of CO_2 from various sources have contributed significantly to climate change and global warming. This approach utilizes algae and other photosynthetic microorganisms to remove pollutants from wastewater while simultaneously converting atmospheric CO_2 into organic biomass, thus helping to mitigate climate change by reducing greenhouse gas emissions.

Numerous studies have demonstrated the efficacy of algae-based wastewater treatment for CO_2 biofixation. Microalgae have proven to be efficient in removing CO_2 through the rapid production of algal biomass. Integrating the biofixation of CO_2 with wastewater treatment has played a significant role in combating the high concentration of CO_2 in the environment, which is the primary contributor to the greenhouse effect. Unlike chemical methods that involve costly processes and potential environmental risks during long-term storage, the use of biological agents, such as algae, offers an economically feasible and environmentally sustainable solution for

CO_2 mitigation (Lackner, 2003). Compared to land plants and trees, which have relatively slower photosynthetic processes, eukaryotic algae and cyanobacteria can fix CO_2 at a much faster rate, approximately 10–50 times faster (Iasimone et al., 2017).

4.6.6   Biomass Valorization

Algae have the remarkable ability to utilize nutrients and organic compounds present in industrial wastewater to produce biomass. The biomass generated during algae-based wastewater treatment via photosynthesis can be utilized for various applications, as discussed below.

4.6.6.1   Biofuel Production

Algae biomass can be transformed into biofuels like biodiesel or bioethanol, which are renewable alternatives to fossil fuels. Because of their rapid growth rates and high lipid content, algae are a prospective feedstock for biodiesel synthesis (Alam et al., 2014). During the wastewater treatment process, algae efficiently absorb nutrients and organic matter, converting them into biomass. This biomass can be further processed to extract lipids, which can be converted into biodiesel through transesterification (Akubude et al., 2019). Li et al. (2021) investigated the lipid productivity of microalgae grown in wastewater and found promising results for biodiesel production. Additionally, algal biomass can also be used for the production of bioethanol. Through fermentation processes, the carbohydrates present in the biomass can be converted into ethanol. Bioethanol serves as a valuable fuel additive and can be blended with gasoline to reduce carbon emissions from transportation. Ashokkumar et al. (2017) explored bioethanol production through the anaerobic fermentation process from brown marine macroalga *Padina tetrastromatica* and reported favorable yields. Harun et al. (2014) also reported efficient conversion of biomass using single-stage fermentation and two-stage gasification/fermentation to bioethanol.

4.6.6.2   Biogas Production

Algae biomass can be anaerobically digested to produce biogas, which is primarily composed of methane (CH_4) and carbon dioxide (CO_2), through the metabolic activities of diverse microorganisms present in the digestion system. The produced biogas has significant potential for various applications. It can be utilized as a renewable energy source for electricity generation and even transportation fuel. Biogas can be used directly in gas engines or can be upgraded to biomethane, a purified form of biogas with a higher methane content, for injection into the natural gas grid or as a vehicle fuel. Research by Zhang et al. (2020b) investigated the anaerobic digestion of microalgal biomass and reported high methane yields. Another study by Vargas-Estrada et al. (2021) explored the potential of utilizing microalgae consortia for biogas production and demonstrated favorable biogas production rates.

4.6.6.3   Animal Feed

Algae biomass can be processed and fed to livestock and aquaculture as a nutrient-rich feed source due to its high protein content and nutrient profile. Algae-based feeds can increase animal growth and overall health by increasing the nutritional value of their diets (Bature et al., 2022). The biomass obtained can be further processed to obtain algal meal or algal protein concentrate, which can serve as a protein-rich ingredient in animal feed formulations (Parisi et al., 2020). Algal biomass offers a sustainable alternative to traditional protein sources, such as soybean meal or fish-meal, which are associated with environmental concerns and limited availability. The use of algal biomass in animal feed provides several benefits. Firstly, it reduces the reliance on land-based protein sources, thus mitigating deforestation and land use pressures (Yong et al., 2022). Secondly, algae-based protein is known to have a balanced amino acid profile and can contribute to improved animal growth, health, and productivity. Chojnacka et al. (2018). Additionally, the inclusion of algae in ani-mal diets can enhance the nutritional value of the feed, as it contains essential vita-mins, minerals, and omega-3 fatty acids (Katiyar and Arora, 2020). Madeira et al. (2017) investigated the potential of using microalgae as a feed ingredient for poultry and reported positive effects on growth performance and egg quality. Another study by Ansari et al. (2021) evaluated the use of algae biomass in aquaculture feed and observed improved feed utilization and fish growth.

4.6.6.4   Fertilizer Manufacturing

Algae, with their ability to efficiently capture and assimilate nutrients from waste-water, offer a valuable resource for the production of organic fertilizers. During the wastewater treatment process, algae effectively absorb nutrients such as nitrogen and phosphorus from the water, converting them into biomass. This biomass can be fur-ther processed to extract nutrient-rich components, which can be used as a raw mate-rial for fertilizer production. Algal biomass can be converted into various forms of fertilizers, including liquid fertilizers, biofertilizers, and nutrient-rich composts. The utilization of algae-based wastewater treatment for fertilizer manufacturing provides several benefits. Firstly, it helps in the recovery of valuable nutrients from wastewater, reducing the need for synthetic fertilizers and minimizing nutrient pollution in water bodies. Secondly, algae-based fertilizers are rich in organic matter and microorgan-isms, which enhance soil fertility, structure, and nutrient availability. Additionally, these fertilizers are known to promote sustainable plant growth, improve crop yield, and contribute to soil health and sustainability (Mahapatra et al., 2022). Dagnaisser et al. (2022) investigated the production of nitrogen- and phosphorus-rich biofertil-izers from microalgae and reported positive effects on soil nutrient content and plant growth. Okoro et al. (2019) explored the utilization of algal biomass in the production of liquid fertilizers and observed improved nutrient availability and crop performance.

4.6.6.5   High-Value Products

Algae biomass contains useful chemicals such as proteins, carbohydrates, pigments, and bioactive molecules that can be extracted and used in a variety of sectors. Pharmaceuticals, cosmetics, nutraceuticals, and food additives all use these molecules.

    a. **Pharmaceuticals industry**: Algal extracts have demonstrated various pharmaceutical applications, including antimicrobial, anti-inflammatory, and anticancer properties. The unique biochemical composition of algae, including the presence of bioactive compounds, makes it a promising source for the development of pharmaceutical products. Algae contain a diverse range of bioactive compounds, such as polysaccharides, polyphenols, pigments, and fatty acids, which exhibit various biological activities, including antioxidant, anti-inflammatory, antimicrobial, and anticancer properties. These bioactive compounds can be isolated and utilized for the development of new drugs, nutraceuticals, and functional food ingredients. Another area of interest is the production of algae-derived pharmaceutical excipients. Algae-based polysaccharides, such as carrageenan and alginate, have been extensively studied and utilized as gelling agents, stabilizers, and drug delivery systems in the pharmaceutical industry. These natural polysaccharides offer advantages such as biocompatibility, biodegradability, and the ability to enhance drug solubility and release. Furthermore, algae-based biomass has shown potential in the production of therapeutic proteins and peptides. Several studies have demonstrated the pharmaceutical potential of algae-based biomass. For instance, a study by Fu et al. (2017) investigated the antimicrobial and antitumor activities of algae-derived compounds and highlighted their potential as sources for new pharmaceuticals.

b. **Cosmetic industry**: Algae-based biomass has gained significant attention in the cosmetic industry due to its unique properties and potential benefits for skincare and beauty products. The bioactive compounds and natural pigments present in algae offer a range of cosmetic applications, making it an attractive ingredient for formulators. Algae extracts are known for their moisturizing, hydrating, and anti-aging properties. The high content of polysaccharides, amino acids, vitamins, and minerals in algae helps to improve skin elasticity, reduce the appearance of fine lines and wrinkles, and enhance overall skin health. Algae-based ingredients are commonly used in moisturizers, serums, masks, and anti-aging creams. Algae-based biomass is also utilized for its skin-soothing and anti-inflammatory effects. Algae extracts have been found to possess calming properties that can help alleviate skin redness, irritation, and sensitivity. They are often included in formulations targeting sensitive or reactive skin types, as well as in products designed to address skin conditions such as acne and rosacea (Ashokkumar et al., 2022). Additionally, the natural pigments derived from algae offer cosmetic applications in terms of color cosmetics. Algae pigments can be used as natural dyes and colorants in makeup products such as lipsticks, eyeshadows, and blushes. These natural pigments provide a sustainable alternative to synthetic colorants, meeting the growing demand for clean and natural beauty products. Kim et al. (2018) investigated the anti-aging effects of an algae extract on human skin cells and observed significant improvements in collagen synthesis and skin elasticity.

c. **As nutraceuticals**: Algae biomass is gaining recognition in the nutraceuticals industry for its rich nutritional profile, including omega-3 fatty acids, vitamins, minerals, and antioxidants. Algal-based supplements and functional foods offer potential health benefits, such as cardiovascular support and immune system enhancement. Spirulina and chlorella, two commonly used algae species, are rich sources of protein, vitamins (such as vitamin B12), iron, and antioxidants, which can contribute to improved energy levels, immune function, and detoxification. Algae-based biomass also shows promise in the functional food industry. Algal extracts and powders are used to fortify food products with essential nutrients and enhance their nutritional value. Algae-derived ingredients are incorporated into various functional foods, such as energy bars, beverages, and snacks, to provide a natural and sustainable source of nutrients. Moreover, algae-derived compounds, such as phycocyanin and astaxanthin, have gained attention as natural colorants and antioxidants in the food industry (Pangestuti and Kim, 2011).

d. As food additives algae biomass is utilized in the food additives industry as a natural and sustainable source of colorants, thickeners, and texturizers. Algal-based additives provide functional properties and enhance the flavor, texture, nutritional value, shelf life, and visual appeal of food products. Algae, such as *Spirulina* and *Chlorella*, contain pigments like chlorophyll, phycocyanin, and carotenoids, which can be extracted and used as natural food dyes. These natural colorants are an attractive alternative to synthetic food dyes, as they provide vibrant colors while offering health benefits in terms of antioxidant properties (Garbary et al., 2017). Algae-derived ingredients are also used as thickening and gelling agents in food products. Certain species of algae, such as carrageenan from red seaweeds, agar from red and green seaweeds, and alginate from brown seaweeds, have gel-forming properties. These hydrocolloids are widely used in the food industry to improve the texture, stability, and mouthfeel of various products, including dairy alternatives, desserts, sauces, and beverages.

4.7 DIFFERENT INDUSTRIAL WASTEWATER TREATED BY AN ALGAE-BASED SYSTEM

Industrial wastewater refers to the contaminated water generated during industrial processes. Common sources of industrial wastewater include dairy, textiles, petroleum, food processing, municipal, pharmaceutical, pulp, and paper industries (Figure 4.3). Proper management of industrial wastewater is essential for sustainable industrial operations and environmental stewardship. Algae-based wastewater treatment has emerged as a promising technology for treating industrial wastewater, offering several advantages over traditional treatment methods.

4.7.1 DAIRY INDUSTRY

The dairy industry is one of the most significant contributors to the food industry, providing dairy products to consumers worldwide. However, the dairy industry also produces large amounts of wastewater, which can be harmful to the environment and human health if not adequately treated. Dairy wastewater is rich in nutrients and organic matter, making it an ideal substrate for algal growth. Algae can utilize these nutrients to grow, producing biomass that can be harvested and used for various applications.

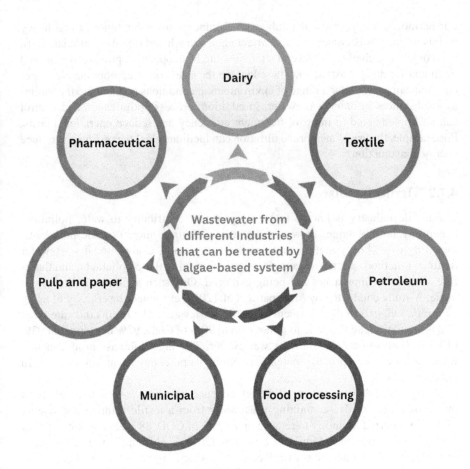

FIGURE 4.3 Showing typical origins of industrial wastewater.

Several species of algae have been investigated for their potential in treating dairy wastewater. A study by Jiang et al. (2021) investigated the use of *Chlorella vulgaris* for treating dairy wastewater. The study reported that the algal biomass produced during the treatment could be used as a potential feedstock for biogas production. Becker (2007) and Spolaore et al. (2006) investigated the use of *Scenedesmus obliquus* and *Nannochloropsis* sp. and reported that the algal biomass produced during the treatment had a high protein content, making it a potential feed source for livestock and biodiesel production, respectively. Chen et al. (2021) showed that a microalgae-based system could remove up to 95% of chemical oxygen demand (COD), 94% of TN, and 96% of total phosphorus (TP) from dairy wastewater. Similarly, Yang et al. (2022) reported that a mixed algal system could achieve removal efficiencies of 80%–90% for COD, TN, and TP from dairy wastewater.

Despite the potential of algae-based wastewater treatment, the variability of dairy wastewater composition can affect algal growth and treatment efficiency. Dairy wastewater can vary significantly depending on the production process, season, and location, making it challenging to optimize algal growth and treatment performance.

Furthermore, the presence of inhibitory substances such as antibiotics and heavy metals in dairy wastewater can also affect algal growth and treatment efficiency. To overcome these challenges, several strategies can be adopted to optimize algal-based treatment for dairy wastewater. These include the selection of appropriate algal species that can grow under a range of environmental conditions and tolerate the inhibitory substances in dairy wastewater. In addition, process optimization and control can be implemented to improve treatment efficiency and reduce operational costs. For example, the use of membrane filtration can facilitate algal harvesting and reduce energy consumption.

4.7.2 Textile Industry

The textile industry is known to be a significant contributor to water pollution, generating large volumes of wastewater that contain a range of toxic pollutants, including dyes, heavy metals, and organic compounds. Conventional wastewater treatment methods are not always effective in removing these pollutants, and therefore alternative approaches are being explored. One such approach is the use of algae. A study conducted by Mahapatra et al. (2013) evaluated the efficacy of using *Chlorella vulgaris* in the treatment of textile wastewater. The results indicated that this particular algae was able to remove up to 90% of COD, 92% of color, and 93% of total dissolved solids from the wastewater. The algal biomass produced was found to have a high protein content, indicating its potential use as animal feed. In another study,

Wang et al. (2020) investigated the effectiveness of using *Scenedesmus obliquus* in the treatment of dye-containing wastewater from a textile printing and dyeing plant and reported its ability to remove up to 95% of COD, 88% of color, and 98% of TN from the wastewater. Similarly, Sun et al. (2021) evaluated the use of *Spirulina platensis* in the treatment of textile dyeing wastewater. The results showed that it was able to remove up to 92% of COD, 94% of TP, and 99% of TN from the wastewater. The algal biomass produced by both algae was rich in lipid content that was suitable to be used for biodiesel production. In another study reported by Shahid et al. (2020), *Chlorococcum* sp. was able to remove up to 95% of COD, 92% of total organic carbon (TOC), and 84% of TN from the wastewater. The algal biomass produced was found to have a high methane yield, indicating its potential use as a feedstock for biogas production. In a study conducted by Ahmad et al. (2021), the efficacy of using mixed algal cultures in the treatment of textile dye wastewater was evaluated. The results indicated the removal efficiency of up to 90% of COD, 70% of TP, and 80% of TN from the wastewater. The algal biomass produced had a high protein content, indicating its potential use as animal feed.

4.7.3 Petroleum Industry Wastewater Treatment

The petroleum industry generates a significant amount of wastewater, which contains high levels of organic and inorganic pollutants. The use of algae for petroleum industry wastewater treatment has been extensively studied, and several studies have reported promising results. For instance, a study conducted by Kalhor et al. (2017)

investigated the potential of *Chlorella vulgaris* for removing petroleum hydrocarbons from wastewater and showed 75% removal efficiency.

Similarly, another study by Song et al. (2021) evaluated the use of *Scenedesmus* sp. and reported that it was able to remove up to 90% of the COD and 80% of the TN from the wastewater. In addition to the above studies, several other studies have reported successful algae-based wastewater treatment for the petroleum industry. For instance, a study by Dai et al. (2016) investigated the use of *Microcystis aeruginosa* for treating petroleum refinery wastewater. The study found that Microcystis aeruginosa was able to remove up to 80% of the COD and 75% of the TP from the wastewater.

Moreover, a study by Singh et al. (2021) revealed that *Chlorella vulgaris* is washable and can remove up to 98% of the cadmium (Cd) and 91% of the lead (Pb) from petroleum wastewater.

4.7.4 Food Processing Industry Wastewater Treatment

Food processing industries generate a substantial amount of wastewater that contains high levels of organic and inorganic pollutants. Algae-based wastewater treatment has the potential to provide a sustainable and cost-effective solution for treating food processing industry wastewater (Li et al., 2019b). A study conducted by Scarponi et al. (2021) reported the potential of *Chlorella vulgaris* and *Scenedesmus obliquus* for treating food processing wastewater by removing nutrients and organic matter from the wastewater. Abd Ellatif et al. (2021) investigated that *Spirulina platensis* was able to remove up to 82% of the COD and 81% of the total suspended solids (TSS) from the wastewater. Also, Aziz et al. (2019) stated that the microalgae were able to remove up to 98% of the COD and 95% of the TN from the dairy wastewater. *Chlorella vulgaris* was also capable of removing organic matter and nutrients from wastewater generated from potato processing (Li et al., 2019a).

4.7.5 Municipal Wastewater Treatment

Several studies have reported successful results using algae-based wastewater treatment for municipal wastewater. A study conducted by Luo et al. (2020) evaluated the use of mixed algal cultures for municipal wastewater treatment. The study found that the mixed algal cultures were able to remove up to 90% of the TN and up to 70% of the TP from the wastewater. Similarly, a study by Jebali et al. (2018) investigated the potential of *Scenedesmus obliquus* for treating municipal wastewater effectively. Chen et al. (2015) also evaluated the use of mixed microalgae for treating municipal wastewater and reported the removal efficiency of ~90% of the TN and up to 70% of the TP. Also, Ahmad et al. (2019) reported that the microalgae were able to remove up to 98% of the TSS from the wastewater. Another study by Lee et al. (2015) reported that *Scenedesmus* sp. was able to remove up to 98% of the TP and up to 80% of the TN from the wastewater. Moreover, a study by Wu et al. (2015) investigated the potential of microalgae for treating domestic wastewater and found that the microalgae were able to remove up to 96% of the COD and up to 91% of the NH_4-N from the wastewater. Similarly, in a study by Li et al. (2019a), it was shown that the microalgae were able to remove up to 85% of the COD and up to 80% of the TN from the wastewater.

4.7.6 Pulp and Paper Industry

Algae-based systems have been applied for the treatment of wastewater from pulp and paper production facilities, which contains high levels of organic matter, lignin suspended solids, and colorants, which can have adverse effects on aquatic ecosystems if discharged untreated. Algae, such as *Chlorella*, *Spirogyra*, and *Scenedesmus*, have demonstrated efficient removal of organic compounds, including lignin, from pulp and paper wastewater (Usha et al., 2016; Sasi et al., 2020).

The complex enzymatic systems in algae enable them to degrade recalcitrant compounds, thereby reducing the COD and color of the wastewater.

4.7.6.1 Pharmaceutical Industry

Algae-based systems have been used for the treatment of wastewater from pharmaceutical production facilities, which can contain various pollutants and pharmaceutical residues, including antibiotics and active pharmaceutical ingredients. Algae-based systems not only remove organic compounds but also play a crucial role in reducing the concentration of nutrients and heavy metals present in pharmaceutical wastewater. The removal of nutrients and pharmaceuticals from wastewater can be achieved through the cultivation and harvesting of *Chlorella sorokiniana* (Escapa et al., 2015). Santos et al. (2017) found that the microalgae species *Chlorella vulgaris* and *Scenedesmus obliquus* exhibited high removal rates of pharmaceutical compounds, indicating their potential for pharmaceutical wastewater treatment.

4.8 CHALLENGES AND FUTURE DIRECTIONS

Algae-based systems have gained increasing attention in recent years due to their potential as a sustainable and cost-effective solution for various applications, including wastewater treatment, biofuel production, and CO_2 capture. However, these systems also have limitations and drawbacks that need to be addressed to ensure their optimal performance and sustainability.

One of the primary limitations of algae-based systems is their susceptibility to environmental stress factors such as high salinity, temperature fluctuations, and nutrient imbalances. These stress factors can lead to reduced algal growth, biomass production, and nutrient removal efficiency, and they can even result in the death of the algae. Therefore, proper control and optimization of environmental parameters are crucial for the successful operation of algae-based systems.

Another limitation of algae-based systems is the potential for algal blooms and the formation of harmful algal toxins. Algal blooms can occur when nutrient concentrations in the wastewater are too high, leading to excessive algal growth. This can result in a depletion of dissolved oxygen levels in the water, leading to adverse effects on aquatic life. Additionally, certain algae species can produce toxins that pose a risk to human and animal health. Therefore, monitoring and control of algal growth and species selection are important considerations in algae-based wastewater treatment systems.

The high capital and operational costs associated with algae-based systems are another major limitation. The cultivation and harvesting of algae require significant energy inputs and infrastructure investment, and the process of algae harvesting and dewatering can be challenging and expensive. In addition, the downstream processing of algal biomass for various applications, such as biofuel production, also requires substantial investment. Therefore, the economic feasibility of algae-based systems needs to be carefully evaluated, and alternative approaches, such as mixed-culture systems, should be considered.

Another drawback of algae-based systems is the potential for wastewater contamination. Algae and other microorganisms used in the treatment process can accumulate heavy metals and other pollutants from the wastewater, which can limit their applicability for certain applications, such as animal feed production. Therefore, the use of algae-based systems for wastewater treatment should be accompanied by careful monitoring and testing to ensure the safety and quality of the resulting products.

Finally, the scalability of algae-based systems is still a significant challenge. While small-scale systems have demonstrated promising results, the application of algae-based systems on a larger scale requires substantial investments and infrastructure development. Furthermore, the integration of algae-based systems into existing wastewater treatment infrastructure can be challenging, requiring significant modifications to the existing infrastructure.

4.9 CONCLUSION

In conclusion, while algae-based systems offer promising solutions for various applications, including wastewater treatment, they also have limitations and drawbacks that need to be carefully considered. The optimization of environmental parameters, careful species selection, and monitoring and testing are necessary to ensure the optimal performance and sustainability of these systems. Furthermore, alternative approaches, such as mixed-culture systems, should also be considered to address the limitations and drawbacks of algae-based systems. With further research and development, algae-based systems have the potential to revolutionize the way we approach wastewater treatment and other applications.

REFERENCES

Abd Ellatif, S., El-Sheekh, M.M. and Senousy, H.H., 2021. Role of microalgal ligninolytic enzymes in industrial dye decolorization. *International Journal of Phytoremediation*, 23(1), pp. 41–52.

Acién, F.G., Molina, E., Fernández-Sevilla, J.M., Barbosa, M., Gouveia, L., Sepúlveda, C., Bazaes, J. and Arbib, Z., 2017. Economics of microalgae production. In R. Muñoz and C. Gonzalez-Fernandez (eds.), *Microalgae-Based Biofuels and Bioproducts* (pp. 485–503). Sawston: Woodhead Publishing.

Ahmad, A., Bhat, A.H., Buang, A., Shah, S.M.U. and Afzal, M., 2019. Biotechnological application of microalgae for integrated palm oil mill effluent (POME) remediation: A review. *International Journal of Environmental Science and Technology*, 16, pp. 1763–1788.

Ahmad, A., Singh, A.P., Khan, N., Chowdhary, P., Giri, B.S., Varjani, S. and Chaturvedi, P., 2021. Bio-composite of Fe-sludge biochar immobilized with *Bacillus* sp. in packed column for bio-adsorption of Methylene blue in a hybrid treatment system: Isotherm and kinetic evaluation. *Environmental Technology & Innovation, 23*, p. 101734.

Akubude, V.C., Nwaigwe, K.N. and Dintwa, E., 2019. Production of biodiesel from microalgae via nanocatalyzed transesterification process: A review. *Materials Science for Energy Technologies, 2*(2), pp. 216–225.

Alam, F., Date, A., Rasjidin, R., Mobin, S., Moria, H. and Baqui, A., 2014. Biofuel from algae: Is it a viable alternative? In B. Gikonyo (ed.), *Advances in Biofuel Production: Algae and Aquatic Plants* (pp. 107–120). Boca Raton, FL: CRC Press.

Alami, A.H., Alasad, S., Ali, M. and Alshamsi, M., 2021. Investigating algae for CO_2 capture and accumulation and simultaneous production of biomass for biodiesel production. *Science of the Total Environment, 759*, p. 143529.

Ali, S., Khoo, K.S., Lim, H.R., Ng, H.S. and Show, P.L., 2023. Smart factory of microalgae in environmental biotechnology. In P.L. Show, W.S. Chai and T.C. Ling (eds.), *Microalgae for Environmental Biotechnology* (pp. 263–292). Boca Raton, FL: CRC Press.

Ankit, Bordoloi, N., Tiwari, J., Kumar, S., Korstad, J. and Bauddh, K., 2020. Efficiency of algae for heavy metal removal, bioenergy production, and carbon sequestration. In R.N. Bharagava (ed.), *Emerging Eco-Friendly Green Technologies for Wastewater Treatment* (pp. 77–101). Berlin: Springer Nature.

Ansari, F.A., Guldhe, A., Gupta, S.K., Rawat, I. and Bux, F., 2021. Improving the feasibility of aquaculture feed by using microalgae. *Environmental Science and Pollution Research, 28*(32), pp. 43234–43257.

Ashokkumar, V., Jayashree, S., Kumar, G., Sharmili, S.A., Gopal, M., Dharmaraj, S., Chen, W.H., Kothari, R., Manasa, I., Park, J.H. and Shruthi, S., 2022. Recent technologies in biorefining of macroalgae metabolites and their industrial applications: A circular economy approach. *Bioresource Technology*, p. 127235.

Ashokkumar, V., Salim, M.R., Salam, Z., Sivakumar, P., Chong, C.T., Elumalai, S., Suresh, V. and Ani, F.N., 2017. Production of liquid biofuels (biodiesel and bioethanol) from brown marine macroalgae *Padina tetrastromatica*. *Energy Conversion and Management, 135*, pp. 351–361.

Aziz, A., Basheer, F., Sengar, A., Khan, S.U. and Farooqi, I.H., 2019. Biological wastewater treatment (anaerobic-aerobic) technologies for safe discharge of treated slaughterhouse and meat processing wastewater. *Science of the Total Environment, 686*, pp. 681–708.

Bature, A., Melville, L., Rahman, K.M. and Aulak, P., 2022. Microalgae as feed ingredients and a potential source of competitive advantage in livestock production: A review. *Livestock Science, 259*, p.104907.

Becker, E.W., 2007. Micro-algae as a source of protein. *Biotechnology Advances, 25*(2), pp. 207–210.

Benner, P., Meier, L., Pfeffer, A., Krüger, K., Oropeza Vargas, J.E. and Weuster-Botz, D., 2022. Lab-scale photobioreactor systems: Principles, applications, and scalability. *Bioprocess and Biosystems Engineering, 45*(5), pp. 791–813.

Boruff, B.J., Moheimani, N.R. and Borowitzka, M.A., 2015. Identifying locations for large-scale microalgae cultivation in Western Australia: A GIS approach. *Applied Energy, 149*, pp. 379–391.

Chen, G., Zhao, L. and Qi, Y., 2015. Enhancing the productivity of microalgae cultivated in wastewater toward biofuel production: A critical review. *Applied Energy, 137*, pp. 282–291.

Chen, S., Xie, J., and Wen, Z., 2021. Microalgae-based wastewater treatment and utilization of microalgae biomass. *Advances in Bioenergy, 6*(1), pp. 165–198.

Cho, D.H., Ramanan, R., Heo, J., Kang, Z., Kim, B.H., Ahn, C.Y., Oh, H.M. and Kim, H.S., 2015. Organic carbon, influent microbial diversity and temperature strongly influence algal diversity and biomass in raceway ponds treating raw municipal wastewater. *Bioresource Technology, 191*, pp. 481–487.

Chojnacka, K., Wieczorek, P.P., Schroeder, G. and Michalak, I. (eds.), 2018. *Algae Biomass: Characteristics and Applications: Towards Algae-Based Products* (Vol. 8). Berlin, Heidelberg: Springer.

Dagnaisser, L.S., dos Santos, M.G.B., Rita, A.V.S., Chaves Cardoso, J., de Carvalho, D.F. and de Mendonça, H.V., 2022. Microalgae as bio-fertilizer: A new strategy for advancing modern agriculture, wastewater bioremediation, and atmospheric carbon mitigation. *Water, Air, & Soil Pollution, 233*(11), pp. 1–23.

Dai, X., Chen, C., Yan, G., Chen, Y. and Guo, S., 2016. A comprehensive evaluation of recirculated bio-filter as a pretreatment process for petroleum refinery wastewater. *Journal of Environmental Sciences, 50*, pp. 49–55.

Dwivedi, S., 2012. Bioremediation of heavy metal by algae: Current and future perspective. *Journal of Advanced Laboratory Research in Biology, 3*(3), pp. 195–199.

Elgarahy, A.M., Elwakeel, K.Z., Mohammad, S.H. and Elshoubaky, G.A., 2021. A critical review of biosorption of dyes, heavy metals and metalloids from wastewater as an efficient and green process. *Cleaner Engineering and Technology, 4*, p. 100209.

Escapa, C., Coimbra, R.N., Paniagua, S., García, A.I. and Otero, M., 2015. Nutrients and pharmaceuticals removal from wastewater by culture and harvesting of Chlorella sorokiniana. *Bioresource Technology, 185*, pp. 276–284.

Fazal, T., Mushtaq, A., Rehman, F., Khan, A.U., Rashid, N., Farooq, W., Rehman, M.S.U. and Xu, J., 2018. Bioremediation of textile wastewater and successive biodiesel production using microalgae. *Renewable and Sustainable Energy Reviews, 82*, pp. 3107–3126.

Fu, W., Nelson, D.R., Yi, Z., Xu, M., Khraiwesh, B., Jijakli, K., Chaiboonchoe, A., Alzahmi, A., Al-Khairy, D., Brynjolfsson, S. and Salehi-Ashtiani, K., 2017. Bioactive compounds from microalgae: Current development and prospects. *Studies in Natural Products Chemistry, 54*, pp. 199–225.

Gao, L., Liu, G., Zamyadi, A., Wang, Q. and Li, M., 2021. Life-cycle cost analysis of a hybrid algae-based biological desalination-low pressure reverse osmosis system. *Water Research, 195*, p. 116957.

Garbary, D.J., Bąk, M., Dąbek, P. and Witkowski, A., 2017. Abstracts of papers to be presented at the 11th International Phycological Congress. *Phycologia, 56*(sup4), pp. 1–224.

Goli, A., Shamiri, A., Talaiekhozani, A., Eshtiaghi, N., Aghamohammadi, N. and Aroua, M.K., 2016. An overview of biological processes and their potential for CO2 capture. *Journal of Environmental Management, 183*, pp. 41–58.

Guo, G., Guan, J., Sun, S., Liu, J. and Zhao, Y., 2020. Nutrient and heavy metal removal from piggery wastewater and CH4 enrichment in biogas based on microalgae cultivation technology under different initial inoculum concentration. *Water Environment Research, 92*(6), pp. 922–933.

Harun, R., Yip, J.W., Thiruvenkadam, S., Ghani, W.A., Cherrington, T. and Danquah, M.K., 2014. Algal biomass conversion to bioethanol: A step by step assessment. *Biotechnology Journal, 9*(1), pp. 73–86.

Iasimone, F., De Felice, V., Panico, A. and Pirozzi, F., 2017. Experimental study for the reduction of CO2 emissions in wastewater treatment plant using microalgal cultivation. *Journal of CO2 Utilization, 22*, pp. 1–8.

Jebali, A., Acién, F.G., Barradas, E.R., Olguín, E.J., Sayadi, S. and Grima, E.M., 2018. Pilot-scale outdoor production of Scenedesmus sp. in raceways using flue gases and centrate from anaerobic digestion as the sole culture medium. *Bioresource Technology, 262*, pp. 1–8.

Jiang, R., Qin, L., Feng, S., Huang, D., Wang, Z. and Zhu, S., 2021. The joint effect of ammonium and pH on the growth of Chlorella vulgaris and ammonium removal in artificial liquid digestate. *Bioresource Technology, 325*, p. 124690.

Kalhor, A.X., Movafeghi, A., Mohammadi-Nassab, A.D., Abedi, E. and Bahrami, A., 2017. Potential of the green alga Chlorella vulgaris for biodegradation of crude oil hydrocarbons. *Marine Pollution Bulletin, 123*(1–2), pp. 286–290.

Katiyar, R. and Arora, A., 2020. Health promoting functional lipids from microalgae pool: A review. *Algal Research, 46*, p. 101800.

Kim, J.H., Lee, J.E., Kim, K.H. and Kang, N.J., 2018. Beneficial effects of marine algae-derived carbohydrates for skin health. *Marine Drugs, 16*(11), p. 459.

Kwon, G., Nam, J.H., Kim, D.M., Song, C. and Jahng, D., 2020. Growth and nutrient removal of *Chlorella vulgaris* in ammonia-reduced raw and anaerobically-digested piggery wastewaters. *Environmental Engineering Research, 25*(2), pp. 135–146.

Lackner, K.S., 2003. A guide to CO2 sequestration. *Science, 300*(5626), pp. 1677–1678.

Ledda, C., Villegas, G.R., Adani, F., Fernández, F.A. and Grima, E.M., 2015. Utilization of centrate from wastewater treatment for the outdoor production of Nannochloropsis gaditana biomass at pilot-scale. *Algal Research, 12*, pp. 17–25.

Lee, C.S., Lee, S.A., Ko, S.R., Oh, H.M. and Ahn, C.Y., 2015. Effects of photoperiod on nutrient removal, biomass production, and algal-bacterial population dynamics in lab-scale photobioreactors treating municipal wastewater. *Water Research, 68*, pp. 680–691.

Li, G., Zhang, J., Li, H., Hu, R., Yao, X., Liu, Y., Zhou, Y. and Lyu, T., 2021. Towards high-quality biodiesel production from microalgae using original and anaerobically-digested livestock wastewater. *Chemosphere, 273*, p. 128578.

Li, K., Liu, Q., Fang, F., Luo, R., Lu, Q., Zhou, W., Huo, S., Cheng, P., Liu, J., Addy, M. and Chen, P., 2019a. Microalgae-based wastewater treatment for nutrients recovery: A review. *Bioresource Technology, 291*, p. 121934.

Li, S., Show, P.L., Ngo, H.H. and Ho, S.H., 2022. Algae-mediated antibiotic wastewater treatment: A critical review. *Environmental Science and Ecotechnology, 9*, p. 100145.

Li, S., Zhao, S., Yan, S., Qiu, Y., Song, C., Li, Y. and Kitamura, Y., 2019b. Food processing wastewater purification by microalgae cultivation associated with high value-added compounds production: A review. *Chinese Journal of Chemical Engineering, 27*(12), pp. 2845–2856.

Liu, X.Y., Hong, Y., Zhao, G.P., Zhang, H.K., Zhai, Q.Y. and Wang, Q., 2022. Microalgae-based swine wastewater treatment: Strain screening, conditions optimization, physiological activity and biomass potential. *Science of the Total Environment, 807*, p. 151008.

Luo, Y., Le-Clech, P., and Henderson, R. K., 2020. Characterisation of microalgae-based monocultures and mixed cultures for biomass production and wastewater treatment. *Algal research, 49*, pp. 101963.

Madeira, M.S., Cardoso, C., Lopes, P.A., Coelho, D., Afonso, C., Bandarra, N.M. and Prates, J.A., 2017. Microalgae as feed ingredients for livestock production and meat quality: A review. *Livestock Science, 205*, pp. 111–121.

Mahapatra, D.M., Chanakya, H.N. and Ramachandra, T.V., 2013. Treatment efficacy of algae-based sewage treatment plants. *Environ Monit Assess, 185*, pp. 7145–7164.

Mahapatra, D. M., Satapathy, K. C. and Panda, B., 2022. Biofertilizers and nanofertilizers for sustainable agriculture: Phycoprospects and challenges. *Science of the total environment, 803*, pp. 149990.

Mahmoud, E.A., Madian, H.R. and El-Sheekh, M.M., 2021. Application of green bioremediation technology for refractory and inorganic pollutants treatment. In M.P. Shah (ed.), *Removal of Refractory Pollutants from Wastewater Treatment Plants* (pp. 485–510). Boca Raton, FL: CRC Press.

Mara, D. and Pearson, H., 1998. *Design Manual for Waste Stabilization Ponds in Mediterranean Countries.* Leeds: Lagoon Technology International.

Nagarajan, D., Lee, D.J., Chen, C.Y. and Chang, J.S., 2020. Resource recovery from wastewaters using microalgae-based approaches: A circular bioeconomy perspective. *Bioresource Technology, 302*, p. 122817.

Narala, R.R., Garg, S., Sharma, K.K., Thomas-Hall, S.R., Deme, M., Li, Y. and Schenk, P.M., 2016. Comparison of microalgae cultivation in photobioreactor, open raceway pond, and a two- stage hybrid system. *Frontiers in Energy Research, 4*, p. 29.

Okoro, V., Azimov, U., Munoz, J., Hernandez, H.H. and Phan, A.N., 2019. Microalgae cultivation and harvesting: Growth performance and use of flocculants: A review. *Renewable and Sustainable Energy Reviews*, 115, p. 109364.

Pangestuti, R. and Kim, S.K., 2011. Biological activities and health benefit effects of natural pigments derived from marine algae. *Journal of Functional Foods*, 3(4), pp. 255–266.

Pankratz, S., Kumar, M., Oyedun, A.O., Gemechu, E. and Kumar, A., 2020. Environmental performances of diluents and hydrogen production pathways from microalgae in cold climates: Open raceway ponds and photobioreactors coupled with thermochemical conversion. *Algal Research*, 47, p. 101815.

Parisi, G., Tulli, F., Fortina, R., Marino, R., Bani, P., Dalle Zotte, A., De Angelis, A., Piccolo, G., Pinotti, L., Schiavone, A. and Terova, G., 2020. Protein hunger of the feed sector: The alternatives offered by the plant world. *Italian Journal of Animal Science*, 19(1), pp. 1204–1225.

Pham, T.L. and Bui, M.H., 2020. Removal of nutrients from fertilizer plant wastewater using *Scenedesmus* sp.: Formation of bioflocculation and enhancement of removal efficiency. *Journal of Chemistry*, 2020, pp. 1–9.

Ramanan, R., Kannan, K., Deshkar, A., Yadav, R. and Chakrabarti, T., 2010. Enhanced algal CO_2 sequestration through calcite deposition by Chlorella sp. and Spirulina platensis in a mini-raceway pond. *Bioresource Technology*, 101(8), pp. 2616–2622.

Sambusiti, C., Bellucci, M., Zabaniotou, A., Beneduce, L. and Monlau, F., 2015. Algae as promising feedstocks for fermentative biohydrogen production according to a biorefinery approach: A comprehensive review. *Renewable and Sustainable Energy Reviews*, 44, pp. 20–36.

Santos, C.E., de Coimbra, R.N., Bermejo, S.P., Pérez, A.I.G. and Cabero, M.O., 2017. Comparative assessment of pharmaceutical removal from wastewater by the microalgae Chlorella sorokiniana, Chlorella vulgaris and Scenedesmus obliquus. In R. Farooq and Z. Ahmad (eds.), *Biological Wastewater Treatment and Resource Recovery* (p. 99–118). BoD – Norderstedt: Books on Demand.

Sasi, P.K.C., Viswanathan, A., Mechery, J., Thomas, D.M., Jacob, J.P. and Paulose, S.V., 2020. Phycoremediation of paper and pulp mill effluent using planktochlorella nurekis and *Chlamydomonas reinhardtii*: A comparative study. *Journal of Environmental Treatment Techniques*, 8(2), pp. 809–817.

Scarponi, P., Ghirardini, A.V., Bravi, M. and Cavinato, C., 2021. Evaluation of *Chlorella vulgaris* and *Scenedesmus obliquus* growth on pretreated organic solid waste digestate. *Waste Management*, 119, pp. 235–241.

Shahid, A., Malik, S., Zhu, H., Xu, J., Nawaz, M.Z., Nawaz, S., Alam, M.A. and Mehmood, M.A., 2020. Cultivating microalgae in wastewater for biomass production, pollutant removal, and atmospheric carbon mitigation: A review. *Science of the Total Environment*, 704, p. 135303.

Singh, A., Pal, D.B., Mohammad, A., Alhazmi, A., Haque, S., Yoon, T., Srivastava, N. and Gupta, V.K., 2022. Biological remediation technologies for dyes and heavy metals in wastewater treatment: New insight. *Bioresource Technology*, 343, p. 126154.

Singh, D.V., Bhat, R.A., Upadhyay, A.K., Singh, R. and Singh, D.P., 2021. Microalgae in aquatic environs: A sustainable approach for remediation of heavy metals and emerging contaminants. *Environmental Technology & Innovation*, 21, p. 101340.

Song, Y., Wang, X., Cui, H., Ji, C., Xue, J., Jia, X., Ma, R. and Li, R., 2021. Enhancing growth and oil accumulation of a palmitoleic acid-rich *Scenedesmus obliquus* in mixotrophic cultivation with acetate and its potential for ammonium-containing wastewater purification and biodiesel production. *Journal of Environmental Management*, 297, p. 113273.

Spolaore, P., Joannis-Cassan, C., Duran, E. and Isambert, A., 2006. Commercial applications of microalgae. *Journal of Bioscience and Bioengineering*, 101(2), pp. 87–96.

Sun, L., Tian, Y., Zhang, J., Cui, H., Zuo, W. and Li, J., 2018. A novel symbiotic system combining algae and sludge membrane bioreactor technology for wastewater treatment and membrane fouling mitigation: Performance and mechanism. *Chemical Engineering Journal, 344*, pp. 246–253.

Sun, T., Zhang, B., Chen, X., Wang, C., Wu, Y. and Zhang, Y., 2021. Spirulina platensis cultivation using poultry slaughterhouse wastewater for lipid production and pollutant removal. *Bioresource Technology, 327*, p. 124991.

Tripathi, R., Gupta, A. and Thakur, I.S., 2019. An integrated approach for phycoremediation of wastewater and sustainable biodiesel production by green microalgae, *Scenedesmus sp.* ISTGA1. *Renewable Energy, 135*, pp. 617–625.

Udaiyappan, A.F.M., Hasan, H.A., Takriff, M.S. and Abdullah, S.R.S., 2017. A review of the potentials, challenges and current status of microalgae biomass applications in industrial wastewater treatment. *Journal of Water Process Engineering, 20*, pp. 8–21.

Ummalyma, S.B., Pandey, A., Sukumaran, R.K. and Sahoo, D., 2018. Bioremediation by microalgae: Current and emerging trends for effluents treatments for value addition of waste streams. In S.J. Varjani, B. Parameswaran, S. Kumar and S.K. Khare (eds.), *Biosynthetic Technology and Environmental Challenges* (pp. 355–375). Berlin, Heidelberg: Springer.

Usha, M.T., Chandra, T.S., Sarada, R. and Chauhan, V.S., 2016. Removal of nutrients and organic pollution load from pulp and paper mill effluent by microalgae in outdoor open pond. *Bioresource Technology, 214*, pp. 856–860.

Vargas-Estrada, L., Longoria, A., Arenas, E., Moreira, J., Okoye, P.U., Bustos-Terrones, Y. and Sebastian, P.J., 2021. A review on current trends in biogas production from microalgae biomass and microalgae waste by anaerobic digestion and co-digestion. *BioEnergy Research, 15*(1), pp. 1–16.

Wang, B., Lan, C.Q. and Horsman, M., 2012. Closed photobioreactors for production of microalgal biomasses. *Biotechnology Advances, 30*(4), pp. 904–912.

Wang, Q., Xie, Y., Zhang, X., Chen, X., Zhao, Y., & Xiong, B., 2020. Simultaneous nutrient removal from dairy wastewater and biomass production by mixed algae cultivation. *Bioresource Technology, 299*, p. 122581.

Wu, K.C., Ho, K.C. and Yau, Y., 2015. Effective removal of nitrogen and phosphorus from saline sewage by *Dunaliella tertiolecta* through acclimated cultivation. *Modern Environmental Science and Engineering, 1*(5), pp. 225–234.

Wuang, S.C., Khin, M.C., Chua, P.Q.D. and Luo, Y.D., 2016. Use of Spirulina biomass produced from treatment of aquaculture wastewater as agricultural fertilizers. *Algal Research, 15*, pp. 59–64.

Yadav, G., Shanmugam, S., Sivaramakrishnan, R., Kumar, D., Mathimani, T., Brindhadevi, K., Pugazhendhi, A. and Rajendran, K., 2021. Mechanism and challenges behind algae as a wastewater treatment choice for bioenergy production and beyond. *Fuel, 285*, p. 119093.

Yang, W., Li, S., Qu, M., Dai, D., Liu, D., Wang, W., Tang, C. and Zhu, L., 2022. Microalgal cultivation for the upgraded biogas by removing CO_2, coupled with the treatment of slurry from anaerobic digestion: A review. *Bioresource Technology, 364*, p. 128118.

Yong, W.T.L., Thien, V.Y., Rupert, R. and Rodrigues, K.F., 2022. Seaweed: A potential climate change solution. *Renewable and Sustainable Energy Reviews, 159*, p. 112222.

Zhang, F., Man, Y.B., Mo, W.Y. and Wong, M.H., 2020a. Application of Spirulina in aquaculture: A review on wastewater treatment and fish growth. *Reviews in Aquaculture, 12*(2), pp. 582–599.

Zhang, L., Li, F., Kuroki, A., Loh, K.C., Wang, C.H., Dai, Y. and Tong, Y.W., 2020b. Methane yield enhancement of mesophilic and thermophilic anaerobic co-digestion of algal biomass and food waste using algal biochar: Semi-continuous operation and microbial community analysis. *Bioresource Technology, 302*, p. 122892.

Zhao, C. and Chen, W., 2019. A review for tannery wastewater treatment: Some thoughts under stricter discharge requirements. *Environmental Science and Pollution Research, 26*, pp. 26102–26111.

Znad, H., Awual, M.R. and Martini, S., 2022. The utilization of algae and seaweed biomass for bioremediation of heavy metal-contaminated wastewater. *Molecules, 27*(4), p. 1275.

5 Microalgae Cultivation Systems

Sudha Sahay, Shailesh R. Dave, and
Vincent Braganza

5.1 INTRODUCTION

Microalgae represent prokaryotic and eukaryotic (true algae) photosynthetic microorganisms that live in an aquatic environment (Vonshak, 1997). Approximately 50,000 microalgal species such as Euglenophyceae (euglenoids), Chrysophyceae (golden-brown algae), Chlorophyceae (green algae), Rhodophyceae (red algae), Phaeophyceae (brown algae), Xanthophyceae (yellow-green algae), and Cyanophyceae (blue-green algae) have been cultivated in different water resources such as fresh, brackish, sea, or marine and wastewaters (Cuellar-Bermudez et al., 2015a,b; Rashid et al., 2014). The microalgal cells are made of carbohydrates, lipids, proteins, fatty acids, vitamins, pigments, etc. (Saratale et al., 2022; Cuellar-Bermudez et al., 2017). Over the past 40 years, microalgal biotechnology has developed and diversified significantly due to its intriguing features that qualify it as feedstock for several biorefinery operations. Microalgae are used in pharmaceuticals, biochemicals, fertilizers, nutraceuticals, health food, and animal feed industries and, more recently, have been proposed as a source of biofuels (Figure 5.1).

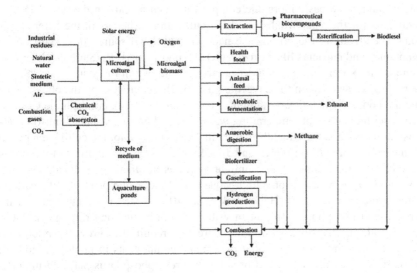

FIGURE 5.1 Applications of microalgae.

DOI: 10.1201/9781003390213-5

Various researchers have identified that microalgae are an excellent third-generation biofuel resource compared to other resources (Chen et al., 2018; Hu et al., 2017). For example, microalgal species such as *Chlorella* sp., *Botryococcus* sp., and *Chlamydomonas* can produce 15–300 times more oil as compared to other viable conventional feedstocks. Microalgal oil can be easily converted into biodiesel. Also, the traditional oil-producing crops are usually seasonal crops that can be harvested once or twice a year, while the microalgae have a very short biomass generation and harvesting cycle (~5–30 days) depending on the type of microalgal species. There are several other advantages of cultivating microalgae. First of all, microalgae cultivation does not require fertile land, a large quantity of fresh water, herbicides, and pesticides, which are generally required to cultivate other traditional oil-producing crops. Therefore, microalgae do not compete with other crops for their survival (Khan et al., 2018). Secondly, microalgae cultivation can also be done using different sources of wastewater, such as industrial, agricultural, domestic, treated, or untreated effluents; moreover, it helps to remove pollutants from such wastewater (phycoremediation) (Posadas et al., 2017; Selmani et al., 2013). Thirdly, the cultivation of microalgae helps to reduce atmospheric carbon dioxide gas through the photosynthesis mechanism and effectively contributes to tackle the greenhouse effect and global warming. Over and above this, microalgal cultivation provides not only biomass and biofuels but also reduces large amounts of greenhouse gas (CO_2) in the atmosphere during the cultivation process. We can use wastewater from various sources as a circular technology, which leads to zero or minimal pollution (Khan et al., 2021; Bhatia et al., 2021). In spite of these several benefits of microalgae cultivation, its production, however, still faces several challenges during the upstream and downstream steps, thus making the microalgae-based biofuel production technology very costly and unviable on a commercial scale.

In nature, the microalgal cell's growth and development processes mainly require sunlight and some basic inorganic components such as carbon dioxide (CO_2), and metal salts for their autotrophic mode of cultivation, whereas in the heterotrophic mode of cultivation, the microalgae need an additional source of certain organic compounds and nutrients like nitrogen (N) and phosphorus (P) (Saratale et al., 2022; Brennan and Owende, 2010). Under natural conditions, in the water reservoirs, the microalgae grow through the oxygenic photosynthesis process by the photoautotrophic method, in which direct sunlight (as an energy source), atmospheric CO_2 (as a carbon source), ambient temperature, water pH, and nutrients such as nitrogen (N), phosphorus (P), and trace metals present in water are used for the cultivation process (Ghimire et al., 2017; Masojidek and Torzillo, 2008; Arnold, 2013). In this process, the solar energy obtained from sunlight activates the metabolic signals and interacts with CO_2 to produce some high-value-added metabolic products like carbohydrates, proteins, and lipids (Hallenbeck et al., 2016). Other basic nutrients, such as nitrogen and phosphorus, present in water (generally supplied through industrial, agricultural, and livestock wastewaters) are also required to grow the microalgae. Nitrogen is required for the synthesis of many components in phytoplankton such as chlorophyll, amino acids, and proteins, whereas phosphorus plays an important role in transferring the energy molecules through the channels during photosynthesis and nucleic acid formation (Adeniyi et al., 2018; Hoh et al., 2016; Cuellar-Bermudez et al., 2015a,b; Brennan and Owende, 2010).

Two major cultivation systems have been developed so far for the cultivation of microalgae: an open system and a closed system. Various modifications in the design of open and closed systems have been done to enhance biomass yield. Microalgae productivity has been shown to be significantly higher in all types of closed systems. This is because, in closed systems, all controlled physical and chemical conditions required by the specific microalgae are controlled. But the capital cost of operation of a closed system is expensive as compared to open pond systems. On the other hand, microalgae cultivation using the open system is more profitable in terms of expenditure costs, but the overall productivity is reduced due to several drawbacks of the system, for example, survival competition with other contaminants for nutrition. Since both open and closed systems have pros and cons, therefore, none of the systems is considered so far viable for the production of microalgae-based biofuels on a commercial scale and has yet to give us an acceptable cost of production and yield that is advantageous.

In the last 30 years, several publications have appeared in the literature on microalgal cultivation for biofuel production, but very few articles have focused on microalgal biorefinery with a cost-effective technological approach. In this context, this chapter focuses on the current developments and future perspectives on large-scale microalgal cultivation strategies for the biorefinery economy and bioremediation of wastewater. In this chapter, various advanced cultivation techniques established by researchers and adopted by a few industries for enhanced biomass production and cost-effective methods for biodiesel production will be discussed.

5.2   APPLICATIONS OF MICROALGAE

Depending on the nature of the requirements of the end user, microalgae have several applications. For example, some specific species have been identified as more appropriate for lipid production, while others are for nutrient supplementation, enzyme production, etc. Some of the potential characters may not be expressed simultaneously. For example, high lipid production is usually indicative of lower growth rates, which is a disadvantage for species with this characteristic. The complete utilization of microalgal biomass can be achieved by producing several by-products in parallel with zero waste. Table 5.1 shows specific species of microalgae and the advantages that were identified with each species.

5.3   VARIOUS CULTURE CONDITIONS OF MICROALGAE

Microalgae are tiny phytoplankton that have a tendency to grow and multiply faster than plants. Their generation time is 1 to 3 days. Due to their high efficiency in photosynthesis, there are possibilities to metabolize and accumulate a large number of phytoconstituents within their cells, making them a suitable candidate to serve as rich industrial raw materials (Randrianarison and Ashraf, 2017). Considering the metabolism of microalgae, they can be classified as phototrophic, heterotrophic, mixotrophic, and photoheterotrophic.

TABLE 5.1
Products Obtained from Microalgae Species

Microalgae	Products	Source
Chlorella vulgaris	Biodiesel, biogas	Babich et al. (2011)
Dunaliella salina	Biodiesel, biogas	Sialve et al. (2009)
Chlamydomonas reinhardtii	Biohydrogen, biogas	Mussgnug et al. (2010)
Spirogyra sp.	Biohydrogen, pigments	Pacheco et al. (2015)
Phaeodactylum ellipsoidea	Pigments, fatty acids	Chew et al. (2017)
Scenedesmus obliquus	Biodiesel, biogas, fatty acids, pigments	Harman-Ware et al. (2013)
Isochrysis galbana	Pigments, fatty acids	Gilbert-López et al. (2015)
Nannocholoropsis sp.	Biodiesel, biogas, pigments	Pan et al. (2010)
Chlorococcum sp., *Chlorella vulgaris*	Bioethanol production	Nguyen et al. (2009) and Choi et al. (2010)

5.3.1 PHOTOTROPHIC/AUTOTROPHIC

Microalgae cultures are said to be phototrophic when they use light. One such example is when microalgae use sunlight, as a source of energy, and CO_2 as a source of inorganic carbon and convert it to chemical energy such as carbohydrates by the process of photosynthesis (Huang et al., 2010). This type of phototrophic cultivation has an environmental advantage because the microalgae utilize direct sunlight coupled with atmospheric carbon dioxide. As we know, CO_2 is the principal contributor to the greenhouse effect, and microalgae's use of atmospheric CO_2 for the production of microalgal biomass and biofuels brings about a favorable energy balance. Phototrophic is also referred to as an autotrophic mode of development. It is one of the most commonly used techniques for the cultivation of microalgae in open systems (Yoo et al., 2010).

5.3.2 HETEROTROPHIC

Some species of microalgae have a tendency to grow not only under phototrophic conditions but also use organic carbon in the absence of light for their photosynthesis mechanism. Such a type of microalgae culture is called heterotrophic. In this case, microalgae use organic carbon both as a source of energy and carbon (Huerlimann et al., 2010; Xiong et al., 2008). The major advantage of heterotrophic cultivation is that it avoids problems associated with the availability of sufficient light and can produce relevant results in the production of microalgal biomass. Researchers have estimated that the yields of microalgal biomass are significantly higher in the case of heterotrophic cultivation as compared to phototrophic cultivation (Huang et al., 2010). However, in heterotrophic metabolism, the need for the addition of organic carbon to the natural water or nutrient medium for microalgae cultivation leads to high costs, making the large-scale production of biofuels from microalgae unviable (Feng et al., 2011; Liang, 2013).

5.3.3 MIXOTROPHIC

In mixotrophic cultivation, the microalgae use both organic and inorganic (CO_2) compounds as a source of carbon for the process of photosynthesis. Thus, microalgae are capable of living under both phototrophic as well as heterotrophic conditions. Here, the microalgae use organic compounds and release CO_2 through the process of respiration, which is later absorbed and utilized under phototrophic metabolism for photosynthesis (Mata et al., 2010).

5.3.4 PHOTOHETEROTROPHIC

In photoheterotrophic cultivation, microalgae require light when they use organic compounds as a source of carbon. The main difference between mixotrophic and photoheterotrophic is that in the case of photoheterotrophic, light is used as a source of energy, but light and other organic sources are necessary at the same time. Since light is the only source of energy here, light intensity regulates the metabolism to increase photoheterotrophic cultivation. However, this type of cultivation for the production of biodiesel is very rare. Scientists prefer to use the mixotrophic mode of cultivation, though both types of cultivation are limited by the risk of contamination. To overcome the risk of contamination, a specially designed closed cultivation system provided with artificial light is the best option, but this again results in a high cost of operation (Suali and Sarbatly, 2012).

For any selected species of microalgae in a specific cultivation system, the microalgae can be produced by using various methods, ranging from closely controlled laboratory methods to less predictable methods in outdoor tanks. The terminology used to describe the type of microalgal culture includes:

- **Indoor/outdoor**: Indoor cultures are maintained under controlled physical/chemical/environmental conditions such as illumination, photoperiod, temperature, and level of nutrients, free from contamination by predators, and other competing microalga, whereas for outdoor microalgae cultures, it is very difficult to grow specific microalgae for extended periods.
- **Open/closed**: When microalgae cultures are grown in an uncovered pond or tank and have direct contact with the surrounding environment, then it is said to be an open cultivation system. Open cultures (indoors or outdoors) are more readily contaminated, whereas microalgae cultures growing in closed systems are not prone to contamination because they are growing in closed vessels such as tubes, flasks, carboys, bags, etc.
- **Axenic/xenic**: Axenic cultures are pure cultures of any single microalgal species that are free of any foreign microorganism, such as bacteria, fungi, or any other microalgal species. Axenic microalgal cultures are maintained in strictly sterilized glassware, culture media, and vessels to avoid contamination. On the other hand, the xenic cultures are non-sterile and mixed with unidentified foreign microorganisms, especially bacteria. In terms of total microalgal biomass and biofuel-producing efficiencies, xenic cultures are weaker than axenic cultures, and therefore, xenic

cultures are impractical for commercial operations where productivity and yield need to be high.
*   **Batch, continuous, and semi-continuous**: These are the three basic types of phytoplankton culture techniques, which are described below:

5.3.5 Batch Culture

When only one cycle of microalgal culture development occurs, which consists of a single inoculation of cells (inoculum) into a container of natural/artificial medium, it is called batch culture. The incubation period continuously runs for several days and finally harvests the microalgal biomass when the microalgal population reaches its maximum or nearly maximum density. In the laboratory, for scale-up of the microalgae biomass, the bench-scale batch cultures are transferred from smaller to larger vessels containing a large volume of sterile nutrient medium before they reach the stationary phase. The large-volume batch cultures are then brought to a maximum density and harvested in one cycle of the cultivation system.

5.3.6 Continuous Culture

The continuous culture method is the one in which a supply of natural/artificial medium is continuously pumped into a growth chamber and the excess culture is simultaneously removed, permitting the maintenance of cultures very close to the maximum growth rate. There are two types of continuous culture systems: (1) turbidostatic culture, in which, by using an automatic system, the microalgal concentration is kept at a pre-set level. To set the initial cell density of microalgae in the culture system, we dilute it with a fresh medium as per our requirement, and (2) chemostatic culture, in which the rate of flow of the addition of fresh medium in the growing microalgal culture remains constant with respect to the rate of harvesting. In a chemostatic continuous culture system, a limited amount of vital nutrients (e.g., nitrate) are added at a fixed rate, and therefore, the rate of growth of microalgal cells remains constant irrespective of the cell density produced.

5.3.7 Semi-Continuous Culture

Semi-continuous cultures are those where large tanks are used to culture the microalgal strain, and periodically, partial harvesting is done during a specific period of growth. At the time of harvesting, the tank is immediately refilled to the original volume with fresh medium to supplement with nutrients at the original level of nutrient enrichment. The refilled culture is once again growing since it was partially harvested, and the remaining microalgae cells in the culture act as an inoculum to grow further in the freshly added nutrient-enriched medium. Semi-continuous cultures can be placed indoors or outdoors, but usually, their duration of growth is unpredictable because the culture condition is not rate-limiting. It has been observed that in outdoor systems, competitors, predators, and contaminants eventually grow, making the culture unsuitable for further use. However, the culture is partially harvested; therefore, the semi-continuous culturing method gives a higher yield in terms of microalgae biomass than the batch method.

5.4 CULTIVATION SYSTEMS OF MICROALGAE

There are several methods for the cultivation of microalgae using both natural and artificial environmental conditions. The selection of a suitable cultivation method and the optimization of various parameters within that cultivation process to enhance the yield of microalgal biomass, biofuel, and by-products are the prime requirements when we work with any specific microalgae species. In the process of optimizing a cultivation system, a few basic features must be considered, such as illumination, photoperiod, pH, temperature, air circulation, and gaseous exchange for the supply of CO_2 and O_2 degassing. Based on the nature of the microalgae culture conditions and applications for a specific microalgae species, different configurations and designs of cultivation systems are used for microalgal cultivation. From this perspective, there are mainly three types of microalgae cultivation systems: open, closed, and hybrid.

5.4.1 OPEN CULTIVATION SYSTEMS

The open cultivation of microalgae is primarily of three types: open ponds, circular ponds, and raceway ponds.

5.4.1.1 Open Ponds

Open-pond cultivation has been the simplest way to cultivate microalgae on a large scale. The open pond is widely used due to its relatively cheaper construction, maintenance, operation cost, low energy demand, and ease of scale-up (Costa and de Morais, 2014). The open pond system includes the use of natural water sources such as lakes, ponds, and seawater. Technically, artificial water bodies such as circular and raceway ponds are also considered open ponds. On a small scale, a container such as a tank can also be used to culture microalgae in an open system (Sun et al., 2016). Despite the large-scale cultivation possibility in the open pond system, the overall production of microalgal biomass has been shown to have a lower microalgal cell density, and it is also difficult to harvest the microalgal biomass with a minimum loss of microalgal cells (Tredici, 2004). In addition, factors such as rainwater runoff, changes in salinity and pH, and erosion of pond banks result in leakage and increased water turbidity. The contamination due to protozoa and bacteria causes toxicity for the growing microalgae and significantly affects the productivity of microalgae in pond-based open cultivation systems (Borowitzka and Borowitzka, 1990). To solve this problem, the researchers advised using those specific microalgae species for cultivation that are capable of surviving in alkaline or saline conditions since only a few contaminants can survive under these conditions (Ugwu and Aoyagi, 2012). Another problem is that it is harder to control certain growth parameters such as temperature, light intensity, and photoperiod in the open cultivation system. These parameters are equally important for microalgae cultivation and directly affect the rate of growth of microalgae and biofuel production (Stark and O'Gara, 2012). However, open cultivation systems mostly depend on naturally occurring water bodies to provide the right conditions and nutrients for the growth of microalgae. Currently, one of the largest commercial cultivations of microalgae in natural water is located at Hutt Lagoon, Australia, which is capable of producing 6 tons of β-carotene every year from *Dunaliella* using its 700-ha ponds (Borowitzka and Borowitzka, 1990).

5.4.1.2   Circular Ponds

Circular ponds are artificial ponds that are used for small and large-scale cultivation of microalgae. The circular-shaped tanks are used in the cultivation system with a depth of about 30–70cm and a diameter of 45m, along with a rotating agitator located at the center of the pond (Shen et al., 2009). The purpose of the rotating agitator is to ensure efficient mixing of the nutrient medium and also prevent sedimentation of microalgae biomass (Show et al., 2017). The design of the circular pond is limited by its size. The larger circular pond may have strong water resistance, and therefore extra pressure will be drawn on the agitator for its constant circular movement (Lee, 2001). Moreover, high energy usage in the agitation process and high construction costs made the circular pond cultivation system more unattractive to end users (Hamed, 2016). It is reported that this cultivation system has been used in Japan and Taiwan for the cultivation of *Chlorella* sp. (Lee, 1997; Shen et al., 2009).

5.4.1.3   Raceway Ponds

In recent years, raceway ponds have become one of the most frequently used open pond systems for the cultivation of microalgae. There are a series of closed-loop channels around 30cm deep, and a paddlewheel is used to construct raceway ponds (Figure 5.2). The paddlewheel runs with the help of an electric motor. There is continuous circulation of the nutrient medium in one direction to ensure an equal distribution of nutrients. It also helps with proper mixing to produce homogenous microalgal culture development. With the movement of the paddlewheel, some bubbles are formed in the microalgae culture medium, which helps to trap atmospheric gases such as CO_2, etc., which are then made available to the microalgae underneath for their metabolism and help in the enhancement of overall microalgal biomass.

Rogers et al. (2014) stated that raceway ponds are considered to be one of the best open pond cultivation designs because of their lower energy consumption due to a

FIGURE 5.2   Schematic diagram of raceway pond.

single paddlewheel. The paddlewheel can be used to properly agitate up to a 5-ha raceway pond. There are several raceway pond systems that have been established all over the world for small and large-scale cultivation of microalgae at the research and commercial levels. One such raceway pond cultivation is established by Sapphire Energy's Columbus Algal Biomass Farm located in Columbus, United States, which has successfully produced 520 metric tonnes of dried microalgae biomass during 2 years of its operation without any technical issues (White and Ryan, 2015). The Energy and Resources Institute (TERI) has set up a 100,000 L algal production system as part of the DBT-TERI Centre of Excellence on Integrated Production of Advanced Biofuels and Bio-commodities at its Airoli site in Navi Mumbai. The algal production system is based on an indigenous sunlight-distributed algal growth system that has been found to give 1.5 times the productivity of the standard raceway pond system. The sunlight-distributed system is a step in realizing the high lab yields of algae in outdoor conditions (Kannan and Devi, 2019).

Outdoor cultures are considered photo-limited systems as they are operated in sunlight. Therefore, they have to maintain the optimum concentration of nutrients rather than a maximum growth rate. However, to improve biomass productivity, all open ponds such as artificial tanks, raceways, and circular ponds have impellers, rotating arms, paddle wheels, or streams of CO_2-enriched air supply systems for the proper mixing of the culture medium so that they can provide proper mixing of the nutrients that make them available to all the microalgal cells homogeneously (Figure 5.3a–d).

In one of the reports, it is estimated that the microalgal culture is typically mixed at 0.25 m/s by a paddlewheel, and microalgal cell densities were low, up to 0.3 g/L (Norsker et al., 2011). The investment cost for the establishment of such a 100-ha open raceway pond plant was 0.37 M€/ha (Norsker et al., 2011). Chisti (2007) stated that open raceway ponds are widely used as low-cost cultivation systems for the commercial production of microalgae.

5.4.2 CLOSED CULTIVATION SYSTEMS

A closed cultivation system for microalgae is performed in a specially designed enclosed glass container, such as tubes, flasks, or vessels. There is no direct contact between the container's inner and outer environments. The microalgal cultures growing inside the closed systems are away from the outer spaces. There is an arrangement of artificial light provided either from inside or outside the system that facilitates the growth of microalgae. Therefore, it is said to be a photobioreactor. Several problems raised by open cultivation systems are overcome by closed cultivation systems or photobioreactors. First, the size of the photobioreactor is more compact compared to the open pond, therefore providing more efficient land usage. Second, the system provides closed and highly controlled growth conditions for the culture and is thus able to produce contamination-free, single-strain microalgae culture development (Posten, 2009). In addition, the highly controlled culture conditions can also translate into higher nutrient and metabolic efficiency, resulting in higher biomass production per unit of substrate. However, the bottleneck of practical usage of the photobioreactor is its limited scalability due to various design flaws, rendering

FIGURE 5.3 Open outdoor systems for the cultivation of microalgae. (a) A walled pond lined with plastic foil and mixed by air (+CO$_2$) bubbling at the Centre for Aquaculture, University of Naples Federico II, Portici, Italy; (b) a circular pond with a rotating arm (100 L) at the Institute for Ecosystem Study of the CNR (Florence, Italy); (c) a raceway pond with a paddle-wheel mixer (600 L) at the Faculty of Marine Sciences and Technology, Canakkale Onsekiz Mart University, Turkey; (d) an inclined-surface system of sloping planes arranged in cascades (90 m^2; 600–1,000 L) at the Institute of Microbiology, Trebon, Czech Republic.

it uneconomical to be used in large-scale production (Gupta et al., 2015). Moreover, the highly controlled growth conditions of photobioreactors always come with high capital and operational costs.

Various types of closed photobioreactors are available, such as tubular photobioreactors, vertical/bubble column photobioreactors, flat panel photobioreactors, stirred-tank photobioreactors, and soft-frame photobioreactors.

5.4.2.1 Tubular Photobioreactor

Tubular photobioreactors are comprised of transparent tubes through which the culture is circulated at liquid velocities of typically 0.5 m/s (Norsker et al., 2011). The photobioreactor is closed; therefore, the oxygen liberated by microalgal cells is either consumed by the microalgal cells themselves in their metabolism or accumulated in the vacuum space of the glass container. To prevent high oxygen concentrations, the transparent tubes are connected to a degasser or stripper vessel, where oxygen is removed by air injection. The diameter of the tubes varies with the system, ranging between 3 and 10 cm. The source of light could be either natural with the help of

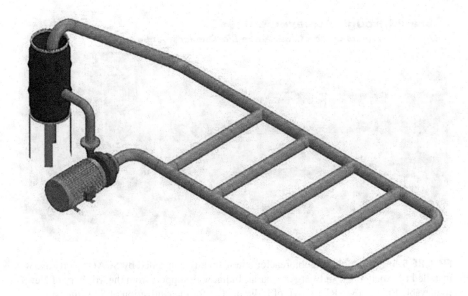

FIGURE 5.4   Schematic diagram of tubular photobioreactor.

sunlight or artificial by using LED tube lights or lamps. Similarly, tubular photobio-reactors can be maintained at outdoor temperatures for the incubation of microalgal biomass or can also be placed indoors at controlled temperatures. Therefore, the advantage of the closed cultivation system is that we can implement it in both natural environmental conditions and artificially controlled conditions as per the require-ments of the specific microalgal species. Also, the orientation of the tubular photo-bioreactor can be horizontal, vertical, helix, or slanted to capture the maximum light (Mishra et al., 2019). The microalgae culture is circulated within the glass container with the help of a mechanical pump or airlift system (Figure 5.4), and there is a con-trolling device to control the air pressure (Xu, 2007).

The construction cost of tubular photobioreactors is higher than that of open race-way ponds, especially vertically oriented tubular photobioreactors. Norsker et al. (2011) estimated an investment cost of 0.51 M€/ha for a tubular photobioreactor, whereas they calculated a flat panel photobioreactor cost of 0.8 M€/ha. In addition, the operating costs for cultivating 1 kg of dry-weight microalgae biomass were found to be 5.55 and 6.02 €/ha for tubular and flat panel systems, respectively.

There are several software programs that are used to operate the closed photo-bioreactor cultivation system automatically, in which we can monitor and control the pH, temperature, air pressure, levels of CO_2 and O_2, rate of flow of the medium, etc. One such type of closed tubular photobioreactor with a capacity of 750 L was installed by Abellon Clean Energy, Gujarat, India, with support from the Ministry of New & Renewable Energy (MNRE), Govt. of India, in 2011. This is the first of its kind in India, a large-scale algae bioreactor system automatically controlled by SCADA software (Figure 5.5).

The problem with the tubular photobioreactor design is the poor movement of the microalgal biomass across the system. Therefore, the long tubular tubes used in the

FIGURE 5.5 A tubular photobioreactor automatically controlled by SCADA software was installed by Abellon Clean Energy, Gujarat, India, with support from the Ministry of New & Renewable Energy (MNRE), Govt. of India in 2011 for the cultivation of *Spirulina* sp.

bioreactor might result in differences in the concentration of substrate and product along the tubes. It has been reported that commercially established large-scale tubular photobioreactors are being used to produce *Haematococcus* sp. and *Chlorella* sp. in Germany and Israel, respectively (Torzillo and Zittelli, 2015).

5.4.2.2   Vertical/Bubble Column Photobioreactor

Vertical column photobioreactors consist of transparent vertical cylindrical tubes and a sparger that pumps in air bubbles that help in the homogenization of the culture and allow the transfer of carbon dioxide and oxygen between air and microalgae culture (Figure 5.6) (Mohan et al., 2019). In this system, the transfer efficiency of gases through liquid is better than that of tubular photobioreactor. This is due to the sparger used in this system to generate small bubbles, which provide a larger surface area for more efficient transfer of substances (Rinanti et al., 2013). In addition, the design of this system is very simple and has a lower energy demand with a simple operating procedure. The only drawback with this system is that cylindrical-shaped, vertically arranged glass tubes do not provide an adequate amount of light in the center of the photobioreactor, which is required by the growing microalgae for efficient photosynthesis. In vertical column photobioreactors, microalgal cells absorb less amount of light energy as a result of lower photon flux densities because of light dilution on the reactor surface, whereas in a horizontal system, the capture of a large number of photons is not a problem. Therefore, higher area productivity was found in the horizontal systems. Besides, there is a high construction cost and difficulty in cleaning the vertical/bubble column photobioreactor tubes, just like any other closed photobioreactor, making it difficult to use for commercial-scale microalgal production (Huang et al., 2017). Though the commercial-scale usage of this photobioreactor is limited, there are examples of large-scale experimental vertical column photobioreactors that have

FIGURE 5.6 Vertical/bubble column photobioreactor: mixing of the microalgae suspension is maintained by bubbling through a mixture of air with 1% CO_2. (Courtesy: Institute of Microbiology, Academy of Sciences, Trebon, Czech Republic.)

been made, including a 40-L vertical column outdoor photobioreactor for the cultivation of *Chlorella zofingiensis* in Guangdong, China (Huo et al., 2018).

5.4.2.3 Flat-Plate Photobioreactor

Flat-plate photobioreactors are rectangular compartments made up of transparent material with a depth of 1–5 cm. There is a recirculating airlift system attached inside the reactor for aeration and to mix the microalgal culture (Tamburic et al., 2011). The flat-plate photobioreactor has a larger surface area that supports maximum capturing of illumination and low oxygen build-up, thus achieving the highest photosynthetic efficiency among all photobioreactor designs (Yan et al., 2016). However, the attached airlift in this photobioreactor causes damage to the microalgae cells (Sierra et al., 2008). Researchers have introduced innovations in the existing design to further improve the efficiency of the flat-plate photobioreactor, e.g., the implementation of twin-layer flat-plate and plastic sheet photobioreactors (Vo et al., 2018). At the commercial level, Algamo, a company located in Krkonose, Czech Republic, has used a flat-plate photobioreactor to produce astaxanthin – a red pigment (carotenoids).

Closed photobioreactors are manageable cultivation systems in which the required parameters can be optimized under controlled physical and chemical conditions, depending on the biological and physiological characteristics of the microalgal species. Masojídek and Torzillo (2008) reported various types of closed cultivation systems that have been established at different institutions and industries for the mass cultivation of microalgae in outdoor conditions by using natural light and temperature (Figure 5.7). The simplest cultivation system is illuminated plastic bags containing the microalgal suspension culture. The supply of a stream of CO_2-enriched air is done using plastic tubes connected to the cultivation bags (Figure 5.7a). This type

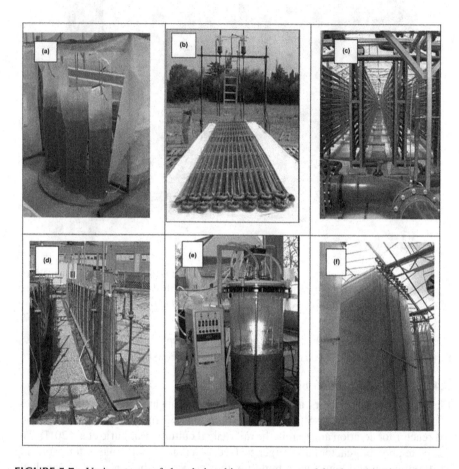

FIGURE 5.7 Various types of closed photobioreactors are used for the cultivation of microalgae. (a) Hanging plastic bags mixed by air (CO_2) bubbling (Faculty of Marine Sciences and Technology, Canakkale Onsekiz Mart University, Turkey); (b) vertically stacked tubular photobioreactor mounted in a greenhouse developed by IGV GmbH (Salata GmbH, Germany); (c) a vertical flat-panel photobioreactor 'Green Wall Panel' (Institute for Ecosystem Study of the CNR, Florence, Italy); (d) a two-plane, horizontal tubular photobioreactor (Institute for Ecosystem Study of the CNR, Florence, Italy); (e) a 100L annular column photobioreactor consisting of two glass cylinders placed one inside the other to form the culture chamber; the LED light source is mounted in the internal cylinder and can be placed outdoors to combine natural and artificial light (Institute of Microbiology, Trebon, Czech Republic); (f) an innovative flat-plate photobioreactor 'Hanging Gardens' developed by ecoduna GmbH (Bruck a/L, Austria) consisting of 12 closely spaced parallel panels ($0.03 \times 2 \times 6 \, m^3$) placed on a movable frame that allows to track the sun movements. The panels are internally partitioned by baffles to allow culture circulation as air and CO_2 are injected from the bottom to generate a gas-lift effect.

of photobioreactor was installed by the Faculty of Marine Sciences and Technology, Canakkale Onsekiz Mart University, Turkey, where the bags were placed horizontally, arranged in a row. Here, the air can be supplied either with the help of a pump or by injecting a stream of compressed air from the lower end of the bags in an upward direction. Several pumps have been introduced for the purpose of bubbling air into the microalgal suspension culture. Peristaltic and membrane pumps are found to be more suitable for microalgal cells than centrifugal pumps, which may cause higher sheer stress. Cooling can be maintained either by submerging the bags in a pool of water or by spraying water onto the surface of the bags (photobioreactor surface). Another closed photobioreactor for microalgal biomass production has been built in a greenhouse in Klotze and Ritscherhausen, Germany, with a capacity of 700 m³. In this system, the tubular glass tubes are arranged vertically in the form of a fence to utilize direct, diffused sunlight. A similar type of arrangement with a slight modification was also implemented by the IGV BiHotech company, where they developed a glass tube photobioreactor with a capacity of 3,000 L. The whole photobioreactor setup is placed in a mobile container for sustainable aquaculture of microalgae-fish cultivation (Figure 5.7b). Figure 5.7c shows a closed photobioreactor, which was installed by the Institute for Ecosystem Study of the CNR, Florence, Italy. They used a two-plane horizontal tubular bioreactor, which led to a high *Spirulina* productivity of 30 g dry weight/m²day. Flat-panel photobioreactors with a short light path and a horizontal, inclined, or vertical orientation have a high surface-to-volume ratio. These photobioreactors are made from plexiglass or polycarbonate alveolar sheets, or flexible polyethylene bags, 2–4 cm thick, enclosed in a rigid framework (Figure 5.7d). A variation of this column system is used to make annular photobioreactors, which consist of two glass or plexiglass cylinders of different diameters placed one inside the other to form a culture chamber some 5–10 cm thick and 50–200 L in volume (Figure 5.7e). Illumination can be provided by either natural or artificial light. In column photobioreactors, sensitive strains with fragile cells or filaments can be grown, e.g., *Nostoc*, *Isochrysis*, *Navicula*, and *Skeletoma*, as the culture mixing is very gentle. A similar principle of cultivation has been used in flat photobioreactors developed as transparent flat-panels connected in series, vertically arranged 20 cm apart (Figure 5.7f).

Compared with tubular systems, flat-panel photobioreactors have one serious disadvantage: fouling up of the channels due to reduced turbulence in their narrow, rectangular-shaped channels. Among all, the vertical-column photobioreactors are relatively simple systems in which mixing is achieved by air and CO_2 bubbling up from the bottom. There is no doubt that the microalgal biomass yield is higher in photobioreactors, yet their high construction and maintenance costs make them unaffordable for the commercial production of microalgal biomass. However, closed photobioreactors are very useful for the production of high-value bioactive substances, which require the adoption of sterile conditions.

5.4.3 Hybrid Cultivation System

The open-culture system is the one in which microalgae are allowed to grow in open raceway ponds, subject to the conditions of the environment. This means that minimum human intervention is needed in cultivating the microalgae, but it also means that some yield potential is lost. However, this is not a problem with closed cultivation systems because the environment is controlled. The advantages of closed cultivation systems may be offset by the higher operational and capital costs required by such facilities. The two-stage hybrid cultivation system attempts to achieve a balance between the advantages of both the open and closed cultivation systems. In a hybrid cultivation system, the microalgae are initially grown in a closed photobioreactor in a nutrient-rich medium. In 8–10 days, when nutrients are exhausted and microalgal cell density is at its peak, the culture is transferred into the raceway pond for lipid induction. The mass cultivation in a closed photobioreactor, followed by the transfer of culture in a raceway pond, can be repeated in four to five cycles. During each of the harvesting events, at least half of the closed photobioreactor culture is transferred to the raceway ponds, where lipid biosynthesis and accumulation are stimulated by nutrient depletion. In a two-stage hybrid cultivation system, cultivating the microalgae in a closed photobioreactor and allowing yields in open ponds leads to less susceptibility to yield loss from contaminants.

5.4.4 MICROALGAE CULTIVATION IN WASTEWATER (PHYCOREMEDIATION)

We have described in detail the demerits of an open system for microalgae cultivation using any natural water resource, which brings about a source of contamination due to open environmental factors. It also reduces the production of microalgal biomass and biofuel. However, if we cultivate the microalgae in any wastewater, such as domestic, industrial, agricultural, etc., then not only can we further reduce the cost of cultivation, but it also leads to the bioremediation of wastewater, which can be further used for irrigation purposes. It is reported that microalgae cultivation using wastewater offers the highest atmospheric carbon fixation rate of 1.83 kg CO_2/kg of biomass and the fastest biomass productivity of up to 40%–50% higher than terrestrial crops among all terrestrial bioremediation with concomitant pollutant removal (80%–100%) (Ayesha et al., 2020). Thomas et al. (2016) used *Chlorococcum humicola*, *Chlorella vulgaris*, and *Selenastrum* sp. for the treatment of sewage water and the generation of microalgal biomass and lipid contents. *C. humicola* showed a reduction in BOD of nearly 60% and total nitrogen of more than 80%. TDS and TS reductions were ~65%. Total phosphorus was reduced to nil by these microalgae. The reduction of all these parameters was in the range of 60%–65% by *C. vulgaris*, except for total N, which was nearly 85%. The treatment efficiency was lower for *Selenastrum* sp. When compared to *C. humicola* and *C. vulgaris*. Similarly, *C. humicola* and *C. vulgaris* showed higher biomass production, and *Selenastrum* sp. showed higher lipid production. Given this and many other studies, the current microalgal cultivation systems need to be significantly improved because they are plagued by impurities added in the form of other microbes present in the wastewater. However, bioremediation by microalgae is accomplished, and the selection of the method of biomass recovery is essential, as it is during this stage that the largest operational and deployment costs are seen. Therefore, to enable the application of microalgae for bioremediation purposes, it is essential that integrated processes, combining cultivation in wastewater and the use of biomass for the production of commercially valuable by-products such as fertilizers and biofuels, be explored.

5.5 MICROALGAE CULTIVATION FOR BEGINNERS

As we mentioned, microalgae can be grown under two conditions: autotrophic conditions, where microalgae cultivation occurs in an open system in which sunlight is used as an energy source and CO_2 as a carbon source, and heterotrophic conditions, where microalgae cultivation occurs in a closed system where microalgae use organic matter by fermentation as a carbon and energy source.

Whether in open systems or closed systems, it is necessary that all the nutrients required for microalgal growth and reproduction be available in a liquid source, either in the form of a natural water reservoir or an artificial or synthetic nutrient medium. The level of each nutrient should fulfill its essential role in meeting the nutritional requirements of the microalgae being produced. The medium should have sufficient mineral ions and potential growth factors, as well as vitamins. In the case of a heterotrophic culture, glucose is added as a source of carbon and energy. Along with nutrients, the culture medium must have a pH close to the optimal pH for microalgal growth (between 7 and 8.5). The ionic strength of the medium must be adjusted in such a way that it does not cause microalgal cell destruction. Generally, synthetic or artificial culture media have a precisely known composition of each chemical nutrient, both qualitatively and quantitatively, whereas natural water resources have a more random composition due to the availability of surrounding raw materials added to them. Therefore, one can say that the synthetic medium is selected as per the requirements of the microalgae, and suitable microalgae are selected for cultivation as per the composition of natural water resources.

5.5.1 PROCUREMENT OF PURE MICROALGAL STRAINS

Axenic cultures of microalgae used for mass cultivation may be obtained from authorized culture collection centers. A list of authorized culture collection centers is provided by Vonshak (1986), viz., (1) The Culture Collection of Algae, Department of Botany, Indiana University, Bloomington, Indiana, U.S.A., (2) The Culture Centre of Algae and Protozoa, Storey's Way, Cambridge, England, (3) The Culture Collection of Algae and Microorganisms, Institute of Applied Microbiology, University of Tokyo, Bun-kyoku, Tokyo, Japan; and (4) Universitaet Goettingen, Botanisches Institut, Algae Sammlung, Goettingen Federal Republic of Germany. In India, we have a National Repository for Microalgae and Cyanobacteria (sponsored by DBT, Govt. of India), Department of Marine Biotechnology, Bharathidasan University, Tamil Nadu, India. So far, around 3,00,000 species of microalgae are known all over the world. They have specific biological and physiological status and applications depending on their native conditions. However, there are many more microalgal strains that still need to be isolated and identified from the natural water reservoirs around the world.

5.5.2 ISOLATION OF PURE MICROALGAL STRAINS

The isolation of an axenic microalgal strain from a natural water reservoir and furthering its mass cultivation ability initially depend on local environmental conditions. The isolation of microalgal species is not a simple task because of the small cell size and the association with other epiphytic microbial species. Several laboratory techniques are available for isolating individual cells, such as serial dilution and successive plating on agar media and separation using capillary pipettes. Bacteria can be eliminated from the phytoplankton culture by washing or plating in the presence of antibiotics. The sterility of the culture can be checked with a test tube containing seawater and 1 g/L bactopeptone. After sterilization, a drop of the culture to be tested is added, and any residual bacteria will turn the bactopeptone solution turbid. The axenic microalgal strain should be carefully protected against contamination during handling and poor temperature regulations. To reduce risks, two series of stocks are often retained, one of which supplies the starter cultures or master cultures for the production system and the other of which is a seed culture or inoculum prepared from the stock or master culture necessary for maintenance and mass cultivation. Stock microalgal cultures are kept in sterile test tubes, glass vials, or Erlenmeyer flasks at a light intensity of about 1,000 lux and a temperature of 16°C–19°C. At these low light intensities and temperatures, there is a slow microalgal cell metabolism leading to retarded growth, and we can maintain the microalgal cells for a long duration with functional stability and reproducibility. The short-term storage of microalga stock cultures can be done at a constant low illumination, which may result in decreased cell size. Such stock cultures are maintained for about a month and then transferred to create a new culture line.

5.5.3 Cryopreservation of Microalgae Stock Cultures

It is crucial to ensure both viability and functionality are retained by stored stock cultures at very low temperatures. Very low-temperature storage, ranging from the use of domestic freezers to storage under liquid nitrogen, is widely being used, but the implications for stability and function are rarely investigated. Rahul et al. (2019) reported the retention of functionality in the stored master stock cultures of an industrially relevant lipid-producing alga under a variety of cryopreservation regimes. Storage in domestic (−15°C) and conventional (−80°C) freezers and deep freezers (−196°C) was tested. There was a rapid reduction in the viability of stored microalga at −15°C, and >50% viability was lost at −80°C within one month of storage. However, no reduction in viability occurred at −196°C. Functional performance after thawing the cryopreserved cultures is also influenced by the cryopreservation approach; e.g., functional performance varies for the pre-treated and untreated microalga before storage in cryopreservation. These results have important implications for microbial biotechnology, especially for those responsible for the conservation of genetic resources.

5.5.4 Maintenance of Stock Cultures

Stock cultures, also known as master cultures of any specific microalgal species, are the basic foundation of the cultivation system. They are normally supplied as axenic or uni-algal cultures from an authorized culture collections center of national institutions or research laboratories. Axenic algal cultures are isolated in the laboratory from water samples and identified authentically. Since they are valuable, they are normally kept in specialized maintenance media or species-specific media such as Hoagland, Bold basal, Chu-10, BG-11, etc. Culture media recipes have been categorized into two broad water types: freshwater media and saltwater media. Table 5.2 shows the various types of nutrient media developed for the cultivation of microalgae found in these two types of water.

The stock cultures can be maintained either on a liquid medium or on nutrient-enriched agar plates or slants under controlled environmental conditions such as temperature, illumination, and photoperiod. Stock cultures are used only to provide pure lines of starter cultures, also known as inoculum. Every effort should be made to minimize the risk of contamination of the stock cultures and starter cultures. Contaminants are in the form of competing microorganisms that can also grow on the same nutrient medium in which the microalgae can grow. Therefore, sterile

TABLE 5.2
Various Types of Media for the Two Broad Categories of Water

Freshwater Media

1/2 CHEV diatom medium	Modified bolds 3N medium
1/3 CHEV diatom medium	Modified COMBO medium
1/5 CHEV diatom medium	Modified desmidiacean medium
1:1 DYIII/PEA + Gr+ medium	N/20 medium
2/3 CHEV diatom medium	Ochromonas medium
2X CHEV diatom medium	P49 medium
Ag diatom medium	Polytomella medium
Allen medium	Proteose medium
BG-11 medium	Snow algae medium
BG-11(-N) medium	Soil extract medium
Bold 1NV medium	Soil extract + sodium metasilicate medium
Bold 3N medium	Soilwater: BAR medium
Bold basal medium	Soilwater: GR-medium
Botryococcus medium	Soilwater: GR − NH$_4$ medium
Bristol medium	Soilwater: GR+ medium
CHEV diatom medium	Soilwater: GR + NH$_4$ medium
Chu's medium	Soilwater: PEA medium
CR1 diatom medium	Soilwater: peat medium
CR1+ diatom medium	Soilwater: VT medium
CR1-S diatom medium	Spirulina medium
Cyanidium medium	TAP medium
Cyanophycean medium	Trebouxia medium
Desmid medium	Volvocacean medium
DYIII medium	Volvocacean-3N medium
DYV medium	Volvox medium
Euglena medium	Volvox-dextrose medium
HEPES medium	Waris medium
J medium	Waris + soil extract medium
Malt medium	WC medium
MES-Volvox medium	WC+ medium

(Continued)

TABLE 5.2 (*Continued*)

Various Types of Media for the Two Broad Categories of Water

Saltwater Media

1/2 Enriched seawater	Bold 1NV: erdshreiber (1:1) medium
1/2 Erdschreiber medium	Bold 1NV: erdshreiber (4:1) medium
1/2 Soil + seawater medium	Bristol-NaCl medium
1/3 Soil + seawater medium	Dasycladales seawater medium
1/4 Erdschreiber's medium	Enriched seawater medium
1/4 Soil + seawater medium	Erdschreiber's medium
1/5 Soil + seawater medium	ES/10 enriched seawater medium
1% F/2 medium	ES/2 enriched seawater medium
20% Allen + 80% erdschreiber medium	ES/4 enriched seawater medium
2/3 Enriched seawater	F/2 medium
2X Erdschreiber's medium	F/2 + NH$_4$ medium
2X Soil + seawater medium	LDM medium
5/3 Soil + seawater agar medium	Modified 2X CHEV medium
5% F/2 medium	Modified 2X CHEV + soil medium
8 ppt F/2 medium	Modified artificial seawater medium
A+ medium	Modified CHEV medium
Artificial seawater medium	Porphryridium medium
BG-11 + 1% NaCl medium	Soil + seawater medium
BG-11 + 0.36% NaCl medium	SS diatom medium

Source: https://utex.org/pages/algal-culture-media

procedures are performed under aseptic conditions inside the laminar air hood to ensure that contamination does not occur. Stock cultures are kept in small, transparent, autoclaved containers. For example, 500-mL Erlenmeyer flasks fitted with a cotton wool plug at the neck, suitable for containing 250 mL of sterile, autoclaved medium, are ideal. Concentrations of microalgal cells in the culture medium are generally much higher than those found in nature. Therefore, the isolated microalgal cultures must be grown in an enriched nutrient medium to increase the microalgal biomass. The composition of a few microalgae cultivation mediums is given in Table 5.3.

Sometimes, additions to suitably treated seawater can also be used according to the manufacturer's instructions for the cultivation of marine microalgae. Stock cultures are also sometimes maintained in seawater agar medium impregnated with suitable nutrients in glass petri-plates or on slants in test tubes.

5.5.5 CULTIVATION OF MICROALGAE AT A LABORATORY SCALE

For laboratory-scale research and development purposes, the microalgae can be cultivated in transparent glass or plastic cylinders or flat flasks, in which the microalgal culture can be grown in various artificial media under controlled environmental conditions in mild-shaking or static conditions. The most important parameters

TABLE 5.3

Various Media Compositions for Microalgae Cultivation

Modified Hoagland's Medium	Acidified Bold's Basal Medium	Bold's Basal Medium	Half Strength Chu 10 Medium	BG 11 Medium
$(NH_4)_2NO_3$-0.115 g	$(NH_4)_2 SO_4$-0.25 g	$NaNO_3$-0.25 g	$Ca(NO_3)_2$-2 g	$NaNO_3$-1.5 g
H_3BO_3-0.003 g	$NaNO_3$-0.75 g	$MgSO_4 \cdot 7H_2O$-0.075 g	K_2HPO_4-0.25 g	K_2HPO_4-0.04 g
$Ca(NO_3)_2$-0.656 g	$CaCl_2 \cdot 2H_2O$-0.025 g	$NaCl$-0.025 g	$MgSO_4 \cdot 7H_2O$-1. 25 g	$MgSO_4 \cdot 7H_2O$-0.075 g
$CuSO_4$-0.08 mg	$MgSO_4 \cdot 7H_2O$-0.075 g	K_2HPO_4-0.075 g	Na_2CO_3-1 g	$CaCl_2 \cdot 2H_2O$-0.036 g
$Fe_2(C_4H_4O_6)_3$-0.005 g	$K_2HPO_4 \cdot 3H_2O$-0.075 g	KH_2PO_4-0.175 g	Na_2SiO_3-1.25 g	$C_6H_8O_7$-0.006 g
$MgCl_2$-0.24 g	KH_2PO_4-0.175 g	$CaCl_2 \cdot 2H_2O$-0.025 g	$FeCl_3$-0.04 g	$Fe(NH_4)_3(C_6H_5O_7)_2$,
$MnCl_2$-0.016 mg	$NaCl$-0.025 g	$ZnSO_4 \cdot 7H_2O$-8.82 mg	H_3BO_3-0.248 mg	Na_2CO_3-0.02 g
KNO_3-0.3 g	Na_3EDTA-0.005 g	$MnCl_2 \cdot 4H_2O$-0.44 mg	$MnSO_4 \cdot H_2O$-0.147 mg	$EDTA$-0.001 g
$ZnSO_4$-0.22 mg	$FeCl_3 \cdot 6H_2O$-0.0006 g	MoO_3-0.71 mg	$ZnSO_4 \cdot 7H_2O$-0.023 mg	Trace metal mix
—	$MnCl_2 \cdot 4H_2O$-0.0002 g	$CuSO_4 \cdot 5H_2O$-1.57 mg	$CuSO_4 \cdot 5H_2O$-0.010 mg	H_3BO_3-2.86 g
—	$ZnCl_2 \cdot 6H_2O$-0.00003 g	$Co(NO_3)_2 \cdot 6H_2O$-0.49 mg	$(HN_4)_6Mo_7O_{24} \cdot 4H_2O$-0.007 mg	$MnCl_2 \cdot 4H_2O$-1.81 g
—	$CoCl_2 \cdot 6H_2O$-0.00001	H_3BO_3-11.42 mg	$Co(NO_3)_2 \cdot 6H_2O$-0.014 mg	$ZnSO_4 \cdot 7H_2O$-0.222 g
	$Na_2MoO_4 \cdot 2H_2O$-0.00002 g	$EDTA$-50 mg	Vit B1-5 mg	$Na_2MoO_4 \cdot 2H_2O$-0.39 g
	Vit B1-0.012 g Vit B12-10 µg	KOH-31 mg	Vit B7-2.5 mg	$CuSO_4 \cdot 5H_2O$-0.079 g
		$FeSO_4 \cdot 7H_2O$-4.98 mg	Vit B12-15 mg	$Co(NO_3)2 \cdot 6H_2O$-0.0494 g
		H_2SO_4-1 µL	—	
		Vit B1-10 µg	—	
		Vit B12-10 µg	—	

TABLE 5.4

Physical Parameters Required for Culturing Microalgae (Anonymous, 1991)

Parameters	Range	Optimal
Temperature (°C)	16–27	18–24
Salinity (g/L)	9–40	20–24
Light intensity (lux)	1,000–10,000 (depends on microalgae culture volume and cell density)	2,500–5,000
Photoperiod (light hours)	10–16	12–16
pH	7–9	8.2–8.7

regulating microalgal growth are nutrient quantity and quality, light, pH, turbulence, salinity, and temperature. The most optimal parameters as well as the tolerated ranges are species-specific, and a broad generalization for the most important parameters is given in Table 5.4. Also, various factors may be interdependent, and a parameter that is optimal for one set of conditions may not necessarily be optimal for another.

5.5.6 INDOOR MICROALGAE CULTURE DEVELOPMENT

As we mentioned, the concentrations of microalgae cells in synthetic/artificial culture mediums are generally higher than in natural water. Therefore, in laboratory conditions, when we cultivate any specific microalgae, the nutrient medium must be enriched with nutrients required by the selected microalgae to make up for the deficiencies actually found in its native natural water. Figure 5.8 shows the indoor cultivation of microalgae in artificial saline medium under sterile and controlled environmental conditions such as 2,000 lux white light illumination, 25°C ± 2°C temperature, 16 h photoperiod, and pH 7.8 (Sahay and Braganza, 2022). In indoor laboratory conditions, researchers may also optimize the physical and chemical parameters to further enhance the rate of growth of the microalgae strain. The specific growth rate (μ) of the microalgae is calculated using the equation $\mu = \ln(N_2/N_1)/t_2 - t_1$, where μ is the specific growth rate and N_1 and N_2 are the biomass at time 1 (t_1) and time 2 (t_2), respectively (Shuler and Kargi, 2002). Microalgae grow fast, and some can double in size in 24–26 h, and some can even reproduce within 8 h. Microalgae growth generally reaches its peak in 30 days or 4 weeks, though one doesn't need to wait that long to harvest the microalgae. Once the microalgal culture has grown to its maximum strength, it needs to be transferred to fresh medium, either indoors or outdoors, for mass cultivation.

If one looks at the growth curve of microalgae, one will find that the basic relationship between microalgal growth rates and nutrient concentrations is an asymptotic relationship in which the nutrient concentration decreases with an increase in microalgal density. Several reports state that the nutrients specifically affect the rate of microalgal growth; e.g., nitrogen and phosphorus are essential macronutrients needed to promote microalgal growth, and they regulate metabolic activities if supplied in an acceptable form. Various nitrogen and phosphorus concentrations in the

FIGURE 5.8 Cultivation of microalgae in the laboratory under sterile and controlled environmental conditions. (Courtesy: Loyola Centre for Research & Development (LCRD), Xavier Research Foundation (XRF), Ahmedabad, 380009, Gujarat, India).

microalgae cultivation medium may influence lipid and fatty acid yield (Yang et al., 2008; Zienkiewicz et al., 2020).

5.5.7 OUTDOOR MICROALGAE CULTURE DEVELOPMENT

The transfer of indoor microalgae cultures to the outdoors is scaled up in a stepwise process, starting with the laboratory culture in a dilution ratio of approximately 1 to 5–10. It is advisable not to expose diluted laboratory cultures directly outdoors to full sunlight during the first few days to avoid the risk of photo-inhibition. However, a minimum biomass concentration corresponding to about 10 gm/m^2 (~200 mg/m^2 chlorophyll) is recommended (Masojıdek and Torzillo, 2008). The cultivation of microalgae at high cell density often encounters a problem of photolimitation because of light shading. The high light intensity at the surface cell layers saturates the photosynthetic process and causes photo-inhibition, whereas excess energy is dissipated through non-photochemical quenching. Meanwhile, the low-light intensity in the lower layer of cells compels them to perform photorespiration instead of photosynthesis.

This uneven distribution of the light intensity results in sub-optimum photosynthetic efficiency, which eventually reduces the biomass yield.

5.5.8 MICROALGAE CULTIVATION IN WASTEWATER

The nutrients present in the form of carbon, nitrogen, and phosphorous in any wastewater can be turned into an economic opportunity by feeding them to microalgae (Wang et al., 2018). Microalgae can utilize the organic carbon in wastewater and convert it into biomass. The use of wastewater for microalgal cultivation in mixotrophic and heterotrophic cultivation modes can balance respiratory losses, improve the energy budget, and give a boost to biomass productivity. Earlier reports have shown that the utilization of nutrients from wastewater together with industrial flue gases (CO_2) helps to decrease the cost of microalgae mass cultivation and makes microalgae-based biofuel production technology commercially viable (Acien Fernandez et al., 2012). Besides biofuel, one can also recover other value-added by-products, which can compensate for the total cost involved in microalgal cultivation. The wastewater used for microalgal cultivation does not require any additional treatment to meet ecological and environmental regulations. Microalgae release oxygen as a by-product during wastewater treatment, and this is utilized by aerobic bacteria to further degrade the remaining organic loads. This reduces the energy cost compared to the cost of mechanical energy for aeration during conventional wastewater treatment.

Municipal wastewater is used to cultivate microalgae in open raceway ponds or closed photobioreactors. At the Algae Systems plant in Daphne, AL, this process was used to treat up to 50,000 gal/day of incoming raw wastewater (Lucie et al., 2016). A combination of algae nutrient uptake, aeration by photosynthetically produced oxygen, and dewatering via suspended air flotation removed 75% of total nitrogen, 93% of total phosphorus, and 92% of BOD from influent wastewater. Researchers used diverse polycultures dominated by genera *Chlorella*, *Cryptomonas*, and *Scenedesmus*. The resulting biomass was suitable for biofuel conversion via hydrothermal liquefaction due to its consistent lipid content, low ash content, and consistent elemental composition. Biomass production rates ranged from 3.5 to 22.7 g/m²/day during continuous operation, with productivity predominantly driven by temperature and frequency of harvest. This is not the only example; there are several other practices reported in the literature to cultivate microalgae in wastewater. Therefore, the utilization of a large quantity of wastewater for microalgal cultivation could promote waste-free, carbon-neutral, and environmentally sustainable technology.

5.6 NUTRITIONAL CONDITIONS OF MICROALGAE CULTIVATION

Microalgae uptake of nutrients takes place via three modes, as mentioned earlier: autotrophic, heterotrophic, and mixotrophic. In the autotrophic mode, inorganic carbon and artificial light/solar are the primary sources of energy for microalgal growth. In the heterotrophic mode, the microalgae use organic carbon and energy through the Krebs cycle. In the mixotrophic mode, the carbon sources for microalgal growth can be supplied by both inorganic and organic forms. In nature, microalgae are evolved

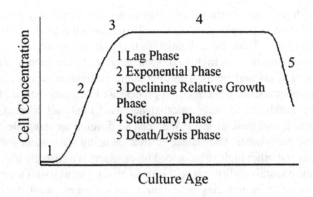

FIGURE 5.9 Growth curve of microalgae.

in fresh, marine, and brackish water, and they utilize solar energy (light sources), air (CO_2), and water for their growth. Microalgal growth occurs in five phases: (1) the lag phase, the initial growth period, where the microalgae take time to adapt themselves to a new environment; (2) the log/exponential phase, where rapid cell division occurs and growth is faster; (3) the decline phase, which contains limiting cell division; (4) the stationary phase, in which the cell density of microalgae is stable because of the limiting factors; and (5) the death phase, in which cell growth is almost stopped due to a lack of nutrients (Figure 5.9). Microalgae are easy to cultivate because they can tolerate a broad range of pH, salinity, and temperature.

5.7   LARGE-SCALE CULTIVATION OF MICROALGAE

The loss of microalgal cell biomass during microalgae cultivation and the harvesting steps affects the final yield. Therefore, while scaling up microalgae cultivation from bench-scale to large-scale, one must optimize the parameters to increase the rate of growth of microalgae to compensate for the low cell density and difficulties in harvesting. Numerous equipment and technologies have been improved over the years to enhance microalgal biomass and lipid production. Closed cultivation systems can be easily used in physically and chemically controlled conditions; however, it is still hard to ensure the high productivity of microalgae on a large scale. An ideal microalgae culturing system should possess a range of characteristics, including (1) an adequate light source, (2) effective transfer of material across the liquid-gas barrier, (3) simple operation procedures, (4) minimal contamination rate, (5) low cost of overall building and production, and (6) high land efficiency (Jegathese and Farid, 2014).

Large-scale microalgal biomass production can be done in an open-raceway pond or in a closed photobioreactor (PBR). Both cultivation systems have their pros and cons. However, the ultimate goal is to overcome the hurdles in both systems to enhance the final yield or compensate for the loss. If we increase the number of fractions obtained from microalgae cultivation, we can compensate for the cultivation loss. For example, we can produce high-value-added products such as pharmaceuticals, nutraceuticals, pigments, enzymes, and biofuels like biodiesel and bioethanol. Another approach is to recycle the nutrients from wastewater sources, which is considered a step in the treatment of industrial wastewater using microalgal species. As the world's population continuously increases day by day, wastewater discharge also increases. Thus, the utilization of this polluted wastewater as a source of microalgal growth nutrients is highly recommended for the environmentally friendly production of biomass and lipids. The microalgal cultivation strategy also helps to mitigate atmospheric CO^2 in the fixation process and helps to reduce global warming. It is very important to make microalgae-based biodiesel production technology scale up to a commercial level. We need to develop an integrated biorefinery approach using wastewater. Microalgae cultivation using wastewater for the production of biofuels and other high-value-added by-products is a strongly influenced technology for future sustainability that will address many global issues together, such as high-energy production, reducing greenhouse gas emissions, wastewater treatment, and lowering production costs.

5.8   RECENT ADVANCEMENTS IN MICROALGAE CULTIVATION

Some researchers have reported that the lipid content of microalgae is usually between 20% and 50% on a dry weight basis (Halder and Azad, 2019), whereas in some microalgal species, e.g., *Botryococcus braunii*, the lipid production can reach up to 75% on a dry weight basis. In recent years, microalgae have not only been cultivated to produce lipids but are also used as a vital source of by-products such as polysaccharides, pigments, proteins, vitamins, bioactive compounds, and antioxidants. Therefore, researchers are making a range of efforts not only to enhance its biomass and lipid accumulations but also to harvest many other useful biorefinery products. Research is being done at the chemical, physical, and molecular levels for the improvement of the microalgae potential.

There are some genetically modified microalgal species that have been developed to promote desirable qualities in microalgae species. In particular, there is much interest in enhancing lipid production due to its association with energy yield. The first nuclear transformation of *Chlamydomonas reinhardtii* using polyethylene glycol or poly-L-ornithine was done in the early 1990s. Until 2008, most of the studies focused on the genetic modification of model microalgal strains such as *Chlamydomonas reinhardtii*, *Phaeodactylum tricornutum*, *Thalassiosira pseudonana*, and more recently *Nannochloropsis* spp., due to the lack of a universal genetic toolbox for all microalgal species (Fu et al., 2019). There are reports on the development of microalgae with highly suitable genetic manipulations and desirable genetic transformations with foreign genes. Figure 5.10 summarizes the advancements made in the last 30 years on the genetic engineering of specific microalgae to enhance lipid production (Camilo et al., 2021).

Recent research has demonstrated that enhancement in lipid production in microalgal cells can be achieved by altering the lipid biosynthesis pathway, the Kennedy pathway, PUFA and TAG metabolism, transcription factors, and nicotinamide adenine dinucleotide phosphate (NADPH) generation. Keeling et al. (2014) reported a transcriptome sequencing project named "Marine Microbial Eukaryote Transcriptome Sequencing Project" that aimed to sequence nearly 700 marine microbial species from 17 phyla. The sequence information for this dataset is

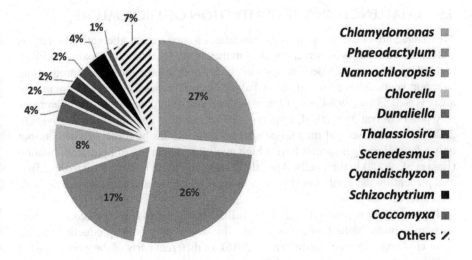

Chlamydomonas
Phaeodactylum
Nannochloropsis
Chlorella
Dunaliella
Thalassiosira
Scenedesmus
Cyanidischyzon
Schizochytrium
Coccomyxa
Others

FIGURE 5.10   Frequency of engineered microalga species for enhancing lipid production.

available at the iMicrobe Project (www.imicrobe.us/#/projects/104) and the Sequence Read Archive (BioProject PRJNA231566). Among the other sequenced transcriptomes, 140 are marine microalgae species. Most of these sequenced species are taxonomically well-defined. This transcriptomic data is very helpful because it provides an extensive reference dataset for novel gene discovery and the construction of computation-based metabolic models (Kumar et al., 2020). Another significant development based on genetic engineering is getting insights into the enhancement of photosynthetic efficiency in model microalgal systems. However, this aspect is yet to be applied to large-scale applications for biofuel production. However, from a commercial point of view, the classical approaches are the main choice of the researchers for the enhancement of lipid productivity, such as the optimization of physical and chemical parameters or the selection of a potential microalgal candidate.

In fact, lipids derived from microalgae play a key role in their commercial utilization for food, feed, or fuel purposes. In the last few years, the research has focused on the enhancement of microalgal lipid content by various means without compromising overall lipid productivity. Research has mainly focused on the environmental, nutritional, and physiological alterations for the cultivation of microalgae and the genetic manipulations for enhanced lipid production (Kumar et al., 2020). The lipid extraction facility at TERI was scaled up based on a novel method of extracting lipids from wet microalgae without the need for drying (Patel and Kannan, 2021). However, genetic engineering of the robust strains for enhanced lipid production remains one of the most viable options to improve the process. In addition, the design of the cultivation system has a significant effect on photosynthetic efficiency and eventually on productivity. There is a need for a more comprehensive and cumulative approach, such as fine-tuning the flux balance of the Calvin cycle toward enhanced CO_2 fixation or perturbing multiple targets at once to get a synergistic effect.

5.9 CHALLENGES IN THE CULTIVATION OF MICROALGAE

Microalgae are one of the promising candidates to produce an alternative biofuel as well as for cleaning wastewater and the environment. Therefore, microalgae-derived technologies are in high demand for biorefinery and sustainable energy production. In recent years, it has been estimated that the cost of biodiesel produced from microalgae is $20.53 and $9.84 per gallon using a tubular photobioreactor and open raceway pond cultivation method, respectively (Veeramuthu and Ngamcharussrivichai, 2020). There are several microalgal species available for biodiesel production, but only a few microalgal species have a high quality and quantity of lipid accumulation (Raja et al., 2008) in their cells. Also, there are many more potential microalgae species present on the earth, but they have not been isolated and identified so far and, therefore, are not in use.

In the current scenario, the ultimate utilization of microalgae feedstock for biomass cultivation, biofuel production, and value-added by-product production faces a lot of challenges (Sahay and Braganza, 2016). In different parts of the world, several research initiatives have been undertaken to overcome the issues related to microalgae-based technologies. However, we are still far away from the ultimate goal of commercializing microalgae technology for biofuel production due to its higher production costs. We need to address the challenge of microalgae-based biodiesel production on a commercial scale by improving the technology from the laboratory scale to the commercial scale. The conversion factor from laboratory to commercial level is not a straight-forward step. There are several intermediate factors that affect drastically large-scale production when scaling up the technology. The most critical steps that need to be addressed are to improve microalgal biomass productivity, harvesting, pre-treatment and oil extraction, and biodiesel production. Though we have several advanced technologies available for large-scale biomass production and lipid conversion into biodiesel, the cost of production is still too high, which makes the final microalgal biodiesel too costly, and we are not able to compete with the cost of petro-diesel. For instance, the design of the microalgae cultivation system requires temperature and growth-limiting controlled conditions, viz., CO_2, water sources, nutrient sources, and optimization of all other parameters that directly or indirectly impact the rate of growth of microalgae. During the microalgae cultivation and harvesting process, biomass dewatering is one of the major obstacles because this process is energy-intensive and therefore costly. The closed photobioreactor system also faces major operating challenges, such as overheating and a foul smell, due to gaseous exchange in a closed system.

It has been observed that the construction of raceway ponds is much cheaper, and easier to operate, and can scale up to several hectares, which makes them the right choice for commercial-scale biomass production. About 95% of commercial microalgal biomass production is performed using open raceway ponds. The microalgae biomass produced in the raceway ponds is used for high-value-added by-products, which sell for high prices in the global market. However, the open cultivation systems also have several limitations, mainly due to frequent contamination by other microalgal species, algal grazers, fungi, amoeba, etc., and changing environmental temperatures with changes in seasons. One study revealed that hundreds of research papers have been published, but there is still no proper information available that can give an accurate estimation of the relevance of a specific cultivation design, operation, yield, and other important aspects that can be used at the commercial level (Zhu et al., 2017). The major problem in microalgal biofuel production is high capital and operating costs, and therefore, alternate approaches have been applied to cut the expenditure involved. There was a vast technological gap during commercialization. For example, in large-scale biomass production, there is a large demand for water, CO_2, nitrogen, and phosphorous. The use of wastewater can fulfill the requirement for nutrients; however, there is a serious concern about contamination by bacteria, pathogens, and chemicals present in wastewater. Venkata et al. (2015) reported that there is a low requirement of nitrogen (0.16 kg) and phosphorous (0.02 kg), but a high amount of CO_2 (3.5–9.3 kg) is required for the production of 1 liter of microalgal oil. Microalgae utilize a large amount of CO_2 for their growth and biomass production. Another major challenge is microalgal lipid extraction prior to biodiesel production. After dewatering the microalgal biomass, several expensive solvents or ultrafiltration techniques are used for the extraction of crude microalgal oil, which significantly increases the production cost. Researchers are trying to develop an advanced technology in which oil is extracted without drying or solvent extraction of the microalgal biomass in order to reduce expenditure. Currently, the microalgae-based biodiesel production process is expensive since it requires a clean microalgal feedstock free from other microbial and protozoan contaminants, free fatty acids, and water or synthetic medium residues. For this kind of extraction technique, the microalgal biomass must be clean and dry, which adds to costs. To reduce the free fatty acid content of microalgal crude oil, the esterification process is carried out via acidic or enzymatic treatment. The esterification via the enzymatic process is usually done with lipases because they can run even at low temperatures. However, during the esterification process, the first component produced is glycerol as a by-product, which inhibits lipase activity, and the reaction is not complete. Baadhe et al. (2014) demonstrated that, using methyl acetate, triacetin is produced as a by-product. Therefore, one can avoid glycerol formation and lipase inhibition by using this method.

5.10 CONCLUSION

Microalgae is a third-generation biofuel-producing feedstock that offers potential characteristics that make it a promising candidate for various industrial and environmental applications. Many efforts have been made to address different challenges for microalgae-based biodiesel production technology, particularly low-cost and high-efficiency microalgal biomass cultivation, biofuel production, wastewater treatment, and CO_2 mitigation. Scientists have given a proof-of-concept for a biorefinery – an integrated and sustainable process of microalgae-based biofuel production. However, the high cost of microalgal biodiesel production is the main hurdle. It may be clearly noted that zero-nutrient cost technology for biomass production, inexpensive large-scale harvesting, and biodiesel conversion processes are yet to be achieved through detailed investigation. The cultivation of microalgae on a commercial scale can play a key role in the present global energy scenario and raise concerns about related environmental issues. Researchers believe that microalgae, as a third-generation candidate, can fulfill the energy demand and its challenges in the future. The paramount challenges in microalgae cultivation techniques and their limitations, with an emphasis on the cost factor, need to be kept in mind. Therefore, establishing a new and innovative biorefinery-based, low-cost technology should be focused on to overcome these problems. Evidently, substantial advancements have been achieved in the field of microalgal biotechnology for the sustainable production of third-generation biofuels and green chemicals. Bioprospecting of more suitable host species and/or metabolic engineering of genetically accessible microalgal strains could accomplish the aim of commercially feasible lipid production. Future research also needs to focus on microalgae that are promising for high-value lipids and are most likely to reach economic feasibility in a short period of time. In recent times, microalgae-based biorefinery has emerged as an emerging technology that aims to address these issues. Besides, it helps wastewater treatment with zero-cost technology, CO_2 mitigation, and attractive value-added by-products. These economic processes could be improved by adopting various cost-cutting activities, such as utilizing wastewater and industrial flue gas as nutrient and carbon sources, respectively. Finally, the microalgae-based biorefinery process does seem to be the most feasible approach in the forthcoming years to compete with fossil fuels in order to develop a sustainable and renewable bioenergy source.

ACKNOWLEDGMENTS

We acknowledge the Gujarat State Biotechnology Mission (GSBTM), Govt. of Gujarat, for funding the microalgae project, and the Xavier Research Foundation (XRF), Ahmedabad, Gujarat, for support.

REFERENCES

Acien Fernandez F.G., GonzalezLopez C.V., Fernández Sevilla J.M., et al. (2012) Conversion of CO_2 into biomass by microalgae: How realistic a contribution may it be to significant CO_2 removal? *Applied Microbiology and Biotechnology*, 96(3), 577–586.

Adeniyi O.M., Azimov U., Burluka, A. (2018) Algae biofuel: Current status and future applications. *Renewable and Sustainable Energy Reviews*, 90, 316–335. doi: 10.1016/j.rser.2018.03.067.

Anonymous (1991) The design and operation of live feeds production systems. In: Fulks W., Main K.L. (Eds.), Rotifer and Micro-Algae Culture Systems: Proceedings of a US-Asia Workshop, Honolulu, Hawaii, January 28–31, 1991 (pp. 3–52). The Oceanic Institute.

Arnold M. (Ed.) (2013) *Sustainable Algal Biomass Products by Cultivation in Wastewater Flows*, (p. 88). VTT Technology Series, No. 147. Espoo: VTT Technical Research Centre of Finland.

Ayesha S., Sana M., Hui Z., et al. (2020) Cultivating microalgae in wastewater for biomass production, pollutant removal, and atmospheric carbon mitigation: A review. *Science of the Total Environment*, 704, 135303. doi: 10.1016/j.scitotenv.2019.135303.

Baadhe R.R., Potumarthi R., Gupta V.K. (2014) Chapter 8 - Lipase-catalyzed biodiesel production: Technical challenges. In: Gupta V.K., Tuohy M.G., Kubicek C.P., Saddler J., Xu F. (Eds.), *Bioenergy Research: Advances and Applications*, (pp. 119–129). Amsterdam: Elsevier.

Babich I.V., Van der Hulst M., Lefferts L., et al. (2011) Catalytic pyrolysis of microalgae to high-quality liquid bio-fuels. *Biomass Bioenergy*, 35, 3199–3207.

Bhatia S.K., Mehariya S., Bhatia R.K. (2021) Wastewater-based microalgal biorefinery for bioenergy production: Progress and challenges. *Science of the Total Environment*, 751, 141599.

Borowitzka L.J., Borowitzka M.A. (1990) Commercial production of β-carotene by *Dunaliella salina* in open ponds. *Bulletin of Marine Science*, 47(1), 244–252.

Brennan L., Owende P. (2010) Biofuels from microalgae: A review of technologies for production, processing, and extractions of biofuels and co-products. *Renewable and Sustainable Energy Reviews*, 14(2), 557–577.

Camilo F.M., Christian S., Mihris I.S.N., et al. (2021) Genetic engineering of microalgae for enhanced lipid production. *Biotechnology Advances*, 52, 107836.

Chen J., Li J., Dong W., et al. (2018) The potential of microalgae in biodiesel production. *Renewable and Sustainable Energy Reviews*, 90, 336–346.

Chew K.W., Yap J.Y., Show P.L., et al. (2017) Microalgae biorefinery: High value products perspectives. *Bioresource Technology*, 229, 53–62.

Chisti Y. (2007) Biodiesel from microalgae. *Biotechnology Advances*, 25, 294–306.

Choi O., Das A., Yu C.P., et al. (2010) Nitrifying bacterial growth inhibition in the presence of algae and cyanobacteria. *Biotechnology and Bioengineering*, 107, 1004–1011.

Costa J.A.V., de Morais M.G. (2014) An open pond system for microalgal cultivation. In: Pandey A., Lee D.J., Chisti Y., Soccol C.R. (Eds.), *Biofuels from Algae*, (pp. 1–22). Elsevier. doi: 10.1016/B978-0-444-59558-4.00001-2.

Cuellar-Bermudez S.P., Aleman-Nava G.S., Chandra R., et al. (2017) Nutrients utilization and contaminants removal: A review of two approaches of algae and cyanobacteria in wastewater. *Algal Research*, 24, 438–449.

Cuellar-Bermudez S.P., Garcia-Perez J.S., Rittmann B.E., et al. (2015a) Photosynthetic bioenergy utilizing CO_2: An approach on flue gases utilization for third generation biofuels. *Journal of Cleaner Production*, 98, 53–65.

Cuellar-Bermudez S.P., Romero-Ogawa M.A., Vannela R., et al. (2015b) Effects of light intensity and carbon dioxide on lipids and fatty acids produced by *Synechocystis* sp. PCC6803 during continuous flow. *Algal Research*, 12, 10–16.

Feng Y., Li C., Zhang D. (2011) Lipid production of *Chlorella vulgaris* cultured in artificial wastewater medium. *Bioresource Technology*, 102, 101–105.

Fu W., Nelson D.R., Mystikou A., et al. (2019) Advances in microalgal research and engineering development. *Current Opinion in Biotechnology*, 59, 157–164. doi: 10.1016/j.copbio.2019.05.013.

Ghimire A., Kumar G., Sivagurunathan P., et al. (2017) Bio-hythane production from microalgae biomass: Key challenges and potential opportunities for algal biorefineries. *Bioresource Technology*, 241, 525–536.

Gilbert-López B., Mendiola J.A., Fontecha, J. (2015) Downstream processing of *Isochrysis galbana*: A step towards microalgal biorefinery. *Green Chemistry*, 17, 4599–4609.

Gupta P.L., Lee S.M., Choi H.J. (2015) A mini review: Photobioreactors for large scale algal cultivation. *World Journal of Microbiology and Biotechnology*, 31(9), 1409–1417.

Halder P., Azad A.K. (2019) Chapter 7 - Recent trends and challenges of algal biofuel conversion technologies. In: Azad A.K., Rasul M. (Eds.), *Advanced Biofuels*, (pp. 167–179). Cambridge: Woodhead Publishing.

Hallenbeck P.C., Grogger M.M., Veverka D. (2016) Solar biofuels production with microalgae. *Applied Energy*, 179, 136–145.

Hamed I. (2016) The evolution and versatility of microalgal biotechnology: A review. *Comprehensive Reviews in Food Science and Food Safety*, 15(6), 1104–1123.

Harman-Ware A.E., Morgan T., Wilson M., et al. (2013) Microalgae as a renewable fuel source: Fast pyrolysis of *Scenedesmus* sp. *Renewable Energy*, 60, 625–632.

Hoh D., Watson S., Kan E. (2016) Algal biofilm reactors for integrated wastewater treatment and biofuel production: A review. *Chemical Engineering Journal*, 287, 466–473.

Hu Z., Ma X., Jiang E. (2017) The effect of microwave pre-treatment on chemical-looping gasification of microalgae for syngas production. *Energy Conversion and Management*, 143, 513–521.

Huang G., Chen F., Wei D., et al. (2010) Biodiesel production by microalgal biotechnology. *Applied Energy*, 87(1), 38–46.

Huang Q., Jiang F., Wang L., et al. (2017) Design of photobioreactors for mass cultivation of photosynthetic organisms. *Engineering*, 3(3), 318–329.

Huerlimann R., Nys R., Heimann K. (2010) Growth, lipid content, productivity, and fatty acid composition of tropical microalgae for scale-up production. *Biotechnology and Bioengineering*, 107, 245–257.

Huo S., Wang Z., Zhu S., et al. (2018) Biomass accumulation of *Chlorella zofingiensis* G1 cultures grown outdoors in photobioreactors. *Frontiers in Energy Research*, 6, 49.

Jegathese S.J.P., Farid M. (2014) Microalgae as a renewable source of energy: A Niche opportunity. *Journal of Renewable Energy*, 2014, 1–10.

Kannan D.C., Devi V. (2019) An outdoor algal growth system of improved productivity for biofuel production. *Journal of Chemical Technology and Biotechnology*, 94, 222–235.

Keeling P.J., Burki F., Wilcox H.M., et al. (2014) The marine microbial eukaryote transcriptome sequencing project (MMETSP): Illuminating the functional diversity of eukaryotic life in the oceans through transcriptome sequencing. *PLoS Biology*, 12, e1001889. doi: 10.1371/journal.pbio.1001889.

Khan M.I., Shin J.H., Kim J.D. (2018) The promising future of microalgae: Current status, challenges, and optimization of a sustainable and renewable industry for biofuels, feed, and other products. *Microbial Cell Factories*, 17(1), 36.

Khan M.J., Harish, Ahirwar A., et al. (2021) Insights into diatom microalgal farming for treatment of wastewater and pre-treatment of algal cells by ultrasonication for value creation. *Environmental Research*, 201, 111550.

Kumar G., Shekh A., Jakhu S., et al. (2020) Bioengineering of microalgae: Recent advances, perspectives, and regulatory challenges for industrial application. *Frontiers in Bioengineering and Biotechnology*, 8(914), 1–31. doi: 10.3389/fbioe.2020.00914.

Lee Y.K. (1997) Commercial production of microalgae in the Asia-Pacific rim. *Journal of Applied Phycology*, 9(5), 403–411.

Lee Y.K. (2001) Microalgal mass culture systems and methods: Their limitation and potential. *Journal of Applied Phycology*, 13(4), 307–315.

Liang Y. (2013) Production of liquid transportation fuels from heterotrophic microalgae. *Applied Energy*, 104, 860–868.

Lucie N.A., Katharina M., Zapata A., et al. (2016) Optimizing microalgae cultivation and wastewater treatment in large-scale offshore photobioreactors. *Algal Research*, 18, 86–94. doi: 10.1016/j.algal.2016.05.033.

Masojídek J., Torzillo G. (2008) Mass cultivation of freshwater microalgae. In: Jorgensen S.E., Fath B.D., (Eds.), *Encyclopedia of Ecology*, (pp. 2226–2235). Cambridge, MA: Academic Press.

Mata T.M., Martins A.A., Caetano N.S. (2010) Microalgae for biodiesel production and other applications: A review. *Renewable and Sustainable Energy Reviews*, 14, 217–232.

Mishra A.K., Kaushik M.S., Tiwari D. (2019) Nitrogenase and hydrogenase: Enzymes for nitrogen fixation and hydrogen production in cyanobacteria. In: Mishra A.K., Tiwari D.N., Rai A.N. (Eds.), *Cyanobacteria*, (pp. 173–191). Cambridge, MA: Academic Press.

Mohan S.V., Rohit M.V., Subhash G.V., et al. (2019) Biofuels from algae. In: Pandey A., Chang J.S., Soccol C.R., Lee D.J., Chisti Y. (Eds.) *Algal Oils as Biodiesel*, (pp. 287–323). Amsterdam, Netherlands: Elsevier.

Mussgnug J.H., Klassen V., Schlüter A., et al. (2010) Microalgae as substrates for fermentative biogas production in a combined biorefinery concept. *Journal of Biotechnology*, 150, 51–56.

Nguyen M.T., Choi S.P., Lee J., et al. (2009) Hydrothermal acid pre-treatment of *Chlamydomonas reinhardtii* biomass for ethanol production. *Journal of Microbiology and Biotechnology*, 19, 161–166.

Norsker N.H., et al. (2011) Microalgal production-a close look at the economics. *Biotechnology Advances*, 29, 24–27.

Pacheco R., Ferreira A.F., Pinto T., et al. (2015) The production of pigments & hydrogen through a *Spirogyra* sp. biorefinery. *Energy Conversion and Management*, 89, 789–797.

Pan P., Hu C., Yang W., et al. (2010) The direct pyrolysis and catalytic pyrolysis of *Nannochloropsis* sp. residue for renewable bio-oils. *Bioresource Technology*, 101, 4593–4599.

Patel S., Kannan D.C. (2021) A method of wet algal lipid recovery for biofuel production. *Algal Research*, 55, 102237.

Posadas E., Alcantara C., Garcia-Encina P.A., et al. (2017) Microalgae-based biofuels and bio-products. In: Gonzalez-Fernandez C., Muñoz R. (Eds.), *Microalgae Cultivation in Wastewater*, (pp. 67–91). Cambridge: Woodhead Publishing.

Posten C. (2009) Design principles of photo-bioreactors for the cultivation of microalgae. *Engineering in Life Sciences*, 9(3), 165–177.

Rahul V.K., María H.O., John G.D., et al. (2019) Effect of cryopreservation on viability and functional stability of an industrially relevant alga. *Scientific Reports*, 9, 2093.

Raja R., Hemaiswarya S., Kumar N.A., et al. (2008) A perspective on the biotechnological potential of microalgae. *Critical Reviews in Microbiology*, 34(2), 77–88.

Randrianarison G., Ashraf M.A. (2017). Microalgae: A potential plant for energy production. *Geology, Ecology, and Landscapes*, 1(2), 104–120.

Rashid N., Rehman M.S.U., Sadiq M., et al. (2014) Current status, issues and developments in microalgae derived biodiesel production. *Renewable and Sustainable Energy Reviews*, 40, 760–778.

Rinanti A., Kardena E., Astuti D.I., et al. (2013) Integrated vertical photobioreactor system for carbon dioxide removal using phototrophic microalgae. *Nigerian Journal of Technology*, 32(2), 225–232.

Rogers J.N., Rosenberg J.N., Guzman B.J., et al. (2014) A critical analysis of paddle-wheel-driven raceway ponds for algal biofuel production at commercial scales. *Algal Research*, 4, 76–88.

Sahay S., Braganza V. (2016) Microalgae - based biodiesel production: Current and future scenario. *Journal of Experimental Sciences*, 7, 31–35.

Sahay S., Braganza V. (2022) Effect of nitrogen on the biomass production and lipid accumulation of freshwater microalgae Micractinium reisseri: A potential strain for bio-fuel production. *Bioscience Biotechnology Research Communications*, 15(3), 442–449.

Saratale R.G., Ponnusamy V.K., Jeyakumar R.B., et al. (2022) Microalgae cultivation strategies using cost-effective nutrient sources: Recent updates and progress towards biofuel production. *Bioresource Technology*, 361, 127691.

Selmani N., Mirghani M.E., Alam M.Z. (2013) Study the growth of microalgae in palm oil mill effluent wastewater. https://iopscience.iop.org/journal/1755-1315 *IOP Conference Series: Earth and Environmental Science*, **16**, 012006.

Shen Y., Yuan W., Pei Z., et al. (2009) Microalgae mass production methods. *Transactions of the ASABE*, 52(4), 1275–1287.

Show P., Tang M., Nagarajan D., et al. (2017) A holistic approach to managing microalgae for biofuel applications. *International Journal of Molecular Sciences*, 18(1), 215.

Shuler M.L., Kargi, F. (2002) *Bioprocess Engineering-Basic Concepts*, 2nd Edition. New Delhi: Prentice-Hall of India Pvt. Ltd.

Sialve B., Bernet N., Bernard O. (2009) Anaerobic digestion of microalgae as a necessary step to make microalgae biodiesel sustainable. *Biotechnology Advances*, 27, 409–416.

Sierra E., Acien F.G., Fernandez J.M., et al. (2008) Characterization of a flat plate photobioreactor for the production of microalgae. *Chemical Engineering Journal*, 138(1–3), 136–147.

Stark M., O'Gara I. (2012) An introduction to photosynthetic microalgae. *Disruptive Science and Technology*, 1(2), 65–67.

Suali E., Sarbatly R. (2012) Conversion of microalgae to biofuel. *Renewable and Sustainable Energy Reviews*, 16, 4316–4342.

Sun Z., Liu J., Zhou Z.G. (2016) Algae for biofuels: An emerging feedstock. In: Luque R., Lin C.S.K., Wilson K., Clark J. (Eds.), *Handbook of Biofuels Production*, (pp. 673–698). Cambridge: Woodhead Publishing. doi: 10.1016/B978-0-08-100455-5.00022-9.

Tamburic B ., Zemichael F.W., Crudge P., et al. (2011) Design of a novel flat-plate photobioreactor system for green algal hydrogen production. *International Journal of Hydrogen Energy*, 36(11), 6578–6591.

Thomas D.G., Minj N., Mohan N., et al. (2016) Cultivation of Microalgae in domestic wastewater for biofuel applications: An upstream approach. *Journal of Algal Biomass Utilization (JABU)*, 7(1), 62–70.

Torzillo G., Zittelli G.C. (2015) Tubular photobioreactors. In: Prokop A., Bajpai R.K., Zappi M.E. (Eds.), *Algal Biorefineries*, (pp. 187–212). Switzerland: Springer International Publishing.

Tredici M.R. (2004) Mass production of microalgae: Photobioreactors. *Handbook Microalgal Culture*, 1, 178–214.

Ugwu C.U., Aoyagi H. (2012) Designs, operation and applications. *Biotechnology*, 11(3), 127–132.

Veeramuthu A.K., Ngamcharussrivichai C. (2020) Potential of microalgal biodiesel: Challenges and applications. In: Taner T., Tiwari A., Ustun T.S. (Eds.) *Renewable Energy: Technologies and Applications*, (pp. 1–14). Norderstedt, Germany: BoD – Books on Demand. doi: 10.5772/intechopen.91651.

Venkata M.S., Rohit M.V., Chiranjeevi P., et al. (2015) Heterotrophic microalgae cultivation to synergize biodiesel production with waste remediation: Progress and perspectives. *Bioresource Technology*, 184, 169–178.

Vo H.N., Ngo H.H., Guo W., et al. (2018) A critical review on designs and applications of microalgae-based photobioreactors for pollutants treatment. *Science of the Total Environment*, 651, 1549–1568.

Vonshak A. (1986) Laboratory techniques for the culturing of microalgae. In: Richmond A. (Ed.), *Handbook for Microalgal Mass Culture*, (pp. 117–146). Boca Raton, FL: CRC Press.

Vonshak, A. (1997) *Spirulina Platensis (Arthrospira) Physiology, Cell-Biology and Biotechnology*. London: Taylor & Francis.

Wang S.K., Wang X., Tao H.H., et al. (2018) Heterotrophic culture of *Chlorella pyrenoidosa* using sucrose as the sole carbon source by co-culture with immobilized yeast. *Bioresource Technology*, 249, 425–430.

White R.L., Ryan R.A. (2015) Long-term cultivation of algae in open-raceway ponds: Lessons from the field. *Industrial Biotechnology*, 11(4), 213–220.

Xiong W., Li X., Xiang J., et al. (2008) High-density fermentation of microalga *Chlorella protothecoides* in bioreactor for microbiodiesel production. *Applied Microbiology Biotechnology*, 78, 29–36.

Xu Z. (2007) Biological production of hydrogen from renewable resources. In: Yang S.T. (Ed.), *Bioprocessing for Value-added Products from Renewable Resources*, (527–558). Amsterdam, Netherlands: Elsevier.

Yan C., Zhang Q., Xue S., et al. (2016) A novel low-cost thin-film flat plate photobioreactor for microalgae cultivation. *Biotechnology and Bioprocess Engineering*, 21(1), 103–109.

Yang L., Chen J., Qin S., et al. (2008) Growth and lipid accumulation by different nutrients in the microalga *Chlamydomonas reinhardtii*. *Biotechnology for Biofuels and Bioproducts*, 11, 40.

Yoo C., Jun S.Y., Lee J.Y., et al. (2010) Selection of microalgae for lipid production under high levels of carbon dioxide. *Bioresource Technology*, 101, 71–74.

Zhu L., Nugroho Y.K., Shakeel S.R. et al. (2017) Using microalgae to produce liquid transportation biodiesel: What is next? *Renewable and Sustainable Energy Reviews*, 78, 391–400.

Zienkiewicz A., Zienkiewicz K., Poliner E., et al. (2020) The microalga *Nannochloropsis* during transition from quiescence to qutotrophy in response to nitrogen availability. *Plant Physiology*, 82, 819–839.

6 Microalgae Cultivation in Industrial Wastewater

Pınar Akdoğan Şirin

ABBREVIATIONS LIST

g	gram
L	liter
mg	milligram
mL	milliliter
d	day

6.1 INTRODUCTION

Microalgae, including prokaryotic cyanobacteria and eukaryotic protists, are single or multicellular microscopic organisms. Interest in microalgae is increasing every day due to their rich protein, carbohydrate, lipid, fatty acid, and pigment content. Today, microalgae species are used in many fields, such as food, energy, cosmetics, wastewater treatment, health, and agriculture. Since microalgae species have different biochemical contents, their uses vary from species to species.

The global demand for freshwater is estimated to increase by 22%–34% by 2050 due to industrial development, population growth, economic development, and socio-economic factors (Boretti and Rosa, 2019; Mao et al., 2021). Industrial wastewater causes many adverse events in the ecosystem, such as climate change, melting glaciers, groundwater depletion, and ozone depletion (Brar et al., 2017). Although the industries that cause wastewater are diverse, the main ones are power plants, metal processing plants, mining industry, steel/iron, textiles, and leather, food, pulp and paper production facilities, industrial laundries, and oil and gas fracturing facilities (Ahmed, Thakur, and Goyal, 2021). Textile, paper, and dyeing industries are the leading industries that cause negative impacts on environmental sustainability and human health in terms of toxic heavy metals, phenolic organic compounds, and other organic pollutants (Fu and Wang, 2011). Metals and heavy metals, asbestos, paint wastes, phenol-containing wastes, pharmaceutical industry wastes, halide solvents, chlorine, sulfur-containing wastes, organic peroxides, polychlorinated biphenyl, pesticides, refinery wastes, and cyanide are the most hazardous substances in industrial wastes (Yıldız Töre and Ata, 2019). In order to treat the various pollutants generated by different industries, a method suitable for each sector should be applied.

Industrial wastewater is generally classified into two categories: inorganic and organic. Inorganic industrial wastewater is mainly produced in the coal and steel industry, in producing non-metallic minerals, and in the surface treatment of metals.

DOI: 10.1201/9781003390213-6

These wastewaters usually contain iron or aluminum salts. Organic industrial wastewater is generated by pharmaceuticals, organic dyestuffs, soaps, synthetic detergents, pesticides, herbicides, cosmetics, tanneries, leather and textile mills, pulp and paper production plants, petroleum refineries, breweries, fermentation plants, and metal processing industries (Hanchang, 2009).

Considering the damages, industrial wastewater should be treated using economical and easily applicable methods. To date, there have been many studies on the treatment of industrial wastewater, depending on the source and content of the waste. Physical, chemical, and biological treatment technologies have been employed in these studies. Physical treatment technologies generally include adsorption, membrane filtration, centrifugal separation, and gravity separation to separate the suspended soluble or insoluble pollutants in the wastewater. Chemical treatment, which includes technologies such as coagulation and flocculation, oxidation and reduction, and electrodialysis, removes dissolved or colloidal pollutants in wastewater by chemical reactions. Biological treatment is the technology that uses microorganisms to decompose dissolved organic matter and reduce or oxidize it to inorganic substances (Mao et al., 2021). Cheaper and environmentally friendly alternative methods, such as bioremediation, have been developed to reduce the cost of chemical, physical, and biological treatment technologies (Malik, 2004; Fikirdeşici-Ergen et al., 2018).

Microalgae play an essential role in biological treatment systems as they have various culture models (phototrophic, mixotrophic, and heterotrophic) (Wollmann et al., 2019). The remediation of wastewater using microalgae cells is called phytoremediation. Through phytoremediation, pollutants (excess nutrients, heavy metals, and acids) in wastewater are removed from the water environment using microalgae cells.

The use of microalgae in different commercial fields, including wastewater treatment, dates back to the 1950s (Li et al., 2019). To date, many microalgae species (*Chlorella vulgaris*, *Scenedesmus* sp., *Chlamydomonas* sp., *Arthrospira platensis*, *Porphyridium purpureum*, and *Botryococcus* sp.) have been used to treat wastewater from different industries (Latiffi et al., 2016; Lin, Nguyen, and Lay, 2017; Arashiro et al., 2020; Behl et al., 2020; Fazal et al., 2021). Microalgae are preferred in wastewater treatment because they can utilize nutrients such as nitrogen and phosphorus in industrial or domestic wastewater (Darvehei, Bahri, and Moheimani, 2018). Using microalgae in treatment reduces the amount of carbon dioxide and removes nitrogen and phosphorus from wastewater. Waste nutrients are assimilated for biomass production, resulting in reduced chemical and biological oxygen demand (Bich, Yaziz, and Bakti, 1999). While water resources are cleaned using microalgae, the resulting microalgae biomass enables the production of economically valuable products (bioplastics, biodiesel) (Lee, Lee, and Sim, 2023). In addition to their ability to use nitrogen and phosphorus in the environment, microalgae can also remove heavy metals from wastewater, which are dangerous for human health.

Toxic compounds contained in different wastewater sources can be removed by microalgae cells through different processes such as biosorption, bioaccumulation, and biodegradation. Biosorption is an environmentally friendly physicochemical process involving the binding of metal ions from industrial wastewater to the surface of a biosorbent (algae, fungi, bacteria, and biopolymer) (Veglio' and Beolchini, 1997;

Shrestha et al., 2021). Biosorption is a passive and energy-free method involving processes such as ion exchange, absorption, adsorption, electrostatic interaction, and precipitation. The binding sites (chemical groups such as carboxyl, hydroxyl, and sulfate) in the cell walls of microalgae enable the adsorption of organic substances from wastewater. Microalgae are very suitable for biosorption due to their high surface/volume ratio. Bioaccumulation is an active metabolic process that requires energy. Bioaccumulation is considerably slower than biosorption, and various substrates in the cell lumen are used during this process. In bioaccumulation, living cells can take organic and inorganic pollutants into the cell to remove them (Abdelfattah et al., 2023). During the bioremediation process, the aquatic environment's chemical structure should be adjusted according to the needs of the microalgae species to be used. Biodegradation, which takes place inside or outside the cell, is an active process that transforms complex compounds in wastewater into simple chemical building blocks. Biodegradation consists of metabolic degradation involving microalgal cells and cometabolism involving non-living substances. In cometabolism, nutrients are added to the aquatic environment to facilitate microalgae cells to biodegrading waste materials.

Many microalgae species and methods have been used to treat wastewater from various industries, such as textile, food, metal plating, aquaculture, paper and pulp, chemical, pharmaceutical, and petroleum factories (Figure 6.1). While biosorption mechanisms, in which living and non-living microalgae species are preferred for removing heavy metals, bioaccumulation and adsorption mechanisms, and biodegradation, in which toxic substances are broken down and rendered harmless, are preferred in the treatment of wastewater from the pharmaceutical industry. In the studies conducted to date, it has been reported that the treatment properties of microalgae have been mostly investigated in domestic wastewater, and their treatment properties in industrial wastewater have been less investigated (Maurya et al., 2022).

This chapter examines the ability of microalgae species to clean wastewater from industrial production areas with different characteristics. In addition, information on the factors affecting microalgae cultures and their nutritional modes is also provided.

6.2　INDUSTRIAL WASTEWATER SOURCES

The main industries that cause pollution of water resources are textile, food and dairy, metal coating, aquaculture, paper and pulp, chemical, pharmaceutical, petroleum and petrochemical, distillery, sugar, mining, battery manufacturing, nuclear power, iron and steel, soap and detergent, electric power plant, pesticide, and biocide (Xiong, Kurade and Jeon, 2018; Cardoso et al., 2021; Dutta, Arya and Kumar, 2021). Wastewater generated by various industrial activity areas has different features and contents. The main heavy metals found in industrial wastewater are aluminum (Al), arsenic (As), cadmium (Cd), chromium (Cr), copper (Cu), mercury (Hg), lead (Pb), nickel (Ni), zinc (Zn), cobalt (Co), iron (Fe), and manganese (Mn) (Lokhande, Singare and Pimple, 2011; Bielen et al., 2017; Tran et al., 2017). Industrial wastewaters cause significant pollution of aquatic and terrestrial ecosystems due to the heavy metals (copper, lead, nickel, zinc, mercury, arsenic, chromium, cadmium, etc.) they contain (Shrestha et al., 2021).

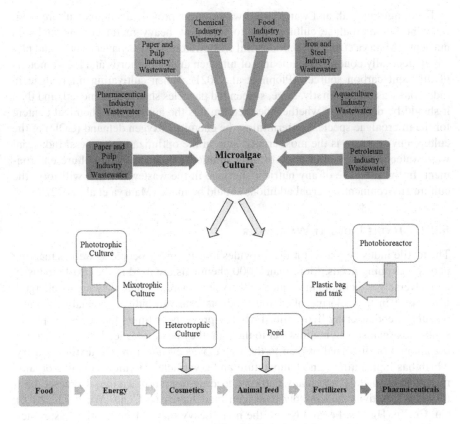

FIGURE 6.1 A schema for microalgae cultivation in industrial wastewater.

Highly toxic heavy metals such as copper, nickel, zinc, and lead can leach into aquatic environments through industrial wastewater and are absorbed by living microorganisms. Thus, these toxic substances are mixed into the food chain and adversely affect human health (Babel and Kurniawan, 2003). Arsenic, which ranks first in the Agency for Toxic Substances and Disease Registry's (ATSDR) ranking, originates from the wastewater of industries such as organic chemistry, fertilizers, petroleum refining, steelmaking, pharmaceuticals, pesticides, fungicides, and metal smelters. Arsenic pollution causes cancer, cardiovascular, endocrine, nervous, respiratory, immune, and fertility diseases in humans (Mohammed Abdul et al., 2015). Lead, which ranks second in the ATSDR ranking, arises from wastewater from paper mills, fertilizers, oil refineries, steel production, pharmaceuticals, dyes, and chemical industries and causes diseases in humans in a wide range of areas, from respiratory, nervous, digestive, and cardiovascular systems to kidneys, teeth, and bones (Boskabady et al., 2018). Mercury, which ranks third in the ATSDR ranking, arises from the wastewater of the chemical, sugar, textile, and plastic production industries and causes neurological and behavioral disorders, rheumatoid arthritis, circulatory and nervous system diseases, and kidney disorders (Tran et al. 2017).

Food, agricultural, and water-based wastewaters provide a suitable culture environment for microalgae cultures due to their low heavy metal content and rich nutrients (Maurya et al., 2022). Industrial wastewaters, such as paper, fabric, and plywood, generally contain low amounts of nitrogen and phosphorus and high amounts of different carbon sources (Plöhn et al., 2021). When cultivating microalgae in industrial wastewater, firstly, large, suspended particles should be removed, and then it should be determined whether the medium has the appropriate chemical content for the microalgae species. Determining the chemical oxygen demand (COD) of the culture environment is the most crucial step in the optimization process. Industrial wastewater can be used directly or after dilution as a microalgae culture environment. In the presence of any nutrient shortage in the wastewater that will form the culture environment, external additions should be made (Maurya et al., 2022).

6.2.1 TEXTILE INDUSTRY WASTEWATER

The textile industry is a sector that provides jobs for many people as well as meeting people's clothing needs. More than 2,000 chemicals are used in the textile industry and adversely affect flora and fauna in the water sources where they are discharged. Dyes used in the coloring of textile products cause turbidity, especially in water resources, endangering living life. Turbidity prevents or limits the access of photosynthetic organisms living in water to the light they need. This leads to a decrease in the amount of dissolved oxygen in the water. Wastewater from the textile industry, which has high salinity and temperature and a variable pH value, contains organic matter, dyes, metals, and many nutrients leading to eutrophication, decreased photosynthetic activity and dissolved oxygen, and increased COD (Wu et al., 2020). Cu, Cr, Zn, Hg, As, Fe, and Ni are the main heavy metals from textile wastewater, which have toxic properties for aquatic organisms and humans. Artificial azo dyes and metabolites cause mutations dangerous for human health (Chu, See and Phang, 2009). Biological methods are preferred for the treatment of dye wastewater, especially due to their complex molecular structure (Lim, Chu and Phang, 2010). Dyes in wastewater are degraded by bioconversion or biosorption processes using microalgae species. Microalgae use nitrogen and phosphorus in textile wastewater as nutrients and remove heavy metals from the environment (El-Kassas and Mohamed, 2014). The high cost required for the production of microalgae in large volumes can be minimized by using textile wastewater as a culture environment (Aslam et al., 2019). At the same time, dye and nutrient pollution are treated by microalgae treatment, while valuable products are obtained from microalgae cells.

In a study using *Chlorella vulgaris* for bioremediation of textile wastewater and biodiesel production at the same time, it was emphasized that diluted and undiluted wastewater could be used in biodiesel production, and undiluted wastewater is more suitable since it eliminates the cost of freshwater by being added to the medium (Fazal et al., 2021). In a study using *Cosmarium* sp. to remove malachite green, a cationic textile dye, it was determined that the microalgae species had the highest decolorization ability at pH 9.0 (Daneshvar et al., 2007). In the phytoremediation study of dry and wet *Chlorella pyrenoidosa* in textile wastewater, it was determined that dry biomass was a better biosorbent for methylene blue dye than wet biomass thanks to its large surface area (Pathak et al., 2015). In the same study, it was determined that the amounts of phosphate, nitrate, and BOD (biochemical oxygen demand) decreased by 87%, 82%, and 63%, respectively, in the medium containing 75% wastewater, and it was emphasized that the biomass obtained could be used as biofuel. While phytoremediation was carried out with *Chlorella* sp. G23 in textile wastewater, an increase in valuable fatty acids was also achieved (Wu et al., 2017). To remove chromium (VI) from the wastewater of the textile dyeing and leather coating industries, two different *Dunaliella* species were cultured for 72 h in a medium with an initial chromium concentration of 100 mg/L (Dönmez and Aksu, 2002). At the end of the experiment, it was found that microalgae species were biosorbent at 58.3 and 45.5 mg/g chromium (VI).

6.2.2 FOOD INDUSTRY WASTEWATER

Food waste generated by the food industry and residential areas such as restaurants, hotels, and households is caused by spoiled, expired, or spilled food. Although food processing wastewater can be treated with chemical and physical techniques, environmentally friendly biological methods are preferred. Thus, high acid usage and high treatment costs are avoided. Unlike other industrial wastewater, food processing wastewater has characteristics such as oil and a bad odor. Food wastes are significant as they contain vital nutrients such as nitrogen and phosphorus, which are essential for microalgae to thrive (Kavitha et al., 2020). Food processing wastewater is a suitable growth medium for microalgae cells due to its organic content, nutrient-rich nature, and low toxicity. Since there are many different types of food processing areas, each food production facility has its own unique wastewater content. The COD value was determined as 10–20 g/L, 2–10 g/L, 6–56 g/L, and 2–32.5 g/L in soybean, milk, starch, and beer processing wastewater, respectively (Li et al., 2019). While the wastewater of soybean processing plants has a 40%–60% carbohydrate content, the wastewater of starch processing plants is rich in minerals.

Nostoc sp., *Arthrospira platensis*, and *Porphyridium purpureum* selected for phycobiliprotein production were grown in food industry wastewater (Arashiro et al., 2020). It was determined that COD, inorganic nitrogen, and phosphate amounts decreased. *Scenedesmus dimorphus*, used in wastewater treatment from breweries, removed more than 99% of the nitrogen and phosphorus in the environment, while the COD decreased by 65% (Lutzu, Zhang and Liu, 2016). It was determined that the microalgae biomass obtained contains high amounts of lipids and is highly suitable for biodiesel. *Chlorella pyrenoidosa* species grown in soybean processing wastewater were found to remove 77.8% of COD, 88.8% of total nitrogen, 89.1% of ammonium, and 70.3% of total phosphorus at the end of 120 h (Hongyang et al., 2011). At the same time, *C. pyrenoidosa* achieved an average biomass productivity of 0.64 g/L/day, an average lipid content of 37%, and a high lipid productivity of 0.40 g/L/day, providing both wastewater cleaning and economically valuable microalgal biomass production. The removal of ammonia and phosphorus by growing *Chlorella vulgaris* and *Scenedesmus dimorphus* species in bioreactors with agro-industrial wastewater from the dairy industry and pig farming was studied by González et al. (1997). In the cylindrical bioreactor, *S. dimorphus* was more effective in ammonia removal, while both microalgae species showed similar characteristics in phosphorus removal. On the other hand, *C. vulgaris* grown in a triangular bioreactor was more successful in ammonia removal. The wastewater caused by sago starch production was first treated by anaerobic fermentation. Then the growth of *Spirulina platensis* was observed in this environment (Phang et al., 2000). COD was reduced by 98%, ammonia-nitrogen by 99.9%, and phosphate by 99.4% in the wastewater. A phytoremediation study was conducted with *Chlamydomonas polypyrenoideum* to treat dairy wastewater (Kothari et al., 2013). In *C. polypyrenoideum* culture, nitrate (90%), nitrite (74%), phosphate (70%), chloride (61%), fluoride (58%), and ammonia (90%) pollution loads decreased on the tenth day, and more crude oil was obtained compared to the control group.

6.2.3 Aquaculture Industry Wastewater

The aquaculture production industry is increasing day by day. In recirculating aquaculture systems, which are preferred for fish farming, some wastewater is treated and reused. The use of these systems not only reduces pollution from fish farming but also saves the amount of water used for aquaculture. However, nutrients such as phosphorus and suspended solids in these systems need to be treated (Tossavainen et al., 2019). It has been determined that aquaculture wastewater has a very suitable environment for microalgae cultivation due to its high nitrogen and phosphorus content. Microalgae species are preferred for feeding aquaculture larvae, molluscs, and crustaceans due to their protein, fatty acids, amino acids, and valuable metabolites. Nutrient-rich wastewater obtained from aquaculture activities is very important in reducing the cost required for the cultivation of microalgae species and evaluating the microalgae biomass obtained in different commercial areas such as biofuel, bio-methane, and fertilizer (Ansari et al., 2017).

   Chaetoceros calcitrans, Nannochloris maculate, and *Tetraselmis chuii* species were grown in wastewater obtained from aquaculture systems, and it was reported that *Nannochloris maculate* increased lipid productivity compared to control conditions (Khatoon et al., 2016). *Spirulina* sp. was cultivated in a closed photobioreactor with aquaculture wastewater or culture media diluted at different ratios (Cardoso et al., 2021). It was determined that *Spirulina* sp. had the highest sulfate (94.01%), phosphate (93.84%), bromine (96.77%), and COD (90.00%) removal rates in the environment containing 25% aquaculture wastewater. The study also suggested that the biomass obtained from the groups containing 25% and 50% wastewater can be used as raw material in biodiesel. It has been determined that mixed microalgae cultures consisting of *Euglena gracilis* and *Selenastrum* sp. species have reduced COD by 45%–67%, total nitrogen by 75%–89%, and total phosphorus by 84%–95% by cultivation in wastewater obtained from aquaculture, and the microalgae biomass obtained can be utilized in eicosapentaenoic acid, docosahexaenoic acid, and tocopherol production (Tossavainen et al., 2019). COD removal rates of *Scenedesmus obliquus, Chlorella sorokiniana*, and *Ankistrodesmus falcatus* species cultivated in wastewater obtained from aquaculture activities were 42%, 69%, and 61%, respectively. *Ankistrodesmus falcatus* provided the highest removal of NO_3^- –N (80%), NO_2^- –N (99%), and total oxidizable nitrogen (75%). *S. obliquus* and *C. sorokiniana* achieved ~100% PO_4^{3-}−P removal from wastewater. The biomass obtained from wastewater treatment with these three microalgae species was used for biofuel and nutrient production. In a study using *Platymonas subcordiformis*, 87%–95% of nitrogen and 82% of phosphorus were removed from wastewater (Gong et al., 2013). In the study where the biomass obtained was used in protein, carbohydrate, and lipid production, *Chlorella vulgaris* was preferred for treating paper-pulp wastewater and aquaculture wastewater. This study conducted by Daneshvar et al. (2018) determined that total nitrogen, total phosphorus, COD, and total organic carbon amounts were removed by 76%, 92%, 75%, and 70%, respectively.

6.2.4 PAPER AND PULP INDUSTRY WASTEWATER

During paper and pulp production processes, a large amount of water is required, and liquid and solid wastes are generated as a result of production (Patel, Mehta and Solanki, 2017). This wastewater is rich in Cd, Cr, Cu, Hg, Pb, Ni, Zn, and Mn heavy metals (Tran et al. 2017). In addition, the color pollution caused by the dyes in the waste creates a thin film on the water layer and causes a decrease in photosynthetic activity. Studies have determined that secondary treatment is insufficient to clean wastewater, which poses a danger to living organisms in the aquatic environment (Sharma et al., 2021). Wastewater from the paper and pulp industry, which contains high amounts of COD, BOD, lignin, sulfur compounds, tannins, and acids, as well as a high phosphorus content, must be treated before discharge. The preference for microalgae species in treating paper and pulp wastewater using physicochemical and biological methods provides both environmental and economic advantages. These advantages include low-cost treatment of wastewater, the use of microalgae biomass as raw material in various sectors, and low-cost microalgae production (Bhatti, Richards and McGinn, 2021).

The study with *Chlorella vulgaris* and *Dictyosphaerium* sp. suggested that pulp and paper mill wastewater could not meet the nitrogen and phosphorus required for microalgae species' growth; mixing with municipal or dairy wastewater may be beneficial. The same study also revealed that wastewaters contain chemicals such as sodium, peroxide, and tannins that inhibit microalgae growth (Bhatti, Richards and McGinn, 2021). Silva et al. (2021) aimed to improve water quality and produce microalgae biomass that can be used for bioenergy by cultivating *C. vulgaris* in paper industry wastewater. As a result of the study, it was determined that while the biomass productivity of *C. vulgaris* increased, the phosphorus content in the environment decreased. *Chlorella vulgaris* was used to remove phosphorus from the secondary treatment water of a Portuguese paper company. In the experiment designed as batch cultures, microalgae cells were grown in wastewater with different dilutions for 11 days. It was found that undiluted wastewater inhibited microalgae cells, while the removal efficiency in diluted wastewater was 54% (Porto et al., 2020). In a study in which *Planktochlorella nurekis* and *Chlamydomonas reinhardtii* species were grown in paper mill wastewater, it was determined that *P. nurekis* provided higher rates of nitrate, phosphate, sulfate, and COD removal than *C. reinhardtii*. The biomass obtained was used in lipid production (Sasi et al., 2020).

6.2.5 PHARMACEUTICAL INDUSTRY WASTEWATER

The treatment of wastewater from the rapidly growing pharmaceutical industry, which is frequently used in areas such as human health, agriculture, and animal husbandry, is important for the sustainability of the ecosystem. Pharmaceutical industry wastewater poses a threat to drinking water and human health as it contains carcinogenic, mutagenic, and other harmful compounds, as well as increasing antibiotic resistance in bacteria and causing differentiation in the endocrine system. The content of wastewater from this sector, where different pharmaceuticals and cosmetic materials are produced, ranges from antibiotics, hormones, polyaromatic hydrocarbons, and phenols (de Jesus Oliveira Santos, Oliveira de Souza and Marcelino, 2023). The mixotrophic growth of microalgae is beneficial for removing pollutants from the pharmaceutical industry from the water environment. Bacteria, fungi, or microalgae cells use mechanisms such as biosorption, bioaccumulation, biodegradation, photo-degradation, and volatilization to remove harmful substances in pharmaceutical industry wastewater from the aquatic environment (Hena, Gutierrez and Croué, 2020).

In a study investigating the removal rates of levofloxacin contaminants in *Chlamydomonas mexicana*, *Chlamydomonas pitschmannii*, *Chlorella vulgaris*, *Ourococcus multisporus*, *Micractinium resseri*, and *Tribonema aequale*, it was determined that *C. vulgaris* had the highest removal capacity of levofloxacin with 12% (Xiong, Kurade and Jeon, 2017). In the same study, it was reported that 1% NaCl added to the medium increased the removal of levofloxacin, and this was reported to be due to simultaneous bioaccumulation and biodegradation within the cell. In a study investigating the removal of florfenicol using two different *Chlorella* sp., it was concluded that 97% of florfenicol could be removed by biodegradation (Song et al., 2019). In a study investigating the removal rates of trimethoprim, sulfamethoxazole, carbamazepine, ciprofloxacin, and triclosan wastes from the water environment with *Nannochloris* sp., it was determined that trimethoprim and carbamazepine could not be degraded by the microalgae cell, while sulfamethoxazole was 40%, ciprofloxacin, and triclosan were 100% removed from the water environment (Bai and Acharya, 2017). In a study investigating the removal of four different steroid hormones (17α-estradiol, 17β-estradiol, estrone, and estriol) using *Scenedesmus dimorphus*, 17α-estradiol and estrone were removed by 85% and 17β-estradiol and estriol were removed by 95% as a result of 8-day culture (Zhang et al., 2014). In a 20-day study on metronidazole treatment with *Chlorella vulgaris*, the microalgae cells removed 100% of metronidazole from the environment using the adsorption mechanism (Hena, Gutierrez and Croué, 2020). In addition, in the study, the removal rate of *C. vulgaris* decreased with the increase in the amount of antibiotics.

6.2.6 HEAVY METAL EFFLUENTS FROM VARIOUS INDUSTRIAL WASTEWATERS

The increase in the industrial sector as a result of industrialization and urbanization has led to an increase in heavy metal waste. Heavy metals, which are not biodegradable, originate from various industrial wastewater and accumulate in the aquatic environment, causing many diseases ranging from respiratory diseases to cancer in humans. Among heavy metals, lead, mercury, and titanium cause brain damage, while tin, lead, mercury, magnesium, argon, and chromium slow down the breathing rate of the parasympathetic system. To treat heavy metals with toxic properties, methods such as membrane filtration, electrochemical methods, chemical precipitation, and ion exchange, which require considerable cost, can be used. Microalgae, fungi, and bacteria can be used for bioremediation, an environmentally friendly method for treating heavy metals. Living and non-living microalgal biomass can be used for the phytoremediation of heavy metals.

Steel industry wastewater contains high amounts of heavy metals (Blanco-Vieites et al., 2022). These heavy metals accumulate in the food chain, increasing their concentration and causing significant toxicity to higher-level organisms (Arora et al., 2008). Blanco-Vieites et al. (2022) emphasized that steel wastewater is difficult to remediate by conventional methods due to its high iron and hydrocarbon content. The same researchers stated that especially iron in steel industry wastewater was used as a nutrient by *Arthrospira maxima* and reduced the iron content in the wastewater by 97.5%, which resulted in an increase in microalgae biomass. At the same time, the total hydrocarbons in the wastewater decreased by 75%. It was determined that *Chlorella variabilis* provided 85% iron and 60% zinc removal in steel industry wastewater, while a 30% reduction in COD amount occurred (Hasanoğlu, Kutluk and Kapucu, 2021). The same study suggested that the microalgal biomass obtained can be used as a raw material in biodiesel production. The ability of *Chlamydomonas reinhardtii, Chlorella vulgaris, Scenedesmus almeriensis,* and *Chlorophyceae* spp. to remove arsenic, copper, boron, manganese, and zinc toxic elements from the aquatic environment was investigated by Saavedra et al. (2018). *C. vulgaris* removed 99% manganese, *Chlorophyceae* spp. removed 91% zinc and 88% copper, and *S. almeriensis* removed 40% arsenic and 38% boron. Battery and accumulator production, cadmium mining, cadmium electroplating, and ceramic industry wastewater contain high amounts of cadmium (II), and *Chlorella vulgaris* has a high biosorption capacity for cadmium (II) ions (Aksu, 2001). Petrochemical plants that produce chemicals such as hydrogen, carbon monoxide, benzene, toluene, and ethylene are rich in nitrogen, sulfur, cyanide, heavy metals, suspended particles, phenolic compounds, aliphatic compounds, and polycyclic aromatic hydrocarbons (Asatekin and Mayes, 2009). The use of microalgae in the treatment of petrochemical wastewater provides lipid production that can be used in bioenergy. Thus, the amount of heavy metals and other pollutants in wastewater is reduced (Santos et al., 2018). In the bioremediation process with *Chlorella vulgaris* in Iranian petrochemical wastewater, BOD and total phosphorus removal were 100%, COD removal was 38%, and total petroleum hydrocarbon removal was 27% in the groups using surfactants (Madadi et al., 2016).

6.3 MICROALGAE CULTURE

6.3.1 Factors Affecting Microalgae Cultures

Microalgae species' growth and biochemical structure vary depending on their environment's physical and chemical properties. Therefore, wastewater content, feeding mode, carbon dioxide content, light intensity, pH, temperature, C/N ratio, and N/P ratio are important in purifying industrial wastewater with microalgae (Hu, 2013).

6.3.1.1   Nutrients

Microalgae species need different nutrients for their growth. Nitrogen and phosphorus are the most important of these nutrients. In laboratory experiments, artificial culture conditions are created for the isolated species, and the cells can multiply this way. The cultural conditions of species isolated from marine and fresh waters differ. The presence of more or less of any nutrient element in the culture environment compared to optimal conditions causes a change in the microalgae species' biochemical structure. Changes in nutrient elements cause an increase or decrease in the protein, carbohydrate, lipid, fatty acid, and pigment substances contained in microalgae cells. In addition, the specific growth rates of microalgae cells are also affected by stress conditions. Industrial wastewaters differ considerably from each other in terms of their nutrient content. While purifying wastewater with microalgae species, the cost should be minimized simultaneously. Therefore, the nutrient composition of the wastewater to be purified should be analyzed in detail, and the microalgae species to be used should be adapted to the nutrient requirements. Some industrial wastewaters are insufficient in nitrogen, phosphorus, and micronutrients required for microalgae growth, and nutrients such as sodium bicarbonate, sodium acetate, carbon dioxide, nitrogen, and phosphorus should be added to the culture environment (Maurya et al., 2022). Microalgae species can be grown entirely in industrial wastewater; in some cases, wastewater can be used by diluting it in certain proportions.

6.3.1.2   pH

Another issue to be considered when creating a culture environment is the pH value of the medium. While some microalgae species can survive at high pH values, others prefer low pH values. The pH value plays a major role in the microalgal surface's functional group charges and binding abilities (Plöhn et al., 2021). Therefore, the pH value of the wastewater in which microalgae will be grown should be calibrated according to the species to be cultured and measured at regular intervals.

6.3.1.3   Light

Light's amount, duration, and color are significant in microalgae cultures as they directly impact photosynthesis and growth. The light-dark photoperiod applied to microalgae cultures affects photosynthetic efficiency. In addition, cells need light for the production of adenosine triphosphate and nicotinamide adenine dinucleotide phosphate-oxidase, while they need darkness for the synthesis of molecules involved in growth (Cheirsilp and Torpee, 2012). The amount of light absorption, which affects photosynthetic efficiency, is closely related to the turbidity and color of wastewater. For example, the dark brown-grayish color of paper and pulp wastewater reduces light absorption and inhibits the growth of microalgae cells (Lee and Lee, 2001). Since the high turbidity of wastewater and highly suspended solids reduce the amount of photosynthesis, methods such as sedimentation, adsorption, and coagulation can be used to remove solids (Amenorfenyo et al., 2019). Another issue to be considered during microalgae culture is that the cells proliferate excessively and shade each other. In such a case, the optimal amount of light cannot be provided to the cells, and a decrease in growth rate occurs (Larsdotter, 2006). When microalgae cells are exposed to very intense light, the photosynthetic apparatus is damaged, the photosynthetic rate decreases, and a condition called photoinhibition occurs (Cirik and Gökpınar, 2008).

6.3.1.4 Temperature

Temperature is one of the most important physical factors affecting biochemical reactions in microalgal cells; temperature change directly affects the cells' biochemical structure (Hu, 2013). The temperature requirements, which vary according to the microalgae species, have a negative effect on enzymatic activities when they exceed a certain level, as in the light factor. If the temperature increase continues, some proteins in the photosystem and electron transport chain denature and cause the cell to die (Serra-Maia et al., 2016). Therefore, attention should be paid to changes in the temperature factor, especially in the purification of wastewater in the outdoor environment.

6.3.2 NUTRITIONAL MODES OF MICROALGAE

Microalgae species' growth and biochemical content are closely related to the content and physical conditions of the culture environment. Microalgae species have three main nutritional modes: photoautotrophic, heterotrophic, and mixotrophic (Pagnanelli et al., 2014). Each culture mode provides different nutrients and energy sources for the development of the species. Microalgae species may prefer different cultural modes depending on environmental conditions (Mata, Martins and Caetano, 2010). Microalgae species produced in photoautotrophic mode have low biomass, and production costs increase accordingly (Perez-Garcia et al., 2015). Therefore, more biomass is produced by benefiting from the ability of microalgae to use organic carbon sources (Abreu et al., 2022).

6.3.2.1 Photoautotrophic Cultivation

In photoautotrophic cultures, the oldest and most well-known method used in microalgae cultivation, light is used as an energy source, while carbon dioxide, an inorganic carbon source, is used as chemical energy through photosynthesis (Chen et al., 2011; Chew et al., 2018). The microalgal biomass produced under these conditions increases up to a certain point in carbon dioxide-rich environments (Cirik and Gökpınar, 2008).

6.3.2.2 Heterotrophic Cultivation

Heterotrophic growth is a culture mode that produces algal biomass faster and at higher densities than autotrophic growth. Although high amounts of biomass can be produced, heterotrophic cultures require costly organic carbon sources (Perez-Garcia et al., 2011). In this growth mode, microalgae cells use organic carbon sources such as glucose, acetate, and glycerol as carbon and energy sources. Heterotrophic growth can be considered economically viable because it yields high amounts of microalgal biomass, is suitable for large-scale production, and is easy to maintain (Chen, 1996).

Furfural, a chemical substance, is used to produce many chemicals and is a widely used extractant in the petrochemical industry. Furfural wastewater, for which conventional treatment methods are unsuitable, was treated with *Chlorella pyrenoidosa*, which is produced as mixotrophic and heterotrophic (Cheng et al., 2022). It was found that COD and total nitrogen decreased the most in the group where the wastewater was diluted ten times.

Galdieria sulphuraria, which has phycocyanin as its primary pigment (Albertano et al., 2000), was grown heterotrophically in a 500 mL Erlenmeyer on wastes from restaurants and bakeries, and it was determined that the microalgae species could use carbohydrates and amino acids in wastes but needs ammonium and organic nutrients to be added externally for phycocyanin synthesis (Sloth et al., 2017). In the study, where the phycocyanin content was determined at 20–22 mg/g, it was emphasized that food waste could be used to produce valuable products.

The growth rate, organic carbon, and nutrient removal of *Chlorella sorokiniana* grown under phototrophic, mixotrophic, and heterotrophic culture conditions were compared (Kim et al., 2013). Nitrogen removal rates were 13.1, 23.9, and 19.4 mg-N/L/day, while phosphorus removal rates were 3.4, 5.6, and 5.1 mg-P/L/day, respectively. At the end of the study, it was emphasized that heterotrophic nutrient modes could be used in wastewater clarification for microalgae cell development and nitrogen and phosphorus removal.

6.3.2.3 Mixotrophic Cultivation

In photoautotrophic mode, light and inorganic carbon sources are necessary for energy and carbon source growth, while in mixotrophic mode, organic carbon sources are used as both energy and carbon sources. In this mode, where photosynthesis and respiration metabolisms are active, light and inorganic carbon additionally support growth (Patel, Choi and Sim, 2020). Roostaei et al. (2018) reported that microalgae grown under mixotrophic culture conditions produced more biomass than photoautotrophic and heterotrophic cultures. In mixotrophic cultures, bacterial contamination is the most crucial problem. In large-volume pond production, organic carbon sources are continuously added to the culture medium and in small amounts during the day. The reason for this is that bacteria multiply very fast in the dark and cause contamination in the environment (Lee, 2001). In the mixotrophic culture mode, which is unsuitable for cultivating all microalgae species, organic carbon sources such as glucose, glycerol, acetic acid, and acetate are added to the culture medium. *Chlorella vulgaris*, *Chlorella sorokiniana*, *Nannochloropsis salina*, *Haematococcus* sp., *Tetraselmis* sp., *Botryococcus braunii*, *Dunaliella salina*, and *Phaeodactylum tricornutum* are some microalgae species showing mixotrophic characteristics.

Agro-industrial wastes are produced in large quantities worldwide. These wastes can be used as a source of organic carbon in the mixotrophic cultivation of microalgae (Lowrey, Brooks and McGinn, 2015; Laraib et al., 2021). Chong et al. (2021) reported that microalgae increase mixotrophic cultures, especially lipids, and biomass in wastewater caused by the food industry. As a result of using waste soy sauce in *Chlorella sorokiniana* culture, it was found that microalgae biomass increased 1.93 times and fatty acids increased 1.76 times (Lee, Lee and Sim, 2023). Nutrient removal (21–22 mg TN/L), lipid productivity (10–11 mg/L/day), and carbohydrate productivity (13–16 mg/L/day) were determined in a 6-day mixotrophic culture of *Scenedesmus obliquus* with 0.5%–1% diluted food wastewater and 10%–14.1% CO_2 (Ji et al., 2015). In addition, palmitic and oleic acid contents were found to increase in the study, and it was emphasized that the method used was cost-effective.

It is considered that the disadvantage caused by adverse weather conditions and insufficient sunlight levels, especially in northern regions, can be overcome by photoautotrophic-mixotrophic microalgae cultivation in industrial wastewater (Park et al., 2012).

6.4 DESIGN OF MICROALGAE CULTURE TREATMENT IN WASTEWATER

The production of microalgae species used as raw materials in commercial areas such as wastewater treatment, food, energy, cosmetics, pharmaceuticals, fertilizers, and animal feed can be done at different scales, from small to large volumes. Microalgae can be produced in small-volume erlenmeyers or large-volume fermenters, plastic bags, tanks, photobioreactors, and ponds. There are some important issues to be considered in producing microalgae species to be used in wastewater treatment. The most important of these is to analyze the physical and chemical properties of the wastewater to be used in detail. The wastewater to be used must first be purified from large particles. This separation can be achieved by filtration (Pathak et al., 2015), centrifugation (Lin, Nguyen and Lay, 2017), or by using different methods such as acetate ultrafiltration membrane discs and glass filter fiber discs (Lutzu, Zhang and Liu, 2016). After the wastewater is purified from suspended particles, water quality analyses are required, including COD, dissolved oxygen, total phosphate, total nitrogen, and total phosphate. In addition to these analyses, the pH value, which is very important for microalgae cells to maintain their vital activities, should also be measured to ensure that it is within the pH range determined during the trials according to the needs of the species. Sterilized wastewater should be used in culture media to prevent bacterial contamination. The chemical content of the effluents can be revised based on the nutrient requirements of the microalgae species to be cultured. In some cases, it is more appropriate to apply different dilution ratios because the wastewater structure inhibits microalgae growth. For the dilution process, the culture medium suitable for the optimal growth conditions of the microalgae species should be preferred. Phytoremediation is initiated by adding a certain amount of cells from the stock microalgae culture to the wastewater medium. Depending on the volume at which the purification process takes place, the culture medium should be aerated, mixed, and illuminated. The design of microalgae culture trials for purification varies in each study.

In the *Chlorella vulgaris* textile wastewater treatment study, glass photobioreactors with a volume of 1 L were used. The treatment properties of synthetic textile wastewater prepared in aerated photobioreactors illuminated with LED lamps were investigated with microalgae cells (Fazal et al., 2021). In treating textile wastewater with *C. pyrenoidosa* in India, the wastewater was first filtered to remove large particles and sterilized in an autoclave to prevent bacterial contamination (Pathak et al., 2015). In the same study, continuously aerated and illuminated 1 L erlenmeyers were used for the phytoremediation of wastewater. In another study in which foreign particles in textile wastewater were settled to the bottom by centrifugation, the wastewater medium was diluted to make it suitable for the growing conditions of *Scenedesmus* sp., the pH was adjusted according to the species, and batch cultures were carried out in glass tubes (Lin, Nguyen and Lay, 2017). Cylindrical photobioreactors are also preferred for the phytoremediation of wastewater. Phycobiliprotein and biogas production were achieved in the experimental setup in which mixed microalgae species grown under controlled conditions in cylindrical reactors with a total volume of 3 L were continuously mixed and aerated (Arashiro et al., 2020).

Microalgae biomass obtained at the end of the treatment process can be used as raw material in the production of biofuels (biodiesel, bioethanol, biomethane, biohydrogen), lipids, proteins, carbohydrates, valuable fatty acids, phycobiliproteins, animal feed, pigments, eicosapentaenoic acid (EPA), docosahexaenoic acid (DHA), Arachidonic acid (ARA), and tocopherol (Table 6.1).

TABLE 6.1
Microalgae Cultivation Studies in Different Industrial Wastewater

Microalgal Species	Wastewater Origin	Application	Utilization of Microalgal Biomass	References
Chlorella vulgaris	Textile wastewater	Reduction of dye and COD (99%)	Biodiesel production	Fazal et al. (2021)
Chlorella pyrenoidosa	Textile wastewater	Reduction of nitrate (63%), phosphate (82%) and BOD (82%)	Biofuel production	Pathak et al. (2015)
Chlorella sp. G23	Textile wastewater	Removal of COD (75%)	FAMEs accumulation	Wu et al. (2017)
Scenedesmus sp.	Textile desizing wastewater	Removal of color (92.4%), COD (89.5%), carbohydrates (97.4%) and organic acids (94.7%)	Production of biogas and algae biomass	Lin et al. (2017)
Chlamydomonas sp. TRC-1	Textile wastewater	Removal of COD (83%), total nitrogen (87%), phosphate (92%)	Lipid production	Behl et al. (2020)
Mixed microalgae species	Textile wastewater	Removal of COD (70%), total phosphorus (99%), total nitrogen (100%)	–	Kumar et al. (2018)
Nostoc sp. *Arthrospira platensis Porphyridium purpureum*	Food wastewater	Reduction of COD, inorganic nitrogen and phosphate (99%)	Production of phycobiliproteins and biogas	Arashiro et al. (2020)

(Continued)

TABLE 6.1 (*Continued*)
Microalgae Cultivation Studies in Different Industrial Wastewater

Microalgal Species	Wastewater Origin	Application	Utilization of Microalgal Biomass	References
Scenedesmus dimorphus	Beer wastewater	Nitrogen and phosphate removal Reduction of COD (65%)	Biodiesel production	Lutzu et al. (2016)
Chlorella pyrenoidosa	Soya bean processing wastewater	Reduction of COD (77.8%) Nitrogen (88.8%) and phosphate (89.1%) removal	Biodiesel production	Hongyang et al. (2011)
Botryococcus sp.	Meat processing industry wastewater	Removal of BOD (97%), COD (94%)	–	Latiffi et al. (2016)
Neochloris sp.	Poultry slaughterhouse wastewater	Removal of COD (96%), nitrite (95%), and phosphate (79%)	Dietary supplements, cosmetics, nutraceuticals, biofuel production	Ummalyma et al. (2023)
Chlorella pyrenoidosa	Anaerobic food processing wastewater	Removal of COD (42%–53%), total nitrogen (82%–88%), total phosphate (59%–67%)	Biodiesel production	Tan et al. (2021)
S. obliquus + *C. vulgaris* + *C. sorokiniana*	Meat processing wastewater	Removal COD (91%), total nitrogen (67%), total phosphate (69%)	–	Hu et al. (2019)
Scenedesmus sp. HXY5	Potato wastewater	Removal of total dissolved nitrogen (59%), total dissolved phosphorus (32%), COD (93%)	Produce pigments (especially lutein)	Yuan et al. (2021)
Chlorella variabilis	Steel industry wastewater	Removal of COD, Fe metal (85%), Zn metal (60%)	Biodiesel production	Sümeyye Hasanoğlu et al. (2021)
Neochloris sp. SK57	River water contaminated with pharmaceutical effluent	Removal of COD (90%), BOD (91%)	Food and fuel production	Singh and Ummalyma (2020)
Chlamydomonas polypyrenoideum	Dairy wastewater	Nitrate (90%), phosphate (96%) removal	Biodiesel production	Kothari et al. (2013)

(Continued)

TABLE 6.1 (*Continued*)
Microalgae Cultivation Studies in Different Industrial Wastewater

Microalgal Species	Wastewater Origin	Application	Utilization of Microalgal Biomass	References
Spriulina sp.	Aquaculture wastewater	Reduction of COD (90%), sulfate (94.01%), phosphate (93.84%), bromine (96.77%)	Biodiesel production	Cardoso et al. (2021)
Euglena gracilis+ Selenastrum sp.	Aquaculture wastewater	Reduction of COD (45%–67%), total nitrogen (75%–89%), total phosphorus (84%–95%)	EPA, DHA, ARA and tocopherol production	Tossavainen et al. (2019)
Platymonas subcordiformis	Aquaculture wastewater	Removal of nitrogen (87%–95%), phosphorus (98%–99%)	Biomass and biofuel production	Guo et al. (2013)
Chlorella vulgaris	Aquaculture wastewater	Removal of total nitrogen (86%), total phosphorus (82%)	–	Gao et al. (2016)
Chlorella vulgaris	Aquaculture and pulp wastewater	Removal of total nitrogen (76%), total phosphorus (92%), COD (75%) and TOC (70%)	Protein, carbohydrate and lipid production	Daneshvar et al. (2018)
Scenedesmus obliquus	Aquaculture wastewater	Removal of COD (42%), NO_3^- - N (77%), NO_2^- - N (73%), total oxidizable nitrogen (68%), NH_4^+ - N (88%), PO_4^{3-} - P (~100%)	Biofuels and feed production	Ansari et al. (2017)
Chlorella sorokiniana		Removal of COD (69%), NO_3^- - N (75%), NO_2^- - N (81%), total oxidizable nitrogen (67%), NH_4^+ - N (98%), PO_4^{3-} - P (~100%)		
Ankistrodesmus falcatus		Removal of COD (61%), NO_3^- - N (80%), NO_2^- - N (99%) total oxidizable nitrogen (75%), NH_4^+ - N (86%), PO_4^{3-} - P (98%)		
Chlorella vulgaris	Paper industry wastewater	Phosphorus removal	Bioenergy production	Silva et al. (2021)

(Continued)

TABLE 6.1 *(Continued)*
Microalgae Cultivation Studies in Different Industrial Wastewater

Microalgal Species	Wastewater Origin	Application	Utilization of Microalgal Biomass	References
Chlorella vulgaris	Paper industry wastewater	Phosphorus removal (54%), (nitrate + nitrite) removal (80%)	–	Porto et al. (2020)
Planktochlorella nurekis	Pulp and paper mill industry wastewater	Removal of nitrate (95%), phosphate (100%), sulfate (80%), COD (92%)	Lipid production (22%)	Sasi et al. (2020)
Chlamydomonas reinhardtii		Removal of nitrate (85%), phosphate (87%), sulfate (60%), COD (91%)	Lipid production (20%)	
Chlorella vulgaris	Pharmaceutical industry wastewater	Removal of levofloxacin 9.5% with 0% NaCl; 91.5% with 1% NaCl	–	Xiong et al. (2017)
Chlorella vulgaris	Pharmaceutical industry wastewater	Removal of florfenicol (97%)	–	Song et al. (2019)
Nannochloris sp.	Pharmaceutical industry wastewater	Removal of sulfamethoxazole (40%), ciprofloxacin (100%), and triclosan (100%)	–	Bai and Acharya (2017)
Scenedesmus dimorphus	Pharmaceutical industry wastewater	Removal of 17α-estradiol and estrone (85%), 17β-estradiol and estriol (95%)	–	Zhang et al. (2014)
Chlorella vulgaris	Pharmaceutical industry wastewater	Removal of metronidazole (100%)	–	Hena et al. (2020)
Arthrospira maxima	Steel industry wastewater	Removal of iron (97.5%), hydrocarbon (75%)	–	Blanco-Vieites et al. (2022)
Chlorella variabilis	Steel industry wastewater	Removal of iron (85%), zinc (60%), COD (30%)	Biodiesel	Sümeyye Hasanoğlu et al. (2021)
Chlorella vulgaris	Manganese	Removal of manganese (99%)	–	Saavedra et al. (2018)
Chlorophyceae spp.	Zinc, copper	Removal of zinc (91%), copper (88%)		
Scenedesmus almeriensis	Arsenic, boron	Removal of arsenic (40%), boron (38%)		
Chlorella vulgaris	Cadmium	Removal of cadmium	–	Aksu (2001)

6.5 CONCLUSION

Microalgae are used in many industrial areas due to their biochemical content, rapid reproduction, resistance to extreme conditions, and easy production processes. In particular, they are widely used in many sectors, such as food, energy, aquaculture, wastewater treatment, and cosmetics.

As a result of the rapid increase in the world population and industrial activities, the cleaning of wastewater with microalgae, which is an environmentally friendly method, is very valuable for our future. Previous studies have shown that excess nutrient input and heavy metals caused by industrial wastewater can be removed from the wastewater environment with microalgae cells. Microalgae biomass, which was obtained from wastewater treatment processes, is used as a raw material in many areas, such as biodiesel, fertilizer, and pigment production. Treatment of industrial wastewater with microalgae cells is important for the production of high value products, a sustainable environment, and reducing the cost of wastewater treatment.

REFERENCES

Abdelfattah, A. et al. (2023) 'Microalgae-based wastewater treatment: Mechanisms, challenges, recent advances, and future prospects', Environmental Science and Ecotechnology. Available at: https://doi.org/10.1016/j.ese.2022.100205.

Abreu, A.P. et al. (2022) 'A comparison between microalgal autotrophic growth and metabolite accumulation with heterotrophic, mixotrophic and photoheterotrophic cultivation modes', Renewable and Sustainable Energy Reviews, 159, p. 112247. Available at: https://doi.org/10.1016/j.rser.2022.112247.

Ahmed, J., Thakur, A. and Goyal, A. (2021) 'Industrial wastewater and its toxic effects'. In: Shah, M.P. (ed.), Biological Treatment of Industrial Wastewater. London: The Royal Society of Chemistry, pp. 1–14. Available at: https://doi.org/10.1039/9781839165399-00001.

Aksu, Z. (2001) 'Equilibrium and kinetic modelling of cadmium (II) biosorption by C. vulgaris in a batch system: Effect of temperature', Separation and Purification Technology, 21(3), pp. 285–294. Available at: https://doi.org/10.1016/S1383-5866(00)00212-4.

Albertano, P. et al. (2000) 'The taxonomic position of Cyanidium, Cyanidioschyzon and Galdieria: An update', Hydrobiologia, 433, pp. 137–143. Available at: https://doi.org/10.1023/A:1004031123806.

Amenorfenyo, D.K. et al. (2019) 'Microalgae brewery wastewater treatment: Potentials, benefits and the challenges', International Journal of Environmental Research and Public Health, 16(11), p. 1910. Available at: https://doi.org/10.3390/ijerph16111910.

Ansari, F.A. et al. (2017) 'Microalgal cultivation using aquaculture wastewater: Integrated biomass generation and nutrient remediation', Algal Research, 21, pp. 169–177. Available at: https://doi.org/10.1016/j.algal.2016.11.015.

Arashiro, L.T. et al. (2020) 'Natural pigments from microalgae grown in industrial wastewater', Bioresource Technology, 303, p. 122894. Available at: https://doi.org/10.1016/j.biortech.2020.122894.

Arora, M. et al. (2008) 'Heavy metal accumulation in vegetables irrigated with water from different sources', Food Chemistry, 111(4), pp. 811–815. Available at: https://doi.org/10.1016/j.foodchem.2008.04.049.

Asatekin, A. and Mayes, A.M. (2009) 'Oil industry wastewater treatment with fouling resistant membranes containing amphiphilic comb copolymers', Environmental Science and Technology, 43(12), pp. 4487–4492. Available at: https://doi.org/10.1021/es803677k.

Aslam, A. et al. (2019) 'Biorefinery of microalgae for nonfuel products', In: Yousuf, A. (ed.), *Microalgae Cultivation for Biofuels Production.* Cambridge, MA: Academic Press, pp. 197–209. Available at: https://doi.org/10.1016/B978-0-12-817536-1.00013-8.

Babel, S. and Kurniawan, T.A. (2003) 'Low-cost adsorbents for heavy metals uptake from contaminated water: A review', *Journal of Hazardous Materials*, 97, pp. 219–243. Available at: https://doi.org/10.1016/S0304-3894(02)00263-7.

Bai, X. and Acharya, K. (2017) 'Algae-mediated removal of selected pharmaceutical and personal care products (PPCPs) from Lake Mead water', *Science of the Total Environment*, 581–582, pp. 734–740. Available at: https://doi.org/10.1016/j.scitotenv.2016.12.192.

Behl, K. et al. (2020) 'Multifaceted applications of isolated microalgae *Chlamydomonas* sp. TRC-1 in wastewater remediation, lipid production and bioelectricity generation', *Bioresource Technology*, 304, p. 122993. Available at: https://doi.org/10.1016/j.biortech.2020.122993.

Bhatti, S., Richards, R. and McGinn, P. (2021) 'Screening of two freshwater green microalgae in pulp and paper mill wastewater effluents in Nova Scotia, Canada', *Water Science and Technology*, 83(6), pp. 1483–1498. Available at: https://doi.org/10.2166/wst.2021.001.

Bich, N.N., Yaziz, M.I. and Bakti, N.A.K. (1999) 'Combination of *Chlorella vulgaris* and *Eichhornia crassipes* for wastewater nitrogen removal', *Water Research*, 33(10), pp. 2357–2362. Available at: https://doi.org/10.1016/S0043-1354(98)00439-4.

Bielen, A. et al. (2017) 'Negative environmental impacts of antibiotic-contaminated effluents from pharmaceutical industries', *Water Research*, 126, pp. 79–87. Available at: https://doi.org/10.1016/j.watres.2017.09.019.

Blanco-Vieites, M. et al. (2022) 'Removal of heavy metals and hydrocarbons by microalgae from wastewater in the steel industry', *Algal Research*, 64, p. 102700. Available at: https://doi.org/10.1016/j.algal.2022.102700.

Boretti, A. and Rosa, L. (2019) 'Reassessing the projections of the World Water Development Report', *NPJ Clean Water*, 2(1), pp. 1–6. Available at: https://doi.org/10.1038/s41545-019-0039-9.

Boskabady, M. et al. (2018) 'The effect of environmental lead exposure on human health and the contribution of inflammatory mechanisms: a review', *Environment International*, 120, pp. 404–420. Available at: https://doi.org/10.1016/j.envint.2018.08.013.

Brar, A. et al. (2017) 'Photoautotrophic microorganisms and bioremediation of industrial effluents: Current status and future prospects', *3 Biotech*, 7(1), pp. 1–8. Available at: https://doi.org/10.1007/s13205-017-0600-5.

Cardoso, L.G. et al. (2021) '*Spirulina* sp. as a bioremediation agent for aquaculture wastewater: Production of high added value compounds and estimation of theoretical biodiesel', *Bioenergy Research*, 14(1), pp. 254–264. Available at: https://doi.org/10.1007/s12155-020-10153-4.

Cheirsilp, B. and Torpee, S. (2012) 'Enhanced growth and lipid production of microalgae under mixotrophic culture condition: Effect of light intensity, glucose concentration and fed-batch cultivation', *Bioresource Technology*, 110, pp. 510–516. Available at: https://doi.org/10.1016/j.biortech.2012.01.125.

Chen, C.Y. et al. (2011) 'Cultivation, photobioreactor design and harvesting of microalgae for biodiesel production: A critical review', *Bioresource Technology*, 102(1), pp. 71–81. Available at: https://doi.org/10.1016/j.biortech.2010.06.159.

Chen, F. (1996) 'High cell density culture of microalgae in heterotrophic growth', *Trends in Biotechnology*, 14(11), pp. 421–426. Available at: https://doi.org/10.1016/0167-7799(96)10060-3.

Cheng, P. et al. (2022) 'Heterotrophic and mixotrophic cultivation of microalgae to simultaneously achieve furfural wastewater treatment and lipid production', *Bioresource Technology*, 349, p. 126888. Available at: https://doi.org/10.1016/j.biortech.2022.126888.

Chew, K.W. et al. (2018) 'Effects of water culture medium, cultivation systems and growth modes for microalgae cultivation: A review', *Journal of the Taiwan Institute of Chemical Engineers*, 91, pp. 332–344. Available at: https://doi.org/10.1016/j.jtice.2018.05.039.

Chong, J.W.R. et al. (2021) 'Advances in production of bioplastics by microalgae using food waste hydrolysate and wastewater: A review', Bioresource Technology, 342, p. 125947. Available at: https://doi.org/10.1016/j.biortech.2021.125947.

Chu, W.L., See, Y.C. and Phang, S.M. (2009) 'Use of immobilised *Chlorella vulgaris* for the removal of colour from textile dyes', *Journal of Applied Phycology*, 21(6), pp. 641–648. Available at: https://doi.org/10.1007/s10811-008-9396-3.

Cirik, S. and Gökpınar, Ş. (2008) Plankton Bilgisi ve Kültürü. İzmir. Available at: https://scholar.google.com/scholar?hl=tr&as_sdt=0%2C5&q=Plankton+bilgisi+ve+kültür ü.+Ege+Üniversitesi+Su+Ürünleri+Fakültesi+Yayınları%2C+47%2C+131-133.&b tnG= (Accessed: 9 March 2023).

Daneshvar, E. et al. (2018) 'Investigation on the feasibility of *Chlorella vulgaris* cultivation in a mixture of pulp and aquaculture effluents: Treatment of wastewater and lipid extraction', *Bioresource Technology*, 255, pp. 104–110. Available at: https://doi.org/10.1016/j.biortech.2018.01.101.

Daneshvar, N. et al. (2007) 'Biological decolorization of dye solution containing Malachite Green by microalgae *Cosmarium* sp.', *Bioresource Technology*, 98(6), pp. 1176–1182. Available at: https://doi.org/10.1016/j.biortech.2006.05.025.

Darvehei, P., Bahri, P.A. and Moheimani, N.R. (2018) 'Model development for the growth of microalgae: A review', *Renewable and Sustainable Energy Reviews*, 97, pp. 233–258. Available at: https://doi.org/10.1016/j.rser.2018.08.027.

de Jesus Oliveira Santos, M., Oliveira de Souza, C. and Marcelino, H.R. (2023) 'Blue technology for a sustainable pharmaceutical industry: Microalgae for bioremediation and pharmaceutical production', *Algal Research*, 69, p. 102931. Available at: https://doi.org/10.1016/j.algal.2022.102931.

Dönmez, G. and Aksu, Z. (2002) 'Removal of chromium(VI) from saline wastewaters by Dunaliella species', *Process Biochemistry*, 38(5), pp. 751–762. Available at: https://doi.org/10.1016/S0032-9592(02)00204-2.

Dutta, D., Arya, S. and Kumar, S. (2021) 'Industrial wastewater treatment: Current trends, bottlenecks, and best practices', *Chemosphere*, 285, p. 131245. Available at: https://doi.org/10.1016/j.chemosphere.2021.131245.

El-Kassas, H.Y. and Mohamed, L.A. (2014) 'Bioremediation of the textile waste effluent by *Chlorella vulgaris*', *Egyptian Journal of Aquatic Research*, 40(3), pp. 301–308. Available at: https://doi.org/10.1016/j.ejar.2014.08.003.

Fazal, T. et al. (2021) 'Integrating bioremediation of textile wastewater with biodiesel production using microalgae (*Chlorella vulgaris*)', *Chemosphere*, 281, p. 130758. Available at: https://doi.org/10.1016/j.chemosphere.2021.130758.

Fikirdeşici-Ergen, Ş. et al. (2018) 'Bioremediation of heavy metal contaminated medium using Lemna minor, Daphnia magna and their consortium', *Chemistry and Ecology*, 34(1), pp. 43–55. Available at: https://doi.org/10.1080/02757540.2017.1393534.

Fu, F. and Wang, Q. (2011) 'Removal of heavy metal ions from wastewaters: A review', *Journal of Environmental Management,* 92(3), pp. 407–418. Available at: https://doi.org/10.1016/j.jenvman.2010.11.011.

Gao, F. et al. (2016) 'Continuous microalgae cultivation in aquaculture wastewater by a membrane photobioreactor for biomass production and nutrients removal', *Ecological Engineering*, 92, pp. 55–61. Available at: https://doi.org/10.1016/j.ecoleng.2016.03.046.

Gong, Y. et al. (2013) 'Triacylglycerol accumulation and change in fatty acid content of four marine oleaginous microalgae under nutrient limitation and at different culture ages', *Journal of Basic Microbiology*, 53(1), pp. 29–36. Available at: https://doi.org/10.1002/jobm.201100487.

González, L.E., Cañizares, R.O. and Baena, S. (1997) 'Efficiency of ammonia and phosphorus removal from a Colombian agroindustrial wastewater by the microalgae *Chlorella vulgaris* and *Scenedesmus dimorphus*', *Bioresource Technology*, 60(3), pp. 259–262. Available at: https://doi.org/10.1016/S0960-8524(97)00029-1.

Guo, Z. et al. (2013) 'Microalgae cultivation using an aquaculture wastewater as growth medium for biomass and biofuel production', *Journal of Environmental Sciences (China)*, 25(S1), pp. S85–S88. Available at: https://doi.org/10.1016/S1001-0742(14)60632-X.

Hanchang, S. (2009) 'Industrial wastewater: Types, amounts and effects', *Point Sources of Pollution: Local Effects and their Control*, I, pp. 191–203.

Hasanoğlu, S., Kutluk, T. and Kapucu, N. (2021) 'Investigation of the characteristics and growth of *Chlorella variabilis* via biosorption of a steel industry wastewater', *Journal of Water Chemistry and Technology*, 43(5), pp. 423–431. Available at: https://doi. org/10.3103/s1063455x21050064.

Hena, S., Gutierrez, L. and Croué, J.P. (2020) 'Removal of metronidazole from aqueous media by *C. vulgaris*', *Journal of Hazardous Materials*, 384, p. 121400. Available at: https:// doi.org/10.1016/j.jhazmat.2019.121400.

Hongyang, S. et al. (2011) 'Cultivation of *Chlorella pyrenoidosa* in soybean processing wastewater', *Bioresource Technology*, 102(21), pp. 9884–9890. Available at: https://doi. org/10.1016/j.biortech.2011.08.016.

Hu, Q. (2013) 'Environmental effects on cell composition', In: Richmond, A. and Hu, Q. (eds.), *Handbook of Microalgal Culture: Applied Phycology and Biotechnology: Second Edition*. Hoboken, NJ: John Wiley & Sons, Ltd, pp. 114–122. Available at: https://doi. org/10.1002/9781118567166.ch7.

Hu, X. et al. (2019) 'Acclimation of consortium of micro-algae help removal of organic pollutants from meat processing wastewater', *Journal of Cleaner Production*, 214, pp. 95–102. Available at: https://doi.org/10.1016/j.jclepro.2018.12.255.

Ji, M.K. et al. (2015) 'Mixotrophic cultivation of a microalga *Scenedesmus obliquus* in municipal wastewater supplemented with food wastewater and flue gas CO2 for biomass production', *Journal of Environmental Management*, 159, pp. 115–120. Available at: https://doi.org/10.1016/j.jenvman.2015.05.037.

Kavitha, S. et al. (2020) 'Introduction: Sources and characterization of food waste and food industry wastes', In: Banu, R., Kumar, G., Gunasekaran, M., and Kavitha, S. (eds.), *Food Waste to Valuable Resources: Applications and Management*. Cambridge, MA: Academic Press, pp. 1–13. Available at: https://doi.org/10.1016/B978-0-12-818353-3.00001-8.

Khatoon, H. et al. (2016) 'Re-use of aquaculture wastewater in cultivating microalgae as live feed for aquaculture organisms', *Desalination and Water Treatment*, 57(60), pp. 29295–29302. Available at: https://doi.org/10.1080/19443994.2016.1156030.

Kim, S. et al. (2013) 'Growth rate, organic carbon and nutrient removal rates of *Chlorella sorokiniana* in autotrophic, heterotrophic and mixotrophic conditions', *Bioresource Technology*, 144, pp. 8–13. Available at: https://doi.org/10.1016/j.biortech.2013.06.068.

Kothari, R. et al. (2013) 'Production of biodiesel from microalgae *Chlamydomonas polypyrenoideum* grown on dairy industry wastewater', *Bioresource Technology*, 144, pp. 499–503. Available at: https://doi.org/10.1016/j.biortech.2013.06.116.

Kumar, G. et al. (2018) 'Evaluation of gradual adaptation of mixed microalgae consortia cultivation using textile wastewater via fed batch operation', *Biotechnology Reports*, 20, p. e00289. Available at: https://doi.org/10.1016/j.btre.2018.e00289.

Laraib, N. et al. (2021) 'Mixotrophic cultivation of *Chlorella vulgaris* in sugarcane molasses preceding nitrogen starvation: Biomass productivity, lipid content, and fatty acid analyses', *Environmental Progress and Sustainable Energy*, 40(4), p. e13625. Available at: https://doi.org/10.1002/ep.13625.

Larsdotter, K. (2006) 'Wastewater treatment with microalgae - a literature review Avloppsrening med mikroalger - en litteraturstudie Algal growth', *Vatten*, 62, pp. 31–38.

Latiffi, N.A.A. et al. (2016) 'Removal of nutrients from meat food processing industry wastewater by using microalgae botryococcus SP', *ARPN Journal of Engineering and Applied Sciences*, 11(16), pp. 9863–9867. Available at: https://www.researchgate.net/publication/309119452 (Accessed: 7 May 2023).

Lee, K. and Lee, C.G. (2001) 'Effect of light/dark cycles on wastewater treatments by microalgae', *Biotechnology and Bioprocess Engineering*, 6(3), pp. 194–199. Available at: https://doi.org/10.1007/BF02932550.

Lee, S.Y., Lee, J.S. and Sim, S.J. (2023) 'Enhancement of microalgal biomass productivity through mixotrophic culture process utilizing waste soy sauce and industrial flue gas', *Bioresource Technology*, 373, p. 128719. Available at: https://doi.org/10.1016/j.biortech.2023.128719.

Lee, Y.K. (2001) 'Microalgal mass culture systems and methods: Their limitation and potential', *Journal of Applied Phycology*, pp. 307–315. Available at: https://doi.org/10.1023/A:1017560006941.

Li, K. et al. (2019) 'Microalgae-based wastewater treatment for nutrients recovery: A review', Bioresource Technology, 291, p. 121934. Available at: https://doi.org/10.1016/j.biortech.2019.121934.

Lim, S.L., Chu, W.L. and Phang, S.M. (2010) 'Use of *Chlorella vulgaris* for bioremediation of textile wastewater', *Bioresource Technology*, 101(19), pp. 7314–7322. Available at: https://doi.org/10.1016/j.biortech.2010.04.092.

Lin, C.Y., Nguyen, M.L.T. and Lay, C.H. (2017) 'Starch-containing textile wastewater treatment for biogas and microalgae biomass production', *Journal of Cleaner Production*, 168, pp. 331–337. Available at: https://doi.org/10.1016/j.jclepro.2017.09.036.

Lokhande, R.S., Singare, P.U. and Pimple, D.S. (2011) 'Toxicity study of heavy metals pollutants in waste water effluent samples collected from Taloja Industrial Estate of Mumbai, India', *Resources and Environment*, 1(1), pp. 13–19. Available at: https://www.researchgate.net/publication/288346888 (Accessed: 3 May 2023).

Lowrey, J., Brooks, M.S. and McGinn, P.J. (2015) 'Heterotrophic and mixotrophic cultivation of microalgae for biodiesel production in agricultural wastewaters and associated challenges: A critical review', *Journal of Applied Phycology*, 27, pp. 1485–1498. Available at: https://doi.org/10.1007/s10811-014-0459-3.

Lutzu, G.A., Zhang, W. and Liu, T. (2016) 'Feasibility of using brewery wastewater for biodiesel production and nutrient removal by *Scenedesmus dimorphus*', *Environmental Technology (United Kingdom)*, 37(12), pp. 1568–1581. Available at: https://doi.org/10.1080/09593330.2015.1121292.

Madadi, R. et al. (2016) 'Treatment of petrochemical wastewater by the Green Algae *Chlorella vulgaris*', *International Journal of Environmental Research*, 10(4), pp. 555–560.

Malik, A. (2004) 'Metal bioremediation through growing cells', Environment International, 30(2), pp. 261–278. Available at: https://doi.org/10.1016/j.envint.2003.08.001.

Mao, G. et al. (2021) 'A bibliometric analysis of industrial wastewater treatments from 1998 to 2019', *Environmental Pollution*, 275, p. 115785. Available at: https://doi.org/10.1016/j.envpol.2020.115785.

Mata, T.M., Martins, A.A. and Caetano, N.S. (2010) 'Microalgae for biodiesel production and other applications: A review', *Renewable and Sustainable Energy Reviews*, 14(1), pp. 217–232. Available at: https://doi.org/10.1016/j.rser.2009.07.020.

Maurya, R. et al. (2022) 'Advances in microalgal research for valorization of industrial wastewater', *Bioresource Technology*, p. 126128. Available at: https://doi.org/10.1016/j.biortech.2021.126128.

Mohammed Abdul, K.S. et al. (2015) 'Arsenic and human health effects: A review', *Environmental Toxicology and Pharmacology*, pp. 828–846. Available at: https://doi.org/10.1016/j.etap.2015.09.016.

Pagnanelli, F. et al. (2014) 'Mixotrophic growth of *Chlorella vulgaris* and *Nannochloropsis oculata*: Interaction between glucose and nitrate', *Journal of Chemical Technology and Biotechnology*, 89(5), pp. 652–661. Available at: https://doi.org/10.1002/jctb.4179.

Park, K.C. et al. (2012) 'Mixotrophic and photoautotrophic cultivation of 14 microalgae isolates from Saskatchewan, Canada: Potential applications for wastewater remediation for biofuel production', *Journal of Applied Phycology*, 24(3), pp. 339–348. Available at: https://doi.org/10.1007/s10811-011-9772-2.

Patel, A.K., Choi, Y.Y. and Sim, S.J. (2020) 'Emerging prospects of mixotrophic microalgae: Way forward to sustainable bioprocess for environmental remediation and cost-effective biofuels', *Bioresource Technology*, p. 122741. Available at: https://doi.org/10.1016/j.biortech.2020.122741.

Patel, S.B., Mehta, A. and Solanki, H.A. (2017) 'Physiochemical analysis of treated industrial effluent collected from ahmedabad mega pipeline', *Journal of Environmental & Analytical Toxicology*, 07(05), p. 5. Available at: https://doi.org/10.4172/2161-0525.1000497.

Pathak, V.V. et al. (2015) 'Experimental and kinetic studies for phycoremediation and dye removal by *Chlorella pyrenoidosa* from textile wastewater', *Journal of Environmental Management*, 163, pp. 270–277. Available at: https://doi.org/10.1016/j.jenvman.2015.08.041.

Perez-Garcia, O. et al. (2011) 'Heterotrophic cultures of microalgae: Metabolism and potential products', *Water Research*, pp. 11–36. Available at: https://doi.org/10.1016/j.watres.2010.08.037.

Perez-Garcia, O. et al. (2015) 'Microalgal heterotrophic and mixotrophic culturing for bio-refining: From metabolic routes to techno-economics', In: Prokop, A., Bajpai, R.K. and Zappi, M.E. (eds.), *Algal Biorefineries: Volume 2: Products and Refinery Design*. Springer International Publishing, pp. 61–131. Available at: https://doi.org/10.1007/978-3-319-20200-6_3.

Phang, S.M. et al. (2000) 'Spirulina cultivation in digested sago starch factory wastewater', *Journal of Applied Phycology*, pp. 395–400. Available at: https://doi.org/10.1023/a:1008157731731.

Plöhn, M. et al. (2021) 'Wastewater treatment by microalgae', *Physiologia Plantarum*, 173(2), pp. 568–578. Available at: https://doi.org/10.1111/ppl.13427.

Porto, B. et al. (2020) 'Microalgal growth in paper industry effluent: Coupling biomass production with nutrients removal', *Applied Sciences (Switzerland)*, 10(9), p. 3009. Available at: https://doi.org/10.3390/app10093009.

Roostaei, J. et al. (2018) 'Mixotrophic microalgae biofilm: A novel algae cultivation strategy for improved productivity and cost-efficiency of biofuel feedstock production', *Scientific Reports*, 8(1), pp. 1–10. Available at: https://doi.org/10.1038/s41598-018-31016-1.

Saavedra, R. et al. (2018) 'Comparative uptake study of arsenic, boron, copper, manganese and zinc from water by different green microalgae', *Bioresource Technology*, 263, pp. 49–57. Available at: https://doi.org/10.1016/j.biortech.2018.04.101.

Santos, S.C.R. et al. (2018) 'Macroalgae biomass as sorbent for metal ions', In: Popa, V.I. and Volf, I. (eds.), *Biomass as Renewable Raw Material to Obtain Bioproducts of High-Tech Value*. Elsevier, pp. 69–112. Available at: https://doi.org/10.1016/B978-0-444-63774-1.00003-X.

Sasi, P.K.C. et al. (2020) 'Phycoremediation of paper and pulp mill effluent using planktochlorella nurekis and chlamydomonas reinhardtii: A comparative study', *Journal of Environmental Treatment Techniques*, 8(2), pp. 809–817. Available at: https://www.dormaj.com/docs/Volume8/Issue%202/Phycoremediation%20of%20paper%20and%20pulp%20mill%20effluent%20using%20Planktochlorella%20nurekis%20and%20Chlamydomonas%20reinhardtii%20%E2%80%93%20a%20comparative%20study.pdf (Accessed:24 March 2024).

Serra-Maia, R. et al. (2016) 'Influence of temperature on *Chlorella vulgaris* growth and mortality rates in a photobioreactor', *Algal Research*, 18, pp. 352–359. Available at: https://doi.org/10.1016/j.algal.2016.06.016.

Sharma, P. et al. (2021) 'Integrating phytoremediation into treatment of pulp and paper industry wastewater: Field observations of native plants for the detoxification of metals and their potential as part of a multidisciplinary strategy', *Journal of Environmental Chemical Engineering*, 9(4), p. 105547. Available at: https://doi.org/10.1016/j.jece.2021.105547.

Shrestha, R. et al. (2021) 'Technological trends in heavy metals removal from industrial wastewater: A review', *Journal of Environmental Chemical Engineering*, p. 105688. Available at: https://doi.org/10.1016/j.jece.2021.105688.

Silva, M.I. et al. (2021) 'Article experimental and techno-economic study on the use of microalgae for paper industry effluents remediation', *Sustainability (Switzerland)*, 13(3), pp. 1–29. Available at: https://doi.org/10.3390/su13031314.

Singh, A. and Ummalyma, S.B. (2020) 'Bioremediation and biomass production of microalgae cultivation in river watercontaminated with pharmaceutical effluent', *Bioresource Technology*, 307, p. 123233. Available at: https://doi.org/10.1016/j.biortech.2020.123233.

Sloth, J.K. et al. (2017) 'Growth and phycocyanin synthesis in the heterotrophic microalga *Galdieria sulphuraria* on substrates made of food waste from restaurants and bakeries', *Bioresource Technology*, 238, pp. 296–305. Available at: https://doi.org/10.1016/j.biortech.2017.04.043.

Song, C. et al. (2019) 'Biodegradability and mechanism of florfenicol via *Chlorella* sp. UTEX1602 and L38: Experimental study', *Bioresource Technology*, 272, pp. 529–534. Available at: https://doi.org/10.1016/j.biortech.2018.10.080.

Tan, X.B. et al. (2021) 'Nutrients recycling and biomass production from *Chlorella pyrenoidosa* culture using anaerobic food processing wastewater in a pilot-scale tubular photobioreactor', *Chemosphere*, 270, p. 129459. Available at: https://doi.org/10.1016/j.chemosphere.2020.129459.

Tossavainen, M. et al. (2019) 'Integrated utilization of microalgae cultured in aquaculture wastewater: Wastewater treatment and production of valuable fatty acids and tocopherols', *Journal of Applied Phycology*, 31(3), pp. 1753–1763. Available at: https://doi.org/10.1007/s10811-018-1689-6.

Tran, T.K. et al. (2017) 'Electrochemical treatment of heavy metal-containing wastewater with the removal of COD and heavy metal ions', *Journal of the Chinese Chemical Society*, 64(5), pp. 493–502. Available at: https://doi.org/10.1002/jccs.201600266.

Ummalyma, S.B. et al. (2023) 'Sustainable microalgal cultivation in poultry slaughterhouse wastewater for biorefinery products and pollutant removal', *Bioresource Technology*, 374, p. 128790. Available at: https://doi.org/10.1016/j.biortech.2023.128790.

Veglio', F. and Beolchini, F. (1997) 'Removal of metals by biosorption: A review', *Hydrometallurgy*, 44(3), pp. 301–316. Available at: https://doi.org/10.1016/s0304-386x(96)00059-x.

Wollmann, F. et al. (2019) 'Microalgae wastewater treatment: Biological and technological approaches', *Engineering in Life Sciences*, pp. 860–871. Available at: https://doi.org/10.1002/elsc.201900071.

Wu, J.Y. et al. (2017) 'Lipid accumulating microalgae cultivation in textile wastewater: Environmental parameters optimization', *Journal of the Taiwan Institute of Chemical Engineers*, 79, pp. 1–6. Available at: https://doi.org/10.1016/j.jtice.2017.02.017.

Wu, J.Y. et al. (2020) 'Immobilized Chlorella species mixotrophic cultivation at various textile wastewater concentrations', *Journal of Water Process Engineering*, 38, p. 101609. Available at: https://doi.org/10.1016/j.jwpe.2020.101609.

Xiong, J.Q., Kurade, M.B. and Jeon, B.H. (2017) 'Biodegradation of levofloxacin by an acclimated freshwater microalga, *Chlorella vulgaris*', *Chemical Engineering Journal*, 313, pp. 1251–1257. Available at: https://doi.org/10.1016/j.cej.2016.11.017.

Xiong, J.Q., Kurade, M.B. and Jeon, B.H. (2018) 'Can microalgae remove pharmaceutical contaminants from water? *Trends in Biotechnology*, pp. 30–44. Available at: https://doi.org/10.1016/j.tibtech.2017.09.003.

Yıldız Töre, G. and Ata, R. (2019) 'Assessment of recovery & reuse activities for industrial waste waters in miscellaneous sectors', *European Journal of Engineering and Applied Sciences*, 2(1), pp. 19–43.

Yuan, S. et al. (2021) 'Purification of potato wastewater and production of byproducts using microalgae Scenedesmus and Desmodesmus', *Journal of Water Process Engineering*, 43, p. 102237. Available at: https://doi.org/10.1016/J.JWPE.2021.102237.

Zhang, Y. et al. (2014) 'Evaluating removal of steroid estrogens by a model alga as a possible sustainability benefit of hypothetical integrated algae cultivation and wastewater treatment systems', *ACS Sustainable Chemistry and Engineering*, 2(11), pp. 2544–2553. Available at: https://doi.org/10.1021/sc5004538.

7 Usability of Microalgaes as Biofuel Raw Material

Meltem Kizilca Çoruh

7.1 INTRODUCTION

Technological developments and industrialization play a significant role in human life. Today, the rapidly increasing world population and the energy requisition due to technology are provided by existing fossil resources. The difference between the energy needed in all areas of the world and that supplied in response to this demand is increasing day by day. Fossil fuels used to meet energy needs are only available in a few countries in the world, and this makes other countries dependent on these countries for energy. In addition, the fact that currently used fossil fuels pollute the environment and increase carbon dioxide emissions, the negative effects of carbon dioxide emissions on the climate, the melting of glaciers and the increase in drought, and the damage to the environment caused by increasing global energy consumption constitute an important problem for the whole world (Oelkers and Cole, 2008; Voumik et al., 2023). As a result of these problems, terrible scenarios will emerge in the next 30–50 years. Due to industrial and economic development, the rapid increase in energy production and consumption activities, together with the world population, causes air, water, and soil pollution. It poses serious risks to the availability and quality of water resources in many areas of the world. Due to the energy crisis affecting almost every part of the world as a result of speedy industrialization and population increases, the search for renewable energy resources in this century is among the most important issues to be overcome to promote more sustainable energy development in the future and has led scientists to develop renewable energy sources (Renuka et al., 2015; Sharma et al., 2020; Sheth and Babu, 2010; Shahbaz et al., 2021). Renewable energy sources are classified as biomass, hydrogen, solar, wind, geothermal, and wave energy.

Recently, interest in biomass energy from renewable energy sources has increased considerably. Studies have focused on the production of renewable, clean, sustainable, efficient, low-cost, and safe alternative energy sources. At the beginning of these energy sources is the energy obtained from biomass, which contains many biofuels such as biodiesel, bioethanol, biogas, bio-oil, and biohydrogen (Demirbas, 2001; He et al., 2022; Malode et al., 2021). They consider environmentally friendly biofuels as an alternative energy source that has the potential to eliminate greenhouse gas emissions and meet the energy demand in the future. Biofuels, which are environmentally friendly compared to traditional energy sources, have become the focus of attention to be obtained from terrestrial plants and microalgae.

DOI: 10.1201/9781003390213-7

Due to the difficulty and high cost of breaking down the lignin complex contained in terrestrial plants, it has led to the search for more effective raw materials. Due to the difficulty and high cost of breaking down the lignin complex contained in terrestrial plants, it has led to the search for more effective raw materials. Among these energy sources, the use of aquatic microorganisms, such as microalgae, which are very rich in oil, carbohydrates, and protein, is a hope for biofuel production with its advantages such as being able to grow in small areas without the need for special areas such as terrestrial plants, rapid growth, high reproduction rates in wastewater, and the ability to absorb carbon dioxide from flue gases (Gupta et al., 2013; Sahu et al., 2013). Algae also have a significant advantage over oilseeds, which photosynthesize carbon dioxide and sunlight very effectively into energy and produce oil in the process. Depending on the content of algae, electricity, ethanol, hydrogen, methane, and biodiesel can be produced by biochemical methods, as well as synthesis gas, biological coal, biodiesel, and electricity can be produced by using thermochemical methods. The most important component of algae is water; accordingly, it plays an important role in wastewater treatment. Thanks to its role in wastewater treatment, both the cleaning of polluted water and its use as biomass play a significant role in the cultivation of algae (Adeniyi et al., 2018; Liu et al., 2010; Sharma et al., 2020; Zabed et al., 2019; Ryu et al., 2006).

Fossil fuels have been formed over hundreds of years, and resources are being depleted day by day. The depletion of fossil resources and the fact that petroleum and its derivatives are produced from limited natural resources constitute an important problem for the whole world. The shortness of the reserve life of fossil fuels, which meet a large part of the energy need, the increase in their use with the increase in population and the development of technology, and the environmental effects it creates pose serious risks regarding the availability and quality of water resources in most parts of the earth (Stone et al., 2010; Beringer et al., 2011). The decrease in clean water resources, which is one of the most important of these effects, causes water scarcity. Today, with the increasing demand for clean water resources, human beings have tended to develop innovative treatment technologies. Countries under water stress have started to use their water resources consciously and to recover their waste water by purifying it. They aim to ensure sustainability with an innovative approach to water by making use of green technology. For this reason, new natural methods have started to attract attention in wastewater purification. In this direction, negative environmental effects can be minimized by using microalgae production and wastewater purification synergistically, and environmental and economic contributions can be made by obtaining high value-added products from algae biomass.

7.2 BIOMASS

Biomass is a mass of non-fossil organic matter of biological origin, containing carbon, hydrogen, oxygen, and nitrogen. The main components of that are carbohydrates, which can be renewed in <100 years, renewable land and aquatic plants, the food industry, animal wastes and urban wastes, and forest by-products. Biomass resources generally have a complex and heterogeneous structure formed as a result of the combination of many components. Apart from major components such as hemicellulose, cellulose, extractive substances, and lignin, many organic and inorganic components are present in different proportions in biomass sources. The main components of biomass are given in Figure 7.1 (Kumar et al., 2012; Vassilev et al., 2013).

Although the sources of biomass are very diverse, they are generally terrestrial plants (forest plants, grasses, energy plants, and plant wastes), aquatic (macro and micro algae living in seas and lakes, grasses, reeds, and some microorganisms), and waste (animal, urban, and industrial) that can be used to produce energy. Biomass is a renewable, environmentally friendly, domestic, and national energy source that provides socio-economic development, can be grown anywhere, can produce electricity, and provides fuel for vehicles. Biomass, one of the renewable energy sources, constitutes about 13% of the earth's energy consumption. The carbon used to form biomass is absorbed by plants from the atmosphere as CO_2, utilizing solar energy. The source of energy, which is stored chemically, especially in the form of cellulose, during the photosynthesis of plants and which can then be used in various forms, is the sun. The plants are then eaten by animals and, thus, can be turned into animal biomass. But primary absorption is carried out by plants. If the plant product is not eaten by animals, it is spoiled or burned by microorganisms. During decomposition, its carbon content is usually released back to the atmosphere as CO_2 or CH_4, depending on the circumstances and processes it contains. During combustion, it returns to the atmosphere as carbon dioxide. This process, known as the carbon cycle, will continue as long as plants exist on earth. Apart from energy, biomass is evaluated in many other areas, such as paper, furniture, and insulation material. When used for energy purposes, various technologies are used to get solid, liquid, and gaseous fuels. One of the most major renewable energy resources is biofuels.

Biofuels are called first-, second-, and third-generation biofuels depending on the raw material source and production process used (Gustavsson et al., 2006; Li et al., 2008; Kaygusuz and Türker, 2002). The first-generation biofuels were obtained from agricultural products such as sugar cane, canola, and corn. However, agricultural biofuels have important disadvantages, such as not being able to compete with fossil fuels economically and risking food security as a result of using fertile lands. In the second generation, biofuel production obtained from lignocellulosic biomass of vegetable origin, although there is no problem of food-biofuel competition, the high cost

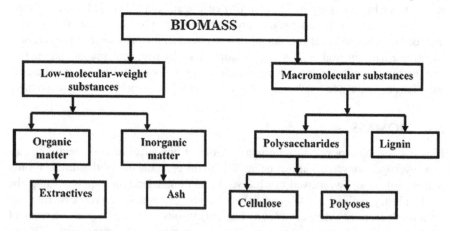

FIGURE 7.1 Biomass basic components.

of separating lignin from lignocellulosic biomass has created an important problem. Third-generation biofuels refer to low-budget biooils. It is the production of biofuels by switching from lignocellulosic sources to cellulosic sources and using genetically modified plants and algae with higher oil and cellulose content within the scope of integrated biorefinery technologies. Generally, algae-based biofuels fall into this category.

Microalgae, which has been the subject of intense research in biofuel production in recent years, is produced in biofuels due to its high photosynthetic efficiency, ability to reproduce on lands that are not suitable for agriculture, wastewater purification by evaluating wastewater, capturing CO_2 gas from flue gases that cause climate change, and offering valuable by-products other than biofuel. It has much more important advantages than oily plants to be used. It has a significant advantage over oil seeds in terms of biomass and biodiesel efficiency. Biomass is used in biofuel technology such as biodiesel, bioethanol, biogas, fertilizer, and hydrogen, which can be easily transported, stored, and used with properties equivalent to existing fuels by directly burning or raising the fuel quality in various processes. The biomass, which has different contents and sources from each other, causes the applied methods to be different.

7.3   BIOMASS CONVERSION PROCESSES

Conversion processes from biomass can be separated into three basic categories: physical, thermochemical, and biochemical. The first is grinding, filtration, drying, and extraction by physical method. The second is direct combustion, gasification, pyrolysis, carbonization, and liquefaction by the thermochemical method. The other is the conversion of high-moisture raw materials into high-energy products like biogas by microbiological fermentation (Demirbas, 2001; Goyal et al., 2008; Griffin and Schultz, 2012; Ryu et al., 2006). The energy stored in the structure of the biomass can be released directly in the form of heat by combustion or converted into solid (charcoal) fuels through carbonization, liquid (bio-oils) through pyrolysis or liquefaction, or gaseous (syngas) fuels through gasification using many different process conditions and reactor setups. Especially, the amount and species of biomass, the desired energy form, environmental standards, economic terms, and the characteristics of the applied work affect the conversion processes. In many cases, the desired energy form is determined primarily by the performed process and then the species and amount of biomass chosen.

When thermochemical and biological methods are compared, it can be concluded that thermochemical methods are quite advantageous (Yadav et al., 2023). Because in thermochemical methods, reactions are completed in minutes and seconds and many complex products are formed, whereas in biochemical methods, reactions can last for hours, days, weeks, or even years and one or more specific products such as ethanol and biogas are formed. The aim of thermochemical conversion processes is to obtain fuels that are easily stored and transported and have stable properties as an alternative to fossil fuels. Basically, thermochemical conversion technologies cover processes applied to produce solid, liquid, and gaseous fuels with high energy content from biomass. These processes are especially applied in order to obtain high thermal value fuels from biomass sources with low calorific value and high moisture content. The transportation, distribution, and burning of the fuels obtained as a result of the thermal conversion processes are easier compared to the biomass source.

Moreover, their high yield, ashless nature, and low damage to the environment are other important advantages. Primary products directly derived from thermochemical processes are more readily available and more valuable than raw biomass. Primary products can also be used by converting them into more useful and valuable secondary fuels or chemical products. Conversion techniques utilizing biomass sources and fuels get handled by these techniques, and their application areas are summed up in Table 7.1 (Panwar et al., 2012; Özbay et al., 2001).

7.3.1 Physical Process

The difference in moisture, ash, and thermal values of different biomass creates some problems in the conversion processes of biomass. The fact that the processes such as size reduction, drying, filtration, extraction, and condensation that are physically performed in the conversion of biomass are carried out before the thermochemical or microbial conversion processes increases the applicability of the conversions realized in this way. In order to increase the amount of energy to be produced from biomass, it may be necessary to dry the biomass source. The drying process can be carried out by the sun at low costs; it can be made in a short period of time and more effectively by industrial drying methods like convection ovens and spray dryers. Size reduction is an important process in the use of biomass as a fuel and in the production of pellets and briquettes. Other benefits of breaking down biomass into minor particles are reducing storage volume and reducing transportation costs. These processes are defined as the pre-preparation process of the biomass. As a result of physical processes, the fuel quality of the biomass increases, and the biomass can be used directly by burning (Zhang et al., 2022; Villagracia et al., 2016).

7.3.2 Combustion

Although the burning of biomass to produce energy is the earliest known thermochemical conversion method for humanity, new combustion systems have been improved in recent years to increase yield. Developing combustion processes is

TABLE 7.1
Biomass Conversion Techniques, Fuels, and Application Areas

Biomass	Conversion Method	Fuels	Application Areas
Forest waste	Airless refutation	Biogas	Electricity generation heating
Agricultural waste	Pyrolysis	Ethanol	Heating, means of transport
Energy crops	Direct combustion	Hydrogen	Heating
Animal waste	Fermentation, airless refutation	Methane	Means of transport, heating
Rubbish	Gasification	Methanol	Aircraft
Algae	Hydrolysis	Synthetic oil	Synthetic oils, rockets
Energy forests	Biophotolysis	Diesel	Product drying
Vegetable and animal oils	Esterification	Diesel	Heating, means of transport greenhouse cultivation

utilized to use biomass for heat and electricity generation. Especially in the construction of thermal power plants working with biomass, fluidized bed systems replace conventional combustion systems. By burning biomass directly in boilers and producing heat and steam, electricity can be produced with an efficiency of 20%–30%. Fluidized bed technology has been modified in order to prevent particle losses in the fluidized bed combustion process. Stable wastes, such as ash left over from incineration, were removed by cementation, melting and solidification, extraction using acid or other solvents, and stabilization using chemicals. Another method of removal is melting at very high temperatures and solidification by cooling again. These solidified wastes can be used for paving roads as cobblestones and for reclamation of land areas. It is possible to directly burn almost any biomass source. Incineration is not economical due to the transportation of forest and agricultural wastes to distant places and the high moisture content.

Heat energy can be obtained by direct burning of dried agricultural waste. It can also be used in thermal power to burn agricultural wastes by mixing them with fossil fuels such as coal and to generate electricity. However, it is not very efficient and is not preferred due to reasons such as low calorific values, transportation and storage difficulties, variations in fuel quality, high humidity, flue gas environmental problems, and low efficiency of direct combustion plants. Although combustion system technology is known and commercially accepted, many problems may be encountered during operation due to fuel-related variability. For this reason, it is necessary to know the fuel properties well. Agglomeration, fouling, and corrosion are the most important fuel-related problems encountered in biomass combustion. These problems cause agglomeration in the bed, fly ash accumulation on the heat transfer surfaces, and high-temperature corrosion.

7.3.3 PYROLYSIS

Pyrolysis is defined as the process of thermally decomposing biomass by heating it in an airless environment to obtain liquid (biodiesel, bioethanol, biomethanol, and biooil), solid (biochar, wood charcoal, biobriquettes, and biopellets) and gaseous (biogas, biohydrogen, biosynthesis gas) products. However, pyrolysis is a rather complex process. It usually occurs through a series of reactions and is influenced by many factors. Consequently, in the pyrolysis process, in addition to solid, liquid, and gas products, some chemicals and water are also formed. Pyrolysis product distribution and quality are affected by parameters such as pyrolysis temperature, heating rate, particle size, pyrolysis medium, reactor geometry, raw material type and properties, pressure, and catalyst. In addition, the organic-inorganic structure of the biomass also affects properties such as moisture content, porosity, ash content, volatile component amount, calorific value, fixed carbon/volatile matter ratio, and cellulose/lignin ratio.

It is necessary to examine the effect of temperature, which is one of the most major factors affecting the thermal decomposition process, on product compositions in order to correctly design the systems that enable the conversion of biomass sources to fuels and to choose the most suitable conditions due to the physical and chemical differences in the structures of the components. For this reason, it is significant to examine the thermal properties of biomass components. In order to ensure the design of biomass pyrolysis processes, it is also of great importance to determine the kinetic mechanism, besides the evaluation of parameters for optimization and control. The most widely used method for this purpose is thermogravimetric analysis. Pyrolysis products can be used as artificial fuel in power plants, refineries, gas turbines, and diesel engines; as solid-liquid fuel in power stations and steam boilers; and as coke in the iron and steel industry. Biomass pyrolysis starts between 350°C and 550°C and lasts up to 700°C. During decomposition, a large number of highly reactive radicals are formed as a result of bond breaks in the structure of the raw materials. These radicals form gas, liquid, and solid products as a result of a series of reactions to become stable. The resulting solid product is called char, and the liquid product is called tar.

Pyrolysis systems are applied using two methods: slow pyrolysis and fast pyrolysis, depending on the working conditions. These two methods differ from each other in terms of product yields and compositions. Each of these methods is more efficient than the other in obtaining a different pyrolysis product. The slow-pyrolysis technique is a cheap and efficient method. Solid, liquid, and gaseous products are formed as pyrolysis products. Higher temperatures are required to increase gas product yield, while lower temperatures are used for liquid products. On the other hand, rapid pyrolysis is an advanced technology used to obtain liquid products at very high temperature rise rates and very short reaction times. In pyrolysis applied at high temperature and low heating rates with a long residence time, the yield of biochar and gaseous products is at its maximum level. While the heating rate at low temperatures has an important impact on product distribution and composition, there is no significant effect at high temperatures. The heating rate is important only at low temperatures. Because the product distribution and composition show a similar trend at high temperatures.

Today, there are many pyrolysis reactors developed for different purposes (Garcia-Nunez et al., 2017). Pyrolysis reactors differ according to the targeted end products (coal, oil, heat, electricity, and gases), the operating principle of the reactor (discrete or continuous), the method of heating (direct or indirect heating, microwave automatic, thermal), the heat source used (electricity, biomass combustion, gas heater), the method used to load the reactor (manual, mechanical), the pressure at which the process operates (atmospheric, vacuum, and pressurized), the material used in its construction (concrete, brick, steel, soil), and the portability of the reactor (fixed, mobile). These reactors can be classified as fixed-bed reactors, entrainment flow reactors, fluid-bed reactors, vacuum pyrolysis reactors, rotary conical reactors, screw/screw reactors, solar-powered reactors, and microwave-powered reactors.

In order to maximize the biofuel yield obtained as a result of pyrolysis, high temperature, low heating rate, and long retention time should be required. Bio-fuel (primary products) obtained from pyrolysis: compounds with high oxygen content have a low calorific value and are unstable and corrosive. The liquid product yield is increased by removing the oxygen from the oxygen-rich biomass and enriching it with hydrogen. It can be used directly, or it can be converted to secondary products after purification processes. Two methods are generally recommended for this. The first method is reaction with hydrogen and carbon monoxide at high pressure or catalytic hydrogenation using a hydrogen donor solvent. The second method is the purification of the low-molecular-weight gasoline-equivalent liquid product obtained using zeolites under atmospheric conditions (Figure 7.2). In the first method, the oxygen in the biofuel is converted into

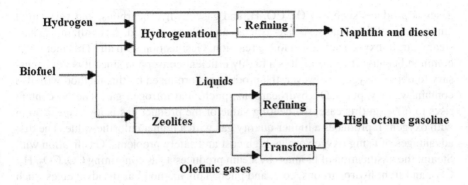

FIGURE 7.2 Purification of biofuel.

water and removed from the environment. In the second method, oxygen is removed from the environment as CO_2 (Demirbas, 2001; Ward et al., 2014). The waste used in the pyrolysis method can be converted into useful products. Pyrolysis is the fundamental thermochemical process used to turn biomass into a more beneficial fuel.

7.3.4 LIQUEFACTION

The principle of liquefaction, where the long molecular structure of the material at low temperature (250°C–500°C) and high pressure (~150 bar) is deteriorated with the help of a catalyst, is to obtain a liquid fuel called bio-oil while being converted into a short-molecule structure. Liquefaction is a process similar to pyrolysis. In both methods, the aim is to convert the biomass into liquid. Pyrolysis occurs in the high temperature range of 923–10,730 K, while the liquefaction process takes in the lower temperature range of 798–8,730 K. The process pressure occurs in the low pressure range of 0.1–0.5 MPa in pyrolysis and in the high pressure range of 5–20 MPa in lique-faction. It is usually performed in subcritical water conditions and at high pressure. The fact that it can be implemented at a lower temperature than processes such as pyroly-sis, high energy efficiency, water being a unique and environmentally friendly solvent, and the fact that the process can be applied to wet biomass without having to dry the biomass make liquefaction attractive. Although it provides many advantages with the liquefaction of biomass, the cost and risk of handling high pressure are the most impor-tant points of the process and should be considered. Thus, oils with high energy content can be produced for the production of other chemicals and fuels, which take up less space, can be pumped, and can be easily stored (Toor et al., 2011; Cocero et al., 2002).

7.3.5 GASIFICATION

Gasification is the process of obtaining combustible gaseous fuel by the high temper-ature decomposition of solid fuels such as carbon-containing biomass. Gasification has been used for years, especially in the gasification of coal, before natural gas was advanced. Concern over the use of this technique in biomass has risen recently as a cleaner fuel and ambidextrous gas can be manufactured from the original biomass. Gasification is a partial thermal oxidation process in which a high percentage of gaseous products such as CO_2, CO, H, and gaseous hydrocarbons, and very small amounts of solid products, such as ash, tar, and oils are formed. As oxidizing agents, steam, air, or oxygen are added to the reaction. Gasification with air: This method is common because it is cheap. It is a highly efficient conversion since it is not necessary to obtain oxygen. However, the product gas produced by this method will also contain water vapor, CO_2, hydrocarbons, pitch, and nitrogen gas. The N_2 content rises to 60%; in this case, the heating value of the product gas is low. Gasification with oxygen: It produces a higher-quality gas with a higher calorific value. The disadvantages of using oxygen are its high cost and safety problems. Gasification with Steam: the gasification of biomass by steam produces a gas containing CO, CO_2, H_2, CH_4, and light hydrocarbons, coal, and pitch. This method has disadvantages, such as corrosion and catalysis poisoning. However, when the aim is to produce hydrogen, not gas, gasification with steam comes to the fore. By making the produced gas more standardized than the original biomass in terms of quality and usage, it can be used as a chemical raw material or in gas engines and gas turbines for liquid fuel manufacture. In order to start the gasification process, a part of the carbon element in the solid waste is burned, and the gasifier first passes the water to the vapor phase, then reaches the pyrolysis phase, and then rises to the gasification reaction temperature.

The reactions occurring in the gasifier are evaporation of water, pyrolysis, reduction (gasification), and combustion. According to the gasifier bed condition, it can be split into a fixed bed, a fluidized bed, and an entrained bed. Plasma gasification, which is mentioned as a new gasification technique, and especially important for aqueous biomass, and where researche is concentrated. When another type of gasification, supercritical water gasification, is added, five different systems emerge. With the gasification process, low-value biomass adds value to the material (Begum et al., 2014; Castello and Fiori, 2011; Islam, 2020).

7.3.6 CARBONIZATION

In carbonization, organic materials such as wood, peat, and coal undergo chemical changes in different temperature regions in an airless environment. The evaporation of water is completed at 170°C, and the degradation reactions of the wood polymer begin to occur at temperatures higher than 180°C. An exothermic reaction occurs by releasing methanol, acetic acid, tar, CO, and water in the temperature range of 200°C–350°C. Dehydration reactions occur at temperatures higher than 500°C. As a result of the carbonization process, 50% CO, 35% CO, 10% CH_4, and 5% other hydrocarbon and H_2 gases are released.

7.3.7 AIRLESS DIGESTION

Airless digestion is a biological process. It is made with microorganisms that can live in an oxygen-free environment. It is the conversion into a fuel and valuable fertilizer that can be used almost everywhere by fermentation in an oxygen-free environment with the help of microorganisms in biomass. Airless digestion method: the efficiency of the conversion process varies depending on the biomass source used, system size, pH value, and temperature. As it is known, biomass undergoes fermentation in an oxygen-free environment with the help of microorganisms, leaving behind valuable fertilizer, methane gas, and carbon dioxide. Biogas is the best-known and most widely used gas fuel produced from biomass by this method (Yelmen and Cakır, 2016; Zhang et al., 2015).

7.3.8 FERMENTATION

It is a chain of reactions in which glucose is converted to alcohol. Sugary and starchy plants are applied to lignocellulosic plants. Biomass contains varying proportions of cellulose, hemicellulose, and lignin. Cellulose can be converted into glucose by applying chemical hydrolysis followed by enzymatic hydrolysis. This process needs to be done with excessive care because chemical hydrolysis can sometimes degrade glucose. With the fermentation of glucose, many chemical products identical to those obtained from ethanol, butanol, crude oil, and acetone can be obtained.

7.3.9 BIOPHOTOLYSIS

Biophotolysis is a process performed with the help of solar energy from microscopic algae in order to obtain oxygen and hydrogen. Algae in the seas and oceans, which act as solar cells, decompose water photosynthetically. It is used in the biological production of hydrogen from renewable energy sources such as the sun and biomass. For this reason, bacteria and microalgae are used (Razu et al., 2019).

7.3.10 ESTERIFICATION

It is the reaction of fatty acids with an alcohol, such as short-chain methanol or ethanol, in an acidic or basic medium. As a result of this reaction, alcohol ester (such as methyl or ethyl ester) and glycerin are formed. Too much glycerin is released and contains impurities, and purification is a very costly process. In biofuel technology, esterification is carried out on oilseed crops such as canola (rapeseed), sunflower, soybean, safflower, and cotton; it is a method applied to animal fats and frying oils used in food. The basis of this method is the reaction of the oils in the biomass, ethanol, methanol, or another type of alcohol, with glycerin, and as a result, biodiesel is obtained (Gomes et al., 2015; Mandari and Devarai, 2022).

7.4 MICROALGAE

Algae are simple microscopic autotrophic (using inorganic compounds like salts, CO_2, and light energy resources to grow) or heterotrophic organisms (using external organic compound nutrients as an energy resource, although not photosynthetic) that grow in aquatic environments and can be in unicellular or multicellular structures. Algae are simple microscopic organisms that grow in an aquatic environment, which can be either unicellular or multicellular, and that are autotrophic (utilization of inorganic compounds such as salts, CO_2, and light energy sources for growth) or heterotrophic (not photosynthetic but using an external organic compound as energy source). Microalgae, which can be based on both prokaryotic and eukaryotic structures, can multiply quickly and survive even under adverse conditions due to their single-celled or easy multi-celled structure. Sizes can range from 3–10 cm to 70 cm long. It is known that they can grow up to 50 cm per day. Ecologically, algae can be found in many areas of the earth. But the main habitat for 70% is water. Algae do not have roots, stems, leaves, or a well-defined vascular system. They live in wetlands, soil, trees, and rocks with bodies or structures with similar functions. Algae can reproduce in three different ways: vegetative, sexual, and asexual reproduction. They usually reproduce through vegetative reproduction. Some species form colonies first and then divide as a result of normal growth. In some species, vegetative reproduction occurs through the growth of the mother plant. Algae are structurally divided into two large groups: prokaryotic (microalgae) and eukaryotic (macroalgae). While microalgae are known as blue-green algae, macroalgae are brown algae, red algae, and green algae according to their flagellum structure and the pigments they contain (Adeniyi et al., 2018; Li et al., 2008; Renuka et al., 2015). There are numerous species of microalgae that live both in aquatic and terrestrial environments and in a wide variety of habitats. It is predicted that there are more than 50,000 species of microalgae on earth, but only about 30,000 of them have been investigated. Microalgae are photosynthetic organisms with relatively simple requirements for growth that need sunlight to transform H_2O and CO_2 into proteins, lipids, amino acids, carotenoids, polysaccharides, and other bioactive compounds.

Microalgae are promising biomass stocks due to their rapid growth, high breeding rate, and low greenhouse gas emissions. Microalgae are microorganisms that include carbohydrates, proteins, lipids, and vitamins. In general, depending on the species, microalgae can contain approximately 15%–77% oil. Microalgae has a high oil rate and growth efficiency compared to other oil crops, which is an advantage for biodiesel and biogas production. The production of these fuels from microalgae has the potential to respond to the increasing global energy desire and to contribute to the interception of global warming, partially by converting the excess carbon dioxide in the atmosphere into an efficient product through photosynthesis. In addition, microalgae that do not pose a great danger, such as terrestrial energy plants, use drinking water resources during their production. It also provides the advantage of removing pollutants such as nitrogen and phosphorus from nature through wastewater treatment. These advantages of microalgae enable them to be identified as a potential raw material with validity for biofuel production. Although the lipid content of microalgae has been identified as the most valuable component in biodiesel biofuel production, their protein and carbohydrate content can also be used in the production of different types of biofuels. For the economics of the biofuel production process, it is necessary to determine the overall chemical composition of algal biomass (Parmar et al., 2011; Liu et al., 2010; Zabed et al., 2019; Ryu et al., 2006).

Microalgae, which prioritize their simple development and energy conservation, have a faster growth rate than plants. The ability to survive in dissimilar environmental circumstances and the production parameters during the growing periods have a prominent effect on microalgae growth. The most important parameters for the living conditions and reproduction of microalgae are nutrient quality and quantity, light, pH, aeration, salinity, and temperature. Microalgae can double their numbers in as little as one day. A wide variety of products can be produced in microalgae due to the high concentrations of native proteins, carbohydrates, lipids, pigments, enzymes, and vitamins. Two types of systems, open and closed, are used in microalgal production technology. In open systems, microalgae culture takes place in inclined, circular, unstirred, and open ponds designed as race tracks, while in closed systems, it takes place in flat plate, tubular, internally illuminated, and fermenter-type photobioreactors. Light source and intensity are important parameters in the regulation of both production systems. While a natural light source from the sun is used as a light source in outdoor production systems for microalgae, different artificial sources, such as optical fibers and light-emitting diodes are used in indoor production systems. In addition, it is possible to use fiber optic systems for the further transfer of solar energy to electrical energy. At the same time, photosynthetically active radiation above saturation can be a growth-limiting factor in both growing systems, thereby minimizing biomass productivity. Growing systems that take place in a laboratory environment can be monitored and edited, while external systems are not.

Microalgae is a versatile raw material with wide energy potential. Most of the microalgae are evaluated in the food industry. It is used in medicine, pharmacy, and cosmetic products due to the pigments and vitamins it contains. It is used as an organic fertilizer in agriculture. In addition, in recent years, different conversion methods have been used depending on the content of algae, and it has started to be evaluated in the production of biomass for the production of fuels like bioethanol, biodiesel, biogas, and biohydrogen from algae. On the other hand, while the high lipid content of some algae species is promising for biodiesel production, proteins, long-chain fatty acids, and pigments are reserved for pharmaceutical and nutritional implementation. Under special cultural conditions such as a high C/N ratio or stressful circumstances, 30%–50% of their weight content of lipid is accumulated by algae. Microalgal lipids are classified in two ways. Fatty acids, including 14–20 carbons, are suitable for biodiesel production. Polyunsaturated fatty acids with more than 20 carbons are used as healthy nutritional additions.

7.5 BIOFUEL

Biofuels are solid, gaseous, and liquid fuels with high energy content obtained through various biochemical and/or thermochemical conversion processes of agricultural and forest products and animal, plant, and municipal residues. Biofuels supplied from biomass can be used alone or mixed with fossil fuels. Since the carbon in biofuels is produced as a consequence of plants breaking down CO_2 in the air, the burning of biofuels does not cause a full rise in CO_2 in the earth's atmosphere. Biodiesel from crude oil of oilseed plants (rapeseed, soybean, safflower, sunflower, etc.) produced in the world, bioethanol from starch, sugar, and cellulose of carbohydrate plants (corn, potato, wheat, sugar beet, etc.), and biogas from vegetable and animal wastes. Depending on the content of algae, electricity, ethanol, hydrogen, methane, and biodiesel can be produced by biochemical methods, as well as syngas, biological coal, biodiesel, and electricity, which can be produced by using thermochemical methods. For example, if it will be used for biodiesel production, species with high oil content should be selected; if it will be used for biohydrogen production, hydrogen-producing species should be selected; and if it will be used for bioethanol production, species with high carbohydrate content should be selected (Parmar et al., 2011; Demirbas, 2001; Li et al., 2008).

7.5.1 BIOHYDROGEN

Hydrogen is a valuable gas with a high energy value (3,042 cal/m^3). Hydrogen gas can be produced by biological, electrochemical, and thermochemical methods. Thermochemical and electrochemical methods use energy and are not always environmentally friendly. Since biological hydrogen production processes are operated at ambient temperature and pressure, there is less energy consumption. Since the main product formed when it is burned is water, it is considered to be a non-polluting fuel. Compared to other gaseous fuels, it is harmless to the environment and humans. Hydrogen energy is renewable, sustainable, environmentally friendly, and less energy-intensive, considering the increasing energy needs and environmental impacts. Today, hydrogen is supplied from fossil fuels, biomass, and water. Hydrogen can be generated by biophotolysis, photofermentation darkfermentation, or a combination of these processes. The most important obstacle to the commercialization of these processes is the low hydrogen production efficiency and speed. Numerous studies have been conducted in the literature on hydrogen production from various organic wastes such as sugar, tofu, red meat, starch factory wastes and food wastes, sewage sludge, and molasses. Studies in which biomass microalgae are used in hydrogen production are very limited. These are generally concentrated on hydrogen production by photofermentation. Some microalgae species can produce biological hydrogen by splitting water into hydrogen and oxygen under anaerobic conditions using solar energy and hydrogenase/nitrogenase enzymes. *Anabaena* sp., *Chlorella pyrenoidosa*, *Spirulina platensis*, *Platymonas subcordiformis*, *Chlorella vulgaris*, and particularly *Chlamydomonas reinhardtii* are commonly used microalgae species for biohydrogen production (Buitrón et al., 2017; Li et al., 2022; Razu et al., 2019).

7.5.2 BIODIESEL

Biodiesel is a product that comes out as a result of the reaction of triglyceride-containing vegetable and animal oils with a short-chain alcohol (methanol, ethanol, butanol, etc.) in the presence of a catalyst and is used as fuel. During the transesterification reaction, which is widely preferred in biodiesel production, fatty acids in the structure of the oil molecule (triglyceride) form new esters with alcohol. Base (NaOH, KOH), acid (H_2SO_4, HCl), or enzyme (biological) catalyst can be used to accelerate the reaction. Biodiesel is one of these esters obtained as a result of this reaction. Vegetable oils, animal fats, algae, or used waste oils can be used as crude oils in the production of biodiesel. Biodiesel is biodegradable and non-toxic. However, due to its low emission rate, it is also beneficial for the environment. Biodiesel's biodegradability, non-toxicity, low emission profile, high cetane number, high oxygen content, sulfur and aromatic-free nature, superior lubricating ability can be listed as its attractive features. Biodiesel does not contain petroleum. Unlike bioethanol, it can be used pure or mixed with petroleum-based diesel at any rate. The cost of the crude oil used is the most important factor determining the price of biodiesel and determining its competition with petroleum-based fuels. On the other hand, the available areas to produce vegetable products such as peanut, sunflower, soybean, rapeseed, and palm oil, which are required for the crude oil to be used in biodiesel production, are very limited. The cost of biodiesel production and the inadequacy of biodiesel processing facilities have limited the use of oilseed plants as energy crops. The search for a cheap and continuous source of crude oil that is not suitable for renewal is inevitable, especially in overpopulated and food-scarce countries, this is of greater importance. Plants with high fatty acid content are generally preferred as raw materials for biodiesel production. It has been determined that algae, as a renewable, non-toxic biodiesel fuel source, has a higher growth rate and is more resistant to changing environmental conditions when compared with agricultural products and other aquatic plants. Oils obtained from microalgae are chemically similar to vegetable oils and are an alternative raw material to biodiesel. One of the major components that allows microalgae to be simply transformed into biodiesel is triglycerides. The biodiesel produced is both energy-friendly and sustainable and can be used in diesel engines by blending it in certain proportions without making any important engine changes (Antolin et al., 2002; Lang et al., 2001; Miao and Wu, 2006).

7.5.3 BIOETHANOL

Bioethanol is an alternative biofuel to fuel types such as gasoline due to its physical and chemical properties. Bioethanol is a fuel produced by fermentation of starch or cellulose-based products from woody plants such as wheat, corn, and sugar beet and used by blending with gasoline with certain mixing ratios. Various starch-sugar plants are used as bioethanol raw materials. Bioethanol, a gasoline-like fuel obtained from the processing of agroindustrial products such as wheat, barley, rye, rice, potatoes, corn, sugar cane, barley, straw, and molasses, has a very important place. Saccharides formed as a result of the enzymatic reaction of starch and sugar can be used to obtain bioethanol, which can be produced by fermentation with the help of microorganisms or as a result of acidic hydrolysis and their distillation from all kinds of cellulosic mass. Bioethanol production can be realized by using the rich carbohydrate content in the structure of microalgae in the fermentation of bacteria as a food source. Generally, yeast (*Saccharomyces cerevisiae, Saccharomyces bayanus, Pichia stipitis*) or bacterial (*Zymomonas mobilis, Escherichia coli*) cells are used in bioethanol fermentation. The type of microorganisms to be used in order to complete the fermentation process, to convert the bioethanol to the desired level, and to prevent the microorganisms from being inhibited during the reaction (cellulose, hemicellulose, lignin, etc.) in the biomass structure. should be selected in accordance with its content. It has been observed that yeast cells rapidly consume the simple sugars obtained from microalgae and convert them into bioethanol, but when there is no simple sugar in the environment, yeast cells begin to consume bioethanol as a carbon source. For this reason, the yeast/biomass ratio and fermentation time to be used in fermentation are very important. Microalgae are cell factories that convert carbon dioxide into carbohydrates depending on sunlight, apart from lipids, for the production of bioethanol (Nevoigt, 2008; Somda et al., 2011).

7.6 RESULT

It is very important to determine the algae species that will be used primarily in the production of biofuel from microalgal biomass. In addition, the nutrient medium to be used, environmental conditions, and the selection of the appropriate bioreactor are also important parameters that affect the biofuel yield. The most important component of algae is water. Therefore, it plays a significant role in wastewater treatment. Thanks to its role in wastewater treatment, both the cleaning of polluted water and its use as biomass play a major role in the cultivation of algae. The biggest obstacle to the commercialization of microalgal biofuels is production costs. For example, production costs can be reduced by using wastewater as a nutrient source in the cultivation of microalgae. Microalgae-based biofuel production will gain more importance in the future, considering the fact that fossil fuel reserves will not be able to meet the needs if environmental effects are taken into account. It is also very important to choose an efficient and economical method for the biomass sample, which constitutes a significant part of the production costs. For the economics of the biofuel production process, reducing algae production costs and using regional power plants or industrial flue gas and wastewater treatment plants can be used effectively and hybridly. Determination of the overall chemical composition of algal biomass is essential for the content of microalgae, their impact on conversion processes, the variation of biofuels obtained, and the economics of the biofuel production process. For the successful development of microalgal biofuels, less expensive and rapid screening for large algae samples is necessary as an alternative to traditional gravimetric-based measurement protocols that only take a few days and require less biomass. Microalgae spend this energy more efficiently than plants that can use sunlight more efficiently. The oil yield of many microalgae species is better than the yield of the best oil plant. The most important problem for microalgae with such high oil productivity is the production parameters during the growing period.

Today, microalgae are used as a raw material source for many types of biofuels, such as methane production (anaerobic degradation of biomass), biodiesel production (microalgal oils), biohydrogen production (photobiological reactions), and biological coal production. Unicellular or multicellular microalgae are attractive candidates for biofuel research. Thermochemical conversion methods turn all types of algal biomass directly into crude oil, which can be converted to biogases, methane, and others. However, algae must be screened for the accumulation of significant amounts of lipids, carbohydrates, and other biomolecules for the production of specific biofuels such as biodiesel, bioethanol, butanol, and biohydrogen. While the majority of microalgae naturally have a high lipid content, it is feasible to increase the concentration by optimizing parameters affecting growth such as light intensity, salinity, nitrogen level control, CO_2 concentration, and temperature. It is necessary to determine these parameters well and to provide optimum conditions.

Algae-based biofuels, which are sustainable and environmentally friendly biofuel raw materials, must have the capacity to compete with petroleum. In the future, the use of microalgal biofuels will gain more importance in the event that fossil fuel reserves are depleted and their environmental effects reach levels that pose a risk to life. The long-term viability of the technologies chosen to produce biofuels is more important than the short-term benefits.

REFERENCES

Adeniyi, O.M., Azimov, U., and Burluka, A. (2018). Algae biofuel: Current status and future applications, *Renewable and Sustainable Energy Reviews*, 90, 316–335, https://doi.org/10.1016/j.rser.2018.03.067.

Antolin, G., Tinaut, F.V., Briceno, Y., Castano, V., Perez, C., and Ramirez, A.I. (2002). Optimisation of biodiesel production by sunflower oil transesterification, *Bioresource Technology*, 83, 111–114, https://doi.org/10.1016/S0960-8524(01)00200-0.

Begum, S., Rasul, M.G., and Akbar, D. (2014). A numerical investigation of municipal solid waste gasification using aspen plus, *Procedia Engineering*, 90, 710–717, ISSN 1877-7058, https://doi.org/10.1016/j.proeng.2014.11.800.

Beringer, T., Lucht, W., and Schaphoff, S. (2011). Bioenergy production potential of global biomass plantations under environmental and agricultural constraints, *GCB Bioenergy*, 3, 299–312, https://doi.org/10.1111/j.1757-1707.2010.01088.x.

Buitrón, G., Carrillo-Reyes, J., Morales, M., Faraloni, C., and Torzillo, G. (2017). Biohydrogen production from microalgae. In: Muñoz, R., and Gonzalez-Fernandez, C. (eds.), *Microalgae-Based Biofuels and Bioproducts: From Feedstock Cultivation to End-Products*, pp. 209–234. Sawston: Woodhead Publishing, https://doi.org/10.1016/B978-0-08-101023-5.00009-1.

Castello, D., and Fiori, L. (2011). Supercritical water gasification of biomass: Thermodynamic constraints, *Bioresource Technology*, 102(16), 7574–7582, ISSN 0960-8524, https://doi.org/10.1016/j.biortech.2011.05.017.

Cocero, M.J., Alonso, E., Sanz, M.T., and Fdz-Polanco, F. (2002). Supercritical water oxidation process under energetically self-sufficient operation, *Journal of Supercritical Fluids*, 24(1), 37–46, https://doi.org/10.1016/S0896-8446(02)00011-6.

Demirbas, A. (2001). Biomass resource facilities and biomass conversion processing for fuels and chemicals, *Energy Conversion and Management*, 42(11), 1357–1378, https://doi.org/10.1016/S0196-8904(00)00137-0.

Garcia-Nunez, J.A., Pelaez-Samaniego, M.R., Garcia-Perez, M.E., Fonts, I., Abrego, J., Westerhof, R.J.M., and Garcia-Perez, M. (2017). Historical developments of pyrolysis reactors: A review, *Energy & Fuels*, 31(6), 5751–5757, https://doi.org/10.1021/acs.energyfuels.7b00641.

Gomes, M.G., Santos, D.Q., De Morais, L.C., and Pasquini, D. (2015). Purification of biodiesel by dry washing, employing starch and cellulose as natural adsorbents, *Fuel*, 155, 1–6, https://doi.org/10.1016/j.fuel.2015.04.012.

Goyal, H., Seal, D., Saxena, R. (2008). Bio-fuels from thermochemical conversion of renewable resources: A review, *Renewable and Sustainable Energy Reviews*, 12(2), 504–517, https://doi.org/10.1016/j.rser.2006.07.014.

Griffin, D.W., and Schultz, M.A. (2012). Fuel and chemical products from biomass syngas: A comparison of gas fermentation to thermochemical conversion routes, *Environmental Progress & Sustainable Energy*, 31, 219–224, https://doi.org/ 10.1002/ep.11613.

Gupta, V., Ratha, S.K., Sood, A., Chaudhary, V., and Prasanna, R. (2013). New insights into the biodiversity and applications of cyanobacteria (blue-green algae)-prospects and challenges, *Algal Research*, 2(2), 79–97, https://doi.org/10.1016/j.algal.2013.01.006.

Gustavsson, L., Madlener, R., Hoen, H.F., Jungmeier, G., Karjalainen, T., KlÖhn, S., Mahapatra, K, Pohjola, J., Solberg B., and Spelter, H. (2006). The role of wood material for greenhouse gas mitigation, *Mitigation and Adaptation Strategies for Global Change*, 1, 1097–1127, https://doi.org/10.1007/s11027-006-9035-8.

He, Z., Kandasamy, S., Zhang, B., Bhuvanendran, N., EL-Seesy AI., Wang, Q., Narayanan, M., Thangavel, P., and Dar, M.A. (2022). Microalgae as a multipotential role in commercial applications: Current scenario and future perspectives, *Fuel*, 308, 122053, https://doi.org/10.1016/j.fuel.2021.122053.

Islam, M.W. (2020). Effect of different gasifying agents (steam, H_2O_2, oxygen, CO_2, and air) on gasification parameters, *International Journal of Hydrogen Energy*, 45(56), 31760–31774, ISSN 0360-3199, https://doi.org/10.1016/j.ijhydene.2020.09.002.

Kaygusuz, K., and Türker, M.F. (2002). Biomass energy potential in Turkey, *Renewable Energy*, 26(4), 661–678, https://doi.org/10.1016/S0960-1481(01)00154-9.

Kumar, A., Pasangulapati, V., Ramachandriya, K.D., Wilkins, M.R., Jones, C.L., and Huhnke, R.L. (2012). Effects of cellulose, hemicellulose and lignin on thermochemical conversion characteristics of the selected biomass, *Bioresource Technology*, 114, 663–669, https://doi.org/10.1016/j.biortech.2012.03.036.

Lang, X., Dalai, A.K, Bakhshi, N.N., Reaney, M.J., and Hertz, P.B. (2001). Preparation and characterization of bio-diesels from various bio-oils, *Bioresource Technology*, 80(1), 53–62, ISSN 0960-8524, https://doi.org/10.1016/S0960-8524(01)00051-7.

Li, S., Li, F., Zhu, X., Liao, Q., Chang, J., and Ho, S. (2022). Biohydrogen production from microalgae for environmental sustainability, *Chemosphere*, 291, Part 1, 132717, ISSN 0045-6535, https://doi.org/10.1016/j.chemosphere.2021.132717.

Li, Y., Horsman, M., Wu N., Lan, C.Q., and Dubois-Calero, N. (2008). Biofuels from microalgae, *Biotechnology Progress*, 24(4), 815–820, https://doi.org/10.1021/bp070371k.

Liu, W., Zhang, Q., and Liu, G. (2010). Lake eutrophication associated with geographic location, lake morphology and climate in China, *Hydrobiologia*, 644, 289–299, https://doi.org/10.1007/s10750-010-0151-9.

Malode, S.J., Prabhu, K.K., Mascarenhas, R.J., Shetti, N.P., and Aminabhavi, T.M. (2021). Aminabhavi, recent advances and viability in biofuel production, *Energy Conversion and Management: X*, 10, 100070, https://doi.org/10.1016/j.ecmx.2020.100070.

Mandari, V., and Devarai, S.K. (2022). Biodiesel production using homogeneous, heterogeneous, and enzyme catalysts via transesterification and esterification reactions: A critical review, *BioEnergy Research*, 15, 935–961. https://doi.org/10.1007/s12155-021-10333.

Miao, X., and Wu, Q. (2006). Biodiesel production from heterotrophic microalgal oil, *Bioresource Technology*, 97(6), 841–846, ISSN 0960-8524, https://doi.org/10.1016/j.biortech.2005.04.008.

Nevoigt, E. (2008). Progress in metabolic engineering of *Saccharomyces cerevisiae*, *Microbiology and Molecular Biology Reviews*, 72(3), 379–412, https://doi.org/10.1128/MMBR.00025-07.

Oelkers, E.H., and Cole, D.R. (2008). Carbon dioxide sequestration a solution to a global problem, *Elements*, 4(5), 305–310, https://doi.org/10.2113/gselements.4.5.305.

Özbay, N., Pütün, A.E., and Pütün, E. (2001). Structural analysis of bio-oils from pyrolysis and steam pyrolysis of cottonseed cake, *Journal of Analytical and Applied Pyrolysis*, 60(1), 89–101, https://doi.org/10.1016/S0165-2370(00)00161-3.

Panwar, N.L., Kothari, R., and Tyagi, V.V. (2012). Thermo chemical conversion of biomass-eco friendly energy routes, *Renewable and Sustainable Energy Reviews*, 16(4), 1801–1816, https://doi.org/10.1016/j.rser.2012.01.024.

Parmar, A., Singh, N.K., Pandey, A., Gnansounou, E., and Madamwar, D. (2011). Cyanobacteria and microalgae: A positive prospect for biofuels, *Bioresource Technololgy*, 102, 10163–10172, https://doi.org/10.1016/j.biortech.2011.08.030.

Razu, M.H., Hossain, F., and Khan, M. (2019). Advancement of bio-hydrogen production from microalgae. In: Alam, M., and Wang, Z. (eds.), *Microalgae Biotechnology for Development of Biofuel and Wastewater Treatment*. Singapore: Springer, pp. 423–462, https://doi.org/10.1007/978-981-13-2264-8_17.

Renuka, N., Sood, A., and Prasanna, R. (2015). Phycoremediation of wastewaters: A synergistic approach using microalgae for bioremediation and biomass generation, *International Journal of Environmental Science and Technology*, 12, 1443–1460, https://doi.org/10.1007/s13762-014-0700-2.

Ryu, C., Yang, Y.B., Khor, A., Yates, N.E., Sharifi, V.N., and Swithenbank, J. (2006). Effect of fuel properties on biomass combustion: Part I. Experiments-fuel type, equivalence ratio and particle size, *Fuel*, 85(7–8), 1039–1046, https://doi.org/10.1016/j.fuel.2005.09.019.

Sahu, A.K., Siljudalen, J., Trydal, T., and Rusten, B. (2013). Utilization of wastewater nutrients for microalgae growth for anaerobic digestion, *Journal of Environmental Management*, 122, 113–120, https://doi.org/10.1016/j.jenvman.2013.02.038.

Shahbaz, M., Topcu, B.A., Sarıgül, S.S., and Vo, X.V. (2021). The effect of financial development on renewable energy demand: The case of developing countries, *Renewable Energy*, 178, 1370–1380, https://doi.org/10.1016/j.renene.2021.06.121.

Sharma, S., Basu, S., Shetti, N.P., and Aminabhavi, T.M. (2020). Waste-to-energy nexus for circular economy and environmental protection: Recent trends in hydrogen energy, *Science of the Total Environment*, 713, 136633, https://doi.org/10.1016/j.scitotenv.2020.136633.

Sheth, P.N. and Babu, B.V. (2010). Production of hydrogen energy through biomass (waste wood) gasification, *International Journal of Hydrogen Energy*, 35(19), 10803–10810, https://doi.org/10.1016/j.ijhydene.2010.03.009.

Somda, M.K., Savadogo, A., Barro, N., Thonart, P., and Traore, A.S. (2011). Effect of minerals salts in fermentation process using mango residues as carbon source for bioethanol production, *Asian Journal of Industrial Engineering*, 3, 29–38, https://scialert.net/abstract/?doi=ajie.2011.29.38.

Stone, K.C., Hunt, P.G., Cantrell, K.B., Ro, K.S. (2010). The potential impacts of biomass feedstock production on water resource availability, *Bioresource Technology*, 101(6), 2014–2025, https://doi.org/10.1016/j.biortech.2009.10.037.

Toor, S.S., Rosendahl, L., and Rudolf, A. (2011). Hydrothermal liquefaction of biomass: A review of subcritical water technologies, *Energy*, 36(5), 2328–2342, https://doi.org/10.1016/j.energy.2011.03.013.

Vassilev, S.V., Baxter, D., and Vassileva, C.G. (2013). An overview of the behaviour of biomass during combustion: Part I. Phase-mineral transformations of organic and inorganic matter, *Fuel*, 112, 391–449, https://doi.org/10.1016/j.fuel.2013.05.043.

Villagracia, A.R.C., Mayol, A.P., Ubando, A.T., Biona, J.B.M.M., Arboleda, N.B., David, M.Y., Tumlos, R.B., Lee, H., Lin, O.H., Espiritu, R.A., Culaba, A.B., and Kasai, H. (2016). Microwave drying characteristics of microalgae (*Chlorella vulgaris*) for biofuel production, *Clean Technologies and Environmental Policy*, 18(8), 2441–2451, https://doi.org/10.1007/s10098-016-1169-0.

Voumik, L.C., Hossain, M.S., Islam, M.A., and Rahaman, A. (2023). Power generation sources and carbon dioxide emissions in BRICS countries: Static and dynamic panel regression, *Strategic Planning for Energy and the Environment*, 41(4), 401–424, https://doi.org/10.13052/spee1048-5236.4143.

Ward, J., Rasul, M.G., and Bhuiya, M.M.K. (2014). Energy recovery from biomass by fast pyrolysis, *Procedia Engineering*, 90, 669–674, https://doi.org/10.1016/j.proeng.2014.11.791.

Yadav, S., Singh, D., Mohanty, P., and Sarangi, P.K. (2023). Biochemical and thermochemical routes of H_2 production from food waste: A comparative review, *Chemical Engineering Technology*, 46, 191–203, https://doi.org/10.1002/ceat.202000526.

Yang, X., Wu, X., Hao, H., and He, Z. (2008). Mechanisms and assessment of water eutrophication, *Journal of Zhejiang University Science B*, 9, 197–209, https://doi.org/10.1631/jzus.B0710626.

Yelmen, B., and Cakir, M.T. (2016). Biomass potential of Turkey and energy production applications, *Energy Sources, Part B: Economics, Planning, and Policy*, 11(5), 428–435, https://doi.org/10.1080/15567249.2011.613443.

Zabed, H.M., Akter, S., Yun, J., Zhang, G., Awad, F.N., Qi, X., and Sahu, J.N. (2019). Recent advances in biological pretreatment of microalgae and lignocellulosic biomass for biofuel production, *Renewable and Sustainable Energy Reviews*, 105, 105–128, https://doi.org/10.1016/j.rser.2019.01.048.

Zhang, A., Shen, J., and Ni, Y. (2015). Anaerobic digestion for use in the pulp and paper indus-
    try and other sectors: An introductory mini review, *BioResources,* 10(4), 8750–8769,
    doi: 10.15376/biores.10.4.Zhang.

Zhang, H., Gong, T., Li, J., Pan, B., Hu, Q., Duan, M., and Zhang, X. (2022). Study on the
    effect of spray drying process on the quality of microalgal biomass: A comprehensive
    biocomposition analysis of spray-dried *S. acuminatus biomass, BioEnerg Research,* 15,
    320–333, https://doi.org/10.1007/s12155-021-10343-8.

8 Phycoremediation of Pesticides

Gargi Sarkar and Ankita Chatterjee

8.1 INTRODUCTION

The world is experiencing various environmental and economic disturbances in the current era related to energy depletion, soil and water contamination, and global warming. Rapid increases in population, urbanization, and industrialization lead to environmental pollution. The absence of appropriate sewage treatment plants in the underdeveloped areas leads to the disposal and dumping of domestic, agricultural, and industrial effluents directly into the water bodies. The existence of certain anthropogenic substances in the natural environment should not be a direct threat to mankind, the biodiverse ecosystem, or any other level. Thus, it is required to confine the use and application of substances that are hazardous to the environment (Priyadharshini et al., 2021). Remediation methods are extremely important in order to remove the undesirable components from the environment. Residual parts of organic compounds, like pharmaceutical drugs and pesticides, create a strong negative impact on the environment. Though wastewater treatment plants have been extensively explored in recent days, they are not appropriately designed and constructed for the elimination of organic compounds, which are complex in nature (Daneshvar et al., 2010). The wastewater treatment plants can, however, be used in the degradation of degradable organic carbon, phosphorus, and nitrogen. In certain cases, it is observed that complex organic contaminants are released from the wastewater plant without being treated properly and thus might cause pollution in the environment. In general, organic contaminants are degraded and eliminated by physicochemical methods, like the application of activated carbon, ozone oxidation, hydrogen peroxide, UV irradiation, and membrane filters (Zaini et al., 2010).

The traditional techniques are associated with certain flaws in the remediation of the organic components, and thus biological methods of remediation are explored by scientists and researchers. Biological remediation techniques are sustainable methods that efficiently remediate organic pollutants, even at low concentrations. Bioremediation is a cost-effective and eco-friendly mechanism that uses biological materials, such as various parts of plants and different microorganisms. Among the microbes, bacteria, algae, and fungi are most commonly used in remediation strategies (Monteiro et al., 2012).

Microalgae are well known for their diverse applications in the fields of environmental remediation and renewable energy. Microalgae are unicellular, eukaryotic organisms that are mixotrophic in nature. These organisms are used in bioremediation techniques due to their extreme adaptability in the ecosystem. Application of

DOI: 10.1201/9781003390213-8

cyanobacteria, microalgae, or macroalgae in partial or complete removal of toxic pollutants in the environment is termed phycoremediation (Aranguren Díaz et al., 2022).

8.2 MICROBIAL BIOREMEDIATION

The transformation of toxic pollutants into non-toxic end products using microorganisms or plants is categorized as bioremediation. The end products after the completion of bioremediation are generally water with either carbon dioxide or methane. The application of microbial bioremediation has gained extreme importance in waste water treatments. The technique has even been innovated and improved for the removal of hazardous components from various effluents. The cost-effectiveness, environmental safety, and reduced or no sludge production of microbial bioremediation are advantages that make the concept much more efficient (Bhardwaj et al., 2019).

Microbial bioremediation of organic pollutants might occur through active or passive phenomena. The active process of bioremediation, often known as bioaccumulation, involves the uptake of contaminants by living microorganisms via their cell membrane. Accumulation or metabolization of the organic compound occurs in the process. However, in the case of passive processes, also termed biosorption, physicochemical properties are observed where chemical structures and functional bonds on the cell wall play a role in the uptake of organic compounds. Various studies have revealed that microalgae are efficient in the biosorption and metabolization of pesticides (Ghasemi et al., 2011; Monteiro et al., 2012). Enzymes released by microbes play an important role in the breakdown of the toxic compound. The active ingredient in the pesticides is degraded by microbial enzymes. Certain enzymes that effectively degrade pesticides belong to the group of dioxygenases, oxidoreductases, hydrolases, monooxygenases, lyases, lipases, cellulases, and proteases. Thus, clearly, the microbial bioremediation follows the principles of oxidation-reduction, hydroxylation, decarboxylation, denitrification, demethylation, ammonification, and desulfurization based on the chemical nature of the target pesticides. Several microbes often use the phenomenon of co-metabolism as well, where microbes would breakdown the pollutants into non-toxic end products in indirect ways rather than directly using them as an energy or nutrient source (Randika et al., 2022).

8.3 PHYCOREMEDIATION

Algae, defined as thallophytic organisms capable of carrying out photosynthesis, can be classified as a diversified group of aquatic life (mostly, as plants), lacking typical leaves, branches, and roots. The cell walls of algae are primarily composed of cellulose. Algal species are further classified into two unique categories, namely, macroalgae and microalgae. As the name suggests, macroalgae are the multicellular algal species usually observed by naked eyes, whereas microalgae, ranging in size from 0.2 to $100\,\mu m$, are the unicellular algal species. Microalgae are extremely efficient in the utilization of carbon dioxide, water, and nutrients due to their uncomplicated cellular structures. Thus, algae are highly prioritized during the conversion of solar energy into biomass and the reduction of carbon dioxide emissions. Earlier studies have revealed that algae can utilize a combination of chicken waste and cow dung to generate biogas, thus fulfilling both the elimination of chicken waste and energy generation (Liu et al., 2020; Pacheco et al., 2020).

The traditional physicochemical techniques used for wastewater treatment are expensive as well as not very efficient, which in turn results in the attraction of scientists and researchers toward biological treatment techniques. Phycoremediation represents the application of algae (macroalgae, microalgae, and *Cyanobacteria*) in the breakdown or degradation of contaminants from the environment (Figure 8.1). Algal cells exist in nature individually or in groups as highly differentiated plants. Algae uses carbon dioxide in the presence of sunlight in order to fix carbon dioxide and release oxygen into the environment. Thus, algal cells are very important in the environment, accounting for almost half of environmental photosynthetic activity (Rani et al., 2021). The efficient generation of oxygen by algae plays a role in reducing the organic contaminants in the environment. In the absence or partial presence of light, the photosynthesis rate is observed to decrease, as is the utilization of organic components as nutrient sources (Al Azad et al., 2017). Algal biomass is explored for energy generation, food, composting, and waste water treatment. The presence of lipids in algal cells adds much value to energy generation (Nur and Buma, 2019). Microalgal cells perform at their highest efficacy based on physical and ecological factors. Thus, the application of microalgae in the restoration of a healthy environment has been explored with much interest in recent days. In certain cases, the rate of degradation of the contaminants by microalgae might be lower when compared to the application of bacteria and fungi; however, the end product in cases where bacteria and fungi are applied often produces more sludge than algae. Thus, the utilization of algae is considered to be safer among the microbial strategies of remediation (Mustafa et al., 2021).

As far as remediation techniques are concerned, algae have many advantages over conventional physicochemical techniques. The advantages are mainly due to their efficient adsorption capacity, immensely available resources, and rapid remediation process. The requirement of a huge quantity of resources in traditional techniques, along with the release of toxic and hazardous sludges, are their major drawbacks.

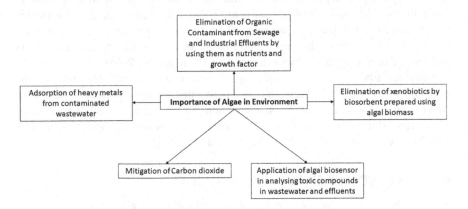

FIGURE 8.1   Implementation and advantages of algal biomass in remediation techniques and improvement of environmental conditions.

However, phycoremediation is an environmentally friendly approach to reducing waste. Among various algal species, *Chlamydomonas* sp., *Chlorella* sp., *Nostoc* sp., *Oscillatoria* sp., *Scenedesmus* sp., and *Spirulina* sp. are the commonly used algae in wastewater treatment. Apart from the above-mentioned algae, filamentous algae like *Oedogonium* sp., *Rhizoclonium* sp., *Microspora* sp., *Spirogyra* sp., *Klebsormidium* sp., *Cladophora* sp., and *Stigeoclonium* sp. also show efficient degradation capacity. In a recent study, microalgae *Pseudokirchneriella subcapitata* is reported to be able to degrade antiretroviral pharmaceutical compounds such as abacavir, darunavir, efavirenz, lamivudine, and telzir in wastewater (Toyama et al., 2018; Deshmukh et al., 2019; Emparan et al., 2019; Reddy et al., 2021).

The preliminary understanding of phycoremediation of pollutants in wastewater occurred in 1960, along with algal biomass harvesting and the generation of biodiesel. However, in recent days, sewage water has been treated in oxidation ponds where microbes, predominantly alae, along with certain bacteria, are harvested to oxidize the waste water. The effective bioremediation of wastewater using algal biomass can simultaneously generate biochar and biogas, which in turn might be a great alternative to fuel. Biocircular economy is targeted with the application of algal biomass as fertilizer (Caines et al., 2014; Huang et al., 2020; Sharma et al., 2021).

Microalgae majorly use three phenomena for the remediation of contaminants, such as biosorption, bioaccumulation, and biodegradation. Pesticides bioremediation are often successfully attempted by biosorption mechanism. The cell wall and extracellular matrix of the microalgae form electrostatic bonds with the contaminants in order to trap them in the biosorbent. Biosorption is greatly affected by various physical factors, like pH and temperature. Biosorbent prepared using microalgal biomass has amino, hydroxyl, carboxyl, and phosphate functional groups, which help in the adsorption of contaminants to the sorbent. The dependence of biosorption on pH is due to the presence of these functional bonds. In cases of decreased pH, the positively charged adsorption sites get activated, which enhances the adsorption of anionic contaminants by electrostatic attraction (Ali, 2010). Bioaccumulation, however, being an active process, involves the treatment of inorganic and organic contaminants by the live algal biomass, followed by transferring them from the aqueous medium to the interior of the microbial cells (Coimbra et al., 2018; Ahmad et al., 2020). The third mechanism observed in biodegradation is that the algal biomass utilizes the contaminants for metabolism and thereby transforms the contaminants into a non-toxic end product. Small molecules of pollutants can be remediated by biodegradation (Narala et al., 2016).

Organic carbon, nitrogen, phosphorus, potassium, and several other nutrients are utilized by microalgae for their growth. These substances are present in the wastewater in an adequate amount. Thus, application of wastewater in algal biomass cultivation can fulfill the purpose of wastewater treatment as well as harvesting biomass simultaneously (Guleri et al., 2020; Sarkar and Dey, 2021).

8.4 HAZARDOUS EFFECTS OF PESTICIDES

To meet the requirements of global agriculture products, the need for pesticides is unavoidable in the current scenario. The development of pesticides and their application in agricultural activities increases crop yield and reduces the chances of crop quality deterioration. In cases where pesticides are not used, it is observed to have a loss of 78%, 54%, and 32% in the production of fruits, vegetables, and cereals, respectively (Tudi et al., 2021). Pesticides, used for horticultural or agricultural practices, can be of various types, such as herbicides, fungicides, insecticides, etc. Although pesticides have a huge impact on the improved economy as well as meeting the world food requirement, they have an adverse effect on human health and the environment.

8.4.1 EFFECTS OF PESTICIDE EXPOSURE ON HUMAN HEALTH

There is ample research going on to study the adverse effects of these chemicals, and significant evidence has been found that proves these chemicals as potential risk factors for humans and other living beings. Some of the pesticides can cause external irritation, while others cause serious poisoning (Aktar et al., 2009). Constant exposure to pesticides can cause modest skin rashes or allergies, as well as severe illnesses like extreme headache, nausea, and dizziness. Certain pesticides, such as organophosphates, for example, have the potential to result in serious illnesses like convulsions, comas, and even death. The level of pesticide toxicity is determined by the medium of exposure, the pathway for exposure, and the portion of the body affected. Some of the toxic effects of pesticides are only short-lived because they can be swiftly reversed and do not result in severe or irreparable harm. Although the damage caused by some pesticides may be reversible, it can take a long time to fully recover. Although exposure to some toxins does not result in death, they may nonetheless cause permanent damage (Damalas and Koutroubas, 2016).

8.4.2 EFFECT OF THE PRESENCE OF PESTICIDES ON FOOD COMMODITIES LIKE FRUITS AND VEGETABLES

Fruits and vegetables are consumed as healthy options due to the presence of vitamins and minerals. These are cultivated in the presence of high concentrations of pesticides to ensure their higher yield and protection against germs, insects, and rodents. According to a current report by the FDA (Food and Drug Administration), a total of 207 different pesticide residues were found in the meals, making them unsafe for ingestion. Pesticide residues have been identified in about 62 percent of vegetables and 82 percent of fruits, which can have adverse effects on human health. According to several investigations, eating food contaminated with a lot of pesticides may increase your risk of developing diseases like cancer, kidney, and lung problems. Children's developing organs make them vulnerable to illness and infection. Children who are exposed to these high levels of chemical residues are at risk of developing cancer as well as mental health issues including autism and attention deficit hyperactivity disorder. The fetus may be harmed if a pregnant woman consumes fruit that has pesticide residue, and she may experience issues during delivery. Other health issues linked to these dangerous compounds include anxiety, dizziness, nausea, cramps in the abdomen, and diarrhea (Gallo et al., 2020).

8.4.3 Effect of Pesticides on the Environment

Pesticides cause an adverse effect on the soil and water adjacent to the area of exposure. These chemicals proved to be toxic for various organisms, like birds, fish, beneficial insects, and non-target plants. Deposition of pesticides in environment leads to the initiation of a process called bio-amplification, where as the pesticides are transferred from one trophic level to another via food chain, the concentration of the pesticides is increased (shown in Figure 8.2) (Baruah and Chaurasia, 2021).

The extensive use of pesticides is primarily responsible for contaminating water resources. Due to the persistent organic compounds found in pesticides, water contamination is on the rise. The migration of pesticide residues to water is influenced by a number of parameters, including rainfall, drainage, microbiological activity, soil texture, and solubility, as well as pesticide half-lives in water. Pesticides are said to be responsible for 5,000–20,000 annual fatalities and 50 lakh to 1 million annual infections. Both groundwater and surface water contain measurable levels of pesticides, which renders them unsafe for use or has detrimental effects on human health (Srivastava et al., 2022).

Populations of beneficial soil microorganisms may drop as a result of heavy pesticide applications to the soil. Dr. Elaine Ingham, a soil scientist, asserts that "soil declines" if both bacteria and fungi are lost. For instance, in order to convert atmospheric nitrogen into nitrates, plants rely on a variety of soil bacteria. Many plants have mycorrhizal fungus growing on their roots, which helps the plants absorb nutrients. Pesticides used on different agricultural lands impede this process (Aftab and Hakeem, 2023). The effects of various chemical pesticides on certain microorganisms are tabulated in Table 8.1.

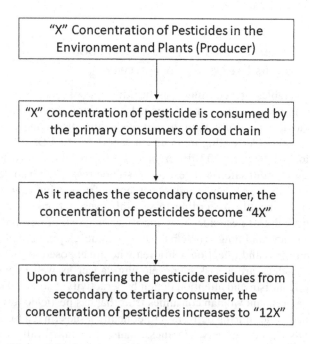

FIGURE 8.2 Bioamplification of pesticides.

TABLE 8.1

Effect of Pesticides on Different Soil Microorganisms

Sl. No.	Compounds in Pesticides	Microorganisms Inhibited	References
1	Triclopyr	Nitrifying bacteria	Pell et al. (1998)
2	Glyphosate	Free-living nitrogen-fixing bacteria	Santos and Flores (1995)
3	2,4-dichlorophenoxyacetic acid (2,4-D)	Growth and activity of nitrogen-fixing blue-green algae	Singh and Singh (1989)
4	Oryzalin and trifluralin	Mycorrhizal fungi	Kelley and South (1978)

8.5 IMPORTANCE OF ALGAE IN PESTICIDE DEGRADATION

Microorganisms are efficient at metabolizing xenobiotic compounds like organic compounds, pesticides, etc. The presence of the cytochrome P450 superfamily of monooxygenase enzymes in these organisms influences the breakdown of these organic compounds. Cytochrome P450 is abundantly found in different types of algae, including Chlorophyta, Rhodophyta and Chromophyta (Lamb et al., 2009). The metabolic activity of several enzymes, including hydrolase, phosphatase, phosphotriesterase, oxygenase, esterase, transferase, and oxidoreductases, is necessary for the biodegradation of pesticides. The herbicide is broken down using three different enzyme- and metabolism-based processes. First, cytochrome P450 activates pesticides in the absence of functional groups by oxidation, reduction, and hydroxylation processes. This is done to obtain more hydrophilic, soluble, intoxicating, and degradable chemicals. Second, the enzymes in the cytosol are transferred to pesticides with functional groups to produce the conjugation with glutathione, glucose, and malonate. Thirdly, the conjugates are moved into vacuoles by glutathione transporters. It was hypothesized that the biosorption, bioaccumulation, and biodegradation processes were crucial in the microalgae remediation of pesticides (Verasoundarapandian et al., 2022). Since algal biomass possesses a higher potential for biosorption, algae are the most effective for the degradation of organic pollutants. The carbohydrate present in the microalgal cell wall aids in the biosorption of harmful pollutants, including pesticides. Their capacity for bioabsorption is dependent on the lipid content of the algal cells, which is influenced by the growing environment and cell distribution. Therefore, by biosorption, microalgae can effectively remove pesticides. Moreover, hazardous substances can accumulate in microalgae (Mustafa et al., 2021). The cleanup procedure uses less energy than other traditional approaches while being more affordable. Microalgae are produced using solar energy; hence, this therapy does not require an external energy source. Microalgae bioprocessing only yields one byproduct in a single stage. This corrective procedure results in a highly pure product. Molecules with added value are easily recoverable. The harvested biomass as a result of this treatment is profitable economically. Microalgal treatment reduces chemical and biological sludge while demonstrating effective pathogen elimination and nutrient recovery (Singhal et al., 2021).

Previous reports show that green algae were used for digesting various organophosphorus pesticides such as methyl parathion, quinalphos, parathion, phorate, malathion, and monocrotophos (Megharaj et al., 1994). Pesticides, namely DTT, chlordimeform, phenol, and lindane, were also degraded by green microalgae (Priyadharshini et al., 2021). The effect of different microalgae on the degradation of chemical compounds present in pesticides has been listed in the table (Ardal, 2014). A study showed that *Scenedesmus quadricauda* effectively removed two fungicides, namely dimethomorph and pyrimethanil, and one herbicide, isoproturon, from their media (Dosnon-Olette et al., 2010). Green alga *Chlamydomonas reinhardtii* showed potential ability to accumulate and break down the herbicide (prometryne) within aquatic ecosystems. The quick elimination of prometryne from the media as a result of this uptake and catabolism can be regarded as an internal tolerance mechanism, indicating that green algae may be helpful in the bioremediation of prometryne-contaminated aquatic habitats (Jin et al., 2012). The pesticide fluroxypyr was accumulated by *Chlamydomonas reinhardtii*. It was severely degraded in the cells, indicating that both accumulation and breakdown took place at the same time (Zhang et al., 2011). Green algae, specifically *Monoraphidium braunii*, appeared to be a viable species for phytoremediating waters from bisphenol A as well as at high contamination levels found in surface water (Gattullo et al., 2012). Recently, it was revealed that bisphenol A was transformed by freshwater microalgae into its mono-glucoside (Nakajima et al., 2007). Three microalgae species were examined for their effect on the herbicides chlortoluron and mesotrione: two chlorophyceae (*Pediastrum tetras* and *Ankistrodesmus fusiformis*) and one diatom (*Amphora coffeaeformis*). While mesotrione caused an increase in cellular density in *A. fusiformis*, chlortoluron significantly inhibited the growth of *A. coffeaeformis* (Moro et al., 2012).

Pesticide biosorption by microalgae is seen as a passive, metabolically impartial method that moves more quickly than bioaccumulation. Microalgae have special cellular walls made of sulfated polysaccharides that can improve the effectiveness of pesticide adsorption from contaminated water. Furthermore, binding sites may be created by the presence of polysaccharides, proteins, or lipids with functional groups like amino, hydroxyl, carboxyl, and sulfate. The parameters that affect absorption capacity are surface-active groups and the attributes of microalgae. The composition of the pesticide, pH, temperature, salinity, nutrients, and the quality and strength of the light are additional elements that influence the elimination and adsorption of the pesticide (Sakurai et al., 2016; Wang et al., 2019). When *Chlorococcum* sp. and *Scenedesmus* sp. degraded α-endosulfan (a cyclodiene insecticide) to endosulfan sulfate, the predominant metabolite, and endosulfan ether, a minor metabolite, in a specific liquid medium, their involvement in this process was clearly confirmed (Sethunathan et al., 2004). When a high density of the algal inoculum was utilized, both metabolites appeared to undergo further breakdown, as indicated by their accumulation only in tiny amounts and the formation of an endosulfan-derived aldehyde. Table 8.2 shows examples of various algal species used for the biodegradation of pesticides. A study was executed to compare the remediation of different pesticides (Atrazine, Molinate, Simazine, Isoproturon, Propanil, Carbofuran, Dimethoate, Pendimethalin, Metoalcholar, and Pyriproxin) using live and lyophilized algae, *Chlorella vulgaris*. The live algal biomass was treated with these pesticides and incubated for 5 days and was able to remove more than 90% of the pesticides, whereas the lyophilized biomass was treated for 60 min only but was able to remove nearly 99% of the same (Hussein et al., 2016). Pesticides can be ingested and eliminated by microorganisms, either actively or passively. In contrast to the passive process, which involves direct physical contact with the chemical structures of the microorganism cell wall, the active process needs externally driven energy. The chemical makeup of the pesticide and the strain of microalgae used have a significant impact on how well pesticides are removed by them. As sources of carbon and nitrogen, microalgae can use cyanide, hydrocarbons, and insecticides. Microalgae use bioaccumulation, biodegradation, and biosorption as methods of removing organic contaminants (Fomina and Gadd, 2014).

8.6 ADVANTAGES AND CHALLENGES IN THE IMPLEMENTATION OF PHYCOREMEDIATION

Phycoremediation offers various benefits, particularly when microalgae are utilized in the process. These benefits can be summed up as rapid remediation due to the high growth rates of microalgae, environmental cleanup with reduced energy and expenses due to autotrophy, and volume reduction of polluted areas due to the straightforward structure and single-celled nature of microalgae. Algal biomass is capable of addressing various concerns that cannot be addressed by the conventional techniques. The simultaneous generation of biomass and production of bioactive compounds during bioremediation of organic components by the microalgae is a huge advantage for mankind. Various bioactive compounds, like phycocyanin, eicosahexanoic acid, and beta-carotene, can be extracted from the harvested algal biomass and utilized for commercial purposes. In recent days, photobioreactors have been an impactful tool for the remediation of contaminants. Algae are used in photobioreactors, and the algal-based reactors can be conveniently used in continuous, semi-continuous, and batch modes. Microalgae, when provided with proper nutrients and important growth requirements, show stable reactions and can be combined with other physicochemical treatments for improved efficiency. The carbon dioxide sequestering ability of algae is a boon against global warming. The oxygen emitted during the process would enhance the quality of air. Since phycoremediation requires no electricity or any toxic chemicals for the treatment, the cost of operation is much lower compared to physicochemical techniques (Sirakov et al., 2013; Young et al., 2017; Daneshvar et al., 2019).

Despite several advantages, certain limitations are faced while implementing the phycoremediation technology on an industrial scale. The major drawback faced in implementation is the requirement of space for the growth of algae. The bioactive compounds produced by algae would be generated only when a huge amount of biomass was produced. However, due to space constraints, enough biomass is not cultivated. The application of genetically modified algae is not acceptable for growth in open systems for phycoremediation. In certain cases, bacterial contamination declines the rate of phycoremediation in open systems. However, in remediation techniques, contamination is not considered a major problem if the microbes are symbiotic with each other. In order to reduce excess contamination, use of low-dose

TABLE 8.2

Remediation of Pesticides Using Microalgae

Sl. No.	Microalgae	Pesticides
1	*Euglena gracilis*	Dichloro-diphenyl-trichloroethane (DDT), parathion, phenol
2	*Chlorococcum sp.*	Mirex, DDT, α-endosulfan, fenamiphos
3	*Scenedesmus obliquus*	DDT, parathion, naphthalene sulfonic acid, α-endosulfan
4	*Cylindrotheca sp.*	DDT
5	*Chlamydomonas* sp.	Lindane, naphthalene, phenol, mirex
6	*Dunaliella* sp.	DDT, naphthalene, mirex
7	*Chlorella vulgaris*	Toxaphene, methoxychlor, lindane, chlordimeform, monocrotophos, quinalphos, methyl parathion, diazinon
8	*Scenedesmus bijugatus*	Monocrotophos, quinalphos, methyl parathion
9	*Chlamydomonas reinhardtii*	Fluroxypyr, isoproturon, prometryne, trichlorfon
10	*Selenastrum capricornutum*	Benzene, toluene, chlorobenzene, 1,2-dichlorobenzene, nitrobenzene naphthalene, 2,6-dinitrotoluene, phenanthrene, di-n-butylphthalate, pyrene
11	*Synechococcus elongatus*	Monocrotophos, quinalphos
12	*Phormidium tenue*	Monocrotophos, quinalphos
13	*Nostoc linckia*	Monocrotophos, quinalphos, methyl parathion, DDT
14	*Oscillatoria animalis*	Methyl parathion
15	*Phormidium foveolarum*	Methyl parathion
16	*Anabaena* sp.	DDT

antibiotics in the open system can be opted for. The application of photobioreactors would help in reducing the problems related to contamination (Brar et al., 2017; Anto et al., 2020; Kumar et al., 2020).

8.7 CONCLUSION AND FUTURE SCOPE

Phycoremediation is drawing the attention of scientists and researchers due to its extreme potential in waste water treatment as an environmentally sustainable approach. The application of algal biomass in remediation opens a great opportunity for introducing a circular economy. Various countries have initiated the process of using microalgae with the aim of restoring environmental quality as well as producing by-products such as biofuel. Although the use of microalgae in the biomonitoring and restoration of aquatic systems favors the extraction and biodegradation of many organic pollutants, other persistent organic pollutants continue to be difficult for the microalgae to degrade. This issue can be resolved, and genetic engineering provides a method for promoting better organic pollutant absorption and bioremediation as well as increased microalgal tolerance to these contaminants. In order to increase the absorption, accumulation, and biodegradation of various pollutants by micro-algae, which speeds up the bioremediation process and shortens the time it takes to decontaminate an aquatic ecosystem, it is also necessary to study and control various aquatic ecosystem parameters such as temperature, pH, nutrient availability, and other environmental parameters.

REFERENCES

Aftab T, Hakeem KR, editors. *Environmental Pollution Impact on Plants: Survival Strategies under Challenging Conditions*. Boca Raton, FL: CRC Press, 2023 May 12.

Ahmad N, Mounsef JR, Abou Tayeh J, Lteif R. Bioremediation of Ni, Al and Pb by the living cells of a resistant strain of microalga. *Water Science and Technology*, 2020 Sep 1;82(5):851–60.

Aktar W, Sengupta D, Chowdhury A. Impact of pesticides use in agriculture: Their benefits and hazards. *Interdisciplinary Toxicology*, 2009 Mar 1;2(1):1.

Al Azad S, Estim A, Mustafa S, Sumbing MV. Assessment of nutrients in seaweed tank from land based integrated multitrophic aquaculture module. *Journal of Geoscience and Environment Protection*, 2017 Aug 3;5(08):137.

Ali H. Biodegradation of synthetic dyes: A review. *Water Air Soil Pollution*, 2010;213:251–73.

Anto S, Mukherjee SS, Muthappa R, Mathimani T, Deviram G, Kumar SS, Verma TN, Pugazhendhi A. Algae as green energy reserve: Technological outlook on biofuel production. *Chemosphere*, 2020 Mar 1;242:125079.

Aranguren Díaz Y, Monterroza Martínez E, Carillo García L, Serrano MC, Machado Sierra E. Phycoremediation as a strategy for the recovery of marsh and wetland with potential in Colombia. *Resources*, 2022 Jan 29;11(2):15.

Ardal E. Phycoremediation of pesticides using microalgae, 2014.

Baruah P, Chaurasia N. Recent perspective on bioremediation of agrochemicals by microalgae: Aspects and strategies. In: Mishra BB, Nayak SK, Mohapatra S, Samantaray D (eds.), *Environmental and Agricultural Microbiology: Applications for Sustainability*. Hoboken, NJ: John Wiley & Sons, 2021 Aug 24, pp. 1–24.

Bhardwaj A, Rajput R, Misra K. Status of arsenic remediation in India. In: Ahuja S (ed.), *Advances in Water Purification Techniques*. Amsterdam, Netherlands: Elsevier, 2019 Jan 1, pp. 219–58.

Brar A, Kumar M, Vivekanand V, Pareek N. Photoautotrophic microorganisms and bioremediation of industrial effluents: Current status and future prospects. *3 Biotech*, 2017 May;7:1–8.

Caines S, Manríquez-Hernández JA, Duston J, Corey P, Garbary DJ. Intermittent aeration affects the bioremediation potential of two red algae cultured in finfish effluent. *Journal of Applied Phycology*, 2014 Oct;26:2173–81.

Coimbra RN, Escapa C, Vázquez NC, Noriega-Hevia G, Otero M. Utilization of non-living microalgae biomass from two different strains for the adsorptive removal of diclofenac from water. *Water*, 2018 Oct 9;10(10):1401.

Damalas CA, Koutroubas SD. Farmers' exposure to pesticides: Toxicity types and ways of prevention. *Toxics*, 2016 Jan 8;4(1):1.

Daneshvar A, Svanfelt J, Kronberg L, Prévost M, Weyhenmeyer GA. Seasonal variations in the occurrence and fate of basic and neutral pharmaceuticals in a Swedish river-lake system. *Chemosphere*, 2010 Jun 1;80(3):301–9.

Daneshvar E, Zarrinmehr MJ, Koutra E, Kornaros M, Farhadian O, Bhatnagar A. Sequential cultivation of microalgae in raw and recycled dairy wastewater: Microalgal growth, wastewater treatment and biochemical composition. *Bioresource Technology*, 2019 Feb 1;273:556–64.

Deshmukh S, Bala K, Kumar R. Selection of microalgae species based on their lipid content, fatty acid profile and apparent fuel properties for biodiesel production. *Environmental Science and Pollution Research*, 2019 Aug 1;26:24462–73.

Dosnon-Olette R, Trotel-Aziz P, Couderchet M, Eullaffroy P. Fungicides and herbicide removal in Scenedesmus cell suspensions. *Chemosphere*, 2010 Mar 1;79(2):117–23.

Emparan Q, Harun R, Danquah MK. Role of phycoremediation for nutrient removal from wastewaters: A review. *Applied Ecology and Environmental Research*, 2019 Jan 1;17(1):889–915.

Fomina M, Gadd GM. Biosorption: current perspectives on concept, definition and application. *Bioresource Technology*, 2014 May 1;160:3–14.

Gallo M, Ferrara L, Calogero A, Montesano D, Naviglio D. Relationships between food and diseases: What to know to ensure food safety. *Food Research International*, 2020 Nov 1;137:109414.

Gattullo CE, Bährs H, Steinberg CE, Loffredo E. Removal of bisphenol A by the freshwater green alga *Monoraphidium braunii* and the role of natural organic matter. *Science of the Total Environment*, 2012 Feb 1;416:501–6.

Ghasemi Y, Rasoul-Amini S, Fotooh-Abadi E. The biotransformation, biodegradation, and bioremediation of organic compounds by microalgae 1. *Journal of Phycology*, 2011 Oct;47(5):969–80.

Guleri S, Singh K, Kaushik R, Dhankar R, Tiwari A. Phycoremediation: A novel and synergistic approach in wastewater remediation. *Journal of Microbiology, Biotechnology and Food Sciences*, 2020 Aug 1;10(1):98–106.

Huang W, Liu D, Huang W, Cai W, Zhang Z, Lei Z. Achieving partial nitrification and high lipid production in an algal-bacterial granule system when treating low COD/NH4-N wastewater. *Chemosphere*, 2020 Jun 1;248:126106.

Hussein MH, Abdullah AM, Eladal EG, El-Din NB. Phycoremediation of some pesticides by microchlorophyte alga, *Chlorella* Sp. *Journal of Fertilizers & Pesticides*, 2016;7(173):2.

Jin ZP, Luo K, Zhang S, Zheng Q, Yang H. Bioaccumulation and catabolism of prometryne in green algae. *Chemosphere*, 2012 Apr 1;87(3):278–84.

Kelley WD, South DB. Effects of selected herbicides on growth and mycorrhizal fungi. *Weed Science*, 1978;28(5):599–602.

Kumar R, Ghosh AK, Pal P. Synergy of biofuel production with waste remediation along with value-added co-products recovery through microalgae cultivation: A review of membrane-integrated green approach. *Science of the Total Environment*, 2020 Jan 1;698:134169.

Lamb DC, Lei L, Warrilow AG, Lepesheva GI, Mullins JG, Waterman MR, Kelly SL. The first virally encoded cytochrome p450. *Journal of Virology*, 2009 Aug 15;83(16):8266–9.

Lavrinovics A, Juhna T. Review on challenges and limitations for algae-based wastewater treatment. *Construction Science*, 2017;20(1):17–25.

Liu J, Pemberton B, Lewis J, Scales PJ, Martin GJ. Wastewater treatment using filamentous algae: A review. *Bioresource Technology*, 2020 Feb 1;298:122556.

Megharaj M, Madhavi DR, Sreenivasulu C, Umamaheswari A, Venkateswarlu K. Biodegradation of methyl parathion by soil isolates of microalgae and cyanobacteria. *Bulletin of Environmental Contamination and Toxicology*, 1994 Aug 1;53(2):292–7.

Monteiro CM, Castro PM, Malcata FX. Metal uptake by microalgae: Underlying mechanisms and practical applications. *Biotechnology Progress*, 2012 Mar;28(2):299–311.

Moro CV, Bricheux G, Portelli C, Bohatier J. Comparative effects of the herbicides chlortoluron and mesotrione on freshwater microalgae. *Environmental Toxicology and Chemistry*, 2012 Apr;31(4):778–86.

Mustafa S, Bhatti HN, Maqbool M, Iqbal M. Microalgae biosorption, bioaccumulation and biodegradation efficiency for the remediation of wastewater and carbon dioxide mitigation: Prospects, challenges and opportunities. *Journal of Water Process Engineering*, 2021 Jun 1;41:102009.

Nakajima N, Teramoto T, Kasai F, Sano T, Tamaoki M, Aono M, Kubo A, Kamada H, Azumi Y, Saji H. Glycosylation of bisphenol A by freshwater microalgae. *Chemosphere*, 2007 Oct 1;69(6):934–41.

Narala RR, Garg S, Sharma KK, Thomas-Hall SR, Deme M, Li Y, Schenk PM. Comparison of microalgae cultivation in photobioreactor, open raceway pond, and a two-stage hybrid system. *Frontiers in Energy Research*, 2016 Aug 2;4:29.

Nur MM, Buma AG. Opportunities and challenges of microalgal cultivation on wastewater, with special focus on palm oil mill effluent and the production of high value compounds. *Waste and Biomass Valorization*, 2019 Aug 1;10:2079–97.

Pacheco D, Rocha AC, Pereira L, Verdelhos T. Microalgae water bioremediation: Trends and hot topics. *Applied Sciences*, 2020 Mar 10;10(5):1886.

Pell M, Stenberg B, Torstensson L. Potential denitrification and nitrification tests for evaluation of pesticide effects in soil. *Ambio*, 1998 Feb 1:24–8.

Priyadharshini SD, Babu PS, Manikandan S, Subbaiya R, Govarthanan M, Karmegam N. Phycoremediation of wastewater for pollutant removal: A green approach to environmental protection and long-term remediation. *Environmental Pollution*, 2021 Dec 1;290:117989.

Randika JL, Bandara PK, Soysa HS, Ruwandeepika HA, Gunatilake SK. Bioremediation of pesticide-contaminated soil: A review on indispensable role of soil bacteria. *Journal of Agricultural Sciences – Sri Lanka*, 2022 Jan 4;17:19–43.

Rani S, Gunjyal N, Ojha CS, Singh RP. Review of challenges for algae-based wastewater treatment: Strain selection, wastewater characteristics, abiotic, and biotic factors. *Journal of Hazardous, Toxic, and Radioactive Waste*, 2021 Apr 1;25(2):03120004.

Rath B. Microalgal bioremediation: Current practices and perspectives. *Journal of Biochemical Technology*, 2012 Sep 30;3(3):299–304.

Reddy K, Renuka N, Kumari S, Bux F. Algae-mediated processes for the treatment of antiretroviral drugs in wastewater: Prospects and challenges. *Chemosphere*, 2021 Oct 1;280:130674.

Sakurai T, Aoki M, Ju X, Ueda T, Nakamura Y, Fujiwara S, Umemura T, Tsuzuki M, Minoda A. Profiling of lipid and glycogen accumulations under different growth conditions in the sulfothermophilic red alga *Galdieria sulphuraria*. *Bioresource Technology*, 2016 Jan 1;200:861–6.

Santos A, Flores M. Effects of glyphosate on nitrogen fixation of free-living heterotrophic bacteria. *Letters in Applied Microbiology*, 1995 Jun;20(6):349–52.

Sarkar P, Dey A. Phycoremediation-an emerging technique for dye abatement: An overview. *Process Safety and Environmental Protection*, 2021 Mar 1;147:214–25.

Sethunathan N, Megharaj M, Chen ZL, Williams BD, Lewis G, Naidu R. Algal degradation of a known endocrine disrupting insecticide, α-endosulfan, and its metabolite, endosulfan sulfate, in liquid medium and soil. *Journal of Agricultural and Food Chemistry*, 2004 May 19;52(10):3030–5.

Sharma GK, Khan SA, Shrivastava M, Bhattacharyya R, Sharma A, Gupta DK, Kishore P, Gupta N. Circular economy fertilization: Phycoremediated algal biomass as biofertilizers for sustainable crop production. *Journal of Environmental Management*, 2021 Jun 1;287:112295.

Singh JB, Singh S. Effect of 2, 4-dichlorophenoxyacetic acid and maleic hydrazide on growth of bluegreen algae (cyanobacteria) Anabaena doliolum and Anacystis nidulans. *Scientific Culture*, 1989;55:459–60.

Singhal M, Jadhav S, Sonone SS, Sankhla MS, Kumar R. Microalgae based sustainable bioremediation of water contaminated by pesticides. *Biointerface Research in Applied Chemistry*, 2021;12:149–69.

Sirakov I, Velichkova K, Beev G, Staykov Y. The influence of organic carbon on bioremediation process of wastewater originate from aquaculture with use of microalgae from genera *Botryococcus* and *Scenedesmus*. *Agricultural Science & Technology*, 2013 Dec 1;5(4):1313–8820.

Srivastava M, Malin B, Srivastava A, Yadav A, Banger A. Role of pesticides in water pollution. *Journal of Agricultural Science and Food Research*, 2022;13:495.

Toyama T, Kasuya M, Hanaoka T, Kobayashi N, Tanaka Y, Inoue D, Sei K, Morikawa M, Mori K. Growth promotion of three microalgae, *Chlamydomonas reinhardtii, Chlorella vulgaris* and *Euglena gracilis*, by in situ indigenous bacteria in wastewater effluent. *Biotechnology for Biofuels*, 2018 Dec;11(1):1–2.

Tudi M, Daniel Ruan H, Wang L, Lyu J, Sadler R, Connell D, Chu C, Phung DT. Agriculture development, pesticide application and its impact on the environment. *International Journal of Environmental Research and Public Health*, 2021 Feb;18(3):1112.

Verasoundarapandian G, Lim ZS, Radziff SB, Taufik SH, Puasa NA, Shaharuddin NA, Merican F, Wong CY, Lalung J, Ahmad SA. Remediation of pesticides by microalgae as feasible approach in agriculture: Bibliometric strategies. *Agronomy*, 2022 Jan 4;12(1):117.

Wang L, Xiao H, He N, Sun D, Duan S. Biosorption and biodegradation of the environmental hormone nonylphenol by four marine microalgae. *Scientific Reports*, 2019 Mar 27;9(1):5277.

Young P, Taylor M, Fallowfield HJ. Mini-review: High rate algal ponds, flexible systems for sustainable wastewater treatment. *World Journal of Microbiology and Biotechnology*, 2017 Jun;33:1–3.

Zaini MA, Amano Y, Machida M. Adsorption of heavy metals onto activated carbons derived from polyacrylonitrile fiber. *Journal of Hazardous Materials*, 2010 Aug 15;180(1–3):552–60.

Zhang S, Qiu CB, Zhou Y, Jin ZP, Yang H. Bioaccumulation and degradation of pesticide fluroxypyr are associated with toxic tolerance in green alga *Chlamydomonas reinhardtii*. *Ecotoxicology*, 2011 Mar;20:337–47.

9 Phycoremediation of Antibiotics

Yousra A. El-Maradny, Dina M. Mahdy, and Mohamed A. Etman

9.1 INTRODUCTION

Antibiotics were extensively used for human, veterinary, and agricultural purposes for the treatment of infections and as growth promoters in livestock farming. However, the abuse and improper use of antibiotics have resulted in a rise in the prevalence of bacteria that are resistant to antibiotics, and the emergence of resistant pathogenic types has made it difficult to use them in therapeutic settings (Rossolini et al., 2014). Antimicrobial resistance is a serious and increasing danger to public health, necessitating a coordinated international effort to create an effective response and stop its rise and spread in both clinical and outdoor contexts (Cantas et al., 2013). Hotspots for antibiotic resistance can also be found in environments under human stress, such as wastewater systems, pharmaceutical production effluents, fishing facilities, and animal production facilities (Berendonk et al., 2015). These locations contribute to the release of antibiotic-resistant bacteria and their genes into the environment and are marked by exceptionally high bacterial loads and subtherapeutic antibiotic amounts. Hospital and industrial effluents are a significant source of pollutants like heavy metals, antibiotics, and disinfectants that are released into receiving public waterways and/or sewage systems, frequently without any previous treatment (Berendonk et al., 2015). One of the major sources of antibiotic and antibiotic-resistant bacteria contamination in soil and aquatic environments is effluent from communities, hospitals, and the pharmaceutical industry (Czekalski et al., 2014; Rizzo et al., 2013). A relationship was detected between the concentration of tetracycline and sulfonamide and the quantity of antibiotic-resistant bacteria and their genes in wastewater (Gao et al., 2012). These pollutants have the potential to access the food chain, water resources used to produce potable water, or therapeutically important areas in the environment (Michael et al., 2013; Rizzo et al., 2013). Wastewater treatment is more complicated than water treatment because of the presence of a wide range of contaminants. When wastewater reuse programs are used for irrigation, these impacts could possibly become even more apparent. The growing scarcity of water, particularly in desert and semi-arid regions, has led to the reuse of water becoming a widespread practice in many parts of the globe. Only a few nations advise pre-treating these wastewaters before releasing them into the environment. As a result, future work should focus on figuring out the processes and circumstances that lead to the development and dissemination of drug resistance, not just in clinical but also in environmental contexts. The majority of wastewater treatment facilities around the globe, especially those

DOI: 10.1201/9781003390213-9

that use mechanical and biological processes, are mainly built to get rid of organic substances, minerals (like nitrogen and phosphorous), and suspended particles. Conventional wastewater treatment focuses on processes that turn wastewater pollutants into harmless materials, allowing the water to be safely disposed of or reused (Acién et al., 2016). However, the effectiveness of the wastewater treatment methods presently in use to eliminate organic micropollutants, such as antibiotics and other antimicrobial agents, is limited. Some antibiotic-resistant bacteria and their genes can withstand wastewater treatment procedures while maintaining (or even increasing) their resistance frequency in comparison to pretreatment levels (Czekalski et al., 2012; Novo et al., 2013; Vaz-Moreira et al., 2014).

Due to the consistent rise in demand, a freshwater shortage has been evident in recent decades. It threatens the long-term, sustainable growth and development of humanity. Water crises are the global risk with the biggest potential effect. The world's expanding population, higher living standards, shifting consumption patterns, and heavy agricultural practices that use irrigation are the main factors contributing to the rising demand for water globally. The goal of wastewater treatment is to convert wastewater into effluent that can be reused by going back through the water cycle. Zhang et al. (2009) analyzed how the wastewater treatment processes affected the frequency of antibiotic resistance in Acinetobacter spp. in the wastewater and the potential dissemination of antibiotic resistance to water sources. The prevalence of antibiotic resistance was discovered to be considerably greater in the outflow samples than in the earlier samples, with trimethoprim (97%), and rifampin (74%).

Phytoplankton, or aquatic organisms that mimic plants, includes algae. They lack roots, branches, or leaves but do contain cell walls made of cellulose. Macroalgae and microalgae are two different varieties of algal cells. Macroalgae are organisms with multiple cells with diameters up to several meters, whereas microalgae are microscopic (unicellular) critters with sizes between 0.2 and 100 μm. Microalgae come in a variety of sizes, colors, and forms, and they are known to generate a wide range of metabolites and enzymes. For protein synthesis (45%–60% of the dry weight of the microalgae), nucleic acids, and phospholipids, microalgae need a lot of phosphorus and nitrogen. In this regard, the use of microalgae for nutrient removal offers significant opportunities for tertiary wastewater treatment that removes ammonia, nitrate, and phosphate. Microalgae are frequently quite susceptible to hazardous substances (Muñoz & Guieysse, 2006). Phycoremediation is the process of removing or transforming pollutants such as nutrients, pharmaceutical products, harmful compounds, poisons, textile dying, heavy metals, and CO_2 from waste air and wastewater using algae for biomass production. Due to their toxicity and persistence, the rising quantity of antibiotics in the environment poses a significant threat. Conventional plant-based treatment methods are ineffective for the purification of effluent-containing antibiotics. Recently, it has been discovered that algae-based technologies are a promising method for antibiotic removal (Li et al., 2022). The most efficient bioremediation will decrease the dissolved nutrient loads in aquaculture effluents by selecting the appropriate algal species and strains. Algae should be highly productive, able to grow rapidly in a variety of environments, and widely available. First, the microbes multiply and produce carbon dioxide for the proliferation of microalgae. Microalgae produce oxygen from carbon dioxide during photosynthesis in the presence of light, which bacteria then use for growth. As a byproduct of photosynthesis, microalgae produce oxygen from water, which microbes use to oxidize organic compounds. Microalgae resolve carbon dioxide, the final product of the bio-oxidation of organic compounds, into cell carbon during photosynthesis. As a result, the pollutants in the effluent are reduced to levels deemed undetectable or acceptable by local authorities. Optimization of temperature, pH, light, and pollutant concentration influences the phytoremediation process. This chapter's objective is to provide a comprehensive overview of the use of microalgae to remove antibiotics from effluent. In addition, the conventional effluent treatment methods and the advantages and disadvantages of these technologies were discussed.

9.2 THE COMPOSITION OF WASTEWATER

Choosing proper wastewater treatment requires knowing the precise makeup of the effluent to be processed, particularly to figure out whether extra carbon, nitrogen, or phosphorus are to be added. Wastewater content might vary significantly depending on the time, source, location, and main activity in the region (agriculture, industry, farms, etc.). Wastewater mainly contains nitrogen, carbon, phosphorus, and other emerging pollutant compounds (Mehrabadi et al., 2015). Furthermore, wastewater may have other contaminants, including heavy metals and other substances (cosmetics, antibiotics, surfactants, etc.) (Muñoz & Guieysse, 2006). The content of wastewater is similar to the culture media used for microalgae, so microalgae could be produced on wastewater and used to clear it at the same time (Acién et al., 2016).

Additionally, three distinct kinds of wastewater are found in the wastewater treatment process: after primary treatment, when the solids and lipids have been eliminated; after secondary treatment, when the majority of the organic matter has been eliminated; and after anaerobic decomposition, when the concentrate, which has a high contaminant content, has been produced (Wagner et al., 2002).

Wastewater contains ammonium, which may be toxic to microalgae at concentrations greater than 100 mg/L when it comes to nitrogen (Collos & Harrison, 2014). It must be diluted before being employed as a growth medium within the reactor since concentrate from anaerobic decomposition often has levels over this limit (Morales-Amaral et al., 2015). Because the bulk of the ammonium over pH 9 is in the form of ammonia, which separates the electron transport in photosystem II and interferes with H_2O in oxidation processes, generating O_2, microalgae are vulnerable to high ammonium concentrations and high pH values (Collos & Harrison, 2014).

The most common antibiotics in the environment that cause a risk of resistance are β-lactams, macrolides, tetracyclines, sulfonamides, and quinolones. These antibiotics are extensively used by humans and animals and can remain in the water, causing the spread of bacterial resistance genes.

9.3 CONVENTIONAL AND ALTERNATIVE TREATMENTS IN WASTEWATER REMOVAL

In order to safely dispose of or repurpose the water, conventional wastewater treatment relies on successive aerobic and anaerobic processes that turn wastewater contaminants into inert substances. These traditional methods are categorized into biological, chemical, and physical methods to remove adequate amounts of carbon, nitrogen, and phosphorus, but at a high cost and with high energy consumption and nutritional loss. Additionally, traditional wastewater treatments are complicated and need trained employees to be managed properly. They also significantly affect the ecosystem because they emit carbon gases (CO_2, CH_4, N_2O, etc.) (Acién et al., 2016). Biological treatment mainly depends on the use of bacteria, algae, and fungi to break down organic pollutants into basic compounds and extra biomass. These biological methods include anaerobic decomposition, aerated wetlands, activated sludge, fungi treatment, trickling filtration, and stabilization (Grady et al., 2011). Chemical processes include catalysis, electrolysis, ion exchange, neutralization, oxidation, and reduction. Physical boundaries are used along with naturally existing forces like van der Waals forces, electrical attraction, and gravity to remove compounds using the physical approach. Typically, physical remediation of polluting compounds does not result in a shift in their chemical makeup (Phoon et al., 2020). Additionally, other conventional physicochemical techniques like coagulation, flocculation, precipitation, sorption, ion-exchange, electrical dialysis, and membrane filtration can be used (Zinicovscaia, 2016).

Conventional and emerging methods for wastewater treatment include those listed below.

Advanced oxidation processes (AOP) are a set of water treatment processes using oxidation, including photochemical and nonphotochemical AOPs (Ameta & Ameta, 2018). AOP mainly acts through the interaction between the organic compound and the hydroxyl radical, forming oxidized products that are less toxic and more susceptible to bioremediation processes (Phoon et al., 2020). AOP of wastewater includes (1) heterogeneous photocatalysis, which depends on a catalyst and light energy, e.g., tetracycline (Demircivi & Simsek, 2019; Kang et al., 2020), ciprofloxacin (Lu et al., 2018), ofloxacin (Wang et al., 2018), and levofloxacin (Zhou et al., 2020); (2) the Fenton reaction, which is a reaction with peroxide, either photo-Fenton (PF) or electro-Fenton (EF). Typically, the presence of Fe ions and H_2O_2 causes a species of active oxygen to develop and oxidize inorganic or organic materials, e.g., tetracycline (Hassan et al., 2020; Kakavandi et al., 2016; Zhao et al., 2020b), sulfamethoxazole (Ranjusha et al., 2020), penicillin (Villegas-Guzman et al., 2017), ciprofloxacin (Li et al., 2017), ofloxacin (Du et al., 2020), and levofloxacin (Liu et al., 2017).

Ozonation is an effective technique to mineralize organic compounds by using ozone, which acts directly by oxidation and indirectly by the production of hydroxyl radicals. Concerning ozone reactivity, phenol, aniline, and amines have very high ozone reactivity, while thioester and anisole have intermediate activity, and amides have no ozone reactivity. This technique could be used in the removal of tetracycline, ampicillin, erythromycin, azithromycin, clarithromycin, sulfamethoxazole-trimethoprim, and ofloxacin (Iakovides et al., 2019). The major drawback of using ozonation is the accumulation of hidden carcinogenic and hazardous byproducts (Ikehata & El-Din, 2004; Liotta et al., 2009).

Adsorption is an advanced method that stands out among the highly effective methods for water treatment and removing contaminants. When chemicals are adsorbed from the liquid phase into the solid phase, a mass transfer process called adsorption takes place. Adsorbates may be attracted to the surface adsorbents through physical or chemical reactions such as activated carbon, silica, and chitosan. Various antibiotics can be removed by adsorption, such as tetracycline (Acosta et al., 2016; Ravikumar et al., 2019), ciprofloxacin (Ma et al., 2020), and other fluoroquinolones (Duan et al., 2019). The increasing capacity degradation after several treatment cycles makes the adsorption method challenging and has some disadvantages. If steam is used for regeneration, the primary disadvantage of adsorption is the secondary waste creation, such as the useless recovered organic substance, spent adsorbent, and organics in the effluent. For the handling of secondary refuse, specialized removal or off-site cleaning is also required (Phoon et al., 2020).

Membrane technology involves the separation of chemical components using synthetic barriers. The quantity of separated antibiotics was found to depend on a variety of variables, including the physio-chemical properties of antibiotics, such as molecular weight, structure, electrical charge, and membrane characteristics, such as material, pore size, charge, and hydrophilicity (Nasrollahi et al., 2022; Vatanpour et al., 2022). Different types of membrane filters are present, including nanofiltration (NF), reverse osmosis (RO), microfiltration (MF), and ultrafiltration (UF). Among these membrane filters, NF and RO are qualified to effectively remove most organic, inorganic, and microorganism complexes from wastewater during treatment. The main disadvantages of membrane technology are its high cost and high energy consumption. Additionally, the concentrated pollutants need a process to be removed (Nasrollahi et al., 2022).

Advanced biological treatment, including aerobic and anaerobic treatments. The anaerobic process is a series of chemical reactions that include hydrolysis, acidogenesis, acetogenesis, and methanogenesis. The developed anaerobic technologies include the anaerobic bio-entrapment membrane bioreactor, the anaerobic sequencing batch reactor, and the anaerobic membrane bioreactor. An anaerobic process is a chemical reaction involving different biochemical steps, including hydrolysis, acid-forming, acetogenesis, and methanogenesis. While in the aerobic process, the chemical pollutants decompose into carbon dioxide and water under aerobic conditions (Dey et al., 2022).

The phytoremediation technique uses plants to remediate polluted water and soil. More research is being done on the application of bioremediation and phytoremediation techniques for the enhancement of new xenobiotics, especially antibiotics and antibiotic resistance genes. Through degradation (either rhizodegradation or phytodegradation), sorption, absorption, translocation, metabolism, and accumulation of pollutants through storage, plants can detoxify and/or remove antibiotics through phytoremediation. The mechanism of phytoremediation differs based on the type of plant used, using a combination of plants, or combining bioremediation with phytore-mediation. Phytoremediation frequently occurs in soil conditions when pollutants are either passively or actively absorbed by plants (Kafle et al., 2022). Aquatic habitats can also experience phytoremediation via aqua-culture plant growth. In addition to surface sorption and rhizodegradation, diffusive transport mechanisms at the roots of plants also allow chemicals to enter the plant. Following absorption by the plant, phytoremediation may proceed through translocation, plant metabolization, and ultimately plant preservation. The plant can pick up or absorb some pollutants, but it won't break them down. Numerous studies have examined antibiotic root absorption by plants in hydroponic systems not only for bioremedia-tion but also to evaluate the danger of antibiotic buildup from irrigation water and waste application in human-consumed crops and products. The main treatment approach aided by engineered pond treatment systems is phytoremediation, which also includes microbial enzymatic biodegradation (McCorquodale-Bauer et al., 2023).

9.3.1 Stages of Conventional Treatment

Conventional wastewater treatment consists of three stages: primary, secondary, and occasionally tertiary, with a variety of biological and physicochemical methods accessible for each step. The primary process aims to reduce the wastewater's solid composition, including oils and lipids, grease, sand, grit, and settleable particles. This process, which is used in all wastewater treatment, is completely mechanically pro-duced through coagulation and flocculation (filtering and sedimentation). However, despite the use of strong coagulants such as $Al_2(SO_4)_3$ and $FeCl_3$, the main procedure did not provide any appreciable results in the removal of medicines, including anti-biotics. Utilizing ozonation and other AOP techniques can be beneficial, but only if their economic and toxicological implications are taken into account. The second-ary treatment, which usually uses a biological procedure to remove organic waste and/or nutrients with aerobic or anaerobic systems, can, however, vary greatly. The most common biological treatment used in contemporary wastewater treatment is standard activated sludge. Less frequently used are membrane bioreactors, movable bed biofilm reactors, and stationary bed bioreactors (Michael et al., 2013). Certain contaminants found in hospital laundry's effluent outflow, along with an abrupt rise in formic acid, may shock the reactor's pH and reduce efficiency by causing sludge to break down. RO, filtration, adsorption, enhanced oxidation, and nanotechnology are all included in the tertiary process.

Antibiotics were removed from the effluent by a number of procedures. These pro-cesses may be non-biotic or abiotic (such as sorption, breakdown, and photolysis) or biotic (biodegradation, which mostly includes bacteria and fungus). Their adsorption on sewage waste and their decomposition or transformation following treatment play a major role in antibiotic removal. Because the compounds are very briefly exposed to light throughout the process of treating wastewater, photolysis is not particularly likely to occur, although hydrolysis may be necessary for certain chemicals to form hydrophobic or hydrophilic byproducts (Michael et al., 2013). The removal rates of β-lactam during biological treatment are >90% because they are unstable because of the hydrolysis of the β-lactam ring (Watkinson et al., 2009).

9.4 TYPES OF MICROALGAE AND MECHANISMS USED IN THE REMOVAL OF ANTIBIOTICS

Microalgae are extremely diverse organisms that live in a variety of habitats and may produce a wide range of potentially beneficial chemicals and metabolic products. Microalgae have more effective access to carbon dioxide, water, and nutrients owing to their simple cellular structure compared to other methods. There are benefits to using microalgae-based cleaning methods, such as the elimination of contaminants and pollutants from wastewater, the release of oxygen (which may be eaten by microbes in the wastewater), and the fixation of carbon dioxide. Different species of microalgae have been used in the phycoremediation of various types of wastewaters (Xiong et al., 2018). High growth rates, high biochemical characteristic content and productivity, higher tolerance to potential pollutants (such as metal ions and toxic compounds present in wastewater), efficient photosynthetic capacity, broad adaptability, high oxygen generation rates, a high carbon dioxide sinking capacity, and robust growth properties with improved tolerance for varied environmental conditions are the features of algae most desired for use in wastewater treatment. These adaptable microalgae species have been used in a range of algal bioreactors and high-rate algal reservoirs to treat a variety of pharmaceutical contaminants as well as water from municipalities, anaerobic digestion effluent, textile pigmentation wastewater, agro-industrial wastewater, metallic wastewater, and industrial wastewater (Amenorfenyo et al., 2019; Mojiri et al., 2021).

Cyanobacteria (blue-green prokaryotes), Chlorophyceae (green eukaryotes), Phaeophyceae (brown eukaryotes), Rhodophyceae (red eukaryotes), and Bacillariophyceae (diatom eukaryotes) are some of the extensively used microalgae. A crucial factor in the treatment of wastewater is the choice of microalgae species based on their capacity for rapid growth and their effectiveness in absorbing nutrients from effluent. For the removal of nutrients (nitrogen and phosphorus), *Scenedesmus, Chlorella,* and *Botryococcus* are often used as microalgae (Emparan et al., 2019).

Microalgae may be used to clean wastewater in both immobilized and suspended free-cell cultures. Microalgae live cells move autonomously inside the bottles carrying media under the suspended free-cell culture condition to achieve uniform cell distribution. Microalgae with immobilized cells cannot easily move from their starting point to any other region of the medium because of this restriction. This method may be carried out by preserving the live microalgae cells in carriers like alginate and chitosan beads. When wastewater is treated using a suspended free-microalgae cell culture, nitrogen and phosphorus are removed from the wastewater, while aerobic microorganisms that cohabit in the culture are given oxygen. Microalgae have a strong capacity for nutrient absorption and have the ability to effectively remove heavy metals, nitrogen, and phosphorus from wastewater (Emparan et al., 2019).

Antibiotic treatment by algae is a multi-way process; how the algae react to the effects of the antibiotics may affect how well they are removed in future treatment runs. Algae's susceptibility to antibiotics generally differs greatly Figure 9.1.

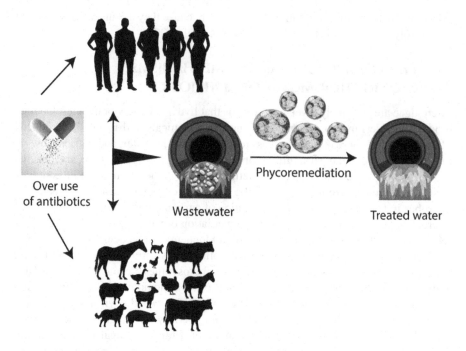

FIGURE 9.1 Diagram demonstrating the overuse of antibiotics by humans and animals and the removal of antibiotics from wastewater by phycoremediation.

The creation of algae-based antibiotic elimination methods must consider the inhibition of microalgae brought on by antibiotics. Antibiotics may have an impact on the development of algae by preventing the production of pigments and compounds like chlorophyll-a, as well as by inhibiting the actions of enzymes like catalase and superoxide dismutase (Bashir & Cho, 2016; Perales-Vela et al., 2016).

Many factors can affect the performance of microalgae in the removal of antibiotics from wastewater. These factors include microalgal species, the growth condition of alga, the concentration and class of antibiotics, and the composition of wastewater (Leng et al., 2020). Each microalgal species is able to remove a certain class of antibiotics at their tolerable level, the concentration of antibiotics they can persist and grow.

9.4.1 MECHANISMS OF PHYCOREMEDIATION OF ANTIBIOTICS FROM WASTEWATER

There are different mechanisms used by algae to remove antibiotics, such as biotic pathways (biodegradation and accumulation) and abiotic pathways (adsorption and photolysis) (Yu et al., 2022), as presented in Table 9.1. Through a variety of methods, biosorption has been employed to remove potentially harmful components. In the past, biosorption was divided into two categories: metabolism-dependent and metabolism-independent. Later, the terms "bioaccumulation" and "biosorption" were used to describe processes that depend on metabolism. **Biodegradation** is an intracellular or extracellular process by which antibiotics are broken down by algae.

TABLE 9.1

Phycoremediation Process Involved in Antibiotics Removal from Wastewater

Microalgae	Antibiotic Removed	Removal (mg/L, %), Cultivation Time	Mechanism	References
Chlamydomonas reinhardti	Chlortetracycline (CTC)	At 1×10^{-3} mol/L CTC, removal efficiency was 99% in 24 h	Biodegradation	Zhao et al. (2020a)
Chlamydomonas sp. Tai-03	Ciprofloxacin and sulfadiazine	>1,000 mg/L day of microalgae removed 65.05% and 35.60% of antibiotics, respectively	Biodegradation and photolysis, respectively	Xie et al. (2020)
Chlorella pyrenoidosa	Ceftazidime and 7-aminocephalosporanic acid	40 mg/mL for 6 and 24 h	Adsorption and biodegradation	Yu et al. (2017)
Chlorella pyrenoidosa	Roxithromycin (ROX)	At 0.1 and 0.25 mg/L ROX, removal efficiency was 45.9–53.3%, in 21 days	Biodegradation, bio-adsorption, and accumulation	Li et al. (2020)
Chlorella pyrenoidosa	Cefradine	60 mg/L of algae removed 76.02%	Sequencing batch reactor algae process (SBAR)	Chen et al. (2015)
Chlorella pyrenoidosa	Sulfamethoxazole (SMX)	At 0.4 μM of SMX, 99.3% removed, after 5 days	Biodegradation	Xiong et al. (2020)
Chlorella sp. Cha-01, *Chlamydomonas* sp. Tai-03, and *Mychonastes* sp. YL-02	Cephalosporin	At 100 mg/L of cephalosporin, 1 mg/mL of Cha-01, YL-02, and Tai-03 removed cephalosporin with 12.0%, 9.6%, and 11.7%, respectively, for 13 days	Photolysis, hydrolysis, and adsorption	Guo et al. (2016)
Chlorella sp. (UTEX1602 and L38)	Florfenicol (FF)	At 46 mg/L of FF, 97% were removed	Biodegradation	Song et al. (2019)
Chlorella sp. CS-436	Sulfonamide	24%–38%, for 30 days	CaO_2 treatment followed by biodegradation	Vo et al. (2021)

(Continued)

TABLE 9.1 (*Continued*)
Phycoremediation Process Involved in Antibiotics Removal from Wastewater

Microalgae	Antibiotic Removed	Removal (mg/L, %), Cultivation Time	Mechanism	References
Chlorella vulgaris	Metronidazole	At 1–50 μM metronidazole and 0.05 and 0.5 g/L of *C. vulgaris*, removal efficiency was 59% in 20 days	Bio-adsorption	Hena et al. (2020)
Chlorella vulgaris and *Scenedesmus obliquus*	Sulfamethazine and enrofloxacin	1 mg/mL of *C. vulgaris* and *S. obliquus* removed 17% and 7.3%, respectively, of sulfamethazine. And 52% and 43.3% of enrofloxacin, respectively, after 15 days	Biodegradation and photolysis	Chen et al. (2020)
Dictyosphaerium	Oxytetracycline (OTC) and ofloxacin (OFLX)	96.3%–100.0% OTC, and 32.8%–60.1% OFLX	Biodegradation	Zhang et al. (2022)
Haematococcus pluvialis, *Selenastrum capricornutum*, *Scenedesmus quadricauda* and Chlorella vulgaris	Sulfamerazine, sulfamethoxazole, sulfamono-methoxine, trimethoprim, clarithromycin, azithromycin, roxithromycin, lomefloxacin, levofloxacin and flumequine	20, 50 and 100 μg/L, for 40 days. *S. capricornutum*, and *C. vulgaris* showed high removel affinity to macrolides and fluoroquinolones, while *H. pluvialis* and *S. quadricauda* showed affinity in removal of sulfonamides	Biodegradation, adsorption, and bioaccumulation	Kiki et al. (2020)
Mainly *C. vulgaris*	Tetracycline	At 0.82 mg/L, 69% of tetracycline removed, after 30 days	Photodegradation and biosorption	De Godos et al. (2012)

(Continued)

TABLE 9.1 (Continued)
Phycoremediation Process Involved in Antibiotics Removal from Wastewater

Microalgae	Antibiotic Removed	Removal (mg/L, %), Cultivation Time	Mechanism	References
Mainly *Chlorella sorokiniana*	Cephalexin and erythromycin	96.54% and 92.38%, respectively	Bio-adsorption, accumulation, and photodegradation	da Silva Rodrigues et al. (2021)
Microalgae-bacteria consortium mainly *Chlorella sorokiniana*	Sulfamethoxazole	52 µg/L of microalgae-bacteria consortium removed 4.34% ± 2.35%	Biodegradation, photodegradation and bio-adsorption	Da Silva Rodrigues et al. (2020)
Microcystis aeruginosa	Amoxicillin	500 ng/L, 31% for 7 days	Biodegradation	Liu et al. (2015)
Microcystis aeruginosa	Tetracycline	At 10–100 mg/L tetracycline, removal efficiency was 98% in 2 days	Biodegradation and hydrolysis	Pan et al. (2021)
Scenedesmus obliquus, *Chlamydomonas mexicana,* *Chlorella vulgaris,* *Our-ococcus multisporus,* *Micractinium resseri*	Enrofloxacin	1 mg/mL of microalgal species removed 18%–26%	Biodegradation	Xiong et al. (2017)
Spirulina platensis	Chlortetracycline (CTC)	At 1.0 mg/L CTC, removal efficiency was 98.63%–99.95% for 13 days	Biodegradation	Zhou et al. (2021)

In the intracellular process, the antibiotics are adsorbed on the algal wall, and then they are broken inside the algae by carboxylation, decarboxylation, hydroxylation, demethylation, etc. using the secreted enzymes (Xiong et al., 2017). However, in the extracellular one, the antibiotics are broken down by the extracellularly secreted metabolites. The processes for antibiotic degradation can either be co-metabolism, in which the existence of non-specific enzymes catalyzing the metabolism of other substrates is required for the breakdown of pollutants, or metabolic degradation, in which organic substances act as the only sources of carbon and energy (Tiwari et al., 2017). Even though biodegradation can completely eliminate antibiotics, some toxic intermediates may be formed. **Accumulation** is an intracellular process to remove antibiotics from wastewater; it can be used alone or as a pre-step before biodegradation (Song et al., 2019). Some antibiotics, such as doxycycline and sulfamethoxazole-trimethoprim, can accumulate in the algae after crossing their cell wall. Antibiotic accumulation can produce reactive oxygen species, resulting in DNA and protein denaturation and cell damage. Bioaccumulation is mainly dependent on passive diffusion and is affected by the concentration and lipophilicity of antibiotics. Bioaccumulation is an energy-independent process that is affected by the type of microalgae used (Prata et al., 2018). **Adsorption** is an extracellular process of removing pollutants removal by their adsorption and passive binding to solid materials. The bio-adsorption of antibiotics to viable or dead algae occurs through the adsorption of antibiotics to specific functional groups on the algal cell wall through hydrogen bonds, electrostatic attraction, surface precipitation, or hydrophobicity (Leng et al., 2020). Because functional groups (such as hydroxyl and carbonyl) are readily available, macromolecules, including proteins, carbohydrates, and lipids found in the cell wall of microalgae, are negatively charged and attract metals and pollutants in the biomass. During the adsorption procedure, the negatively charged cell wall would bind to the oppositely charged pollutants via electrostatic contact and an ion exchange mechanism. Bio-adsorption is preferred in hydrophobic antibiotics with opposite charges to the used algae, as the reduction in antibiotic negativity reduces the adsorption effect. In addition to the previous factors that affected biosorption, the change of other environmental factors such as pH, temperature, incubation time, and the presence of heavy metals can also affect sorption (Xiong et al., 2018). High pH, low temperature, and a decrease inmetal concentrations are factors influencing sorption. **Photodegradation** is the direct and indirect photolysis of antibiotics by light or other reactive metabolites produced from algae in the presence of light, respectively. Some antibiotics, such as tetracycline, ciprofloxacin, and cefazolin, can be removed by direct and indirect photolysis (Jiang et al., 2018; Leng et al., 2020). In indirect photodegradation, hydroxyl radicals are produced from algae in the presence of light, resulting in the breakdown of antibiotics. The main drawback of this method is its instability and shading effect (Zhang et al., 2012). These mechanisms are illustrated in Figure 9.2.

The removal of antibiotics may produce more toxic byproducts than the original ones. A number of antibiotic degradation processes and byproducts, along with the complexities of wastewater, frequently make it difficult for microalgae to remove all antibiotics. Future research must conduct systematic studies on removal processes and the byproducts that come from those pathways. The coexistence of several antibiotics

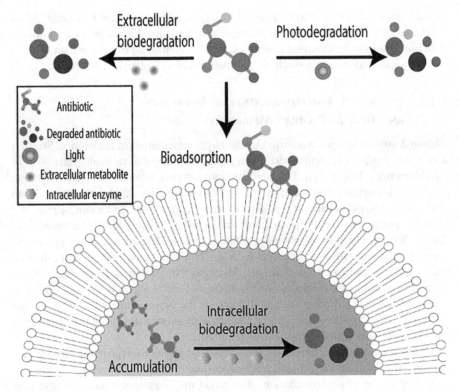

FIGURE 9.2 Schematic presentation of the phycoremediation for the removal of antibiotics from wastewater. The mechanism of phycoremediaiton including intracellular and extracellular bioremediation, accumulation, adsorption, and photodegradation.

and other pollutants in real wastewater is probably the case. The algae species and antibiotic class generally affect how well a medicine is eliminated (Ribeiro et al., 2018). The method by which microalgal biomass is utilized after receiving antibiotic medication is another issue (Leng et al., 2018).

A novel strategy for the efficient treatment of antibiotics in the aqueous phase involves combining microalgae with microbial consortiums. For the improved remediation of organic pollutants like antibiotics, integrated treatment techniques (algae-based technologies paired with advanced oxidation processes (AOPs) and constructed wetlands) and genetic changes would be practical (Xiong et al., 2018). AOP could be used as a pretreatment method, helping in the transformation of complicated compounds into biodegradable ones. However, more efforts are required to overcome the problem, such as in treatment adaptability, overoxidation, and the relation between photoreactor and bioreactor compositions (Marsolek et al., 2014). Constructed wetlands have been widely used because of their cost-effectiveness and ease of maintenance. The integration of microalgae as pretreatment before constructed wetlands can overcome the problem of low oxygen availability, enhance nitrogen removal performance, and improve wastewater treatment quality (Ding et al., 2016). However, these integration methods are still at the laboratory scale and require extensive research to be applied to large-scale wastewater treatment. Algal genetic engineering is a significant systemic technique to address the issue of microalgal biomass in manufacturing processes, to alter the process of metabolism for high product yield, and to create synthetic photoautotrophs (Bajhaiya et al., 2017).

9.4.2 EXAMPLES OF PHYCOREMEDIATION OF ANTIBIOTICS AND THEIR BY-PRODUCT METABOLITES

Chlamydomonas species were used in the phycoremediation of antibiotics. Starting with 8×10^5 m/L of *C. reinhardti* green algae, they are able to biodegrade 99% of the chlortetracycline in 24 h. The degraded byproducts contain iso-chlortetracycline, 4-epi-iso-chlortetracycline, and non-toxic low-molecular-weight products (Zhao et al., 2020a). Ciprofloxacin is a positively charged antibiotic, and it can be removed by biosorption followed by photocatalytic degradation using *Chlamydomonas* sp. Tai-03. The mechanism of photodegradation was achieved by hydroxylation, followed by defluorination and oxidation. Algal biodegradation results in the formation of a byproduct free from fluorine, indicating the detoxification process and the formation of non-toxic products of ciprofloxacin. Ciprofloxacin is ten-fold higher in zeta potential than sulfadiazine, and this affects the biosorption and removal of ciprofloxacin, showing the selectivity of *Chlamydomonas* sp. Tai-03 in the removal of ciprofloxacin over sulfadiazine (Xie et al., 2020).

    Chlorella sp. is one of the most extensively studied microalgae in the phycoremediation process of antibiotics and pharmaceutical products. *Chlorella pyrenoidosa* was used in the phycoremediation of ceftazidime, 7-aminocephalosporanic acid (7-ACA), roxithromycin, cefradine, and sulfamethoxazole by bio-adsorption and biodegradation. Δ-3 ceftazidime and trans-ceftazidime were the non-toxic biodegraded metabolites of ceftazidime, while 4-chlorocinnamic acid is the by-product of the 7-ACA (Yu et al., 2017). Chen et al. (2015) demonstrated that *Chlorella pyrenoidosa* can utilize the cefradine-degraded products as a carbon source to increase their photosynthetic capabilities and better adapt to prolonged exposure to antibiotics and their metabolites. The biodegradation of sulfamethoxazole by *Chlorella pyrenoidosa* was enhanced by the addition of co-metabolites such as sodium acetate. This degradation was held in place by two-step mechanisms. The first pathway was the breakdown of the N–S bond and the formation of byproduct metabolites, including 4-amino-benzenesulfinic acid and 3-amino-5-methylisoxazole. The second pathway was oxidation, hydroxylation, and pterin-related conjugation, which can be held by the cleavage effect of the amino group on the benzene ring (Xiong et al., 2020).

    The primary factor in the adsorption of metronidazole was the formation of exopolymeric compounds, which increased in *C. vulgaris*. Additionally, the zeta potential of *C. vulgaris* in the test cultures was considerably affected by metronidazole, pointing to a change in surface properties. At a stationary phase, this drop in negatively charged surfaces led to auto-flocculation occurrences (Hena et al., 2020). In addition to the previous antibiotics, *C. vulgaris* is able to remove sulfamethazine and enrofloxacin by bioremediation and photolysis. The main degradation pathways of sulfamethazine consist of hydroxylation, methylation, and oxidation. However, the degradation pathways of enrofloxacin are dealkylation, decarboxylation, and defluorination (Chen et al., 2020). The main metabolic pathways of trimethoprim in the presence of algal consortiums are hydroxylation, demethylation, oxidation, and bond cleavage (Kiki et al., 2020). In the former study and other studies, they showed that the same antibiotic can have multiple degradation pathways, either with the same microalgal species or with different species. The mechanisms of degradation are change in the functional group, either by addition or deletion, and ring cleavage (de Godos et al., 2012).

9.4.3 Role of Microalgal Nanoparticles (NPs) for Degrading Toxic Substances

The breakdown of toxic components into less toxic ones allows for the use of biological resources, or their parts and extracts, in bioremediation processes. Nanoformulations are actively used in remediation processes for treating and eliminating contaminants like heavy metals and dye contaminants that can cause serious issues for both terrestrial and aquatic biota. They act mainly by restraining uptake and consumption of dissolved oxygen and diminishing photosynthetic capability (e.g., Fe-based NPs can be used for the removal of heavy metals from soil) (Agarwal et al., 2019).

Algal-produced NPs are employed in a variety of industries because of their toxic profile, simplicity of handling, affordability, and environmental friendliness. Numerous studies have emphasized the crucial role that NPs play in cleaning polluted wastewater.

9.4.3.1 Degradation of Metals and Organic Dyes

Turbinaria conoides and *Sargassum tenerrimum* macroalgae are extracted from water as reducers and caps for the manufacture of gold (Au) NPs. In the presence of sodium borohydride as a reducing agent, the Au-NPs show excellent reduced catalytic performance in the breakdown of organic dyes (Rhodamine B and Sulforhodamine 101). Additionally, nitrogenous substances (4-nitrophenol and p-nitroaniline) were reduced using Au-NPs. Compounds containing nitrogen and organic dyes are also typical pollutants in wastewater. *Ulva lactuca*, a green macrophyte, was extracted to make silver (Ag) NPs, which were then used to catalyze the breakdown of the methyl orange color. Zinc acetate dehydrated from the extract of many algal species, particularly the brown alga *Sargassum muticum*, was used to make zinc oxide NPs. The photo-desulfurization of a dibenzothiophene pollutant and the selective photodegradation of cationic dyes were both successfully accomplished (97%) using the microalgae *Chlorella* zinc oxide NPs. In order to reduce the blooming of toxic algal wastewaters, iron oxide NPs are used for the bioremediation of N and P. These NPs are produced from the water-based extracts of three brown seaweeds (*Petalonia fascia*, *Colpomenia sinuosa*, and *Padina pavonica*). The carbonyl radicals and amines of the polysaccharides and glycoproteins in the algae served as the reducing and stabilizing agents for the formation of iron oxide NPs. Due to their ability to speed up the redox process that results in the production of superoxide and hydroxide radicals, TiO_2 NPs dramatically reduce the amount of light stress (Abideen et al., 2022).

9.4.4 Limitations in the Use of Microalgae in Antibiotic Removal

Temperature, light, pH, and oxygen concentration are just a few of the environmental factors that have an impact on algal biomass production and development. Providing nutrients and recycling them, transferring and changing gases, providing photosynthetically active radiation, keeping the culture's integrity, managing the environment, getting a supply of land and water, and the harvesting procedure are some of the difficulties associated with producing algae on a large scale. Essential nutrients for algal growth include carbon, nitrogen, and phosphate ions. Other less important nutrients that should be present in trace quantities are calcium, zinc, iron, silica, magnesium, and potassium. Microalgae are tiny organisms, and one of the limitations of phycoremediaiton includes the removal of these organisms from water. The issue of collecting and recovering algal biomass from the treated wastewater can be solved by immobilizing microalgae. Immobilized live cells on an appropriate support, as opposed to suspended cells, can contribute to enhanced cell retention time in the reactor, which can simplify the treatment procedure. High rates of hydrogen synthesis and the elimination of inorganic fertilizers from wastewater have both been achieved using a hollow fiber-immobilized cyanobacterial system (Gondi et al., 2022; Ramesh et al., 2023).

Contamination and the presence of other organisms competing with the algae's growth in water are two of the limitations of large-scale microalgae production. These contaminant organisms, such as protozoa and viruses, can infect the microalgae and produce toxic metabolites. In addition, the ability of microalgae to remove antibiotics is mainly affected by the species of algae and the type of antibiotic. For that reason, choosing the proper type of algae strain is a challenging step in phycoremediation. Some wastewater-based microalgae can be genetically engineered to grow faster and survive longer. Understanding algae's metabolic pathways and using genetic engineering to modify various wild-type algal strains can significantly enhance algae's capacity to filter out pollutants (Gondi et al., 2022; Ramesh et al., 2023).

The microalgal production method has certain significant drawbacks, including laborious and time-consuming tasks and microalgae harvesting. Microalgal NPs are often a novel study topic for improved removal efficiency to avoid these effects. The term "nanotechnology" refers to a technology that makes it possible to create, utilize, and comprehend material structures, systems, and devices with essentially unique features and functions as a result of their tiny structure. Nanotechnology is a fantastic field to concentrate on because of how effective NP physiochemical and crystallographic characteristics are (Agarwal et al., 2019; Vargas-Estrada et al., 2020).

9.5 CONCLUSION AND FUTURE PERSPECTIVE OF GREEN TECHNOLOGY

The removal of pollutants from wastewater, the treatment of wastewater that contains significant levels of heavy metals, the storage and capture of CO_2, the conversion and breakdown of antibiotics, and the use of biosensors based on algae to detect toxic compounds are just a few of the many applications for phytoremediation. Because of the potential threats it poses to the environment and public health, antibiotic contamination has received more attention in recent years. Despite the relatively low concentrations of microbial traces found in the environment, mounting evidence suggests that these traces have adverse ecological effects on organisms, limiting the growth of helpful microorganisms and changing the makeup and activity of microbial communities. In the meantime, incorrect or excessive use of antibiotics may encourage the emergence and spread of bacteria and genes that are resistant to them, placing considerable selection pressure on human and other microbial systems. This type of microalgae has a significant impact on the rate of antibiotic clearance. For instance, the Chlorella species are efficient in breaking down cephalosporins, ceftazindin, cephradine, cephalexin, amoxicillin, azithromycin, enrofloxacin, florfenicol, and levofloxacin (Emparan et al., 2019; Leng et al., 2018; Xiong et al., 2018).

Three main methods are used by microalgae to break down antibiotics: (1) fast passive adsorption through physicochemical interactions between pollutants and the cell surface; (2) molecular transportation through the membrane of a cell; and (3) bioaccumulation and biodegradation within the cell. The fundamental biodegradation mechanisms for antibiotics, however, can be divided into two groups: (1) metabolic degradation, in which the antibiotic functions as a source of carbon, an electron donor, or acceptor for microalgae; and (2) co-metabolism, in which organic compounds support the growth of microalgae while also acting as electron donors (Díaz et al., 2022; Xiong et al., 2021).

9.5.1 Advantages and Disadvantages of Conventional and Advanced Methods of Antibiotic Treatment from Wastewater

AOPs are oxidation processes that show high removal efficiency. Heterogeneous photocatalysis is a destructive process for the removal of organic compounds by using semiconductors under light. This process is effective, non-toxic, stable, and low-cost, but it showed poor treatment with high concentrations of chemical compounds. Fenton is a method of using iron ions and hydrogen peroxide for the destruction of antibiotics. It is a rapid and effective method with no sludge production. However, further treatment of the by-products is required.

Adsorption is a fast and effective method that is able to separate different components and could be used in the pretreatment process. However, adsorption is highly affected by the pH, has a low selectivity, is a non-destructive method, and mostly requires further treatment.

Membrane filtration is a fast, simple, and selective process that has various kinds of membranes that can be adopted based on the type and composition of wastewater. However, it is an expensive process with high energy consumption.

Biological therapies might be anaerobic, aerobic, or hybrid. These techniques produce little biomass and require tiny reactors. However, they are costly, use a lot of energy, and are not appropriate for all antibiotics. The phycoremediation using microalgae to remove antibiotics shows many benefits in that microalgae biomass growth depends on a carbon source and photons to perform photosynthesis, microalgae excrete many nontoxic metabolites, they are more resistant to antibiotics than bacteria, and they limit the growth of competitors. Additionally, it is simple, cost-effective, eco-friendly, and safe. Microalgae detoxify and remove pollutants and contaminants from toxic wastewater more efficiently than conventional and traditional treatment procedures. Some limitations have arisen with the study of microalgae in wastewater treatment, such as the fact that most of the research conducted on the use of microalgae was conducted on a small and laboratory scale, that few or single antibiotics were used, that antibiotic pollutants should be in the aqueous phase, and that phycoremediation is highly affected by the algal species and the concentration and type of antibiotics. Furthermore, some antibiotics exert a toxic effect on algae and interfere with the production of chloroplasts, the generation of chlorophyll, and the synthesis of proteins, which has a negative impact on the ability of microalgal cells to proliferate and expand (Yu et al., 2022).

The application of microalgae-based technology in wastewater treatment appears to have a promising prospect, according to the overview of microalgal bioremediation. Biodegradation, bioaccumulation, and bioadsorption are the three main removal processes for antibiotics that are mediated by microalgae. The performance of microalgae in the antibiotic removal process needs to be improved by the pretreatment of wastewater sludge using an adsorbent or catalyst. To evaluate the toxicity of antibiotics on microalgae, other pollutants already in the water should be added. Finally, removing antibiotics by microalgae carries the danger of horizontal antibiotic resistance genes movement, which will require more study. Combinations could have an impact. Most of the time, removing antibiotics was carried out using just one method. But wastewater is a complicated system that includes several types of antibiotics. Methods that are combined may be more effective than those used alone. As a result, effective combining and using of the current approaches is important for antibiotic elimination.

This chapter discusses the development of approaches and methodologies and their advantages and disadvantages in the removal of antibiotics from wastewater. Phycoremediation offers high-efficiency wastewater treatment, a reduction in carbon dioxide emissions, and value-added uses for the biomass generated. The use of collected biomass for energy and biofertilizers should determine the best harvesting technique based on the algae strains being employed. In conclusion, the employment of algae in the biomonitoring and restoration of aquatic systems encourages bioremediation, even though some antibiotics are still difficult for algae to remove.

REFERENCES

Abideen, Z., Waqif, H., Munir, N., El-Keblawy, A., Hasnain, M., Radicetti, E., Mancinelli, R., Nielsen, B. L., & Haider, G. (2022). Algal-mediated nanoparticles, phycochar, and biofertilizers for mitigating abiotic stresses in plants: A review. *Agronomy, 12*(8), 1788. MDPI. https://doi.org/10.3390/agronomy12081788.

Acién, F. G., Gómez-Serrano, C., Morales-Amaral, M. M., Fernández-Sevilla, J. M., & Molina-Grima, E. (2016). Wastewater treatment using microalgae: How realistic a contribution might it be to significant urban wastewater treatment? *Applied Microbiology and Biotechnology, 100*(21), 9013–9022. Springer Verlag. https://doi.org/10.1007/s00253-016-7835-7.

Acosta, R., Fierro, V., Martinez de Yuso, A., Nabarlatz, D., & Celzard, A. (2016). Tetracycline adsorption onto activated carbons produced by KOH activation of tyre pyrolysis char. *Chemosphere, 149*, 168–176. https://doi.org/10.1016/j.chemosphere.2016.01.093.

Agarwal, P., Gupta, R., & Agarwal, N. (2019). Advances in synthesis and applications of microalgal nanoparticles for wastewater treatment. *Journal of Nanotechnology, 2019*. Hindawi Limited. https://doi.org/10.1155/2019/7392713.

Amenorfenyo, D. K., Huang, X., Zhang, Y., Zeng, Q., Zhang, N., Ren, J., & Huang, Q. (2019). Microalgae brewery wastewater treatment: Potentials, benefits and the challenges. *International Journal of Environmental Research and Public Health, 16*(11). https://doi.org/10.3390/ijerph16111910.

Ameta, S., & Ameta, R. (eds.), (2018). *Advanced Oxidation Processes for Wastewater Treatment: Emerging Green Chemical Technology*. Cambridge, MA: Academic Press.

Bajhaiya, A., Moreira, J., & Pittman, J. (2017). Transcriptional engineering of microalgae: Prospects for high-value chemicals. *Trends in Biotechnology, 35*(2), 93–99. Elsevier Ltd. https://doi.org/10.1016/j.tibtech.2016.06.001.

Bashir, K. M. I., & Cho, M. G. (2016). The effect of kanamycin and tetracycline on growth and photosynthetic activity of two chlorophyte algae. *BioMed Research International, 2016*. https://doi.org/10.1155/2016/5656304.

Berendonk, T. U., Manaia, C. M., Merlin, C., Fatta-Kassinos, D., Cytryn, E., Walsh, F., Bürgmann, H., Sørum, H., Norström, M., Pons, M. N., Kreuzinger, N., Huovinen, P., Stefani, S., Schwartz, T., Kisand, V., Baquero, F., & Martinez, J. L. (2015). Tackling antibiotic resistance: The environmental framework. *Nature Reviews Microbiology, 13*(5), 310–317. Nature Publishing Group. https://doi.org/10.1038/nrmicro3439.

Cantas, L., Shah, S. Q. A., Cavaco, L. M., Manaia, C. M., Walsh, F., Popowska, M., Garelick, H., Bürgmann, H., & Sørum, H. (2013). A brief multi-disciplinary review on antimicrobial resistance in medicine and its linkage to the global environmental microbiota. *Frontiers in Microbiology, 4*(May). https://doi.org/10.3389/fmicb.2013.00096.

Chen, J., Zheng, F., & Guo, R. (2015). Algal feedback and removal efficiency in a sequencing batch reactor algae process (SBAR) to treat the antibiotic cefradine. *PLoS One, 10*(7). https://doi.org/10.1371/journal.pone.0133273.

Chen, Q., Zhang, L., Han, Y., Fang, J., & Wang, H. (2020). Degradation and metabolic pathways of sulfamethazine and enrofloxacin in *Chlorella vulgaris* and *Scenedesmus obliquus* treatment systems. *Environmental Science and Pollution Research, 27*(22), 28198–28208. https://doi.org/10.1007/s11356-020-09008-4.

Collos, Y., & Harrison, P. J. (2014). Acclimation and toxicity of high ammonium concentrations to unicellular algae. *Marine Pollution Bulletin, 80*(1–2), 8–23. https://doi.org/10.1016/j.marpolbul.2014.01.006.

Czekalski, N., Berthold, T., Caucci, S., Egli, A., & Bürgmann, H. (2012). Increased levels of multiresistant bacteria and resistance genes after wastewater treatment and their dissemination into Lake Geneva, Switzerland. *Frontiers in Microbiology, 3*(Mar). https://doi.org/10.3389/fmicb.2012.00106.

Czekalski, N., Gascón Díez, E., & Bürgmann, H. (2014). Wastewater as a point source of antibiotic-resistance genes in the sediment of a freshwater lake. *ISME Journal, 8*(7), 1381–1390. https://doi.org/10.1038/ismej.2014.8.

da Silva Rodrigues, D. A., da Cunha C. C. R. F., do Espirito Santo, D. R., de Barros, A. L. C., Pereira, A. R., de Queiroz Silva, S., et al. (2021). Removal of cephalexin and erythromycin antibiotics, and their resistance genes, by microalgae-bacteria consortium from wastewater treatment plant secondary effluents. *Environmental Science and Pollution Research, 28*(47), 67822–67832. https://doi.org/10.1007/s11356-021-15351-x.

da Silva Rodrigues, D. A., da Cunha, C. C. R. F., Freitas, M. G., de Barros, A. L. C., e Castro, P. B. N., Pereira, A. R., de Queiroz Silva, S., da Fonseca Santiago, A., & de Cássia Franco Afonso, R. J. (2020). Biodegradation of sulfamethoxazole by microalgae-bacteria consortium in wastewater treatment plant effluents. *Science of the Total Environment, 749*. https://doi.org/10.1016/j.scitotenv.2020.141441.

de Godos, I., Muñoz, R., & Guieysse, B. (2012). Tetracycline removal during wastewater treatment in high-rate algal ponds. *Journal of Hazardous Materials, 229–230*, 446–449. https://doi.org/10.1016/j.jhazmat.2012.05.106.

Demircivi, P., & Simsek, E. B. (2019). Visible-light-enhanced photoactivity of perovskite-type W-doped BaTiO$_3$ photocatalyst for photodegradation of tetracycline. *Journal of Alloys and Compounds, 774*, 795–802. https://doi.org/10.1016/j.jallcom.2018.09.354.

Dey, R., Maarisetty, D., & Baral, S. S. (2022). A comparative study of bioelectrochemical systems with established anaerobic/aerobic processes. *Biomass Conversion and Biorefinery.* Springer Science and Business Media Deutschland GmbH. https://doi.org/10.1007/s13399-021-02258-3.

Díaz, Y. A., Martínez, E. M., García, L. C., Serrano, M. C., & Sierra, E. M. (2022). Phycoremediation as a strategy for the recovery of marsh and wetland with potential in Colombia. *Resources, 11*(2). MDPI. https://doi.org/10.3390/resources11020015.

Ding, Y., Wang, W., Liu, X., Song, X., Wang, Y., & Ullman, J. L. (2016). Intensified nitrogen removal of constructed wetland by novel integration of high rate algal pond biotechnology. *Bioresource Technology, 219*, 757–761. https://doi.org/10.1016/j.biortech.2016.08.044.

Du, Z., Li, K., Zhou, S., Liu, X., Yu, Y., Zhang, Y., He, Y., & Zhang, Y. (2020). Degradation of ofloxacin with heterogeneous photo-Fenton catalyzed by biogenic Fe-Mn oxides. *Chemical Engineering Journal, 380*. https://doi.org/10.1016/j.cej.2019.122427.

Duan, W., Li, M., Xiao, W., Wang, N., Niu, B., Zhou, L., & Zheng, Y. (2019). Enhanced adsorption of three fluoroquinolone antibiotics using polypyrrole functionalized Calotropis gigantea fiber. *Colloids and Surfaces A: Physicochemical and Engineering Aspects, 574*, 178–187. https://doi.org/10.1016/j.colsurfa.2019.04.068.

Emparan, Q., Harun, R., & Danquah, M. K. (2019). Role of phycoremediation for nutrient removal from wastewaters: A review. *Applied Ecology and Environmental Research, 17*(1), 889–915. https://doi.org/10.15666/aeer/1701_889915.

Gao, P., Munir, M., & Xagoraraki, I. (2012). Correlation of tetracycline and sulfonamide antibiotics with corresponding resistance genes and resistant bacteria in a conventional municipal wastewater treatment plant. *Science of the Total Environment, 421–422*, 173–183. https://doi.org/10.1016/j.scitotenv.2012.01.061.

Gondi, R., Kavitha, S., Yukesh Kannah, R., Parthiba Karthikeyan, O., Kumar, G., Kumar Tyagi, V., & Rajesh Banu, J. (2022). Algal-based system for removal of emerging pollutants from wastewater: A review. *Bioresource Technology, 344*, 126245. Elsevier Ltd. https://doi.org/10.1016/j.biortech.2021.126245.

Grady, J. C., Daigger, G., Love, N., & Filipe, C. (2011). *Biological Wastewater Treatment.* Boca Raton, FL: CRC Press.

Guo, W. Q., Zheng, H. S., Li, S., Du, J. S., Feng, X. C., Yin, R. L., Wu, Q. L., Ren, N. Q., & Chang, J. S. (2016). Removal of cephalosporin antibiotics 7-ACA from wastewater during the cultivation of lipid-accumulating microalgae. *Bioresource Technology, 221*, 284–290. https://doi.org/10.1016/j.biortech.2016.09.036.

Hassan, M., Ashraf, G. A., Zhang, B., He, Y., Shen, G., & Hu, S. (2020). Energy-efficient degradation of antibiotics in microbial electro-Fenton system catalysed by M-type strontium hexaferrite nanoparticles. *Chemical Engineering Journal, 380*. https://doi.org/10.1016/j.cej.2019.122483.

Hena, S., Gutierrez, L., & Croué, J. P. (2020). Removal of metronidazole from aqueous media by *C. vulgaris*. *Journal of Hazardous Materials, 384*. https://doi.org/10.1016/j.jhazmat.2019.121400.

Iakovides, I. C., Michael-Kordatou, I., Moreira, N. F. F., Ribeiro, A. R., Fernandes, T., Pereira, M. F. R., Nunes, O. C., Manaia, C. M., Silva, A. M. T., & Fatta-Kassinos, D. (2019). Continuous ozonation of urban wastewater: Removal of antibiotics, antibiotic-resistant *Escherichia coli* and antibiotic resistance genes and phytotoxicity. *Water Research, 159*, 333–347. https://doi.org/10.1016/j.watres.2019.05.025.

Ikehata, K., & El-Din, M. G. (2004). Degradation of recalcitrant surfactants in wastewater by ozonation and advanced oxidation processes: A review. *Ozone: Science and Engineering, 26*(4), 327–343. https://doi.org/10.1080/01919510490482160.

Jiang, L., Yuan, X., Zeng, G., Wu, Z., Liang, J., Chen, X., Leng, L., Wang, H., & Wang, H. (2018). Metal-free efficient photocatalyst for stable visible-light photocatalytic degradation of refractory pollutant. *Applied Catalysis B: Environmental, 221*, 715–725. https://doi.org/10.1016/j.apcatb.2017.09.059.

Kafle, A., Timilsina, A., Gautam, A., Adhikari, K., Bhattarai, A., & Aryal, N. (2022). Phytoremediation: Mechanisms, plant selection and enhancement by natural and synthetic agents. *Environmental Advances, 8.* Elsevier Ltd. https://doi.org/10.1016/j.envadv.2022.100203.

Kakavandi, B., Takdastan, A., Jaafarzadeh, N., Azizi, M., Mirzaei, A., & Azari, A. (2016). Application of Fe3O4@C catalyzing heterogeneous UV-Fenton system for tetracycline removal with a focus on optimization by a response surface method. *Journal of Photochemistry and Photobiology A: Chemistry, 314*, 178–188. https://doi.org/10.1016/j.jphotochem.2015.08.008.

Kang, J., Jin, C., Li, Z., Wang, M., Chen, Z., & Wang, Y. (2020). Dual Z-scheme MoS2/g-C$_3$N$_4$/ Bi$_{24}$O$_{31}$Cl$_{10}$ ternary heterojunction photocatalysts for enhanced visible-light photodegradation of antibiotic. *Journal of Alloys and Compounds, 825.* https://doi.org/10.1016/j.jallcom.2020.153975.

Kiki, C., Rashid, A., Wang, Y., Li, Y., Zeng, Q., Yu, C. P., & Sun, Q. (2020). Dissipation of antibiotics by microalgae: Kinetics, identification of transformation products and pathways. *Journal of Hazardous Materials, 387.* https://doi.org/10.1016/j.jhazmat.2019.121985.

Leng, L., Li, J., Wen, Z., & Zhou, W. (2018). Use of microalgae to recycle nutrients in aqueous phase derived from hydrothermal liquefaction process. *Bioresource Technology, 256*, 529–542. Elsevier Ltd. https://doi.org/10.1016/j.biortech.2018.01.121.

Leng, L., Wei, L., Xiong, Q., Xu, S., Li, W., Lv, S., Lu, Q., Wan, L., Wen, Z., & Zhou, W. (2020). Use of microalgae based technology for the removal of antibiotics from wastewater: A review. *Chemosphere, 238.* Elsevier Ltd. https://doi.org/10.1016/j.chemosphere.2019.124680.

Li, J., Min, Z., Li, W., Xu, L., Han, J., & Li, P. (2020). Interactive effects of roxithromycin and freshwater microalgae, *Chlorella pyrenoidosa*: Toxicity and removal mechanism. *Ecotoxicology and Environmental Safety, 191.* https://doi.org/10.1016/j.ecoenv.2019.110156.

Li, S., Show, P. L., Ngo, H. H., & Ho, S. H. (2022). Algae-mediated antibiotic wastewater treatment: A critical review. *Environmental Science and Ecotechnology, 9.* https://doi.org/10.1016/j.ese.2022.100145.

Li, Y., Han, J., Mi, X., Mi, X., Li, Y., Zhang, S., & Zhan, S. (2017). Modified carbon felt made using CexA1-xO2 composites as a cathode in electro-Fenton system to degrade ciprofloxacin. *RSC Advances, 7*(43), 27065–27078. https://doi.org/10.1039/c7ra03302h.

Liotta, L. F., Gruttadauria, M., Di Carlo, G., Perrini, G., & Librando, V. (2009). Heterogeneous catalytic degradation of phenolic substrates: Catalysts activity. *Journal of Hazardous Materials, 162*(2–3), 588–606. https://doi.org/10.1016/j.jhazmat.2008.05.115.

Liu, X., Yang, D., Zhou, Y., Zhang, J., Luo, L., Meng, S., Chen, S., Tan, M., Li, Z., & Tang, L. (2017). Electrocatalytic properties of N-doped graphite felt in electro-Fenton process and degradation mechanism of levofloxacin. *Chemosphere, 182*, 306–315. https://doi.org/10.1016/j.chemosphere.2017.05.035.

Liu, Y., Wang, F., Chen, X., Zhang, J., & Gao, B. (2015). Cellular responses and biodegradation of amoxicillin in Microcystis aeruginosa at different nitrogen levels. *Ecotoxicology and Environmental Safety, 111*, 138–145. https://doi.org/10.1016/j.ecoenv.2014.10.011.

Lu, X., Wang, Y., Zhang, X., Xu, G., Wang, D., Lv, J., Zheng, Z., & Wu, Y. (2018). NiS and MoS2 nanosheet co-modified graphitic C3N4 ternary heterostructure for high efficient visible light photodegradation of antibiotic. *Journal of Hazardous Materials, 341,* 10–19. https://doi.org/10.1016/j.jhazmat.2017.07.004.

Ma, J., Jiang, Z., Cao, J., & Yu, F. (2020). Enhanced adsorption for the removal of antibiotics by carbon nanotubes/graphene oxide/sodium alginate triple-network nanocomposite hydrogels in aqueous solutions. *Chemosphere, 242.* https://doi.org/10.1016/j.chemosphere.2019.125188.

Marsolek, M. D., Kirisits, M. J., Gray, K. A., & Rittmann, B. E. (2014). Coupled photocatalytic-biodegradation of 2,4,5-trichlorophenol: Effects of photolytic and photocatalytic effluent composition on bioreactor process performance, community diversity, and resistance and resilience to perturbation. *Water Research, 50,* 59–69. https://doi.org/10.1016/j.watres.2013.11.043.

McCorquodale-Bauer, K., Grosshans, R., Zvomuya, F., & Cicek, N. (2023). Critical review of phytoremediation for the removal of antibiotics and antibiotic resistance genes in wastewater. *Science of the Total Environment, 870.* Elsevier B.V. https://doi.org/10.1016/j.scitotenv.2023.161876.

Mehrabadi, A., Craggs, R., & Farid, M. M. (2015). Wastewater treatment high rate algal ponds (WWT HRAP) for low-cost biofuel production. *Bioresource Technology, 184,* 202–214. Elsevier Ltd. https://doi.org/10.1016/j.biortech.2014.11.004.

Michael, I., Rizzo, L., McArdell, C. S., Manaia, C. M., Merlin, C., Schwartz, T., Dagot, C., & Fatta-Kassinos, D. (2013). Urban wastewater treatment plants as hotspots for the release of antibiotics in the environment: A review. *Water Research, 47*(3), 957–995. Elsevier Ltd. https://doi.org/10.1016/j.watres.2012.11.027.

Mojiri, A., Baharlooeian, M., & Zahed, M. A. (2021). The potential of chaetoceros muelleri in bioremediation of antibiotics: Performance and optimization. *International Journal of Environmental Research and Public Health, 18*(3), 1–13. https://doi.org/10.3390/ijerph18030977.

Morales-Amaral, M. del M., Gómez-Serrano, C., Acién, F. G., Fernández-Sevilla, J. M., & Molina-Grima, E. (2015). Production of microalgae using centrate from anaerobic digestion as the nutrient source. *Algal Research, 9,* 297–305. https://doi.org/10.1016/j.algal.2015.03.018.

Muñoz, R., & Guieysse, B. (2006). Algal-bacterial processes for the treatment of hazardous contaminants: A review. *Water Research, 40*(15), 2799–2815. Elsevier Ltd. https://doi.org/10.1016/j.watres.2006.06.011.

Nasrollahi, N., Vatanpour, V., & Khataee, A. (2022). Removal of antibiotics from wastewaters by membrane technology: Limitations, successes, and future improvements. *Science of the Total Environment, 838.* Elsevier B.V. https://doi.org/10.1016/j.scitotenv.2022.156010.

Novo, A., André, S., Viana, P., Nunes, O. C., & Manaia, C. M. (2013). Antibiotic resistance, Antimicrobial residues and bacterial community composition in urban wastewater. *Water Research, 47*(5), 1875–1887. https://doi.org/10.1016/j.watres.2013.01.010.

Pan, M., Lyu, T., Zhan, L., Matamoros, V., Angelidaki, I., Cooper, M., & Pan, G. (2021). Mitigating antibiotic pollution using cyanobacteria: Removal efficiency, pathways and metabolism. *Water Research, 190.* https://doi.org/10.1016/j.watres.2020.116735.

Perales-Vela, H. V., García, R. V., Gómez-Juárez, E. A., Salcedo-Álvarez, M. O., & Cañizares-Villanueva, R. O. (2016). Streptomycin affects the growth and photochemical activity of the alga *Chlorella vulgaris. Ecotoxicology and Environmental Safety, 132,* 311–317. https://doi.org/10.1016/j.ecoenv.2016.06.019.

Phoon, B. L., Ong, C. C., Mohamed Saheed, M. S., Show, P. L., Chang, J. S., Ling, T. C., Lam, S. S., & Juan, J. C. (2020). Conventional and emerging technologies for removal of antibiotics from wastewater. *Journal of Hazardous Materials, 400.* https://doi.org/10.1016/j.jhazmat.2020.122961.

Prata, J. C., Lavorante, B. R. B. O., Montenegro, M.C. B. S. M., & Guilhermino, L. (2018). Influence of microplastics on the toxicity of the pharmaceuticals procainamide and dox-ycycline on the marine microalgae *Tetraselmis chuii*. *Aquatic Toxicology, 197*, 143–152. https://doi.org/10.1016/j.aquatox.2018.02.015.

Ramesh, B., Saravanan, A., Senthil Kumar, P., Yaashikaa, P. R., Thamarai, P., Shaji, A., & Rangasamy, G. (2023). A review on algae biosorption for the removal of hazardous pollut-ants from wastewater: Limiting factors, prospects and recommendations. *Environmental Pollution, 327*. Elsevier Ltd. https://doi.org/10.1016/j.envpol.2023.121572.

Ranjusha, V. P., Matsumoto, K., Nara, S., Inagaki, Y., & Sakakibara, Y. (2020). Application of phyto-Fenton process in constructed wetland for the continuous removal of antibiotics. *IOP Conference Series: Earth and Environmental Science, 427*(1). https://doi.org/10.10 88/1755-1315/427/1/012006.

Ravikumar, K. V. G., Sudakaran, S. V., Ravichandran, K., Pulimi, M., Natarajan, C., & Mukherjee, A. (2019). Green synthesis of NiFe nano particles using Punica granatum peel extract for tetracycline removal. *Journal of Cleaner Production, 210*, 767–776. https://doi.org/10.1016/j.jclepro.2018.11.108.

Ribeiro, A. R., Sures, B., & Schmidt, T. C. (2018). Ecotoxicity of the two veterinarian antibi-otics ceftiofur and cefapirin before and after photo-transformation. *Science of the Total Environment, 619–620*, 866–873. https://doi.org/10.1016/j.scitotenv.2017.11.109.

Rizzo, L., Manaia, C., Merlin, C., Schwartz, T., Dagot, C., Ploy, M. C., Michael, I., & Fatta-Kassinos, D. (2013). Urban wastewater treatment plants as hotspots for antibiotic resistant bacteria and genes spread into the environment: A review. *Science of the Total Environment, 447*, 345–360. https://doi.org/10.1016/j.scitotenv.2013.01.032.

Rossolini, G. M., Arena, F., Pecile, P., & Pollini, S. (2014). Update on the antibiotic resis-tance crisis. *Current Opinion in Pharmacology, 18*, 56–60. Elsevier Ltd. https://doi.org/10.1016/j.coph.2014.09.006.

Song, C., Wei, Y., Qiu, Y., Qi, Y., Li, Y., & Kitamura, Y. (2019). Biodegradability and mechanism of florfenicol via *Chlorella* sp. UTEX1602 and L38: Experimental study. *Bioresource Technology, 272*, 529–534. https://doi.org/10.1016/j.biortech.2018.10.080.

Tiwari, B., Sellamuthu, B., Ouarda, Y., Drogui, P., Tyagi, R. D., & Buelna, G. (2017). Review on fate and mechanism of removal of pharmaceutical pollutants from wastewater using biological approach. *Bioresource Technology, 224*, 1–12. Elsevier Ltd. https://doi.org/10.1016/j.biortech.2016.11.042.

Vargas-Estrada, L., Torres-Arellano, S., Longoria, A., Arias, D. M., Okoye, P. U., & Sebastian, P. J. (2020). Role of nanoparticles on microalgal cultivation: A review. *Fuel, 280*. https://doi.org/10.1016/j.fuel.2020.118598.

Vatanpour, V., Yavuzturk Gul, B., Zeytuncu, B., Korkut, S., İlyasoğlu, G., Turken, T., Badawi, M., Koyuncu, I., & Saeb, M. R. (2022). Polysaccharides in fabrication of membranes: A review. *Carbohydrate Polymers, 281*. Elsevier Ltd. https://doi.org/10.1016/j.carbpol.2021.119041.

Vaz-Moreira, I., Nunes, O. C., & Manaia, C. M. (2014). Bacterial diversity and antibi-otic resistance in water habitats: Searching the links with the human microbiome. *FEMS Microbiology Reviews, 38*(4), 761–778. Blackwell Publishing Ltd. https://doi.org/10.1111/1574-6976.12062.

Villegas-Guzman, P., Silva-Agredo, J., Florez, O., Giraldo-Aguirre, A. L., Pulgarin, C., & Torres-Palma, R. A. (2017). Selecting the best AOP for isoxazolyl penicillins degrada-tion as a function of water characteristics: Effects of pH, chemical nature of additives and pollutant concentration. *Journal of Environmental Management, 190*, 72–79. https://doi.org/10.1016/j.jenvman.2016.12.056.

Vo, H. N. P., Ngo, H. H., Guo, W., Nguyen, K. H., Chang, S. W., Nguyen, D. D., Cheng, D., Bui, X. T., Liu, Y., & Zhang, X. (2021). Effect of calcium peroxide pretreatment on the remediation of sulfonamide antibiotics (SMs) by Chlorella sp. *Science of the Total Environment, 793*. https://doi.org/10.1016/j.scitotenv.2021.148598.

Wagner, M., Loy, A., Nogueira, R., Purkhold, U., Lee, N., & Daims, H. (2002). Microbial community composition and function in wastewater treatment plants. *Antonie van Leeuwenhoek, 81*, 665–680. http://doi.org/10.1023/a:1020586312170.

Wang, Y., Wang, F., Feng, Y., Xie, Z., Zhang, Q., Jin, X., Liu, H., Liu, Y., Lv, W., & Liu, G. (2018). Facile synthesis of carbon quantum dots loaded with mesoporous g-C3N4 for synergistic absorption and visible light photodegradation of fluoroquinolone antibiotics. *Dalton Transactions, 47*(4), 1284–1293. https://doi.org/10.1039/c7dt04360k.

Watkinson, A. J., Murby, E. J., Kolpin, D. W., & Costanzo, S. D. (2009). The occurrence of antibiotics in an urban watershed: From wastewater to drinking water. *Science of the Total Environment, 407*(8), 2711–2723. https://doi.org/10.1016/j.scitotenv.2008.11.059.

Xie, P., Chen, C., Zhang, C., Su, G., Ren, N., & Ho, S. H. (2020). Revealing the role of adsorption in ciprofloxacin and sulfadiazine elimination routes in microalgae. *Water Research, 172*. https://doi.org/10.1016/j.watres.2020.115475.

Xiong, J. Q., Kurade, M. B., & Jeon, B. H. (2017). Ecotoxicological effects of enrofloxacin and its removal by monoculture of microalgal species and their consortium. *Environmental Pollution, 226*, 486–493. https://doi.org/10.1016/j.envpol.2017.04.044.

Xiong, J. Q., Kurade, M. B., & Jeon, B. H. (2018). Can microalgae remove pharmaceutical contaminants from water? *Trends in Biotechnology, 36*(1), 30–44. Elsevier Ltd. https://doi.org/10.1016/j.tibtech.2017.09.003.

Xiong, Q., Hu, L. X., Liu, Y. S., Zhao, J. L., He, L. Y., & Ying, G. G. (2021). Microalgae-based technology for antibiotics removal: From mechanisms to application of innovational hybrid systems. *Environment International, 155*. Elsevier Ltd. https://doi.org/10.1016/j.envint.2021.106594.

Xiong, Q., Liu, Y. S., Hu, L. X., Shi, Z. Q., Cai, W. W., He, L. Y., & Ying, G. G. (2020). Co-metabolism of sulfamethoxazole by a freshwater microalga Chlorella pyrenoidosa. *Water Research, 175*. https://doi.org/10.1016/j.watres.2020.115656.

Yu, C., Pang, H., Wang, J. H., Chi, Z. Y., Zhang, Q., Kong, F. T., Xu, Y. P., Li, S. Y., & Che, J. (2022). Occurrence of antibiotics in waters, removal by microalgae-based systems, and their toxicological effects: A review. *Science of the Total Environment, 813*. Elsevier B.V. https://doi.org/10.1016/j.scitotenv.2021.151891.

Yu, Y., Zhou, Y., Wang, Z., Torres, O. L., Guo, R., & Chen, J. (2017). Investigation of the removal mechanism of antibiotic ceftazidime by green algae and subsequent microbic impact assessment. *Scientific Reports, 7*(1). https://doi.org/10.1038/s41598-017-04128-3.

Zhang, J., Fu, D., & Wu, J. (2012). Photodegradation of Norfloxacin in aqueous solution containing algae. *Journal of Environmental Sciences, 24*(4), 743–749. https://doi.org/10.1016/S1001-0742(11)60814-0.

Zhang, J., Xia, A., Yao, D., Guo, X., Lam, S. S., Huang, Y., Zhu, X., Zhu, X., & Liao, Q. (2022). Removal of oxytetracycline and ofloxacin in wastewater by microalgae-bacteria symbiosis for bioenergy production. *Bioresource Technology, 363*. https://doi.org/10.1016/j.biortech.2022.127891.

Zhang, Y., Marrs, C. F., Simon, C., & Xi, C. (2009). Wastewater treatment contributes to selective increase of antibiotic resistance among *Acinetobacter* spp. *Science of the Total Environment, 407*(12), 3702–3706. https://doi.org/10.1016/j.scitotenv.2009.02.013.

Zhao, F., Zhang, D., Xu, C., Liu, J., & Shen, C. (2020a). The enhanced degradation and detoxification of chlortetracycline by *Chlamydomonas reinhardtii*. *Ecotoxicology and Environmental Safety, 196*. https://doi.org/10.1016/j.ecoenv.2020.110552.

Zhao, J., Ji, M., Di, J., Zhang, Y., He, M., Li, H., & Xia, J. (2020b). Novel Z-scheme heterogeneous photo-Fenton-like g-C_3N_4/FeOCl for the pollutants degradation under visible light irradiation. *Journal of Photochemistry and Photobiology A: Chemistry, 391*. https://doi.org/10.1016/j.jphotochem.2019.112343.

Zhou, L., Liu, Z., Guan, Z., Tian, B., Wang, L., Zhou, Y., Zhou, Y., Lei, J., Zhang, J., & Liu, Y. (2020). 0D/2D plasmonic Cu_2-xS/g-C_3N_4 nanosheets harnessing UV-vis-NIR broad spectrum for photocatalytic degradation of antibiotic pollutant. *Applied Catalysis B: Environmental, 263*. https://doi.org/10.1016/j.apcatb.2019.118326.

Zhou, T., Cao, L., Zhang, Q., Liu, Y., Xiang, S., Liu, T., & Ruan, R. (2021). Effect of chlortetracycline on the growth and intracellular components of Spirulina platensis and its biodegradation pathway. *Journal of Hazardous Materials, 413*. https://doi.org/10.1016/j.jhazmat.2021.125310.

Zinicovscaia, I. (2016). Conventional methods of wastewater treatment. In: Zinicovscaia, I., & Cepoi, L. (eds.), *Cyanobacteria for Bioremediation of Wastewaters* (pp. 17–25). Springer International Publishing. https://doi.org/10.1007/978-3-319-26751-7_3.

10 Application of Phycoremediation in Lakes and Reservoirs

Meltem Çelen

10.1 INTRODUCTION

Ensuring the quality of lake and reservoir water holds paramount significance for sustaining healthy ecosystems and providing safe water resources. Although the importance of freshwater sources has been well recognized for an extended period of time, many countries across the globe treat both treated and untreated freshwater bodies as receiving environments. With an increasing awareness of the significance of water resources, wastewater treatment technologies aimed at preserving these resources have been continually advancing.

Several in situ measures, incorporating external interventions to enhance the self-purification (Ali et al., 2020) capacities of lakes and reservoirs, have also been developed. Among these methods, various physical measures such as water column aeration, chemical interventions including precipitation and oxidation of pollutants, and biological strategies like bioremediation have been implemented (Huang, 2016). Phycoremediation, which involves the use of algae and aquatic plants for the removal of pollutants from water bodies, is also a prominent biological remediation method.

The application of phycoremediation was initially introduced for wastewater treatment purposes, and subsequently, trials with various types of pollutants using different strains of microalgae have been conducted. Different algal species used for phycoremediation are obtained from natural water sources such as seas, lakes, or sediment environments. Phycoremediation can be applied for pollutant removal from various environments, including soil, water bodies, and rice fields (Cuypers and Vangronsveld, 2017; Wójcik et al., 2017; Ali et al., 2020; Arantza et al., 2022). Furthermore, algae, in addition to their pollutant removal capabilities, also possess nutrient assimilation properties, specifically nitrogen and phosphorus compounds. They have demonstrated efficacy in remediating acid mine drainage from mining sites, as well as in removing pollutants like heavy metals (HMs), persistent organic pollutants (POPs), and pharmaceuticals (Chen et al., 2020; Chirwa et al., 2019; Das et al., 2019; Krishna Samal et al., 2020; Zhou et al., 2023).

Although microplastic pollution is more commonly associated with marine environments, its presence extends to freshwater sources through point-source pollution (Erdogan, 2020; Samavi et al., 2022; Szymańska and Obolewski, 2020). In line with

DOI: 10.1201/9781003390213-10

this, algae have also shown potential for microplastic degradation (Karalija et al., 2022). However, the implementation of in-lake phycoremediation has received limited scientific attention (Touliabah et al., 2022; Cuypers and Vangronsveld, 2017; Arantza et al., 2022). Phycoremediation simultaneously provides a multitude of advantageous processes. Algae utilize excess nutrients present in their surrounding environment, removing targeted pollutants while also generating algal biomass that can serve as a raw material for various value-added products such as biofuels (e.g., biodiesel, bio-alcohol, and bio-oil), biochar, glycerol, functional food, and pigments (Razaviarani et al., 2023; Leong et al., 2021; Singh et al., 2023). Furthermore, it can contribute to the recovery of valuable metals. Additionally, utilizing algae as decontamination agents in drinking water sources, such as lakes and reservoirs, presents significant advantages, including low costs, easy manipulation, minimal environmental impact, relatively straightforward recovery of metal pollutants, and minimal secondary waste generation (Topal et al., 2020; Tufail et al., 2022).

This study aims to emphasize the innovative role of phycoremediation in addressing various types of pollutants affecting water bodies, specifically lakes and reservoirs. The study highlights the position of phycoremediation within the spectrum of available methods and presents detailed mechanisms within these environmental contexts. Since lakes and reservoirs are open systems, in situ investigations can vary significantly. Consequently, the compilation of current scientific research within the literature seeks to establish a foundation for future studies. Moreover, this study is expected to enhance the understanding of the role of micro and macroalgae in biogeochemical processes within lake and reservoir basin systems.

10.2 METHODS USED TO IMPROVE LAKE AND RESERVOIR WATER QUALITY

In response to the contamination challenges faced by lakes and reservoirs, interventions can be undertaken to accelerate their natural purification processes and restore water quality once their assimilation capacities are exceeded. Foremost among the initial steps to enhance surface water quality is the reduction of pressure from point and non-point pollution sources on water resources. Neglecting this primary step may render efforts to improve the water quality of lakes and reservoirs futile, resulting in time and budgetary losses. However, as the focus of this review does not encompass external measures, also known as remediation techniques, they shall not be elaborated upon.

For internal measures aimed at in situ water quality improvement in lakes and reservoirs, a holistic approach considering the entire basin is essential. This approach necessitates a well-defined understanding of the water ecosystem, including its geographic, geological, and topographic characteristics. Despite categorizing lakes and reservoirs as still water bodies within the classification of surface water resources, their formation, morphology, and basins exhibit significant variations. Although detailed classifications exist, lakes can be categorized based on their formation into glacial lakes, tectonic lakes, uplifted seabeds, volcanic lakes, and reservoirs (Holdren et al., 2001). These distinct formations can influence various characteristics such as depth, surface area, shoreline length, and basin-specific features like geology, soil type, and slope. Consequently, these factors directly or indirectly impact the hydrological, hydrodynamic, and ecological attributes of lakes. For instance, a lake within a basin composed of granite will receive fewer nutrients compared to a lake situated in an alluvial basin (Holdren et al., 2001). Another noteworthy feature that substantially varies with lake morphology and geographical location is thermal stratification. This stratification in sufficiently deep lakes leads to seasonal mixing during spring and autumn, causing the transfer of nutrients from sediments to the water column and altering the distribution of algal species (Holdren et al., 2001).

Figure 10.1 provides a general overview of restoration techniques applicable to lakes and reservoirs, encompassing both in-lake and preventive methods. When considering in-lake methods, a range of physical processes like aeration, similar to wastewater treatment facilities, as well as chemical methods such as oxidation and precipitation and biological strategies like bioremediation, phytoremediation, and phycoremediation are discernible (Jilbert et al., 2020; Huang, 2016; Pereira and Mulligan, 2023; Nürnberg, 2017). The sediment layer within lakes and reservoirs also functions as an internal source of several elements, primarily phosphorus. Consequently, efforts to enhance water quality in the water column should also include the control of phosphorus transfer from sediment (Qin et al., 2014). It's important to note that the methods employed to improve lake water quality differ based on the types of parameters causing degradation. The unique hydrodynamic structures of lakes engender distinct self-cleansing mechanisms. Thus, interventions to enhance water quality in lakes should be tailored to these characteristics. In a study evaluating lake improvement methods, Chortek (2017) concluded that the hypolimnetic oxygenation method was most suitable for mercury-contaminated lakes, while phycoremediation was deemed inappropriate for such cases.

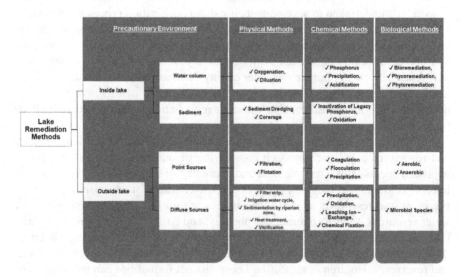

FIGURE 10.1 Methods applicable for managing water quality in lake and reservoir systems.

10.2.1 PHYCOREMEDIATION

Phycoremediation is defined as the removal of pollutants or nutrients from wastewater or aquatic environments using micro and/or macroalgae (Shanmuganathan et al., 2023). It has extended beyond aquatic environments to encompass the remediation of diverse settings, such as soils and rice fields (Prasad, 2022; Tiodar et al., 2021). While microalgae are employed for pollutant removal, the production of valuable products like biofuels from the harvested microalgae adds a significant advantage to this method (Aransiola et al., 2019). The utilization of algae for wastewater treatment began in the 1950s (Oswald et al., 1957), yet attempts at remediating water resources are relatively recent. The use of higher plants and bacteria for the bioextraction and bioremediation of heavy metals and organic pollutants has been extensively explored. Algae play a pivotal role in the control and biomonitoring of organic pollutants in aquatic ecosystems. However, the application of microalgae for the restoration of organically polluted aquatic environments has been investigated only within the last decade (Paranjape et al., 2016; Chekroun et al., 2014; Zhang et al., 2019).

10.2.2 POLLUTANT REMOVAL THROUGH PHYCOREMEDIATION IN LAKE/RESERVOIR WATERSHED SYSTEMS

Micro and macroalgae are commonly found in a variety of environmental settings, including terrestrial, marine, freshwater, wetlands, and sediments. Due to their ecological roles in the carbon cycle, they utilize atmospheric CO_2 and release O_2. Simultaneously, they utilize macronutrients such as N and P and micronutrients like iron and silicon to varying degrees, based on the elemental species present in their stoichiometries. During their life cycles, algae also remove pollutants from their environment through processes like bioaccumulation, biodegradation, and biosorption (Marella et al., 2020). Consequently, the implementation of phycoremediation involving micro and macroalgae in lakes and reservoirs necessitates the consideration of studies conducted in soil, water column, and sediment in tandem with the reservoir watersheds.

Alsamhary (2023) conducted experiments with the use of a cyanobacterial species, *Cylindrospermum stagnale*, and earthworms (*Eisenia fetida*), coupled with rice husk biochar, to remediate Cd-contaminated soils, yielding positive results. Another study employing a cyanobacterial species indicated that the presence of cyanobacterial crusts in arsenic-contaminated mine site soils could mitigate As toxicity and enhance soil productivity (Mao et al., 2023). In a study aimed at increasing yield in saline-alkali soils, the combination of *Azotobacter beijerincki* and *Chlorella pyrenoidosa* effectively lowered soil pH, increasing the availability of phosphorus for plants (Zhou et al., 2023). An experiment in mercury-contaminated rice fields involving algae revealed that algae-derived organic matter altered the properties of soil-dissolved organic matter and significantly regulated methylmercury production. This highlights the importance of considering Hg removal from rice fields through algal assistance (Hu et al., 2023).

Lakes and reservoirs can be exposed to pollutants of diverse origins. The removal mechanisms observed when micro and macroalgae are employed for remediation vary depending on the pollutants. Commonly targeted pollutants for algae-based removal include POPs, HMs, and nutrients. Recent studies have also examined the removal of microplastics by algae.

POPs originate from sources such as insecticides, industrial processes, or combustion, persisting in the environment for extended periods (Jones and Voogt, 1999). These chemicals can resist sunlight and atmospheric oxidation, concentrate in water, sediment, or air, and can be transported over long distances. POPs tend to accumulate in the fatty tissues of living organisms, leading to bioconcentration and long-term accumulation. Consequently, even low levels of exposure can lead to toxic concentrations over time, posing a global threat (Kodavanti et al., 2014). Phycoremediation employs algae to break down POPs like xenobiotics, pesticides, and heavy metals from soil and wastewater. Algae, cyanobacteria, and protozoa are commonly used for this purpose. Algal species such as Dermarestia, Fucus, Rhodococcus, and Ascophyllum nodosum are examples of those employed for POP degradation. Algae produce enzymes that facilitate hydrocarbon degradation and chlorophyll pigments that support photosynthesis, a crucial element in contaminant degradation. The degradation process involves three phases: transformation of compounds into water-soluble forms, addition of polar groups, and incorporation into cell structures. Enzymes produced by algae also contribute to POP degradation. In summary, phycoremediation effectively transforms harmful toxic POPs into less toxic forms (Kumar et al., 2022).

Kumar et al. (2015) provided a detailed description of the mechanisms involved in heavy metal removal using algae. Both living and non-living microalgae can be employed for heavy metal removal. The most studied heavy metals in microalgae-based removal include Cu, Cd, Ni, Pb, Zn, Hg, and Cr. Among the mechanisms, bioremoval, particularly biosorption, is the most rapid process for heavy metal removal. Biosorption encompasses reversible extracellular adsorption and irreversible intracellular adsorption. While both living and non-living microalgae facilitate extracellular adsorption, only living microalgae can perform irreversible intracellular adsorption. Besides biosorption, other heavy metal removal mechanisms result from microalgae's metabolic activities. Research involving microalgae has highlighted that heavy metal removal is optimal within the pH range of 4–9 (Kumar et al., 2015).

Nitrogen enters lake ecosystems through atmospheric deposition, fixation by cyanobacteria (e.g., N_2 fixation), as well as surface and subsurface flows of both inorganic and organic forms from watershed sources. Internal loading of nitrogen occurs due to the microbial decay of N stored in aquatic vegetation, algal biomass, and lake sediments. The primary factor governing nitrogen processes in freshwater ecosystems is the residence time of water. In deep lakes with a low littoral/profundal ratio, the free-floating plankton community dominates and exerts significant control over nitrogen cycling (Durand et al., 2011). In surface waters, nitrogen is present in forms such as NO_2^-, NO_3^-, NH_4^+, NH_3, and also in organic forms such as dissolved organic nitrogen and particulate organic nitrogen. Nitrate, nitrite, ammonium, and dissolved organic nitrogen are nitrogen forms available for utilization by algae.

Phosphorus plays a vital role in biological metabolism. Compared to other necessary macronutrients, phosphorus is the least abundant in nature, often limiting biological productivity. Orthophosphate (PO_4) is the direct, bioavailable form of soluble inorganic phosphorus. Highly reactive, phosphate can interact with many cations (e.g., Fe and Ca), facilitating its easy removal from water. Unlike nitrogen, phosphorus lacks an atmospheric source (Wetzel, 2001).

Plastics production started intensively in the 1950s. The legacy of early plastic production, in the form of discarded plastics, constitutes a significant contemporary environmental challenge in air, water, and soil. Synthetic plastics produced since Bakelite's introduction in 1909 include polyethylene, polypropylene, polystyrene, polyurethane, and polyvinyl chloride. Plastics exist in nano, micro, and macro forms, distributed across the environment in the air, water, soil, and sediment. Due to their structural components and potential adsorption of toxic compounds like HMs and POPs, plastics exhibit toxic attributes. Research is underway to address the removal of microplastics from environmental settings using biological processes, with algae and bacteria among the prominent approaches (Jung et al., 2022; Osman et al., 2023).

In Table 10.1, current applications are provided concerning the utilization of microalgae in diverse environmental contexts for the removal of various pollutants across different species. As evident, the scope extends beyond aquatic environments to encompass terrestrial habitats as well, where contemporary research explores the employment of microalgae for phytoremediation purposes.

TABLE 10.1
Bioremediation Applications in Various Environmental Contexts

Species name	Factor	Pollutants	References
Cyanobacteria (Cylindrospermum stagnale)	Soil	Cd	Alsamhary (2023)
Cylindrospermum stagnale (micro-algae) ve earthworm (Eisenia fetida)	Soil	As	Mao et al. (2023)
Chlorella vulgaris, Scenedesmus dimorphus, and Phormedium sp.	River	Pb, Cr, Mn, Fe, Co, Ni, Cu, Zn, and Cd	Raj et al. (2017)
Azotobacter beijerinck (bacteria) i ve Chlorella pyrenoidosa (micro-algae)	Soil	Alkalinity, salt	Zhou et al. (2023)
Cyanophyceae, Chlorophyceae and Bacillariophyceae	River	turbidity, EC, SO_4, Alkalinity, klorür, TDS, TSS, NO_3^-, KOİ ve BOİ, Cd, Ni,ve Pb	Ugya et al. (2021)
Monoraphidium sp. SL4A, Chlorella sp. SL7A, Selenastrum sp., SL7, Neochloris sp. SK57, ve Chlorococcum sp. SL7B	River	pH, DO, TDS, COD, and BOD	Ummalyama and Singh (2022)
Cyanobacteria	Lake	COD, NH_4^+, NO_3^-, NO_2	Ni et al. (2018)
Chlorella sp.	River	(EC), turbidity, total hardness, (BOD), (COD), Ca, SO_2, NH_3, NO_3^-, NO_2, PO_4, Mg, F ve Cl	Narayanan et al. (2021)
Cryptomonas erosa	Lake	Hg^+_2	Diéguez et al. (2013)
Diatom Sp. (Cylotella meneghiniana, Gomphonema lanceolatum, Nitzschia palea etc.)	Lake	N, P, COD and BOD	Kiran et al. (2016)

10.3 PHYCOREMEDIATION MECHANISMS IN LAKES AND RESERVOIRS

10.3.1 OVERVIEW OF THE PHYCOREMEDIATION PROCESS AND ITS APPLICATION IN LAKES AND RESERVOIRS

The growth of microalgae is dependent on a carbon source and photons for photosynthesis. They can modify their internal structures through both biochemical and physiological adaptations. Autotrophic microalgae utilize inorganic compounds and sunlight for photosynthesis. During this process, CO_2 and water are converted into carbohydrates (glucose), which are subsequently metabolized to generate energy. This energy is used to convert adenosine diphosphate into adenosine triphosphate. The energy in adenosine triphosphate is then utilized to drive various processes within the cell, converting back to adenosine diphosphate in the process and becoming available to acquire more energy for growth (Brennan and Owende 2010). On the other hand, heterotrophic microalgae rely on organic compounds and external nutrients as their carbon and energy source, particularly under dark conditions (Amaro et al., 2011; Huang et al., 2010).

Microalgae can remediate environmental pollutants through mechanisms such as biodesorption, biouptake, and biodegradation. Biodesorption involves the attachment of pollutants to cell wall components or organic molecules secreted by cells. Biouptake, on the other hand, refers to the transportation of pollutants into cells through diffusion, facilitated diffusion, or active transport, often followed by binding to cellular proteins and other chemicals. Microalgae achieve biodegradation by catalytically breaking down pollutant molecules into simpler metabolic components. Biodegradation is a key technology in bioremediation, as it transforms contaminants into less toxic compounds. This process can occur within cells, outside cells, or as a combination of both (Sutherland and Ralph, 2019). While biodesorption can be accomplished by both living and non-living cells, living cells are involved in other mechanisms.

In lakes and reservoirs, the application of phycoremediation involves harnessing the metabolic capabilities of microalgae to mitigate various pollutants. The diverse mechanisms by which microalgae interact with pollutants make them effective tools in environmental restoration strategies. The following sections delve into specific aspects of phycoremediation in lakes and reservoirs, including pollutant removal mechanisms and factors influencing their efficiency.

10.3.1.1 Mechanisms of Heavy Metal Removal

In studies focusing on the removal of heavy metals using microalgae, particular attention has been given to elements such as Cu, followed by Cd, Ni, Pb, Zn, Hg, and Cr. Bioremediation mechanisms have emerged as prominent aspects in heavy metal removal studies, revealing that live cells tend to adsorb a greater amount of heavy metals compared to dead cells. The data indicate that heavy metal removal is notably higher at a pH of 5 (Kumar et al., 2015).

The exceptional metal-sorption capacity of several algal species is attributed to the presence of proteins, lipids, and polysaccharides with heavy metal-binding functional groups on their cell wall surfaces (Priatni et al., 2018). These functional groups, including hydroxyl, amino, sulfate, and carboxyl groups, serve as robust binding sites for heavy metals. The process of heavy metal biosorption by microalgae involves the formation of covalent bonds between heavy metals and ionized cell walls, followed by redox reactions or crystallization on the cell surface. This is followed by the transfer of heavy metals through the cell membrane into the cytosol, followed by diffusion and binding to internal binding sites of proteins and peptides (Ding et al., 2020; Pradhan et al., 2019).

Extracellular metabolites secreted by microalgae possess the capability to chelate metal ions. Moreover, the pH elevation associated with microalgal growth promotes the precipitation of heavy metals (Leong and Chang, 2020). Various pretreatment techniques enhance the metal-sorption capacity of algae, with pretreatment using $CaCl_2$ proving to be the most effective and cost-efficient method for activating algal biomass.

Heavy metal bioremediation constitutes a two-step process involving the initial biosorption of heavy metals onto various metal-binding ligands on cell surfaces, followed by an intracellular process termed bioaccumulation. This process is carried out by cellular mechanisms composed of inorganic molecules and related enzymes (Hernández-Ávila et al., 2017). Biosorption is a more rapid process in which heavy metals primarily attach to the surface of cells, whereas bioaccumulation encompasses the transport of a fraction of metal ions to cells for intracellular metabolic activities, serving various metabolic functions. Initially, metal ions are physically adsorbed onto the cell surface in a matter of seconds or minutes – a process termed physical adsorption. Subsequently, these ions undergo slow transport into the cytoplasm, a phenomenon known as chemisorption (Dhir, 2013). The presence of peptide and polysaccharide polymers (such as cellulose and alginate) on the cell wall of microalgae provides numerous nonspecific adsorption sites, facilitating metal biosorption (Gendy et al., 2022). Bioaccumulation represents a metabolism-dependent process characterized by gradual intracellular diffusion and accumulation. Upon active transport through the cell membrane, metal-binding peptides and proteins, such as glutathione, metallothionein proteins, oxidative stress-reducing agents, and phytochelatins, interact with the metals (Leong and Chang, 2020). In the slow and often irreversible process of bioaccumulation, heavy metals accumulate within cells, binding to intracellular compounds such as polyphosphate bodies and/or vacuoles (Suresh Kumar et al., 2015). Algal polyphosphate bodies facilitate the storage of additional nutrients in freshwater unicellular algae (Dhir, 2013). Numerous researchers have demonstrated the sequestration of metals such as Ti, Pb, Mg, Zn, Cd, Sr, Co, Hg, Ni, and Cu within polyphosphate bodies in green algae.

10.3.1.2 Mechanisms of POP and Organic Matter Removal

Biological processes predominantly facilitate the transformation and degradation of organic pollutants, often leading to mineralization. Algae contribute to the elimination of organic pollutants through two essential mechanisms: bioaccumulation and biodegradation.

POPs are synthetic chemicals capable of long-range transport in the environment, exhibiting persistence, and possessing the potential for bioaccumulation and biomagnification within ecosystems. The most commonly encountered POPs in aquatic systems arise from agricultural runoff (pesticides), industrial activities (polychlorinated biphenyls or PCBs), fire-resistant materials, surfactants, and urban wastewater (products based on perfluorooctane sulfonic acid, or PFOS), polychlorinated dibenzo-p-dioxins, or PCDDs, and dibenzofurans, or PCDFs, often collectively referred to as "dioxins").

Single-celled green algae such as *Chlorella fusca* var. vacuolata and *Chlamydomonas reinhardtii* demonstrate the capability to bioaccumulate herbicides like metfl uorazon and prometryn, undergo biotransformation, and biologically break down these compounds. Additionally, species like *Ankistrodesmus* and *Scenedesmus* show potential for the biotransformation of organic compounds such as naphthalene.

Qiu et al. (2017) analyzed the bioaccumulation of polybrominated diphenyl ethers and organochlorine pesticides in algae. Their findings revealed that both polybrominated diphenyl ethers and organochlorine pesticides exhibited levels within phytoplankton that were ten times higher than those within Ulva, showcasing that phytoplankton with a larger surface area displayed higher uptake efficiency for POPs compared to Ulva.

Sediments in aquatic environments often possess substantial microbial biomass and diversity due to their inherent anaerobic conditions and ample carbon and energy sources (Himmelheber et al., 2007). This microbial biomass and diversity harbor the potential for the breakdown of organic pollutants.

Algae and cyanobacteria play a significant role in polycyclic aromatic hydrocarbon degradation. It has been reported that red, green, and brown algae, along with cyanobacteria, perform naphthalene metabolism under phototrophic conditions (Cerniglia, 1993). Furthermore, various algae species such as *Dunaliella* spp., *Chlamydomonas* spp., *Selenastrum capricornutum*, *S. costatum*, and *Nitzschia* spp., as well as bacterial species like *Pseudomonas migulae*, *Sphingomonas yanoikuyae*, and *Scenedesmus obliquus*, exhibit the capability to degrade several polycyclic aromatic hydrocarbons.

10.3.1.4 Mechanisms of Nutrient Removal (N, P)

Nutrient pollution in surface waters leads to a well-known process called eutrophication, which triggers widespread excessive algal growth. Within this context, phycoremediation assumes a pivotal role as the most effective method for nutrient removal by sustaining the natural nutrient cycle of lakes (Touliabah et al., 2022). Macronutrients such as carbon, nitrogen, phosphorus, and sulfur, as well as trace elements in small amounts like sodium, calcium, and iron, are essential for algae growth. Among these, nitrogen and phosphorus stand out as the most critical nutrients (Cai et al., 2013).

The conversion of inorganic nitrogen into organic forms is accomplished by eukaryotic microalgae through assimilation (Cai et al., 2013). Initially, the translocation of inorganic nitrogen across the cell membrane occurs through the reduction of nitrate to nitrite by nitrate reductase and further reduction of nitrite to ammonia along the plasma membrane of algal cells. The subsequent step involves the conversion of ammonia to amino acids (glutamine) (Cai et al., 2013).

According to Martinez et al. (1999), algal metabolism heavily relies on inorganic phosphorus in the forms of hydrogen phosphate (HPO_4) and dihydrogen phosphate (H_2PO_4), which are subsequently incorporated into organic compounds through a phosphorylation process involving ADP-derived adenosine triphosphate production and energy input. Phosphates are translocated across algal cell membranes. pH can directly influence enzyme activity, the permeability of the cell membrane, or the degree of phosphate ionization, thus affecting the rate of phosphate absorption. Phosphate uptake rates are directly associated with factors such as the presence of potassium in water, the availability of micronutrients, and the presence of organic compounds.

Due to the rise in pH associated with algal photosynthesis, the reduction of nitrogen and phosphorus can be further enhanced through the precipitation of phosphorus and the release of ammonia (Oswald, 2003).

10.3.1.5   Mechanisms of Microplastic Removal

Microalgae have the capability to remove plastics from water sources by utilizing them as a carbon source (Priyadharshini). The biodegradation of plastics by microalgae involves several stages, including biodeterioration (formation of biofilm, pore and crack formation, pH modification), biofragmentation (assimilation of polymers through extracellular enzymes), assimilation (utilization of plastics as a carbon source), and ultimately mineralization (conversion into CO_2, N_2, CH_4, and/or H_2O) (Dussud and Ghiglione, 2014). While not all microalgal species can perform all these steps, similarly, not all types of plastics are susceptible to degradation. When plastics interact with algae, plastic particles can hinder photosynthesis by reducing the exposure of algae to light. This hindrance can lead to decreased survival of the organism and increased oxidative stress (Dussud and Ghiglione, 2014).

10.4   CURRENT APPLICATIONS OF PHYCOREMEDIATION

10.4.1   Remediation Studies in Diverse Aquatic Environments

Several studies have been conducted to explore the efficacy of using algae for the remediation of water bodies contaminated with heavy metals, organic pollutants, microplastics, and other contaminants. In an endeavor to address heavy metal pollution, a noteworthy investigation was carried out by Raj et al. (2021) in India. They examined the remediation potential of various algae species, including Chlorella vulgaris, Scenedesmus dimorphus, and Phormedium sp., in water samples exhibiting elevated levels of heavy metals, such as Pb, Cr, Mn, Fe, Co, Ni, Cu, Zn, and Cd, collected from the Coom River. Following a 15-day incubation period, the microscopic analysis of the algae revealed no detrimental morphological changes, while it was discerned that these algae efficiently removed heavy metals from the water. The outcomes indicated promising prospects for the scalability of this approach for larger remediation initiatives (Raj et al., 2017).

In China's Yangtze Lake, an innovative ecological dam (Eco-dam) system was employed to investigate the improvement in water quality. This system incorporated submerged biofilters and plant floating beds, constructed using plastic materials that facilitated both biofilm formation and the growth of aquatic plants. The influence of different aquatic plants and the bacteria thriving within the biofilm were examined. The upper layers of the lake's SBF contained photosynthetic bacteria alongside algae, with Cyanobacteria predominating but decreasing in abundance with depth. The findings demonstrated that the created biofilm effectively contributed to the reduction of chemical oxygen demand (COD), nitrite, nitrate, and ammonium compounds (Narayanan et al., 2021).

A study in India focused on exploring suitable indigenous algae species, particularly Chlorella sp., for the purification of water contaminated with domestic and medical wastewater from the Thirumanimutharu River. Both laboratory and in-situ trials were conducted, with three different biomass densities investigated over a 15-day period to assess their potential in removing parameters such as electric conductivity, turbidity, total hardness, biochemical oxygen demand (BOD), COD, and various ions. The experiments, conducted both in controlled settings and within a pond created within the river, underscored the efficiency of *Chlorella* sp. in mitigating the assessed parameters. Morphological analyses of *Chlorella* sp. using Fourier-transform infrared spectroscopy and scanning electron microscopy revealed the involvement of absorption mechanisms in the remediation process (Narayanan et al., 2021).

In another investigation, conducted using water samples from four Patagonian lakes, the effects of dissolved organic matter (DOM) on the bioaccumulation of Hg^{+2} by the algae *Cryptomonas erosa*, zooplankters *Brachionus calyciflorus*, and *Brachionus calyciflorus* were examined. The results indicated that Cryptomonas efficiently facilitated the rapid uptake of Hg^{+2}, predominantly through passive adsorption processes that were significantly influenced by the composition of DOM. In cases where DOM exhibited protein-like or small phenolic signatures, the highest levels of bioaccumulation were observed. Conversely, compounds associated with humic or fulvic signatures, linked to high molecular weight molecules, hindered Hg^{+2} uptake by keeping it in the solution phase (Diéguez et al., 2013).

A study by Kiran et al. (2016) emphasized the significance of diatoms in preventing eutrophication in stagnant waters, showcasing their importance through two case studies. As diatoms rely on silicon (Si), its availability in the form of Nualgi, containing iron and other elements, was investigated in laboratory experiments. The results revealed increased numbers and activity of diatom species, such as *C. clostridium* and *C. fusiformis*, in samples collected from Hussain Sagar Lake. The introduction of Nualgi to a pond with in Nualgi Park yielded substantial removal efficiencies of N, P, COD, and BOD, reaching 95.1%, 88.9%, 91%, and 51%, respectively (Kiran et al., 2016).

The phycoremediation of polluted water from the Kaduna River in Nigeria was explored using algae collected from a petrochemically contaminated river. The algae composition in the polluted water consisted of Cyanophyceae, Chlorophyceae, and Bacillariophyceae species. A photobioreactor was employed in a week-long phycoremediation process, during which significant reductions in turbidity, conductivity, sulfate, alkalinity, chloride, total dissolved solids (TDS), total suspended solids (TSS), nitrate, COD, BOD, Cd, Ni, and Pb were recorded. The use of various analytical techniques, including scanning electron microscopy, X-ray fluorescence spectroscopy, X-ray diffraction, Fourier-transform infrared spectroscopy, and

Gas chromatography/Mass spectrometry, revealed the involvement of bioadsorption mechanisms and phytoremediation-related breakdown processes (Ugya et al., 2021).

Finally, in a study conducted in India, the potential of indigenous algae species, namely *Monoraphidium* sp. SL4A, *Chlorella* sp. SL7A, **Selenastrum** sp. SL7, *Neochloris* sp. SK57, and *Chlorococcum* sp. SL7B, was investigated for the purification of water contaminated with domestic wastewater from the Nambur River. The study demonstrated that the obtained biomass was suitable for the production of fatty acid methyl esters for both nutrient and green fuel production. Furthermore, the presence of these algae facilitated the effective removal of both organic and inorganic pollutants (Ummalyma and Singh, 2022).

10.5 CONCLUSION

The utilization of algae for wastewater treatment dates back to the 1950s, and its effectiveness has been demonstrated through numerous studies (Guleri et al., 2020; Kiran et al., 2016). Considering lake ecosystems, the application of algae for improving water quality in lakes has gained traction, involving both standalone algal methods and combinations with other remediation techniques. The adoption of phycoremediation in wastewater treatment has been extensively explored in the literature, particularly in the context of revitalizing polluted freshwater environments. However, research focused on in situ phycoremediation remains scarce, possibly due to the dominance of engineering projects in this realm, which might not have translated into scientific publications.

Extensive efforts have been directed toward the removal of heavy metals. In the future, endeavors could be directed toward developing algae that exhibit lower costs and higher efficiency in removing specific metal types. Research in the domain of microplastic pollution has focused on both the impact of microplastics on algae and the biodegradation of microplastics facilitated by algae. The development of algal species capable of efficiently degrading microplastics holds promising potential. Microplastics tend to adsorb chemicals present in their environment (such as persistent organic pollutants), and thus, findings from marine environments might not directly apply to freshwater contexts (Wagner et al., 2014). Consequently, dedicated investigations are needed to comprehend the mechanisms of microplastic biodegradation through algae in freshwater environments.

Climate change is expected to impact bioremediation through direct and indirect pathways (Ram Prasad, 2022). Predictive studies using ecosystem modeling to estimate the potential alterations in phycoremediation processes under changing climate parameters are crucial (Jilbert et al., 2020). Climate change can influence lake stratification, seasonal circulations, and temperature-dependent biogeochemical processes. The implementation of in situ phycoremediation and open system approaches like constructed wetlands should be mindful of climate change effects (Gendy et al., 2022). Furthermore, the application of phycoremediation in soil should be holistically integrated with lake/reservoir remediation strategies. Algae in the form of biocrust can help mitigate soil moisture loss, which can be advantageous in preventing both climate-driven drought-induced soil desiccation and excessive runoff carrying soil particles into lake ecosystems.

Phycoremediation involving both micro and macroalgae in the purification of lake and reservoir waters faces the challenge of the unique nature of each water body. Factors such as the distribution of pollutants and microorganisms are not uniform across different ecological settings (Kochhar et al., 2022). Due to the inherent characteristics of biological processes, optimal results require a sufficient presence of microorganisms, nutrients, and the targeted pollutants. The duration for which water remains stagnant, which can be referred to as hydraulic residence time, plays a critical role in the nitrogen cycle of freshwater ecosystems. As a dynamic parameter influenced by location and time, hydraulic residence time is also susceptible to climate change impacts (Durand et al., 2011).

It is important to note that certain pollutants may not be completely removed through phycoremediation. In such cases, strategies like genetic engineering and the screening of new strains for bioremediation should be developed to address organic pollution issues (Baghour et al., 2019). Additionally, it's crucial to recognize that the effectiveness of phycoremediation methods can vary depending on the characteristics of the specific lake or reservoir and the severity of pollution. Often, a combination of different approaches is necessary to achieve optimal results. Continuous monitoring and assessment of water quality are essential for ensuring the success of remediation efforts.

The application of algal methods ensures minimal disruption to ecological balance and can even serve as a source of drinking water supply. Chemical methods may produce byproducts that hinder their use for drinking water. Since lakes and reservoirs are open systems, controlled environments like those found in wastewater treatment facilities cannot be replicated. This also implies that direct observation of phycoremediation in open systems may not always be feasible. The presence of macro and microalgae within ecological cycles, along with other organisms, is a reality in open systems. Therefore, for large-scale applications, it may be beneficial to monitor and obtain contour values of biochemical products or conditions (e.g., redox potential) specific to the pollutant being targeted, considering the unique biogeochemical products of each polluted water mass (Rhodes, 2013; Lürling and Mucci, 2020). Additionally, these monitoring activities should ensure that the desired water quality standards are achieved. The diverse physical, chemical, and ecological makeup of each lake and reservoir necessitates methodological recommendations tailored to each setting. Thus, the further proliferation of diverse lake and reservoir applications similar to those in the literature will be instrumental in formulating overarching guidelines for phycoremediation practices in natural aquatic systems.

REFERENCES

Ali, S. A., Rizwan, M., Zaheer, I. E., Yavas, I., Ünay, A., Abdel-Daim, M. M., …, Kalderis, D. (2020). Application of floating aquatic plants in phytoremediation of heavy metals polluted water: A review. *Sustainability (Switzerland)*, *12*(5). https://doi.org/10.3390/su12051927.

Alsamhary, K. (2023). Vermi-cyanobacterial remediation of cadmium-contaminated soil with rice husk biochar: An eco-friendly approach. *Chemosphere*. https://doi.org/10.1016/j.chemosphere.2022.136931.

Amaro, H. M., Catarina, G., & Malcata, X. (2011). Advances and perspectives in using microalgae to produce biodiesel. *Applied Energy*, 3402–3410. doi: 10.1016/j. apenergy.2010.12.014.

Aransiola, S. A., Abioye, O. P., & Bala, J. D. (2019). Microbial-aided phytoremediation of heavy metals contaminated soil: A review. *European Journal of Biological Research*, 9(2), 104–125.

Arantza, S. J., Hiram, M. R., Erika, K., Chávez-Avilés, M. N., Valiente-Banuet, J. I., & Fierros-Romero, G. (2022). Bio- and phytoremediation: Plants and microbes to the rescue of heavy metal polluted soils. *SN Applied Sciences*, 4(2). https://doi.org/10.1007/s42452-021-04911-y.

Baghour, M., Gálvez, F. J., Sánchez, M. E., & Venema, K. (2019). Overexpression of LeNHX2 and SlSOS2 increases salt tolerance and fruit production in double transgenic tomato plants. *Plant Physiology and Biochemistry*, 77–86. https://doi.org/10.1016/j. plaphy.2018.11.028.

Brennan, L., & Owende, P. (2010). Biofuels from microalgae: A review of technologies for production, processing, and extractions of biofuels and co-products. *Renewable & Sustainable Energy Reviews*, 14(2), 557–577.

Cai, Y., Zheng, H., Ding, S., Kropachev, K., Schwaid, A., Tang, Y., …, Wang, S. (2013). Free energy profiles of base flipping in intercalative polycyclic aromatic hydrocarbon-damaged DNA duplexes: Energetic and structural relationships to nucleotide excision repair susceptibility. *Chemical Research in Toxicology*, 1115–1125. https://doi.org/10.1021/tx400156a.

Cerniglia, C. E. (1993). Biodegradation of polycyclic aromatic hydrocarbons. *Current Opinion in Biotechnology*, 4(3), 331–338.

Chekroun, K., Sanchez, E., & Baghour, M. (2014). The role of algae in bioremediation of organic pollutants. *International Research Journal of Public and Environmental Health*, 1, 19–32.

Chen, T., Lei, M., Wan, X., Zhou, X., & Yang, J. (2020). *Phytoremediation of Arsenic Contaminated Sites in China Theory and Practice*. China: Springer.

Chirwa, E. M., Tikilili, P. V., & Lutsinge, T. B. (2019). Bioremediation of chlorinated and aromatic petrochemical pollutants in multiphase media and oily sludge. In: Bharagava, R.N. (ed.), *Recent Advances in Environmental Management*. Boca Raton, FL: CRC Press, pp. 373–390.

Chortek, E. (2017). *Remediation Strategies for Mercury Contaminated Lakes and Reservoirs within the State of California*. California: Master's Projects and Capstones. Retrieved from https://repository.usfca.edu/capstone/691.

Cuypers, A., & Vangronsveld, J. (eds.), (2017). *Advances in Botanical Research*. London, UK: Academic Press.

Das, N., Mandal, S. K., & Selvi, A. (2019). Petroleum hydrocarbons environmental contamination, toxicity, and bioremediation approaches. In: Bharagava, R. N. (ed.), *Recent Advances in Environmental Management*. Boca Raton, FL: CRC Press, pp. 351–372.

Dhir, B. (2013). *Phytoremediation: Role of Aquatic Plants in Environmental Clean-Up*. New Delhi: Springer India. https://dx.doi.org/10.1007/978-81-322-1307-9_1.

Diéguez, M. C., Queimaliños, C. P., Guevara, S. R., & Marvin-DiPasquale, M. C. (2013). Influence of dissolved organic matter character on mercury incorporation by planktonic organisms: An experimental study using oligotrophic water from Patagonian lakes. *Journal of Environmental Sciences*. https://doi.org/10.1016/S1001-0742(12)60281-2.

Ding, J. P., Yang, M. I., Mishchenko, M. I., & Nevels, R. D. (2020). Identify the limits of geometric optics ray tracing by numerically solving the vector Kirchhoff integral. *Optics Express*, 10670–10682. doi: 10.1364/OE.389097.

Durand, P., Breuer, L., & Johnes, P. (2011). *Nitrogen Processes in Aquatic Ecosystems*. England: Cambridge University Press.

Dussud, C., & Ghiglione, J.-F. (2014). Bacterial degradation of synthetic plastics. *Marine Litter in the Mediterranean and Black Seas*. CIESM Workshop Monograph n° 46 [F. Briand, ed.], 180 p., CIESM Publisher, Monaco.

Erdogan, S. (2020). Microplastic pollution in freshwater ecosystems: A case study from Turkey. *Ege Journal of Fisheries and Aquatic Sciences*, 213–221. https://doi.org/10.12714/egejfas.37.3.02.

Gendy, E. A., Oyekunle, D. T., Ifthikar, J., Jawad, A., & Chen, Z. (2022). A review on the adsorption mechanism of different organic contaminants by covalent organic framework (COF) from the aquatic environment. *Environmental Science and Pollution Research International*, 32566–32593. https://doi.org/10.1007/s11356-022-18726-w.

Guleri, S., Abhishek, S., Singh, J., Dhanker, R., Rinku, N., & Kapoor, N. (2020). Phycoremediation: A novel and synergistic approach in wastewater remediation. *The Journal of Microbiology, Biotechnology and Food Sciences*, *10*(1), 98–106.

Hernández-Ávila, J., Salinas-Rodríguez, E., Cerecedo-Sáenz, E., Reyes-Valderrama, M. I., Arenas-Flores, A., Román-Gutiérrez, A. D., & Rodríguez-Lugo, V. (2017). Diatoms and their capability for heavy metal removal by cationic exchange. *Metals*. https://doi.org/10.3390/met7050169.

Himmelheber, D. W., Pennell, K. D., & Hughes, J. B. (2007). Natural attenuation processes during in situ capping. *Environmental Science & Technology*, 5306–5313. https://doi.org/10.1021/es0700909.

Holdren, C., Jones, W., & Taggart, J. (2001). *Managing Lakes and Reservoirs*. Washington, DC: North American Lake Management Society and Terrene Institute.

Hu, J., Yang, N., He, T., Zhou, X., Yin, D., Wang, Y., & Zhou, L. (2023). Elevated methylmercury production in mercury-contaminated paddy soil resulted from the favorable dissolved organic matter variation created by algal decomposition. *Environmental Pollution*. https://doi.org/10.1016/j.envpol.2023.121415.

Huang, G., Chen, F., Wei, D., Zhang, X., & Chen, G. (2010). Biodiesel production by microalgal biotechnology. *Applied Energy*, 38–46. https://doi.org/10.1016/j.apenergy.2009.06.016.

Huang, T. (ed.), (2016). *Water Pollution and Water Quality Control of Selected Chinese Reservoir Basins: The Handbook of Environmental Chemistry 38 Series*. Cham: Springer International Publishing AG, pp. 265–277.

Jilbert, T., Courture, R. M., & Huser, B. J. (2020). Preface: Restoration of eutrophic lakes: Current practices and future challenges. *Hydrobiologia*, 4343–4357. https://doi.org/10.1007/s10750-020-04457-x.

Jones, K., & Voogt, P. (1999). Persistent organic pollutants (POPs): State of the science. *Environmental Pollution*, 209–221. https://doi.org/10.1016/s0269-7491(99)00098-6.

Jung, Y. S., Sampath, V., Prunicki, M., Aguilera, J., Allen, H., LaBeaud, D., …, Erny, B. (2022). Characterization and regulation of microplastic pollution for protecting planetary and human health. *Environmental Pollution*. https://doi.org/10.1016/j.envpol.2022.120442.

Karalija, E., Carbó, M., Coppi, A., Colzi, I., Dainelli, M., Gašparović, M., …, Pilić, S. (2022). Interplay of plastic pollution with algae and plants: Hidden danger or a blessing? *Journal of Hazardous Materials*. https://doi.org/10.1016/j.jhazmat.2022.129450.

Kiran, M. T., Bhaskar, M. V., & Tiwari, A. (2016). Phycoremediation of eutrophic lakes using diatom algae. In: Rashed, M. N. (ed.), *Lake Sciences and Climate Change*. BoD – Books on Demand. doi: 10.5772/64111.

Kochhar, N., Shrivastava, S., Ghosh, A., Rawat, V. S., Sodhi, K. K., & Kumar, M. (2022). Perspectives on the microorganism of extreme environments and their applications. *Current Research in Microbial Sciences*. https://doi.org/10.1016/j.crmicr.2022.100134.

Kodavanti, P. R., Royland, J. E., & Sambasiva Rao, K. R. (2014). Toxicology of persistent organic pollutants. *Reference Module in Biomedical Sciences*. https://doi.org/10.1016/B978-0-12-801238-3.00211-7.

Krishna Samal, D. P., Sukla, L. B., Pattanaik, A., & Pradhan, D. (2020). Role of microalgae in treatment of acid mine drainage and recovery of valuable metals. *Materials Today: Proceedings*, 346–350. https://doi.org/10.1016/j.matpr.2020.02.165.

Kumar, J., Krithiga, T., Sathish, S., Renita, A. A., Prabu, D., Lokesh, S., …, Namasivayam, S. K. (2022). Persistent organic pollutants in water resources: Fate, occurrence, characterization and risk analysis. *The Science of the Total Environment*. https://doi.org/10.1016/j.scitotenv.2022.154808.

Kumar, S., Dahms, H. U., Won, E. J., & Lee, E. J. (2015). Microalgae: A promising tool for heavy metal remediation. *Ecotoxicology and Environmental Safety*, 329–352. https://doi.org/10.1016/j.ecoenv.2014.12.019.

Leong, Y. K., & Chang, J. S. (2020). Bioremediation of heavy metals using microalgae: Recent advances and mechanisms. *Bioresource Technology*. https://doi.org/10.1016/j.biortech.2020.122886.

Leong, Y. K., Chew, K. W., Chen, W. H., Chang, J. S., & Show, P. L. (2021). Reuniting the biogeochemistry of algae for a low-carbon circular bioeconomy. *Trends in Plant Science*, 729–740. doi.org/10.1016/j.tplants.2020.12.010.

Lürling, M., & Mucci, M. (2020). Mitigating eutrophication nuisance: In-lake measures are becoming inevitable in eutrophic waters in the Netherlands. *Hydrobiologia*, 4447–4467. https://doi.org/10.1007/s10750-020-04297-9.

Mao, Q., Xie, Z., Pei, F., Irshad, S., Issaka, S., & Randrianarison, G. (2023). Indigenous cyanobacteria enhances remediation of arsenic-contaminated soils by regulating physicochemical properties, microbial community structure and function in soil microenvironment. *The Science of the Total Environment*. https://doi.org/10.1016/j.scitotenv.2022.160543.

Marella, K., Saxena, A., & Tiwari, A. (2020). Diatom mediated heavy metal remediation: A review. *Bioresource Technology*. https://doi.org/10.1016/j.biortech.2020.123068.

Martinez, M., Jimenez, J. M., & Yousfi, F. E. (1999). Influence of phosphorus concentration and temperature on growth and phosphorus uptake by the microalga Scenedesmus obliquus. *Bioresource Technology*, 67(3), 233–240.

Narayanan, M., Prabhakaran, M., Natarajan, D., Kandasamy, S., Raja, R., Carvalho, I. S., …, Chinnathambi, A. (2021). Phycoremediation potential of *Chlorella* sp. on the polluted Thirumanimutharu river water. *Chemosphere*. https://doi.org/10.1016/j.chemosphere.2021.130246.

Ni, Z., Wu, X., Li, L., Lv, Z., Zhang, Z., Hao, A., Iseri, Y., Kuba, T., Zhang, X., Wu, W. M., Li, C. (2018). Pollution control and in situ bioremediation for lake aquaculture using an ecological dam. Journal of Cleaner Production, 172, 2256–2265. https://doi.org/10.1016/j.jclepro.2017.11.185.

Nürnberg, G. K. (2017). Attempted management of cyanobacteria by phoslock (lanthanum-modified clay) in Canadian lakes: Water quality results and predictions. *Lake and Reservoir Management*, 163–170. doi: 10.1080/10402381.2016.1265618.

Osman, A. I., Hosny, M., & Eltawil, A. S. (2023). Microplastic sources, formation, toxicity and remediation: A review. *Environmental Chemistry Letters*, 2129–2169. https://doi.org/10.1007/s10311-023-01593-3.

Oswald, W. J. (2003). My sixty years in applied algology. *Journal of Applied Phycology*, 99–106. https://doi.org/10.1023/A:1023871903434.

Oswald, W. J., Gotaas, H. B., Golueke, C. G., Kellen, E. F., & Gloyna. (1957). Algae in waste treatment [with discussion]. *Sewage and Industrial Wastes*. https://www.jstor.org/stable/25033322.

Paranjape, K., Leite, G. B., & Hallenbeck, P. C. (2016). Effect of nitrogen regime on microalgal lipid production during mixotrophic growth with glycerol. *Bioresource Technology*, 778–786. https://doi.org/10.1016/j.biortech.2016.05.020.

Pereira, A. C., & Mulligan, C. N. (2023). Practices for eutrophic shallow lake water remediation and restoration: A critical literature review. *Water*. https://doi.org/10.3390/w15122270.

Pradhan, D., Sukla, L. B., Mishra, B., & Devi, N. (2019). Biosorption for removal of hexavalent chromium using microalgae *Scenedesmus* sp. *Journal of Cleaner Production*, *17*, 907–916.

Prasad, R. (2022). *Phytoremediation for Environmental Sustainability*. Singapore: Springer.

Priatni, S., Ratnaningrum, D., Warya, S., & Audina, E. (2018). Phycobiliproteins production and heavy metals reduction ability of *Porphyridium* sp. *IOP Conference Series: Earth and Environmental Science*. Jakarta: 2nd International Symposium on Green Technology for Value Chains 2017 (GreenVC 20170). doi: 10.1088/1755-1315/160/1/012006.

Qin, H., Su, Q., Khu, S., & Tang, N. (2014). Water quality changes during rapid urbanization in the shenzhen river catchment: An integrated view of socio-economic and infrastructure development. *Sustainability*, 7433–7451. https://doi.org/10.3390/su6107433.

Qiu, Y., Zeng, E. Y., Qiu, H., Yu, K., & Cai, S. (2017). Bioconcentration of polybrominated diphenyl ethers and organochlorine pesticides in algae is an important contaminant route to higher trophic levels. *The Science of the Total Environment*, 1885–1893. https://doi.org/10.1016/j.scitotenv.2016.11.192.

Raj, A., Kumar, A., & Dames, J. F. (2021). Tapping the role of microbial biosurfactants in pesticide remediation: An eco-friendly approach for environmental sustainability. *Frontiers in Microbiology*. https://doi.org/10.3389/fmicb.2021.791723.

Raj, D., Chowdhury, A., & Maiti, S. K. (2017). Ecological risk assessment of mercury and other heavy metals in soils of coal mining area: A case study from the eastern part of a Jharia coal field, India. *Human and Ecological Risk Assessment: An International Journal*, *23*(4), 767–787.

Razaviarani, V., Arab, G., & Lerdwanawattana, N. (2023). Algal biomass dual roles in phycoremediation of wastewater and production of bioenergy and value-added products. *International Journal of Environmental Science and Technology*, 8199–8216. https://doi.org/10.1007/s13762-022-04696-6.

Rhodes, C. J. (2013). Applications of Bioremediation and Phytoremediation. *Science Progress*, 417–427. https://doi.org/10.3184/003685013X13818570960.

Samavi, M., Kosamia, N. M., Vieira, E. C., Mahal, Z., & Rakshit, S. K. (2022). Occurrence of MPs and NPs in freshwater environment. In: Tyagi, R.D., Pandey, A., Drogui, P., Yadav, B., & Pilli, S. (eds.), *Current Developments in Biotechnology and Bioengineering: Microplastics and Nanoplastics: Occurrence, Environmental Impacts and Treatment Processes*. Elsevier Science, pp. 125–150. https://doi.org/10.1016/B978-0-323-99908-3.00011-7.

Shanmuganathan, R., Sibtain Kadri, M., Mathimani, T., Hoang Le, Q., & Pugazhendhi, A. (2023). Recent innovations and challenges in the *eradication of emerging contaminants* from aquatic systems. *Chemosphere*. https://doi.org/10.1016/j.chemosphere.2023.138812.

Singh, A., Rana, M. S., Tiwari, H., Kumar, M., Saxena, S., & Anand, V. (2023). Anaerobic digestion as a tool to manage eutrophication and associated greenhouse gas emission. *The Science of the Total Environment*. https://doi.org/10.1016/j.scitotenv.2022.160722.

Sutherland, D. L., & Ralph, P. J. (2019). Microalgal bioremediation of emerging contaminants: Opportunities and challenges. *Water Research*. https://doi.org/10.1016/j.watres.2019.114921.

Szymańska, A., & Obolewski, K. (2020). Microplastics as contaminants in freshwater environments: A multidisciplinary review. Ecohydrology & Hydrobiology, *20*, 333–345.

Tiodar, E. D., Văcar, C. L., & Podar, D. (2021). Phytoremediation and microorganisms-assisted phytoremediation of mercury-contaminated soils: Challenges and perspectives. *International Journal of Environmental Research and Public Health*, https://doi.org/10.3390/ijerph18052435.

Topal, M., Topal, I. A., & Öbek, E. (2020). Remediation of pollutants with economical importance from mining waters: Usage of Cladophora fracta. *Environmental Technology & Innovation*. https://doi.org/10.1016/j.eti.2020.100876.

Touliabah, H. E., El-Sheekh, M. M., Ismail, M. M., & El-Kassas, H. (2022). A review of microalgae- and cyanobacteria-based biodegradation of organic pollutants. *Molecules*. doi: 10.3390/molecules27031141.

Tufail, M. A., Iltaf, J., Zaheer, T., Tariq, L., Amir, M. B., Fatima, R., ..., Umar, W. (2022). Recent advances in bioremediation of heavy metals and persistent organic pollutants: A review. *The Science of the Total Environment*. https://doi.org/10.1016/j.scitotenv.2022.157961.

Ugya, A. Y., Ajibade, F. O., & Hua, X. (2021). The efficiency of microalgae biofilm in the phycoremediation of water from River Kaduna. *Journal of Environmental Management*. https://doi.org/10.1016/j.jenvman.2021.113109.

Ummalyma, S. B., & Singh, A. (2022). Biomass production and phycoremediation of microalgae cultivated in polluted river water. *Bioresource Technology*. https://doi.org/10.1016/j.biortech.2022.126948.

Wagner, M., Scherer, C., & Alvarez-Muñoz, D. (2014). Microplastics in freshwater ecosystems: What we know and what we need to know. *Environmental Sciences Europe*. https://doi.org/10.1186/s12302-014-0012-7.

Wetzel, R. G. (2001). *Limnology Lake and River Ecosystems*. San Diego, CA: Academic Press.

Wójcik, M., Gonnelli, C., Selvi, F., Dresler, S., Rostański, A., & Vangronsveld, J. (2017). Metallophytes of serpentine and calamine soils: Their unique ecophysiology and potential for phytoremediation. *Advances in Botanical Research*, *83*, 1–42.

Zhang, F., Peng, W., Yang, Y., Dai, W., & Song, J. (2019). A novel method for identifying essential genes by fusing dynamic protein-protein interactive networks. *Protein Interactive Networks*. https://doi.org/10.3390/genes10010031.

Zhou, L., Liu, W., Duan, H., Dong, H., Li, J., Zhang, S., ..., Xu, T. (2023). Improved effects of combined application of nitrogen-fixing bacteria Azotobacter beijerinckii and microalgae Chlorella pyrenoidosa on wheat growth and saline-alkali soil quality. *Chemosphere*. https://doi.org/10.1016/j.chemosphere.2022.137409.

11 Algal Biofilm and Phycoremediation

Kaniye Güneş, Berat Batuhan Kaplangı,
Altan Özkan, and Mehmet Ali Kucuker

11.1 INTRODUCTION

Using microalgae or macroalgae to eradicate harmful substances in wastewater is called "phycoremediation". This term comes from "Phyco" meaning algae and "remediation" meaning treating or bringing back to the initial state. It is accepted as a subdivision of bioremediation. Phycoremediation conforms with microalgae-based wastewater treatment, which has growing interest due to its capability to remove nutrients from municipal, agricultural, and industrial wastewater. Its application in commercial and industrial wastewater treatment dates back to the 1950s (Li et al., 2019; Phang et al., 2015).

One of the most significant concerns associated with urbanization, industrialization, and unawareness of water consumption is the freshwater scarcity problem. Proper utilization of wastewater treatment technologies can help alleviate this problem through the elimination of substances like nitrogen and phosphorus and chemicals of emerging concern (CECs) such as industrial chemicals, pharmaceuticals, and personal care products (Ding, 2023; Golovko et al., 2021).

Excessive nitrogen (N) and phosphorus (P) in aquatic systems result in eutrophication. However, both having high concentrations in wastewater would, in the meantime, be a cost-free nutrient source for microalgae cultivation. Algal biomass produced with nitrogen and phosphorus waste can serve as a sustainable energy resource (Hoh et al., 2016). Phycoremediation can be an applicable method for removing nutrients and diverse contaminants given the correct selection of species, culture conditions, and wastewater contents (Gani et al., 2016). Phycoremediation offers a range of advantages as a wastewater treatment method, such as having low operational costs and a high potential for generating valuable biomass.

Additionally, the handling and treatment needs of the generated wastewater sludge and the effluent oxygenation needs before its discharge at the conclusion of the treatment procedure are significantly diminished owing to phycoremediation. In addition, algae biomass can be used as fertilizer without causing any secondary contamination, which is another environmentally friendly feature of the process (Koul et al., 2022). Bioremediation involves the presence of a range of microorganism groups such as bacteria, fungi, and algae. Since algae possess photosynthetic efficiency and high tolerance to extreme conditions in nature, they can be used as a

DOI: 10.1201/9781003390213-11

platform for remarkably efficient removal of salinity, toxic metal ions, and nutrients from water streams (Gondi et al., 2022).

Based on growth mode, phycoremediation can be divided into two systems, namely, attached system and suspended system. The system where the algae grow attached to a surface is called the attached system, which includes microalgal biofilms, algal turf scrubbers (ATSs), and revolving algal biofilm (RAB)-based systems.

On the contrary, the suspension culture system includes naturally suspended sedimentation ponds and artificial systems, such as waste stabilization ponds (WSPs) and high-rate algal ponds (HRAPs). In addition to standard techniques, various algae-based treatment approaches have been extensively studied and documented in the literature. These include photo-bioreactors, algae-based membrane bioreactors, microbial fuel cells, immobilized systems (Li et al., 2022; Lin-Lan et al., 2018; Razak & Sharip, 2020), and a combination of algal cultivation systems for phycoremediation. For example, the integration of algal-biofilm system into conventional high-rate algal ponds for wastewater treatment reduces the limitations of harvesting and recycling of biomass in high-rate algal ponds (Leong et al., 2021).

Integration of the algae biorefinery approach, which directly contributes to the idea of resource recovery, into the wastewater treatment techniques to produce value-added products helps us to understand how effective phycoremediation is in the circular economy. The concept of "circular bioeconomy" is gaining attraction as a way to convert toxic waste into clean energy. To achieve this objective, phycoremediation-derived algal biomass has been identified as a valuable resource that can be employed in the production of valuable biofuels through biological or thermochemical conversion processes (i.e. anaerobic digestion, direct combustion, pyrolysis, gasification,), and this contributes to the production of affordable algal biorefinery products. Figure 11.1 illustrates the value-added products, which are animal feed,

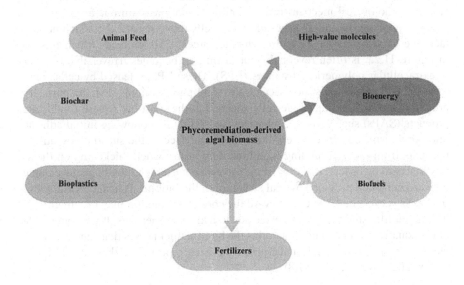

FIGURE 11.1   Value-added products based on algal biomass (Leong et al., 2021).

biochar, bioplastic, fertilizer, high-value molecules, biofuels, and bioenergy, obtained from phycoremediation (Leong et al., 2021).

Phycoremediation should be considered as a holistic process consisting of three main steps: algal biofilm formation and specific techniques for this goal, pollution treatment, and valorization of the resulting biomass.

11.2 ALGAL BIOFILM FORMATION

The term "biofilm" refers to a consortium of embedded microorganisms in a matrix of extracellular polymeric substances (EPS). Under favorable conditions in terms of light, moisture, and nutrients, algae adhere to surfaces and form biofilms, called "algal biofilms" (Carpentier & Cerf, 1999; Kesaano & Sims, 2014). In addition, microalgal biofilm is identified as autotrophic biofilm (cyanobacteria) and heterotrophic microorganism (fungi, bacteria, and protozoa). The bacteria generate carbon dioxide (CO_2) by heterotrophy while microalgae may use this carbon dioxide effectively to produce algal biomass and oxygen for bacteria. Furthermore, it is believed that the consortium between microalgae and bacteria promotes a relief for organic matter in wastewater. This is particularly crucial as microalgae are unable to effectively tolerate such high levels of organic matter. In non-axenic systems, bacteria by itself accelerate initial adhesion in algal biofilm (Mantzorou & Ververidis, 2019; Moreno Osorio et al., 2021; Qian et al., 2023; Wang et al., 2018).

11.2.1 FORMATION MECHANISM

Biofilm formation can be a complicated process and may be influenced by various factors associated with environmental conditions and microbial properties. However, the overall mechanism for biofilm formation is similar for all microorganisms. During the initial adhesion, algal cells move toward the attachment surface due to gravitational or hydrodynamic forces. This process, which is shown in Figure 11.2a, is often reversible, but in time, it becomes irreversible by means of extracellular polymeric substances (EPS). These EPS, released by cells, consist of proteins, nucleic acids, polysaccharides, and phospholipids. For the purpose of providing strong bonding, algae cells attached to the substrate release soluble algal products (SAPs) such as polysaccharides and proteins. Following initial adhesion, microorganisms continue to reproduce on the surface of the substrate by utilizing nutrients from the surrounding liquid medium, after which thickening on the surface starts to be observed. Co-existing bacteria contribute to the bond development between cells and provide physical structure to the biofilm. When a certain biofilm thickness is exceeded, the cells toward the core of the biofilm may lose their viability, and biofilm sloughing takes place due to hydraulic shear resulting from the flow of the aquatic medium. This constrains the biofilm thicknesses that can be achieved; values ranging from 22 μm and 2 mm have been reported (Ozkan, 2012; Tong & Derek, 2021; Wang et al., 2018).

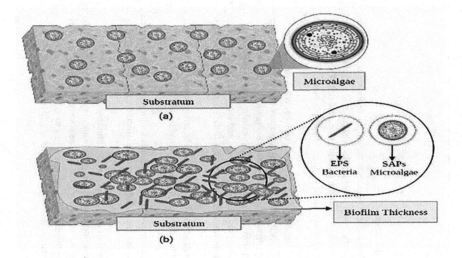

FIGURE 11.2 The illustration of microalgae attachment (a) initial attachment, (b) biofilm growth. (Adapted from Wang et al. (2018).)

11.2.2 ALGAL GROWTH CONDITIONS

Biofilms cultivated for wastewater treatment contain several species, whereas in manufacturing processes, such as fermentation and biotransformation, single species-based biofilms are utilized for process optimization (Li et al., 2007). Even though algal biofilms perform well in wastewater treatment, several factors reduce the efficiency of the biofilm formation and the associated performance (Zhang et al., 2018).

Biological factors, physico-chemical properties, environmental factors, and cellular factors may affect the algal biofilm formation and maturation duration (Moreno Osorio et al., 2021). Table 11.1 summarizes the conditions and factors that positively contribute to biofilm formation.

11.3 APPLICATIONS OF ALGAL BIOFILMS AND PHYCOREMEDIATION

Algal biofilms have mainly been used to treat wastewater with municipal and agricultural origins. Treatment efficiencies have also been assessed for synthetic wastewater and effluents from activated sludge processes. In tertiary treatment, algal biofilms aid in the removal of nutrients, specifically nitrogen and phosphorus, which were not effectively eliminated during secondary treatment. These efforts resulted in the generation of various reactor designs (Dayana Priyadharshini et al., 2021; Hoh et al., 2016). In this section, the treatment efficiency of algal biofilms is presented and discussed.

TABLE 11.1

Factors Affecting the Algal Biofilm Formation

Factors Affecting the Algal Biofilm Formation	Property	Explanation
Physico-chemical properties	Hydrophobicity	Hydrophobicity affects biofilm formation by enabling algae to overcome the attractive forces of the media and adhere to hydrophobic surfaces to minimize contact with water bodies (Osorio et al., 2021). A number of studies investigated the impact of cell hydrophobicity on biofilm growth characteristics. The productivity of mixotrophic biofilms exhibited a significant correlation with their surface hydrophobicity (Roostaei et al., 2018).
	Cell surface energy	Surface energy can be defined as the work that has to be done to create a new surface. When the surface energy of a microorganism is elevated, it is more likely to develop biofilm structures that exhibit a greater degree of homogeneity in comparison to microorganisms with lower surface energy levels. When the surface energy of the cell was between 40 and 50 mJ/m^2, the microalgae cells exhibited the formation of heterogeneous biofilms characterized by a significant presence of open voids and increased porosity. Conversely, a flat and homogeneous biofilm structure was observed within a surface energy range of 50–65 mJ/m^2 (Zhang et al., 2020).
	Nanoparticles	The presence of nanoparticles can also affect microalgae biofilm formation. The study found that low concentrations of nanoparticles (<400 μg/mL) can facilitate cellular growth and biofilm formation (Taghizadeh et al., 2022).
Biological factors	Bacteria	The studies indicate that the presence of the bacteria improves the microalgal biofilm formation by producing growth-promoting organics such as vitamins and other supplemental growth structures (Medipally et al., 2015)
	Photosynthesis	Light conditions of the medium affect the photosynthesis rate of microalgae. Photosynthesis, which is a process in which algae convert light, CO_2, and water into organic compounds and oxygen, has a huge role in microalgal biofilm formation due to maintaining the necessary growing rate since this process provides energy for the series of phototrophic and heterotrophic metabolic processes in a total biofilm community Thus, photosynthetic performance can affect microalgae formation (Zhang et al., 2018; Polizzi et al., 2022).

(Continued)

TABLE 11.1 (*Continued*)
Factors Affecting the Algal Biofilm Formation

Factors Affecting the Algal Biofilm Formation	Property	Explanation
	Nutrient availability	Nutrient availability in the media constrains the formation of algal biofilm. Insufficient availability of nutrients can impose constraints on the growth of microalgae, consequently leading to diminished biomass productivity in microalgae cultivation systems (Lu et al., 2015).
Environmental factors	Liquid flow and composition	It should be noted that flow condition effects vary from species to species. For instance, turbulent flow conditions have a better impact on forming filamentous green algae (*Spirulina platensis*) while laminar flow conditions favor the single-celled green algae (*Chlorella vulgaris*) growth (Osorio et al., 2021).
	Temperature and pH	Different microalgae species have specific temperature and pH ranges in which they thrive and form biofilms (Li et al., 2021). Deviations from the optimal temperature and pH conditions can negatively impact growth and stability.
	Light intensity	The photosynthetic performance, biomass yield, and cellular composition of microalgae biofilms can be influenced by the intensity and duration of light exposure (Li et al., 2021). Furthermore, algal species' optimum light wavelength for biofilm formation varies (Moreno Osorio et al., 2021).
Cellular factors	Genome	Genome is one of the most important factors determining microalgae biofilm formation ability. Genetic factors control biofilm formation and growth processes such as micro-colony formation and EPS production (Moreno Osorio et al., 2021).

11.3.1 INORGANIC AND ORGANIC POLLUTANTS

Sharma et al. (2023) reported promising results from the phycoremediation of X-ray developer (XD) solution using *Desmodesmus armatus* (full name) strains. Food waste (FW) and agricultural compost media (ACM) were used as nutrient sources. The removal efficiencies of biological oxygen demand (BOD) and chemical oxygen demand (COD) were determined as 74.5% and 81.69%, respectively, when using food waste. In the case of agricultural waste medium, the removal efficiencies were measured as 83.05% for BOD and 88.88% for COD. Ugya studied the microalgal biofilm-based phycoremediation of petroleum-contaminated water. This study added to the literature the microalgae previously cultivated in a photobioreactor using nutrient medium (Ugya et al., 2021) in petroleum-contaminated water. Pollutant removal efficiencies were 81% for turbidity, 17.5% for sulfate, 28.4% for alkalinity, 14.6% for chloride, 33% for nitrate, 23.4% for salinity, 28.2% for phosphate, 8% for COD, 16.7% for BOD, and 15% in total petroleum hydrocarbon. The authors attributed the petroleum-based contamination to their adsorption of extracellular polymeric substances. Zhang et al. (2020) utilized pine sawdust as an algal biofilm biocarrier and *Chlorella vulgaris* (*C. vulgaris*) as the treatment agent. Results showed that removal efficiency in synthetic water is 95.95% for COD and 97.16% for total phosphorus (TP), while removal efficiency in real wastewater is 95.94% for total nitrogen (TN) and 96.10% for NH_4^+. Moreover, *C. vulgaris* biofilm has also shown promising lipid productivity. Qian et al. (2023) studied the efficiency of bacterial-algal co-cultures for the treatment of raw and sterilized soy sauce wastewater. *Acinetobacter* and *Comamonas* species dominated the biofilms, and the resulting co-cultivation improved the efficiency of nutrient removal from wastewater.

11.3.2 HEAVY METAL REMOVAL

Heavy metals, such as copper (Cu), cadmium (Cd), arsenic (As), chromium (Cr), mercury (Hg), lead (Pb), and zinc (Zn), commonly exist in domestic wastewater. While certain trace elements such as copper (Cu), manganese (Mn), nickel (Ni), cobalt (Co), and zinc (Zn) play crucial roles as micronutrients for the growth of plants, their excessive accumulation beyond specific thresholds in the food chain can lead to adverse consequences for humans and animals. Heavy metals cannot be degraded and can directly disrupt metabolic processes or undergo transformation into more harmful forms, resulting in detrimental impacts on human health (Kaloudas et al., 2021).

There are various pathways for removing heavy metals, such as sedimentation, flocculation, absorption, cation and anion replacement, microbiological behavior, complexation, accumulation, oxidation/reduction, and adsorption. Algae possess two primary mechanisms through which they can directly eliminate heavy metals from polluted water: (1) absorption into their cells dependent on metabolic activities at low concentrations and (2) a non-active biosorption adsorption process (Kandasamy et al., 2021).

The biosorption abilities of algae are affected by several factors, including the species of algae, the type of contaminant, and whether the algae are in free or immobilized form. Besides free algae cells, immobilized algae cells have been used for heavy metal removal. Immobilization methods, such as covalent binding, adsorption, encapsulation, and entrapment, have been employed to enhance metal removal efficiency. Immobilized algae cells have shown promising results in the removal of heavy metals like chromium (Cr) and uranium (U) from aqueous solutions (Kaloudas et al., 2021).

Chlorella spp., *Cladophora* spp., *Scenedesmus* spp., and *Chlamydomonas reinhardtii* have been investigated for their ability to remove heavy metals from wastewater. Algae can absorb heavy metals through bioaccumulation and biosorption, making them effective in the removal of heavy metals. Different algae species have different heavy metal removal rates, with some achieving high removal efficiencies for metals like Cu, Zn, Mn, and Cd (Ankit et al., 2022).

The study by Chen et al. (2023) shows that the phycoremediation of acid mine drainage (ADM) is a proposed Fe/Mn mineralization method. Rezasoltani and Champagne (2023) reported that *Anabaena* sp. and *Nostoc muscorum* can be utilized for Pb (II) removal; at the end of the phycoremediation process, biomass would be used as a biofertilizer. Another study with Pb (II) conducted using *Scenedesmus obliquus* shows 85.5% removal efficiency. Arsenic removal efficiency obtained with *Coelastrella* sp. was 66.26% in 1 ppm, 49.19% in 2 ppm, and 19.20% in 3 ppm, after which biodiesel is produced as a physio-based product (Angelaalincy et al., 2023). Algae have shown potential for removing heavy metals from wastewater due to their ability to bioaccumulate and biosorb these contaminants. Different algae species have different heavy metal removal rates, and immobilization and growth techniques can enhance their efficiency. Genetic modification, which also controls algal growth characteristics, is also being explored as a tool to improve their heavy metal removal capabilities. These advancements in phycoremediation offer promising insights for developing efficient and sustainable wastewater treatment methods.

11.3.3 EMERGING POLLUTANTS

Emerging contaminants (EC) (i.e., contaminants of emerging concerns "CEC") encompass a wide range of compounds and chemicals that adversely affect the environment. Emerging contaminants from anthropogenic activities, pharmaceutical and personal care products (PPCP), pesticides, and surfactants are common examples. Furthermore, significant amounts of hospital (medical) waste have been discharged into the environment during the COVID-19 pandemic period (Chen et al., 2022; Gupta et al., 2015). Microalgae-based removal technology can be an effective option for ECs that cannot be removed during secondary and tertiary treatments in conventional wastewater treatment plants. However, the efficiency of algae can be hindered by the emerging contamination itself, as these chemicals can inhibit growth. On the other hand, several articles have been published that investigated the EC removal mechanism of algae, particularly for *Chlorella, Scenedesmus, Chlamydomonas, Tetradesmus,* and *Coelastrum* species, and methods for the improvement of EC removal efficiencies (Zhou et al., 2022).

11.4 TECHNIQUES FOR ALGAL CULTIVATION AND POLLUTANT REMOVAL

Phycoremediation can be accomplished through attached growth systems or suspended growth systems. In attached growth systems, microalgae link to a support material, such as polystyrene foam or mesh-type substrates; in designed reactors, these substrates can be stationary (as in algal turf scrubbers) or rotating (revolving biofilm) (Chaturvedi et al., 2013; Lee et al., 2014; Loupasaki & Diamadopoulos, 2013). Attached growth systems provide numerous advantages compared to suspended growth systems, such as increased availability of sunlight due to the more transparent water resulting in higher areal productivity and more biomass produced in a specific area. Suspended growth systems may experience light limitation due to the accumulated microalgal concentration and, therefore, suboptimal growth if there is an insufficient supply of light (Lee et al., 2014). These kinds of systems can be designed in a natural environment (waste stabilization ponds) or can be constructed as an open pond system (High-Rate Algal Ponds). On the other hand, suspended growth systems are used more commonly because of their economic advantages and non-complex setup (Valchev & Ribarova, 2022).

11.4.1 ATTACHED GROWTH SYSTEMS

11.4.1.1 Algal Turf Scrubber

Algal Turf Scrubber (ATS) has been tested for the treatment of wastewater generated as part of industrial, household, and agricultural activities. In these systems, algae grow attached and get harvested by scraping following the draining of the reactor. The harvested algae-rich biomass can be used as a soil conditioner, fertilizer, or biogas feedstock material (Gan et al., 2023).

Reinecke et al. (2023) considered ATS as an affordable solution for small-scale agricultural and industrial sources. An ATS system of a 1,760 m² area was operated with municipal, agricultural, and biogas production plant effluents for 1 year. Based on the results, the fertilizer production cost was estimated at 2.22 €/kg for an optimized ATS. Gan et al. (2023) conducted a study involving the treatment of artificial wastewater over a period of 7 days. The researchers achieved an average nitrogen removal efficiency of approximately 51% and a phosphorus removal efficiency of 90%. These results demonstrate the effectiveness level of the treatment process in reducing nitrogen and phosphorus levels in the wastewater.

Various microbial groups have been found in algal biofilms, including green algae (*Chlorophyceae*), diatoms (*Diatomeae*), gold algae (*Chrysophyceae*), fungi, and ciliates (*Chromista*) as well as prokaryotic cyanobacteria (*Cyanophyceae*). Liu et al. (2016) investigated the use of three benthic filamentous green algae belonging to *Klebsormidium*, *Stigeoclonium*, and *Stigeoclonium* genera for nutrient removal from horticultural wastewater. *Stigeoclonium* showed the highest performance for growth as well as nutrient removal, and *Klebsormidium* strain was not suitable for use to treat horticultural wastewater (Liu et al., 2016).

Ray et al. utilized an ATS of 1 m² area to treat the effluent from a commercial oyster aquaculture facility. An average biomass productivity of 88.8 g/m²-day was achieved. Green filamentous algae dominated the biofilm throughout the test period. Pennate and filamentous diatoms were also significant contributors. An experimental setup comprising Algal Turf Scrubber™ (ATS) systems with a scale of 1 m² was constructed at a commercial oyster aquaculture facility (Ray et al., 2015). The purpose of the study was to examine the rate of algal productivity and explore the possibility of using these facilities to mitigate the release of inorganic nutrients. This system has removed 7.8 kg of nitrogen and 151.6 g of phosphorus, while generating 56.9 kg of algal biomass. *Ulva intestinalis*, green filamentous algae, pennate diatoms, and *Polysiphonia* sp. were the dominant species on ATS biofilm.

11.4.1.2 Revolving Algal Biofilm (RAB)

The RAB reactor uses vertically oriented surfaces that rotate around drive shafts to grow algae. The materials rotate out of the water to expose them to sunlight and facilitate CO_2/O_2 exchange. This design offers a higher solar exposure surface area within a limited space when compared to traditional suspended growth systems. Additionally, the motion of the cylinder generates aerated bubbles within the system. The conveyor belt is usually constructed with polyvinyl chloride (PVC) (Gupta et al., 2021). The productivity of biomass in the RAB reactor is significantly greater, reaching five to ten times higher than that in an open pond. Moreover, the biomass can be conveniently harvested by gently scraping it off the attachment material, resulting in reduced costs in comparison to the more expensive centrifugation harvesting method (Zhou et al., 2018) that uses centrifugal force to enhance the separation of a suspension that is commonly used to harvest microalgae cells efficiently, ~100%, over a short period (Upadhyay et al., 2019).

The nutrient removal efficiency of this kind of system depends on the media's surface area, wastewater's pollution level, and algal-microbial consortia. For example, extracellular polysaccharides (EPS) play an adsorbent role, stabilize the external environment, and promote interactions. In addition to these, EPS act as a carbon and energy source for algae. Thereby, EPS significantly affect the treatment (Gupta et al., 2021).

11.4.2 Suspended Growth Systems

11.4.2.1 Waste Stabilization Ponds

Waste stabilization ponds (WSPs) or open pond systems are commonly used techniques for wastewater treatment in rural areas, especially for those with tropical climates (Kumar et al., 2018). WSPs are used for wastewater treatment, which has high organic content and relies on gravitational settling for harvesting the generated biomass (Ghazy et al., 2008).

The algal accumulation and growth in waste stabilization ponds are attributed to the presence of sunlight, water, and nutrients in wastewater. Algal assimilation and nitrogen consumption play an essential role in nitrogen removal from wastewater in this context. However, algae also deplete phosphorus, typically present in an inorganic form, through enzyme release and secretion of metal-binding compounds like siderophores. Algal cell walls contain various functional groups (amino ($-NH2$), phosphoryl $\left(-PO_3^{2-}\right)$, carboxyl ($-COOH$), sulfhydryl ($-SH$), hydroxyl ($-OH$), and others) that can effectively bind metal ions through weak interactions, as metals are typically present as cations. These functional groups possess a negative charge, facilitating the adsorption of metals onto the cell surface (Koul et al., 2022).

In a study evaluating the performance of a WSP in an arid climate region, it was found that the addition of maturation ponds and optimization of flow rate and retention time improved the treatment efficiency of the system (Achag et al., 2021). Additionally, in situ phycoremediation using microalgae in waste stabilization pond systems has shown promise in reducing physicochemical parameters, including electric conductivity, hardness, and turbidity in polluted water (Venkatesan et al., 2020).

11.4.2.2   High-Rate Algal Ponds

High-rate algal pond (HRAP) (i.e., raceway pond) is a frequently used stirred pond system utilized for wastewater treatment (Randrianarison & Ashraf, 2017). It offers ecological benefits like minimal carbon emissions and decreased greenhouse gas release. The algal biomass produced in HRAPs can be utilized for various purposes, such as the production of biofuels, animal feedstock, and nutraceuticals. However, the harvesting cost is usually a significant challenge for widespread use. Sedimentation with flocculation is a commonly used method for biomass harvesting but it applies only to species that settle under the influence of gravity. *Spirulina*, *Dunaliella*, *Haematococcus*, and *Chlorella* are common algal species cultivated in raceway ponds for commercial purposes (Ferreira de Oliveira & Bragotto, 2022).

HRAPs are large shallow water reservoirs with a depth of up to 25–30 cm. They are equipped with a paddlewheel mechanism that stirs the water at a controlled velocity ranging from 0.15 to 0.30 m/s. The shallow depth and stirring ensure a homogeneous distribution of light, CO_2 diffusion, and nutrients within the algae suspension. CO_2-enriched air can be pumped to the suspensions to enhance algal productivity and increase nutrient uptake rate. Algal photosynthesis can increase the dissolved oxygen levels and assist bacteria in breaking down organic compounds. These kinds of ponds are also more effective at disinfecting wastewater than regular treatment ponds. This is because they allow the establishment of higher average light intensity in the suspensions, which kills bacteria (Kaloudas et al., 2021).

The organic loading rate in HRAPs typically ranges from 100 to 150 kg BOD_5 per hectare per day (Randrianarison & Ashraf, 2017). Algal settling ponds are utilized in conjunction with HRAPs to allow for the gravitational settling of algal biomass. This method can remove colonial microalgae, such as *Micractinium* sp., *Actinastrum* sp., *Pediastrum* sp., and *Dictyosphaerium* sp., without the requirement of immense settling tank volumes. These ponds are also designed to incorporate lamella plates, which enhance gravitational settling efficiencies and secondary thickening of the settled algae. After harvesting, the water goes through a sterilization process, such as membrane filtering, ultraviolet light treatment, ozone (ozonation) or chlorine addition (Kaloudas et al., 2021).

Many researchers reported that biofilm reactors exhibit greater biomass productivity when compared to suspended systems (Hoh et al., 2016). Compared to suspended growth, the reasons why algal biofilm systems are reported to be a much better option for wastewater treatment are the high density/concentration of algal biomass, the effectiveness of cultivation and harvesting processes, and the absence of mixing equipment requirements (Gan et al., 2023). Furthermore, traditional open and closed photobioreactor systems generate low biomass concentrations, usually below 0.5 and 6 g/L, respectively. This necessitates the utilization of energy-intensive harvesting, dewatering, and drying processes. Therefore, commercial utilization of these suspended cultures has been limited because of their low biomass density and capital-intensive harvesting system utilization requirements.

Attached growth systems can offer the solutions to these chronic problems by (1) their capability to generate high biomass density (up to 12%–16% (unit)), (2) enabling the use of simple and efficient harvesting methods, and (3) reducing the energy input requirements for biomass generation (Wang et al., 2018).

HRAPs, on the other hand, are effective systems for wastewater treatment, offering advantages such as a low carbon footprint and the potential for the generation of commercially valuable biomass. Flocculation-induced sedimentation is a widely used method for biomass harvesting, and raceway ponds can be used for cultivating diverse types of algae. HRAPs provide optimal conditions for algae growth and productivity. Settling ponds and algal harvest tanks are used for biomass settling and removal. The treated water then goes through a sterilization process.

In conclusion, attached and suspended growth-based systems have advantages and disadvantages for phycoremediation. Attached growth systems offer higher availability of sunlight, higher areal productivity, and easier biomass harvesting and dewatering. On the other hand, suspended growth systems may face light limitations and lower biomass productivity. Further investigation is required to optimize the efficiency of both types of systems and determine the most suitable method for specific phycoremediation applications.

11.5 ALGAL BIOREFINERIES BASED ON THERMOCHEMICAL AND BIOLOGICAL PROCESSES

Algal biorefinery is essential to simultaneously produce bioactive compounds and biofuel feedstock material. Thermochemical and biological techniques are the two main techniques used in algal biorefineries to utilize algal biomass fully.

Thermochemical techniques involve the use of high temperatures and chemical reactions to convert algal biomass into a range of products. The thermochemical techniques commonly used include (1) direct combustion; (2) pyrolysis; (3) gasification, torrefaction, and their modifications such as supercritical water gasification; (4) hydrothermal liquefaction; and (5) transesterification. These techniques offer a sustainable and efficient route for establishing a bioeconomy (Choudhary et al., 2022). In contrast, biological methods utilize enzymes, microorganisms, and other biological agents to extract and transform algal biomass into desired products. This chapter will discuss microbial fermentation, anaerobic digestion, and photohydrogen production techniques, which have been covered and applied in recent years as a biological algal biomass valorization technique (Osman et al., 2022).

11.5.1 THERMOCHEMICAL TECHNIQUES

11.5.1.1 Direct Combustion of Biomass

This process is among the most accessible thermochemical processes. The main products generated from the combustion process are CO_2, H_2O, and heat (Vaniyankandy et al., 2022). The entire biomass is burned as a solid fuel, typically at temperatures of 800°C–1,000°C with an excess of air present. This process is preferred to extract biodiesel and residual biomass from microalgal biomass containing high lipid content. While researchers have shown preference for alternative biorefinery methods like liquid-targeted liquefaction in the production of biodiesel, there is a growing argument that lipid extraction methods fail to fully capture the algal biomass energy potential. This perspective challenges the common belief that biodiesel is exclusively produced from the lipid content found in algal biomass (Razaviarani et al., 2023).

In fact, experiments have shown that using the entire biomass is much more energy efficient than using only the lipid fraction. In the process of direct combustion, all the components present in the biomass, such as lipid derivatives, proteins, and carbohydrates, are utilized as a source of energy, resulting in a high energy yield and efficiency. This process is straightforward and does not entail any downstream processes that emit carbon. The algal solid fuel can be used in traditional coal boilers without requiring any specific modifications or equipment. However, direct combustion has limitations; white fume emissions may arise where the biomass exhibits a moisture content exceeding 50% by weight (wt.%). To prevent such emissions and maintain optimal process efficiency, energy-intensive dehydration procedures are necessary. In addition, this process involves pretreatment processes such as cutting, dehydrating, and pulverizing, which consume energy and may result in increased expenses. The presence of impurities in biomass, such as sodium, potassium, sulfur, and nitrogen, can also cause issues like corrosion or obstruction (Vaniyankandy et al., 2022). Lastly, due to their lower calorific values compared to fossil fuels, algal solid fuels are not considered suitable as a standalone fuel option. As a result, in industrial settings, coal-biomass co-firing is commonly applied with a small proportion of biomass (Razaviarani et al., 2023).

11.5.1.2 Transesterification of Algal Lipids

Algal lipids have the potential to be transformed into biodiesel using a thermochemical process known as transesterification. This reaction produces glycerol and fatty acid methyl esters (FAMEs) from triglyceride molecules using alcohol and a catalyst. The process reduces the viscosity of the FAME but does not modify its fatty acid composition (Mohan et al., 2018). Regarding algal lipids, the optimal temperature was determined to be 50°C, and the optimal alcohol-to-oil (mostly methanol is used as alcohol) volume ratio was found to be 7.5:1 without ultrasonic pretreatment and 5:1 while applying the ultrasonic pretreatment (Razaviarani et al., 2023). Different catalysts, including acids, bases, or enzymes, can be used in transesterification. Supercritical methanol can also be used to perform transesterification without a catalyst, but it requires elevated temperatures up to 200°C–350°C and pressures in a range of 20–50 MPa (Mohan et al., 2018).

Ultrasound is one of the effective pretreatment technologies for transesterification. It is known to improve emulsification by reducing the size of the droplets in the dispersed phase. With increased ultrasound intensity, the micro-mixing between the oil and methanol phases, as well as the transesterification efficiency, increases. Ultrasonic irradiation has been effectively utilized for lipids extracted from various feedstocks, including algae (Yu et al., 2017).

11.5.1.3 Pyrolysis of Algal Biomass

Pyrolysis is a process based on the controlled decomposition of organic compounds under low atmospheric pressures and the absence of oxygen gas. Pyrolysis is an endothermic process that takes place within a temperature range of 300°C–700°C, varying according to the properties of the material undergoing pyrolysis (López-Aguilar et al., 2022; Vargas e Silva & Monteggia, 2015). This method utilizes the biomass as a whole. Higher temperatures and extended retention times result in the production of bio-oil and volatile biogases as the main output of pyrolysis. On the other hand, lower temperatures and longer residence times cause the formation of solids, known as biochar.

Biomass pyrolysis is a complex process influenced by several factors such as temperature, heating rate, residence time, and inert gas flow rate (N_2, Ar, etc.). The physicochemical properties of the biomass, such as its moisture content, control the relative yield of phases generated. Fast pyrolysis with high heating rates (1,000°C/minutes) is preferred for high yields of liquid products, while slow pyrolysis having arrangements including lower heating rates (5–80°C/minutes) and retention times within 5–30 minutes favors solid char formation. This technology generates the most effective results when utilized in fluidized bed reactor configuration owing to its thermal efficiency and fast devolatilization rates. The pyrolysis bio-oil is generally cleaner than fossil-based alternatives, primarily because of its reduced levels of nitrogen and sulfur contents (Vargas e Silva & Monteggia, 2015).

Pyrolysis is divided into two categories: conventional pyrolysis and modified pyrolysis. Conventional pyrolysis refers to direct pyrolysis without the need for additional techniques or materials. There are two types of conventional pyrolysis: slow pyrolysis and rapid pyrolysis. Slow pyrolysis is performed at relatively lower temperature levels and is favorable for the production of biochar, while fast pyrolysis is conducted at higher temperature levels and is beneficial for the production of bio-oil. Modified pyrolysis, on the other hand, includes (1) catalytic pyrolysis, (2) hydro-pyrolysis, and (3) microwave-induced pyrolysis. Hydro-pyrolysis and catalytic pyrolysis employ a catalyst. Electromagnetic radiation is used to generate necessary heat during microwave-induced pyrolysis (Razaviarani et al., 2023).

11.5.1.4 Hydrothermal Liquefaction of Algal Biomass

Hydrothermal liquefaction (HTL) is a method that involves converting complex molecules found in solid biomass into smaller and more reactive substances. These substances then recombine to form oil compounds with a relatively high molecular weight. This process occurs in a solvent medium at temperatures ranging from 200°C to 350°C and pressures from 5 to 20 MPa and can usually utilize a catalyst. During HTL, lipids get converted to glycerol and fatty acids and broken down into organic compounds that dissolve in water, such as organic acids, benzenes, aldehydes, alcohols, and cyclic ketones. These products then combine again to form aromatic ring structures and long-chain hydrocarbons. Proteins are the main contributor to nitrogen in biocrude. Microalgae are suggested to yield a more significant amount of crude due to their high lipid contents. The liquid phase formed during hydrothermal liquefaction (HTL) is also significant, as it contains dissolved nitrogen compounds and carbon dioxide (Djandja et al., 2020).

Biocrude is an intermediate product that can be used to produce biodiesel. While not compulsory, the integration of a catalyst can enhance the biocrude yield and the associated quality. In one case, the use of a catalyst during HTL of algal biomass resulted in a significant enhancement of the maximum biocrude yield, reaching up to 64%. Sodium carbonate (Na_2CO_3) is a commonly used catalyst in HTL as it is a water-soluble alkali salt. Studies demonstrated that utilizing sodium carbonate at 350°C resulted in the production of biocrude, with yields ranging from 39.9 to 51.6 wt.%, derived from the microalgae *Chlorella pyrenoidosa* (Razaviarani et al., 2023).

11.5.1.5 Gasification and Torrefaction of Algal Biomass

Gasification is a flexible thermochemical technology that converts various types of biomass feedstocks into a high-energy gaseous mixture, including syngas (H_2, CO), CO_2, CH_4, and tar. The process takes place under controlled inputs of steam, oxygen, carbon dioxide, or air at temperatures between 700°C and 1,000°C and can transform biomass in solid, liquid, or gaseous form into a valuable product (Raheem et al., 2021).

Torrefaction is a widely used technique for preparing algal biomass for further processing. This technique involves heating the biomass to temperatures between 200°C and 300°C in an oxygen-free environment, which causes it to lose moisture and become more brittle, increasing biochar yield. So far, the highest biochar yield was 93.9 wt.% using microalga *Chlamydomonas* sp. JSC4 at 200°C. However, torrefaction can be time-consuming, requiring significant energy inputs and long residence times of up to 60 minutes (Razaviarani et al., 2023).

Catalytic gasification is another promising method for converting algal biomass to syngas. Microalgae such as *Chlorella vulgaris* has been shown to be a suitable raw material for syngas production. An optimization study with this species tested a temperature range of 700°C–900°C, reaction time range of 15–40 minutes, and a catalyst loading ratio of 5%–20%. Syngas conversion ratio ranged from 64 to 83 wt.% and that of biochar and tar ranged from 10 to 22 wt.% and from 6 to 13 wt.%, respectively. Hydrogen gas accounted for 27 to 46 mol% of the generated syngas (Raheem et al., 2018; Razaviarani et al., 2023). Similarly, an optimization study was conducted on the catalytic gasification process targeted toward hydrogen using the algae species *Cladophora glomerata* at a process temperature of 700°C. The findings revealed that the syngas constituted approximately 82–99.7 wt.% of the product composition, with hydrogen gas (H_2) comprising 38–44 mol% of the syngas (Ebadi et al., 2019). Although the study conducted by Ebadi et al. (2019) yielded relatively lower amounts of hydrogen gas and biochar, it exhibited significantly reduced molar percentages of tar (0.1–6.5 mol%) (Razaviarani et al., 2023). Gasification is a widely explored alternative method for thermally converting algal biomass. However, the presence of tar and the requirement for gas purification pose challenges to achieving commercial viability. This is primarily due to the susceptibility of the syngas produced to contamination by ammonia and acids (Razaviarani et al., 2023).

11.5.1.6 Supercritical Water Gasification (SCWG)

Supercritical water gasification (SCWG) is the process used to treat hazardous waste materials (Leong et al., 2021) such as domestic solid waste, sludge, and wastewater sludge, including algal biomass generated after phycoremediation. SCWG of biomass is carried out at high temperatures and pressures, with temperatures usually ranging between 600°C and 650°C and with pressures of around 30 MPa. At temperatures exceeding 600°C, water exhibits significant oxidizing properties, leading to the preferential formation of CO_2 through the oxidation of carbon atoms. Additionally, the use of water in this process results in high yields of hydrogen, as hydrogen atoms are released from both the biomass and the water itself (Heidenreich et al., 2016). This technology has the capability to produce high yields of gaseous fuel rich in H_2, CH_4, CO, and CO_2, with high thermal efficiencies (in the range of 65%–77%) and short residence times (<30 minutes).

Having higher lipid content and low activation energy makes the algae, specifically *Spirulina platensis*, the most efficient raw material for SCWG. Produced syngas can be used for the production of power generation, Fischer-Tropsch liquid fuel, bio-H_2, CH_4, and NH_3 (Leong et al., 2021).

11.5.2 BIOLOGICAL TECHNIQUES

11.5.2.1 Microbial Fermentation

Microbial fermentation processes play a crucial role in generating hydrogen and methane through biological means in the biosphere (Kabaivanova et al., 2022). This process converts carbohydrates into biofuels such as bioethanol and biohydrogen. Solvent pretreatment is typically necessary to break down the cell walls of algae, and sulfuric acid has been found to enhance hydrolysis efficiency. Among the various microalgae species, *C. vulgaris* has been identified as the most promising candidate for bioethanol production. However, the yield from microbial fermentation can be low, and acid solvent pretreatment may be needed to increase the bioethanol yield. In addition to bioethanol, microbial fermentation can also produce other valuable by-products, such as biohydrogen, acetic acid, lactic acid, and pyruvate, through different techniques. Among those techniques, dark fermentation has emerged as a promising and feasible method (Razaviarani et al., 2023).

Numerous studies have investigated strategies to enhance ethanol production through fermentation. The factors that control the efficiency of microbial fermentation include (1) the composition of bacterial growth media; (2) temperature; (3) pH; (4) the presence of reducing agents; (5) trace elements; (6) substrate pressure; (7) mass transfer constraints; (8) microbe types; and (9) inhibitory compound formation (Harun et al., 2014).

11.5.2.2   Anaerobic Digestion of Algal Biomass

Anaerobic digestion (AD) is a biological process that entails a sequence of microbial reactions wherein organic matter is decomposed in the absence of oxygen. This process is used mainly for industrial or domestic purposes, primarily for waste management, as well as the generation of biofuels and the production of food and beverages. AD is a stable, viable, and sustainable technology that can help address the issues related to waste and wastewater management (Veerabadhran et al., 2021). AD technology can be enhanced by optimizing different aspects and parameters including the correct selection of pretreatment processes and methods, reactor types, and co-digestion materials (Kabaivanova et al., 2022).

Anaerobic digestion is an efficient method of breaking down algal biomass resulting in the production of biogas consisting of methane, CO_2, hydrogen, and hydrogen sulfide. It can generate up to $20\,m^3$ of methane per ton of algae and is particularly suitable for macroalgae due to its lower lipid content. However, algal anaerobic digestion faces challenges such as the requirement for an energy-intensive drying process and possession of high biomass carbon to nitrogen (C/N) ratios detrimental for establishing process conditions favoring anaerobic microorganisms. Solvent extraction is widely used as a pretreatment method that aims to reduce the carbon present in the biomass. Anaerobic digestion can potentially be a viable conversion process for algae; further research is required to optimize the process (Razaviarani et al., 2023).

The high potential for AD technology to produce valuable products, such as biogas and microalgae liquid biofuel using biomass generated after phycoremediation, makes it an attractive solution for the valorization of algal biomass (Veerabadhran et al., 2021).

11.5.2.3   Photobiological Hydrogen Production

The separation of the water molecules into oxygen and hydrogen ions by algae is called photobiological hydrogen production. First of all, algae have been grown photosynthetically and then cultured by maintaining anaerobic conditions. By this way, hydrogen production is stimulated. After that, hydrogen and oxygen ions will be produced photosynthetically in gas forms and separated spatially (Ferreira & Gouveia, 2020).

Photobiological hydrogen production from photosynthetic algae under various conditions has gained significant attention in the last decades. Recent improvements in valorization techniques, including dark fermentation and anaerobiosis, have increased the commercialization of this technique. Using a suitable support matrix to immobilize microalgal cells is an effective method for producing hydrogen from algal biomass on a pilot-scale. This method enhances the hydrogen production yield, mitigates oxygen sensitivity, and optimizes light utilization (Ferreira & Gouveia, 2020). The studies show that *Chlamydomonas reinhardtii* is a flexible organism to genetic modifications, making it available to produce various forms of bioenergy and other valuable by-products (Ferreira & Gouveia, 2020). *Chlamydomonas reinhardtii* can benefit from light and CO_2 as an energy resource in order to oxidize water to organic carbon compounds and nicotinamide adenine dinucleotide phosphate (NADPH). On the other hand, photosynthesis generates oxygen, inhibiting hydrogen gas production by inactivating the hydrogenase enzyme by oxygen. At this point, dark periods (dark fermentation or anaerobiosis), when the oxygen ions dissolved in the media are consumed by mitochondrial respiration, are adapted to the process to boost hydrogen ion production (Batyrova & Hallenbeck, 2017).

Table 11.2 provides a list of thermochemical and biological techniques to valorize algal biomass after phycoremediation, along with their respective advantages and disadvantages.

Both thermochemical and biological methods have advantages and disadvantages. Thermochemical methods are fast and economically viable for turning algal biomass into biofuels and other various products. They can also handle high moisture content and do not require extensive biomass drying. However, they may require high energy inputs and can degrade some biologically active compounds. Biological methods offer more environmentally friendly and sustainable approaches to algal biorefinery. They can preserve the integrity of biomolecules and have high removal efficiency for contaminants. However, they may require longer processing times and may be limited by the availability of suitable enzymes or microorganisms.

11.5.3 VALUE ADDED PRODUCTS BASED ON ALGAL BIOMASS

Algae have gained popularity in the renewable energy market, especially for being a source of third-generation biofuels. Algae can be considered ideal agents for producing feed, food, and health supplements. They also have a carbon-capturing and concentrating mechanism that enables their potential use for simultaneous greenhouse gas capturing (Rambabu et al., 2023).

In addition, microalgae can grow in highly pollutant media, making them attractive for biorefinery practices. In accordance with the latest studies, algal biomass can intake noxious emissions such as SO_x and NO_x and convert them into valuable by-products. It should be underlined that algal biomass can be used as a feedstock for the generation of alternatives to fossil fuels such as biodiesel, biogas, and biomethane (Uma et al., 2022).

Microalgae are composed of highly bioactive compounds, proteins, amino acids, carotenoids, lipids, fatty acids, auxins, and polysaccharides. These molecules are used in food and pharmaceutical industries as essential materials. Additionally, recent studies have shown that algal metabolites have antibacterial, antioxidant, anti-inflammatory, antitumor, and antiviral properties, and these properties make them suitable for being used in the production of cosmetics, serving as anti-aging agents, sun-protection agents, moisturizers, and emulsifiers. Moreover, certain algal species biosynthesize certain pigments of therapeutic value, such as β-carotene and astaxanthin, and these are used in the treatment of conditions such as cancer, inflammation, and metabolic disorders (Uma et al., 2022).

TABLE 11.2

Advantages and Disadvantages of Thermochemical and Biological Techniques for Valorization of Algal Biomass (Leong et al., 2021; Razaviarani et al., 2023; Ferreira & Gouveia, 2020)

Biorefinery Technique	Advantages	Disadvantages
Direct combustion of biomass	• Energy yield and efficiency are high, • The process is relatively simple, • Utilizes all biomass components as an energy source. • Algal solids become easily grindable and can be utilized, requiring no extra equipment.	• If the moisture content of the biomass is more than 50 wt.%, white fume is generated, and therefore, an energy-intensive dehydration process is needed. • Algal fuels in solid form have lower calorific densities in comparison with fossil fuels, so they are not classified as standalone fuels, and co-fired with coal-biomass is needed.
Transesterification of algal lipids	• Dimethyl carbonate is more expensive than methanol; however, it can lead to glycerol-to-glycerolcarbonate conversion, which can reduce glycerol as an impurity and thus improve the biodiesel quality. • Application of ultrasonic pretreatment can significantly reduce the reaction time by up to nine-fold, increasing the process efficiency.	• The process requires dry algal biomass, which drastically increases the production cost and energy intensity. • The optimum moisture content for the biomass is extremely low at 1% v/v; exceeding amounts may lead to soap formation. • Direct transesterification can be time-consuming, with yields achieved in 36 hours without ultrasonic pretreatment.
Pyrolysis of algal biomass	• Products can be generated in all three phases: (1) solid, (2) liquid, and (3) gas. • Pyrolysis produces higher-quality products compared to direct-combustion and gasification. • Pyrolysis works under less severe conditions, producing high net calorific value syngas. • The bio-oil generated from pyrolysis has a heating value similar to fossil fuels.	• Pyrolysis becomes less efficient without pretreatment, and the yield of the process is maximized when the size of the biomass particles is reduced. • Conventional pyrolysis has high initial costs. • Finding a suitable catalyst is hard for hydro-pyrolysis. • High catalyst expenses when integrated into the process. • Scale up and deciding a cost-effective and appropriate microwave receptor is a difficulty for microwave-induced pyrolysis.

(Continued)

TABLE 11.2 (*Continued*)
Advantages and Disadvantages of Thermochemical and Biological Techniques for Valorization of Algal Biomass (Leong et al., 2021; Razaviarani et al., 2023; Ferreira & Gouveia, 2020)

Biorefinery Technique	Advantages	Disadvantages
	• Conventional pyrolysis is flexible, produces low-sulfur products, and is efficient at large-scales. • Catalytic pyrolysis requires lower energy inputs which allows for separation of solids and impurities and higher product selectivity. • Microwave-induced pyrolysis ensures rapid penetration, uniform heating, improved mass transfer, and comparatively low energy requirement for activation.	• Energy and time are required to reach maximum efficiencies, with residence times of up to 60 minutes.
Gasification and torrefaction of algal biomass	• Torrefaction eliminates the rigid structure of algal biomass, making it more amenable to gasification reactions. • Under certain conditions, such as dry torrefaction at 200°C, biochar yields of up to 93.9 wt.% have been achieved with algae. • The process can be applied to a wide range of algal feedstocks.	• Gasification results in the formation of tar, a complex blend of organic substances. Tar can be problematic as it can clog equipment and affect downstream processes. • The composition of the syngas can vary based on factors such as temperature, reaction time, and catalyst loading. This variability can impact the quality and usability of the syngas. • The hydrogen content in the syngas can vary significantly, thereby impacting its energy content and applicability.
Hydrothermal liquefaction of algal biomass	• Energy conservation through biocrude production for biodiesel. • Possibility of nutrients recovery from algal biomass residue. • High potential for cost-effectiveness.	• Biocrude (liquid biofuel produced using high temperature) quality issues (moisture, viscosity, heteroatoms). • Alkali salt catalyst leads to solid residue and separation problems. • Biocrude may not meet biofuel standards due to heteroatom content.

(Continued)

TABLE 11.2 (*Continued*)

Advantages and Disadvantages of Thermochemical and Biological Techniques for Valorization of Algal Biomass (Leong et al., 2021; Razaviarani et al., 2023; Ferreira & Gouveia, 2020)

Biorefinery Technique	Advantages	Disadvantages
Supercritical water gasification (SCWG)	• Reduces the energy need for energy - intensive drying step. • Minimizes char and tar formation. • Syngas (final product) can produce biofuels and power generation. • Achieves a net negative carbon balance. • Acts as both a reactant and its medium • Allows pre-removal of minerals. • Offers a different approach to enhance biocrude oil.	• Limited exploration of the practical implementation of SCWG and other advanced gasification technologies in the utilization of algae resource
Anaerobic digestion of algal biomass	• Efficient process, generating significant amounts of biomethane. • Utilization of a large portion of the organics present in algal biomass. • Can recover and reuse nitrogen and phosphorus for further nutrient recycling	• Energy-intensive pretreatment required (drying) • High C:N ratio can be problematic • Solvent extraction may be necessary to reduce carbon content in biomass
Microbial fermentation	• It can be performed in various reactors, including bioreactors and fermenters. • Can be scaled up to industrial levels	• Requires careful management to prevent process blockages (pH, temperature monitoring, etc.) • Requires pretreatment with solvents or acids. • Can be energy-intensive. • Low yields of biofuels can be a challenge. • Requires careful monitoring and control of conditions such as temperature and pH
Photobiological hydrogen production	• Flexibility for genetic modification and strain development • Simultaneously producing biomass rich in valuable compounds • Producing clean and versatile energy	• Sensitivity of hydrogenase enzyme to oxygen molecules • Competition from other metabolic pathways for photosynthetic reductants • Challenges in scaling up the process for larger production (high cost)

11.6 CONCLUSION

This chapter defines the term "phycoremediation", which represents the contamination treatment technique using microalgae or macroalgae. The properties of the algal biofilm structures and mechanisms of the algal biofilm formation have been described. Detailed information on the algal biofilm formation as an essential structure to treat inorganic and organic contaminants and emerging pollutants has been underlined. Furthermore, a comprehensive analysis of biomass production and the kinetics of nutrient removal have been described for the commonly used attached growth and suspended growth systems. Finally, thermochemical and biological biorefinery techniques and valorization alternatives of algal biomass have been discussed.

It can be concluded that phycoremediation is an effective method for removing a variety of pollutants, as described by the numerous references and scientific studies cited in this chapter. Microalgae usage during the waste treatment processes improves the waste treatment capacity and efficiency in converting the waste into value-added products through a biorefinery approach. Algal biofilms have proven to be effective tools for phytoremediation. These remediation studies are mostly focused on nutrient removal as algae are known for their high nitrogen and phosphorus uptake rate. Emerging contaminant removal has also been studied and showed promising results. Future research should be targeted toward the analyses of pollutant absorption and utilization mechanisms as well as the concomitant synthesis of commercially relevant metabolites.

Selecting the optimum algal biomass cultivation technique and biorefinery strategy during the phycoremediation depends on the factors such as geographic location and climate conditions of the place where treatment is held, amount of the wastewater, the target contaminant, environmental effects, and desired final bioproducts. It is worth noting that phycoremediation is an integrated process that consists of the cultivation, treatment, and biorefinery processes, and therefore, during the decision period for the development of a phycoremediation strategy, advantages and disadvantages should be considered, and the most cost-effective and environmentally friendly strategy should be selected.

REFERENCES

Achag, B., Mouhanni, H., & Bendou, A. (2021). Improving the performance of waste stabilization ponds in an arid climate. *Journal of Water and Climate Change*, 12(8), 3634–3647.

Angelaalincy, M., Nishtha, P., Ajithkumar, V., Ashokkumar, B., Muthu Ganesh Moorthy, I., Brindhadevi, K., Thuy Lan Chi, N., Pugazhendhi, A., & Varalakshmi, P. (2023). Phycoremediation of Arsenic and biodiesel production using green microalgae Coelastrella sp. M60: An integrated approach. *Fuel*, 333, 126427. https://doi.org/10.1016/j.fuel.2022.126427.

Ankit, Bauddh, K., & Korstad, J. (2022). Phycoremediation: Use of algae to sequester heavy metals. *Hydrobiology*, 1(3), 288–303. https://doi.org/10.3390/hydrobiology1030021.

Batyrova, K., & Hallenbeck, P. C. (2017). Hydrogen production by a *Chlamydomonas reinhardtii* strain with inducible expression of photosystem II. *International Journal of Molecular Sciences*, 18(3), 647. https://doi.org/10.3390/ijms18030647.

Carpentier, B., & Cerf, O. (1999). Biofilms. In R. K. Robinson (Ed.), *Encyclopedia of Food Microbiology* (pp. 252–259). Academic Press: London. https://doi.org/10.1006/RWFM.1999.0205.

Chaturvedi, V., Chandravanshi, M., Rahangdale, M., & Verma, P. (2013). An integrated approach of using polystyrene foam as an attachment system for growth of mixed culture of cyanobacteria with concomitant treatment of copper mine waste water. *Journal of Waste Management*, 2013, 1–7. https://doi.org/10.1155/2013/282798.

Chen, D., Wang, G., Chen, C., Feng, Z., Jiang, Y., Yu, H., Li, M., Chao, Y., Tang, Y., Wang, S., & Qiu, R. (2023). The interplay between microalgae and toxic metal(loid)s: Mechanisms and implications in AMD phycoremediation coupled with Fe/Mn mineralization. *Journal of Hazardous Materials*, 454, 131498. https://doi.org/10.1016/j.jhazmat.2023.131498.

Chen, Y., Lin, M., & Zhuang, D. (2022). Wastewater treatment and emerging contaminants: Bibliometric analysis. *Chemosphere*, 297, 133932. https://doi.org/10.1016/j.chemosphere.2022.133932.

Choudhary, S., Tripathi, S., & Poluri, K. M. (2022). Microalgal-based bioenergy: Strategies, prospects, and sustainability. *Energy and Fuels*, 36(24), 14584–14612. https://doi.org/10.1021/acs.energyfuels.2c02922.

Dayana Priyadharshini, S., Suresh Babu, P., Manikandan, S., Subbaiya, R., Govarthanan, M., & Karmegam, N. (2021). Phycoremediation of wastewater for pollutant removal: A green approach to environmental protection and long-term remediation. *Environmental Pollution*, 290. https://doi.org/10.1016/j.envpol.2021.117989.

Ding, G. K. C. (2023). Wastewater treatment, reused and recycling: A potential source of water supply. In M. A. Abraham (Ed.), *Reference Module in Earth Systems and Environmental Sciences* (pp. 676–693). Elsevier. https://doi.org/10.1016/B978-0-323-90386-8.00062-0.

Djandja, O. S., Wang, Z., Chen, L., Qin, L., Wang, F., Xu, Y., & Duan, P. (2020). Progress in hydrothermal liquefaction of algal biomass and hydrothermal upgrading of the subsequent crude bio-oil: A mini review. *Energy and Fuels*, 34(10), 11723–11751. https://doi.org/10.1021/acs.energyfuels.0c01973.

Ebadi, A. G., Hisoriev, H., Zarnegar, M., & Ahmadi, H. (2019). Hydrogen and syngas production by catalytic gasification of algal biomass (*Cladophora glomerata* L.) using alkali and alkaline-earth metals compounds. *Environmental Technology (United Kingdom)*, 40(9), 1178–1184. https://doi.org/10.1080/09593330.2017.1417495.

Ferreira de Oliveira, A. P., & Bragotto, A. P. A. (2022). Microalgae-based products: Food and public health. *Future Foods*, 6. https://doi.org/10.1016/j.fufo.2022.100157.

Ferreira, A., & Gouveia, L. (2020). Microalgal biorefineries. In E. Jacob-Lopes, M. M. Maroneze, M. I. Queiroz, & L. Q. Zepka (Eds.), *Handbook of Microalgae-Based Processes and Products* (pp. 771–798). Elsevier. https://doi.org/10.1016/b978-0-12-818536-0.00028-2.

Gan, X., Klose, H., & Reinecke, D. (2023). Stable year-round nutrients removal and recovery from wastewater by technical-scale algal Turf Scrubber (ATS). *Separation and Purification Technology*, 307, 122693. https://doi.org/10.1016/J.SEPPUR.2022.122693.

Gani, P., Sunar, N. M., Matias-Peralta, H., Jamaian, S. S., & Latiff, A. A. A. (2016). Effects of different culture conditions on the phycoremediation efficiency of domestic wastewater. *Journal of Environmental Chemical Engineering*, 4(4), 4744–4753. https://doi.org/10.1016/j.jece.2016.11.008.

Ghazy, M. M., Senousy, W. M., Aatty, A. M. & Kamel, M. (2008). Performance Evaluation of a Waste Stabilization Pond in a Rural Area in Egypt. *American Journal of Environmental Sciences*, 4(4), 316–325. https://doi.org/10.3844/ajessp.2008.316.325.

Golovko, O., Örn, S., Sörengård, M., Frieberg, K., Nassazzi, W., Lai, F. Y., & Ahrens, L. (2021). Occurrence and removal of chemicals of emerging concern in wastewater treatment plants and their impact on receiving water systems. *Science of the Total Environment*, 754, 142122. https://doi.org/10.1016/j.scitotenv.2020.142122.

Gondi, R., Kavitha, S., YukeshKannah, R., Parthiba Karthikeyan, O., Kumar, G., Kumar Tyagi, V., & Rajesh Banu, J. (2022). Algal-based system for removal of emerging pollutants from wastewater: A review. *Bioresource Technology*, 344, 126245. https://doi.org/10.1016/j.biortech.2021.126245.

Gupta, P., Bishoyi, A. K., Rajput, M. S., Trivedi, U., & Sanghvi, G. (2021). Biotechnological advances for utilization of algae, microalgae, and cyanobacteria for wastewater treatment and resource recovery. In P. Verma, & M. P. Shah (Eds.), *Phycology-Based Approaches for Wastewater Treatment and Resource Recovery* (pp. 1–24). CRC Press. https://doi.org/10.1201/9781003155713-1.

Gupta, S. K., Shriwastav, A., Kumari, S., Ansari, F. A., Malik, A., & Bux, F. (2015). Phycoremediation of emerging contaminants. In B. Singh, K. Bauddh, and F. Bux (Eds.), *Algae and Environmental Sustainability* (pp. 129–146). Springer India. https://doi.org/10.1007/978-81-322-2641-3_11.

Harun, R., Yip, J. W. S., Thiruvenkadam, S., Ghani, W. A. W. A. K., Cherrington, T., & Danquah, M. K. (2014). Algal biomass conversion to bioethanol-a step-by-step assessment. *Biotechnology Journal*, 9(1), 73–86. https://doi.org/10.1002/biot.201200353.

Heidenreich, S., Müller, M., & Foscolo, P. U. (2016). *Advanced Biomass Gasification: New Concepts for Efficiency Increase and Product Flexibility*, First Edition. Academic Press: Cambridge, MA.

Hoh, D., Watson, S., & Kan, E. (2016). Algal biofilm reactors for integrated wastewater treatment and biofuel production: A review. *Chemical Engineering Journal*, 287, 466–473. Elsevier. https://doi.org/10.1016/j.cej.2015.11.062.

Kabaivanova, L., Ivanova, J., Chorukova, E., Hubenov, V., Nacheva, L., & Simeonov, I. (2022). Algal biomass accumulation in waste digestate after anaerobic digestion of wheat straw. *Fermentation*, 8(12). https://doi.org/10.3390/fermentation8120715.

Kaloudas, D., Pavlova, N., & Penchovsky, R. (2021). Phycoremediation of wastewater by microalgae: A review. *Environmental Chemistry Letters*, 19(4), 2905–2920. https://doi.org/10.1007/s10311-021-01203-0.

Kandasamy, S., Narayanan, M., He, Z., Liu, G., Ramakrishnan, M., Thangavel, P., Pugazhendhi, A., Raja, R., & Carvalho, I. S. (2021). Current strategies and prospects in algae for remediation and biofuels: An overview. *Biocatalysis and Agricultural Biotechnology*, 35. https://doi.org/10.1016/j.bcab.2021.102045.

Kesaano, M., & Sims, R. C. (2014). Algal biofilm based technology for wastewater treatment. *Algal Research*, 5(1), 231–240. https://doi.org/10.1016/J.ALGAL.2014.02.003.

Koul, B., Sharma, K., & Shah, M. P. (2022). Phycoremediation: A sustainable alternative in wastewater treatment (WWT) regime. *Environmental Technology and Innovation*, 25. https://doi.org/10.1016/j.eti.2021.102040.

Kumar, S., Mal, G., & Sharma, H. R. (2018). Waste stabilization ponds: a technical option for liquid waste management in rural areas in Haryana under Swachh Bharat Mission-Gramin. *International Journal of Environmental Science and Technology*, 13(2018), 177–187.

Lee, S. H., Oh, H. M., Jo, B. H., Lee, S. A., Shin, S. Y., Kim, H. S., Lee, S. H., & Ahn, C. Y. (2014). Higher biomass productivity of microalgae in an attached growth system, using wastewater. *Journal of Microbiology and Biotechnology*, 24(11), 1566–1573. https://doi.org/10.4014/jmb.1406.06057.

Leong, Y. K., Huang, C. Y., & Chang, J. S. (2021). Pollution prevention and waste phycoremediation by algal-based wastewater treatment technologies: The applications of high-rate algal ponds (HRAPs) and algal turf scrubber (ATS). *Journal of Environmental Management*, 296. https://doi.org/10.1016/j.jenvman.2021.113193.

Li, K., Liu, Q., Fang, F., Luo, R., Lu, Q., Zhou, W., Huo, S., Cheng, P., Liu, J., Addy, M., Chen, P., Chen, D., & Ruan, R. (2019). Microalgae-based wastewater treatment for nutrients recovery: A review. *Bioresource Technology*, 291, 121934.

Li, M., Zamyadi, A., Zhang, W., Dumée, L. F., & Gao, L. (2022). Algae-based water treatment: A promising and sustainable approach. *Journal of Water Process Engineering*, 46, 102630. https://doi.org/10.1016/j.jwpe.2022.102630.

Li, S. F., Fanesi, A., Martin, T., & Lopes, F. (2021). Biomass production and physiology of Chlorella vulgaris during the early stages of immobilized state are affected by light intensity and inoculum cell density. *Algal Research*, 59, 102453.

Li, X. Z., Hauer, B., & Rosche, B. (2007). Single-species microbial biofilm screening for industrial applications. *Applied Microbiology and Biotechnology*, 76(6), 1255–1262. https://doi.org/10.1007/s00253-007-1108-4.

Lin-Lan, Z., Jing-Han, W., & Hong-Ying, H. (2018). Differences between attached and suspended microalgal cells in ssPBR from the perspective of physiological properties. *Journal of Photochemistry and Photobiology B: Biology*, 181, 164–169. https://doi.org/10.1016/j.jphotobiol.2018.03.014.

Liu, J., Danneels, B., Vanormelingen, P., & Vyverman, W. (2016). Nutrient removal from horticultural wastewater by benthic filamentous algae Klebsormidium sp., Stigeoclonium spp. and their communities: From laboratory flask to outdoor Algal Turf Scrubber (ATS). *Water Research*, 92, 61–68. https://doi.org/10.1016/j.watres.2016.01.049.

López-Aguilar, H. A., Quiroz-Cardoza, D., & Pérez-Hernández, A. (2022). Volatile compounds of algal biomass pyrolysis. *Journal of Marine Science and Engineering*, 10(7). https://doi.org/10.3390/jmse10070928.

Loupasaki, E., & Diamadopoulos, E. (2013). Attached growth systems for wastewater treatment in small and rural communities: A review. *Journal of Chemical Technology and Biotechnology*, 88(2), 190–204. https://doi.org/10.1002/jctb.3967.

Lu, Q., Zhou, W., Min, M., Ma, X., Chandra, C., Doan, Y. T. T., Ma, Y., Zheng, H., Cheng, S., Griffith, R., Chen, P., Chen, C., Urriola, P. E., Shurson, G. C., Gislerød, H. R., & Ruan, R. (2015). Growing Chlorella sp. on meat processing wastewater for nutrient removal and biomass production. *Bioresource Technology*, 198, 189–197. https://doi.org/10.1016/j.biortech.2015.08.133.

Mantzorou, A., & Ververidis, F. (2019). Microalgal biofilms: A further step over current microalgal cultivation techniques. *Science of the Total Environment*, 651, 3187–3201. https://doi.org/10.1016/j.scitotenv.2018.09.355.

Medipally, S. R., Yusoff, F. M., Banerjee, S., & Shariff, M. (2015). Microalgae as sustainable renewable energy feedstock for biofuel production. *BioMed Research International*, 2015. https://doi.org/10.1155/2015/519513.

Mohan, S. V., Rohit, M. V., Subhash, G. V., Chandra, R., Devi, M. P., Butti, S. K., & Rajesh, K. (2018). Algal oils as biodiesel. In A. Pandey, J.-S. Chang, C. R. Soccol, D.-J. Lee, & Y. Chisti (Eds.), *Biomass, Biofuels, Biochemicals: Biofuels from Algae*, Second Edition (pp. 287–323). Elsevier. https://doi.org/10.1016/B978-0-444-64192-2.00012-3.

Moreno Osorio, J. H., Pollio, A., Frunzo, L., Lens, P. N. L., & Esposito, G. (2021). A review of microalgal biofilm technologies: Definition, applications, settings and analysis. *Frontiers in Chemical Engineering*, 3. https://doi.org/10.3389/fceng.2021.737710.

Osman, A. I. et al. (2022) Biochar for agronomy, animal farming, anaerobic digestion, composting, water treatment, soil remediation, construction, energy storage, and carbon sequestration: A review. *Environmental Chemistry Letters*, 20, 2385–2485. https://doi.org/10.1007/s10311-022-01424-x.

Ozkan, A. (2012). Development of a novel algae biofilm photobioreactor for biofuel production. Dissertation, The University of Texas at Austin.

Phang, S.-M., Chu, W.-L., & Rabiei, R. (2015). Phycoremediation. In D. Sahoo, & J. Seckbach (Eds.), *The Algae World* (pp. 357–389). Springer Dordrecht. https://doi.org/10.1007/978-94-017-7321-8_13.

Polizzi, B., Fanesi, A., Lopes, F., Ribot, M., & Bernard, O. (2022). Understanding photosynthetic biofilm productivity and structure through 2D simulation. *PLoS Computational Biology*, 18(4), e1009904.

Qian, J., Wan, T., Ye, Y., Li, J., Toda, T., Li, H., Sekine, M., Takayama, Y., Koga, S., Shao, S., Fan, L., Xu, P., & Zhou, W. (2023). Insight into the formation mechanism of algal biofilm in soy sauce wastewater. *Journal of Cleaner Production*, 394. https://doi.org/10.1016/j.jclepro.2023.136179.

Raheem, A., Abbasi, S. A., Mangi, F. H., Ahmed, S., He, Q., Ding, L., Memon, A. A., Zhao, M., & Yu, G. (2021). Gasification of algal residue for synthesis gas production. *Algal Research*, 58. https://doi.org/10.1016/j.algal.2021.102411.

Raheem, A., Ji, G., Memon, A., Sivasangar, S., Wang, W., Zhao, M., & Taufiq-Yap, Y. H. (2018). Catalytic gasification of algal biomass for hydrogen-rich gas production: Parametric optimization via central composite design. *Energy Conversion and Management*, 158, 235–245. https://doi.org/10.1016/j.enconman.2017.12.041.

Rambabu, K., Avornyo, A., Gomathi, T., Thanigaivelan, A., Show, P. L., & Banat, F. (2023). Phycoremediation for carbon neutrality and circular economy: Potential, trends, and challenges. *Bioresource Technology*, 367, 128257. https://doi.org/10.1016/j.biortech.2022.128257.

Randrianarison, G., & Ashraf, M. A. (2017). Microalgae: A potential plant for energy production. *Geology, Ecology, and Landscapes*, 1(2), 104–120. https://doi.org/10.1080/24749508.2017.1332853.

Ray, N. E., Terlizzi, D. E., & Kangas, P. C. (2015). Nitrogen and phosphorus removal by the Algal Turf Scrubber at an oyster aquaculture facility. *Ecological Engineering*, 78, 27–32. https://doi.org/10.1016/j.ecoleng.2014.04.028.

Razak, S. B. A., & Sharip, Z. (2020). The potential of phycoremediation in controlling eutrophication in tropical lake and reservoir: A review. *Desalination and Water Treatment*, 180, 164–173. https://doi.org/10.5004/dwt.2020.25078.

Razaviarani, V., Arab, G., Lerdwanawattana, N., & Gadia, Y. (2023). Algal biomass dual roles in phycoremediation of wastewater and production of bioenergy and value-added products. *International Journal of Environmental Science and Technology*, 20(7), 8199–8216. https://doi.org/10.1007/s13762-022-04696-6.

Reinecke, D., Bischoff, L.-S., Klassen, V., Blifernez-Klassen, O., Grimm, P., Kruse, O., Klose, H., & Schurr, U. (2023). Nutrient recovery from wastewaters by algal biofilm for fertilizer production part 1: Case study on the techno-economical aspects at pilot-scale. *Separation and Purification Technology*, 305, 122471. https://doi.org/10.1016/j.seppur.2022.122471.

Rezasoltani, S., & Champagne, P., (2023). An integrated approach for the phycoremediation of Pb(II) and the production of biofertilizer using nitrogen-fixing cyanobacteria. *Journal of Hazardous Materials*, 445, 130448. https://doi.org/10.1016/j.jhazmat.2022.130448.

Roostaei, J., Zhang, Y., Gopalakrishnan, K., & Ochocki, A. J. (2018). Mixotrophic microalgae biofilm: A novel algae cultivation strategy for improved productivity and cost-efficiency of biofuel feedstock production. *Scientific Reports*, 8(1). https://doi.org/10.1038/s41598-018-31016-1.

Sharma, S., Kant, A., Sevda, S., Aminabhavi, T. M., & Garlapati, V. K. (2023). A waste-based circular economy approach for phycoremediation of X-ray developer solution. *Environmental Pollution*, 316, 120530. https://doi.org/10.1016/j.envpol.2022.120530.

Taghizadeh, S. M., Ebrahiminezhad, A., Raee, M. J., Ramezani, H., Berenjian, A., & Ghasemi, Y. (2022). A study of l-lysine-stabilized Iron Oxide Nanoparticles (IONPs) on microalgae biofilm formation of Chlorella vulgaris. *Molecular Biotechnology*, 64(6), 702–710. https://doi.org/10.1007/s12033-022-00454-8.

Tong, C. Y., & Derek, C. J. C. (2021). The role of substrates towards marine diatom Cylindrotheca fusiformis adhesion and biofilm development. *Journal of Applied Phycology*, 33(5). https://doi.org/10.1007/s10811-021-02504-1.

Ugya, Y. A., Hasan, D. B., Tahir, S. M., Imam, T. S., Ari, H. A., & Hua, X. (2021). Microalgae biofilm cultured in nutrient-rich water as a tool for the phycoremediation of petroleum-contaminated water. *International Journal of Phytoremediation*, 23(11), 1175–1183. https://doi.org/10.1080/15226514.2021.1882934.

Uma, V. S., Usmani, Z., Sharma, M., Diwan, D., Sharma, M., Guo, M., Tuohy, M. G., Makatsoris, C., Zhao, X., Thakur, V. K., & Gupta, V. K. (2022). Valorisation of algal biomass to value-added metabolites: emerging trends and opportunities. *Phytochemistry Reviews*. https://doi.org/10.1007/s11101-022-09805-4.

Upadhyay, A. K., Singh, R., Singh, J. S., & Singh, D. P. (2019). Microalgae-assisted phycoremediation and energy crisis solution: Challenges and opportunity. In Singh, J. S., & Singh, D. P. (Eds.), *New and Future Developments in Microbial Biotechnology and Bioengineering: Microbial Biotechnology in Agro-Environmental Sustainability* (pp. 295–307). Elsevier. https://doi.org/10.1016/B978-0-444-64191-5.00021-3.

Valchev, D., & Ribarova, I. (2022). A review on the reliability and the readiness level of microalgae- based nutrient recovery technologies for secondary treated effluent in municipal wastewater treatment plants. *Processes*, 10(2), 399. https://doi.org/10.3390/pr10020399.

Vaniyankandy, A., Ray, B., Karthikeyan, S., & Rakesh, S. (2022). Thermochemical conversion of algal based biorefinery for biofuel. In A. Tiwari (Ed.), *Cyanobacteria - Recent Advances and New Perspectives*. IntechOpen. https://doi.org/10.5772/intechopen.106357.

Vargas e Silva, F., & Monteggia, L. O. (2015). Pyrolysis of algal biomass obtained from high-rate algae ponds applied to wastewater treatment. *Frontiers in Energy Research*, 3(Jun). https://doi.org/10.3389/fenrg.2015.00031.

Veerabadhran, M., Gnanasekaran, D., Wei, J., & Yang, F. (2021). Anaerobic digestion of microalgal biomass for bioenergy production, removal of nutrients and microcystin: Current status. *Journal of Applied Microbiology*, 131(4), 1639–1651. https://doi.org/10.1111/jam.15000.

Venkatesan, S., Narayanan, M., Prabakaran, M., Anusha, P., Srinivasan, R., Kandasamy, S., Natarajan, D., & Paulraj, B. (2020). In-situ and ex-situ phycoremediation competence of innate scenedesmus sp. on polluted river water.

Wang, J.-H., Zhuang, L.-L., Xu, X.-Q., Deantes-Espinosa, V. M., Wang, X.-X., & Hu, H.-Y. (2018). Microalgal attachment and attached systems for biomass production and wastewater treatment. *Renewable and Sustainable Energy Reviews*, 92, 331–342. https://doi.org/10.1016/j.rser.2018.04.081.

Yu, G. W., Nie, J., Lu, L. G., Wang, S. P., Li, Z. G., & Lee, M. R. (2017). Transesterification of soybean oil by using the synergistic microwave-ultrasonic irradiation. *Ultrasonics Sonochemistry*, 39, 281–290. https://doi.org/10.1016/j.ultsonch.2017.04.036.

Zhang, Q., Wang, L., Yu, Z., Zhou, T., Gu, Z., Huang, Q., Xiao, B., Zhou, W., Ruan, R., & Liu, Y. (2020). Pine sawdust as algal biofilm biocarrier for wastewater treatment and algae-based byproducts production. *Journal of Cleaner Production*, 256, 120449. https://doi.org/10.1016/j.jclepro.2020.120449.

Zhang, Q., Yu, Z., Zhu, L., Ye, T., Zuo, J., Li, X., Xiao, B., & Jin, S. (2018). Vertical-algal-biofilm enhanced raceway pond for cost-effective wastewater treatment and value-added products production. *Water Research*, 139, 144–157. https://doi.org/10.1016/j.watres.2018.03.076.

Zhou, H., Sheng, Y., Zhao, X., Gross, M., & Wen, Z. (2018). Treatment of acidic sulfate-containing wastewater using revolving algae biofilm reactors: Sulfur removal performance and microbial community characterization. *Bioresource Technology*, 264, 24–34. https://doi.org/10.1016/j.biortech.2018.05.051.

Zhou, J.-L., Yang, L., Huang, K.-X., Chen, D.-Z., & Gao, F. (2022). Mechanisms and application of microalgae on removing emerging contaminants from wastewater: A review. *Bioresource Technology*, 364, 128049. https://doi.org/10.1016/j.biortech.2022.128049.

12 Microalgal-Bacterial Biomass System for Wastewater Treatment
A Case Study on Real Wastewater Treatment

Günay Yıldız Töre, Mehmet Ali Gürbüz, and Enes Özgenç

12.1 PHYCOREMIDATION ALGAE AND MICROALGAE

Water is a universal solvent, playing an important role in the functioning of ecosystems and human health, and is polluted by industrial and urban development and many pollutants (Chaudhry and Malik, 2017). So, wastewater treatment (WWT) is the process applied in order to remove these pollutants generated as a result of domestic and industrial uses. With urbanization and the industrial revolution, the level of contaminants as well as the types of pollutants increased in the wastewater (Saravanan et al., 2021). Excessive use of detergents, soaps, cleansing agents with new formulations, chemical fertilizers, and pesticides has added to wastewater (Mousavi and Khodadoost, 2019). Conventional methods of WWT were successful in removing conventional contaminants from the water, but these techniques require a longer time and more energy input (Crini and Lichtfouse, 2019). As an alternative to conventional treatment technologies, phycoremediation, which includes the use of algae and microalgae for WWT, is an environmentally friendly and sustainable technology. In phycoremediation, compounds such as nitrogen, phosphorus, sulfur, and minerals act as "food" for the algae, not pollutants (Kafle et al., 2022).

Phycoremediation technology has aided the sustainability of WWT plants (Yuliasni et al., 2023). In phycoremediation, compounds such as nitrogen, phosphorus, sulfur, and minerals act as "food" for algae, not pollutants, and allow the removal of phosphates, nitrates, heavy metals, pesticides, hydrocarbons, nitrogen, and phosphorus (Kaloudas et al., 2021). Conditions that support algae growth in phycoremidation also favor water and wastewater disinfection, as they also inhibit the growth of pathogenic bacteria (Singh et al., 2021).

The most important benefits that can be reaped by implementing the phycoremediation technology can be sorted as effective carbon sequestration, oxygenation of the environment due to photosynthetic activity, and generation of biomass for the production of carbon-neutral biofuels, respectively. Hence, this technology offers one

DOI: 10.1201/9781003390213-12

of the most promising solutions for tackling global climate change and the acidification of oceans, thereby offering extreme environmental sustainability (Pandey, 2006).

Algae are polyphyletic (Srivastava et al., 2022), an artificial community of photosynthetic organisms (and, secondarily, non-photosynthetic evolutionary descendants) capable of producing O_2 (Kosourov et al., 2020), including a wide variety of microorganisms known as seaweeds (macroalgae) and microalgae. Eukaryotic algal groups represent at least five distinct evolutionary lineages, some of which include protists traditionally recognized as fungi and protozoa (Naranjo-Ortiz and Gabaldón, 2019). Microalgae is a simple microscopic heterotrophic (Yousuf, 2020) and/or autotrophic photosynthetic organism that grows in the aquatic environment and can be in unicellular or multicellular structures (Zhu, 2015; Jalilian et al., 2020). Microalgae, which can be found in prokaryotic or eukaryotic structures, can reproduce rapidly and can live even under adverse conditions, thanks to their single-celled or simple multicellular structures (Mata et al., 2010; Del Mondo et al., 2022).

The cultivation of microalgae in the laboratory takes ~140 years, while commercial cultivation takes ~60 years (Borowitzka and Moheimani, 2013; Antony and Ramanan, 2021). The first culture study was made by Cohn with Haematococcus pluvialis species (Aydin-Sisman, 2019), then by Famintzin with the green algae *chlorococcum infusionum* and *Desmococcus olivaceus*, which were developed in a simple inorganic culture medium (Borowitzka and Moheimani, 2013; Raposo, 2017; Nowicka-Krawczyk et al., 2022). Modern microalgae culture studies were started by Beijerinck (1890) with *chlorella vulgaris* (Borowitzka and Moheimani, 2013). It was first suggested by Harder and von Witsch that microalgae with rich lipid contents could be used for food and fuel production. Detailed research on this subject was done by Aach with *chlorella pyrenoidosa,* and it was reported that microalgae can accumulate more than 70% (dry weight) lipids in their bodies in a nitrogen-limiting environment (Borowitzka and Moheimani, 2013). In the 1948–1950s, large-scale studies for algae-growing systems began at the Stanford Research Institute (Burlew, 1953; Borowitzka and Moheimani, 2013). In 1951, studies on the production of *chlorella*-type microalgae using pilot-scale open systems were carried out in Cambridge, USA. In the 1960s, William (Bill) Oswald and his team at the University of California focused on large-scale algae development for biomass production (BP) and WWT (Oswald et al., 1957; Oswald and Golueke, 1960). Although the use of microalgae as a food source in Mexico (Farrar, 1966; Sosa-Hernández et al., 2019), Africa, and Asia dates back centuries (Johnston, 1970; Ciferri, 1983; Ullmann and Grimm, 2021), microalgae cultivation has only recently begun to develop (Cadoret et al., 2012; Borowitzka and Moheimani, 2013; Morocho-Jácome et al., 2020).

The chemical composition of microalgae is not an intrinsic constant but varies with a wide variety of factors, depending on the species and growing conditions. Some microalgae have the capacity to adapt to changes in environmental conditions by changing their chemical composition in response to environmental variability (Little et al., 2021). It is possible to accumulate the desired products in microalgae to a large extent by changing environmental factors such as temperature, lighting, pH, CO_2 supply, salt, and nutrients (Sathinathan et al., 2023). Microalgae play an important role in nutrient cycling, fixing inorganic carbon to organic molecules, and producing oxygen in the marine biosphere (Prasad et al., 2021). Fish oil is known for its omega-3 fatty acid content, but in fact, fish do not produce omega-3s but instead accumulate omega-3 reserves by consuming microalgae. These omega-3 fatty acids can be obtained in the human diet directly from the microalgae that produce them (Charles et al., 2019). In addition, microalgae can accumulate significant amounts of protein, depending on the species and growing conditions. Because of their ability to grow on non-arable land, microalgae can provide an alternative protein source for human consumption or animal feed. Microalgae proteins can also be used in the food industry as thickening agents, emulsions, and foam stabilizers in place of animal-based proteins. Some microalgae can also accumulate chromophores such as chlorophyll, carotenoids, or phycobiliproteins, which can be extracted and used as colorants (Hu et al., 2018; Mu et al., 2019; Zheng et al., 2022).

Microalgae generally live autotrophically, perform photosynthesis using their pigments, and convert carbon dioxide, water, and sunlight into biomass (Pragya et al., 2013; Zhu, 2015; Abiusi et al., 2020). However, according to the specific characteristics of the species, they can also develop as heterotrophic and mixotrophic conditions, apart from autotrophic conditions (Lacroux et al., 2022). Autotrophic microalgae use inorganic carbon and raise the pH by producing hydroxyl, while heterotrophic ones retain organic carbon and produce CO_2, lowering the pH (Nair and Chakraborty, 2020). On the other hand, mixotrophic species use both organic and inorganic carbon simultaneously, which results in pH value variation (Cheng et al., 2022).

The use of micro or macroalgae to remove or bioconvert nutrients, heavy metals, and other toxins, including xenobiotics, from WWT plants appears to be a sustainable technique (Priyadharshini et al., 2021). Microalgae are basic, mixotrophic microorganisms that treat distinct wastewater and use nutrients in the environment as food sources (Goswami et al., 2022). Also, microalgae that are unicellular and eukaryotic fix carbon dioxide in the atmosphere, grow in various aquatic climates, and have a size range of nearly 100 μm. Also, their biomass does not cause any adverse effects on the ecosystem due to its potential to contain organic substances (including lipids (7%–23%), proteins (6%–52%), and carbohydrates (5%–23%)) and metabolites that can be used to produce biofuels and products with high added value (Kiran et al., 2017; Priyadharshini et al., 2021).

The main components in algae are carbon, nitrogen, phosphorus, and micronutrients such as cobalt, iron, and zinc, which play an important role in their metabolic activities (Ghosh et al., 2020). Microalgae-based WWT is the most convenient, environmentally friendly, and cost-efficient alternative to conventional treatment techniques that can treat wastewater in a single step. They have the potential for a nutrient removal efficiency of ~90% from wastewater (Schagerl et al., 2022). Phycoremediation, which is one of the treatment techniques, is an interesting stage for treating wastewater because it makes tertiary biotreatment possible while generating precious biomass that may be used for a variety of practices. However, nitrates and other organic matter degraded by tertiary biotreatment are converted into primary, harmless, and fresh wastewater (Goswami et al., 2022). Therefore, wastewater should be treated before being discharged to the receiving environment. There are many different methods for removing organic matter, but it is very costly to produce large quantities of activated sludge.

As indicated in Table 12.1, algae have been proposed as a biological remediation option for removing such nutrients from the aquatic environment. Furthermore, the treatment of wastewater by algae can be done with an open pool system,

TABLE 12.1
Organic Matter Removal Using Algae in the Real Aquatic Environment (Kiran et al., 2017)

Wastewater Treatment	Type of Wastewater	N Removal (%)	P Removal (%)
Algal-bacterial symbiosis (Chlorella + Nitzchia)	Settled domestic sewage	92.0	74.0
Settable algae-bacterial culture (filamentous Blue-Green Algae + Flavobacteria) Gammaproteobacteria, Bacteroidia, and Beta-Proteobacteria)	Pretreated municipal wastewater	88.3	64.8
Cyanobacteria	Secondary treated effluent + swine wastewater	95.0	62.0
Chlorella pyrenoidosa	Settled domestic sewage	93.9	80.0
Chlorella vulgaris	Diluted pig slurry (suspended solids upto 0.2%)	54.0–98.0	42.0–89.0
Chlorella pyrenoidosa	Domestic sewage and industrial wastewater from pig farms and palm oil mill	60.0–70.0	50.0–60.0
Chlorella sp.	Primary settling	68.4	83.2
	Wastewater after primary settling	68.5	90.6
	Wastewater after activated sludge tank	50.8	4.69
	Centrate	82.8	85.6
Chlorella Vulgaris	Untreated urban wastewater	60.1	80.3
Chlorella sp.	Municipal wastewater (raw center)	89.1	80.9
Chlorella sp. 227	Pretreated wastewater (0.2 lim filter)	92.0	86.0
Scenedesmus obliquus	Untreated urban wastewater	100	83.3
Chlorella minutissima	Primary treated wastewater	91.49	87.63
Scenedesmus sp. ZTY1	Primary and secondary treated wastewater	90.0	97.0
Chlorella sp. IM-01	Municipal wastewater after settling	97.81	89.39
Scenedesmus obliquus	Wastewater	92.0–94.0	61.0–99.0
G. sulphurariaia	Primary effluent	99.42	97.8
Mixed algae culture (Microspora sp., Diatoms, Lyngbya sp., Cladophora sp., Spirogyra sp., and Phixoclonium sp.)	Municipal wastewater	97.0	93.0
Algal biofilm wastewater treatment : Algal biofilm	Tertiary treatment of wastewater	NA	97.0

photobioreactors, algae cover, hyper-concentrated culture, dialysis culture, and high-rate algae pools (Kiran et al., 2017; Galès et al., 2019).

As a result, studies have proven that microalgae-based nitrate and phosphate removal methods are beneficial. it is suggested that microalgae can improve nitrate and phosphate removal in wastewater with more than 90% removal efficiency, according to several different studies (Table 12.1).

12.1.1 Cell Structure of Microalgae

Known as the free-living microorganisms of the kingdom Protista, microalgae can be found in many different aquatic environments (Varshney et al., 2015). It is similar to plant cells in that it has photoautotrophs and chloroplasts. Microalgae do not contain roots, stems, and leaves in terms of their structure, but they show some features similar to cellular organelles when compared to high plants (Singh and Saxena, 2015). Algal cells, most of which are known as single-celled organisms, that differentiate from bacteria and other unicellular microorganisms contain nuclei containing genetic information, membrane-bound organelles, and lipid bodies (Schleyer and Vardi, 2020). It consists of polysaccharides and has a cell wall that lets materials pass through, and biochemical compounds function as selective obstacles. It also has a layered structure, so it can survive on its own and is vital due to its capability to generate food thanks to photosynthesis (Singh and Saxena, 2015). Microalgae have photon-converting ability (Shuba and Kifle, 2018) and thus have the potential to produce and accumulate large amounts of carbohydrate biomass. But microalgae can convert nearly 6% of the total radiation energy into fresh biomass (Koberg et al., 2011). Aquatic algal cells are floating and do not include constitutional biopolymers like hemicellulose and lignin, which are crucial for preferable herb growth in earthed media.

12.1.2 Biochemical Composition of Microalgae

Microalgae have a faster growth rate than plants, thanks to their simple growth and energy conservation structure. In addition to the ability to live in different environmental conditions, the production parameters during the growing periods have a significant effect on microalgae growth. The literature data on the parameters required for the living conditions and reproduction of microalgae are given in Table 12.2.

Microalgae can double their numbers in a short time, such as 24 h, and they are also very rich in fat, carbohydrate, and protein and provide the production of different types of biofuels through various conversion processes. The amounts of fat, carbohydrate, and protein contained in some microalgae species are given in Table 12.3.

Microalgae is a versatile raw material with wide energy potential. A wide variety of products can be produced in microalgae due to the high concentrations of natural proteins, lipids, carbohydrates, vitamins, pigments, and enzymes (Salla et al., 2016). In addition, it has been seen that many fuels, such as biodiesel, bioethanol, biogas, and biohydrogen, can be produced with conversion methods from algae (Jones and Mayfield, 2012). Under special cultural conditions such as high C/N ratio or stress

TABLE 12.2

Necessary Parameters for Growing Microalgae (Eleren and Burak, 2019)

Parameters	Literature Data
pH	5.8–9
Temperature (°C)	25–36
Salinity (g/L)	15–150
Light Intensity (μmol/m²/s)	90–360
Lighting time (day : night) (h)	12/12 24/0

TABLE 12.3

Protein, Carbohydrate and Fat of Some Algae Species Contents (Eleren and Burak, 2019)

Algae	Lipid Content (%)	Protein (%)	Carbohydrate (%)
Chlorella protothecoides	50.5	46.3	15.43 ± 0.17
Botryococcus braunii	27.37	–	20–76
Spirulina platensis	29.5	75.76	41.52
Dunaliella tertiolecta	36–42	–	–
Phaeodactylum tricornutum	47	–	–
Chlorella vulgaris	51.41	44.3	58
Scenedesmus obliquus	54.6	50–56	76.6
Chlamydomonas rheinhardii	23–62	48	17

conditions, large amounts of lipids are deposited by algae (30%–50% of their weight content). Microalgal lipids are classified in two ways. Fatty acids containing 14–20 carbons are suitable for biodiesel production. Polyunsaturated fatty acids with more than 20 carbons are used as healthy nutritional supplements. Microalgae with a high content of cellular lipids, soluble polysaccharides, and proteins can be easily pyrolyzed into biogas and bio-oils. In other words, not only fat but also proteins and water-soluble carbohydrate components can be easily converted into gases and fuels by thermochemical methods (Yen et al., 2013; Eleren and Burak, 2019) (Table 12.4).

12.1.3 NEW STRAINS OF MICROALGAE AND CYANOBACTERIA

Microalgae and cyanobacteria, also known as phototrophic microorganisms, can produce biomass efficiently using carbon dioxide, solar energy, and water. However, some of them can survive heterotrophically in very large (freshwater, marine, and terrestrial) habitats in nature. Theoretically, it represents a very diverse group and includes many different species in the range of 200,000–800,000, and 35,000 of these are known to be found in freshwater environments and seas, salty waters,

TABLE 12.4
Oil Content and Productivity of Some Microalgae Species (Aydin-Sisman, 2019)

Microalgae Species	Fat Content (% Dry Weight Biomass)	Oil Productivity (mg/L) day	Volumetric Productivity of Biomass (g/L) days	The Areal Productivity of Biomass (g/m³) day
Ankistrodesmus sp	24.0–31.0	–	–	11.5–17.4
Botryococcus braunii	25.0–75.0	–	0.02	3.0
Chlorella emersonii	25.0–63.0	10.3–50.0	0.036–0.041	0.91–0.97
Chlorella sorokiniana	19.0–22.0	44.7	0.23–1.47	–
Chlorella vulgaris	5.0–58.0	11.2–40.0	0.02–0.20	0.57–0.95
Chlorella sp	10.0–48.0	42.1	0.02–2.5	1.61–16.47/25
Chlorella protothecoides	14.6–57.8	1,214	2–7.7	–
Dunaliella saline	6.0–25.0	116.0	0.22–0.34	1.6–3.5/20–38
Haematococcus pluvialis	25.0	–	0.05–0.06	10.2–36.4
Nannochloropsis oculata	22.7–29.7	84.0–142.0	0.37–0.48	–
Nannochloropsis sp	12.0–53.0	37.6–90.0	0.17–1.43	1.9–5.3
Oocystis pusilla	10.5	–	–	40.6–45.8
Phaeodactylum tricornutum	18.0–57.0	44.8	0.003–1.9	2.4–21
Scenedesmus sp	19.6–21.1	40.8–53.9	0.03–0.26	2.43–13.52
Spirulina platensis	4.0–16.6	–	0.06–4.3	1.5–14.5/24–51
Spirulina maxxima	4.0–9.0	–	0.21–0.25	25
Tetraselmis suecica	8.5–23.0	27.0–36.4	0.12–0.32	19

acidic lakes, and desert areas (Ebenezer et al., 2012). Microalgae and cyanobacteria have been used for thousands of years as a natural source of high-value compounds for the pharmaceutical, food, and aquaculture industries, and large-scale cultivation of microalgae has been done for over half a century. Recently, new microalgae and cyanobacteria species have been identified, and commercial-scale systems have been introduced to grow algal biomass for various crops.

Although microalgae were used especially in aquaculture in the past, today they are preferred in studies due to their ability to contain various high-value molecules such as protein, chlorophyll, carotenoids, and lipids. Commercially, microalgae and cyanobacteria are among the promising organisms in the near future due to the development of sustainable options in the food, feed, pharmaceutical, and energy sectors.

The nutritional components of algae and cyanobacteria species, whose cultivation is becoming widespread day by day, are being examined, and the search for new species to be grown is increasing day by day. In addition to their biological and ecological roles in the aquatic ecosystem, microscopic algae contain important substances for both human health and the nutrition of living things growing in aquaculture (Duru and Yılmaz, 2013).

By making intensive production of algae under controlled conditions, they are also used in the nutrition of terrestrial and aquatic creatures, in the production of powder feed and live feed, in the purification of water, in the food industry, and as a fertilizer source due to the pigments, proteins, vitamins, and minerals they contain.

Algae and cyanobacteria can be produced by biotechnological methods independent of climatic conditions, compared to agricultural production. Since they can show a very rapid increase in biomass thanks to their reproduction by dividing, a much higher amount of pigment can be obtained compared to plants, whose pigments are phycocyanin, phycoerythrin, astaxanthin, canthaxanthin, β-carotene, lutein, and fucoxanthin (Vonshak and Richmond, 1988). Pigments from algae and their extraction studies are shown in Table 12.5.

Today, the main products produced from algae are agar, carrageenan, and alginate. In addition, the investigation of microalgae has become important thanks to the organic acids, saccharides, proteins, and vitamins they have.

Cyanobacteria are photosynthetic prokaryotes that have chlorophyll-a. Generally, water is an electron donor during photosynthesis and provides oxygen oxidation. Until recently, cyanobacteria were also characterized by their ability to form phycocyanin, the pigment of phycobilin. Two of the commonly known names for organisms are cyanobacteria, or blue-green algae, as high concentrations of this pigment cause the bluish color of organisms under certain conditions. However, it has become clear that, in addition to chlorophyll-a, at least oxygen-evolving prokaryotes, such as *prochlorothrix*, which lack phycocyanin but form chlorophyll b, are quite closely related to phycocyanin-containing organisms (Tomitani et al., 1999). Cyanobacteria are also the most important group of organisms ever seen on our planet. Their capacity to use water as an electron source in photosynthesis has removed the limits imposed by the primary production, distribution, and abundance of alternative electron sources such as H_2S, H_2, and Fe^{+2} (Kharecha et al., 2005). The ability to link primary production to nitrogen fixation is known to further increase the biological carrying capacity of the oceans.

TABLE 12.5
Pigments from Algae and Their Extraction Studies (İlter et al., 2017)

Pigment	Algae Obtained	Extraction Method
Phycocyanin Phycoerythrin	Spirulina platensis Spirulina maxima, Spirulina fusiformis, Anabaena sp., Synechococcus sp., Aphanothece halophytica, Nostoc sp., Oscillatoria quadripunctulata ve Phormidium ceylanicum Porphyridium spp.	Freeze-thaw (Abalde et al., 1998; Soni et al., 2006), homogenization (Boussiba and Richmond, 1979; Abalde et al., 1998; Schmidt et al., 2005), high pressure application (Patil et al., 2006; Patil and Raghavarao, 2007), sonication (Abalde et al., 1998), acid application (Sarada et al., 1999), lysozyme application (Boussiba and Richmond, 1979), microorganism extraction (Zhu et al., 2017), nitrogen cavitation method (Viskari and Colyer, 2003)
Astaxanthin	Haematococcus pluvialis spp.	Solvent extraction (Sarada et al., 2006), ultrasonic assisted microwave extraction (Ruen-ngam et al., 2011), enzyme assisted solvent extraction (Kobayashi et al., 1997), soxhlet extraction (Wang et al., 2012)
Canthaxanthin	Haematococcus lacustris Bradyrhizobium Halobacterium	Solvent extraction (Papaioannou et al., 2008), ultrasonic assisted extraction (Macias-Sanchez et al., 2009), acid treatment (Ni et al., 2008), supercritical CO_2 extraction (Macias-Sanchez et al., 2009)
Beta-carotene	Dunaliella salina, Dunaliella bardawil	Supercritical extraction (Mendes et al., 2003), pressurized liquid extraction, ultrasonic assisted extraction, accent electric field extraction
Lutein	Chlorella pyrenoidosa, Scenedesmus obliquus, Chlorella ellipsoidea	Microwave assisted extraction, ultrasonic assisted extraction, classical extraction (Pasquet et al., 2011; Abo-Hashesh and Hallenbeck, 2012)
Fucoxanthin	Undaria pinnatifida, Hijikia Fusiformis, Sargassum fulvellum, Chaetoseros sp., Eisenia bicyclis, Kjellmaniella crassifolia, Alaria crassifolia, Sargassum horneri, Cystoseira hakodatensis, Laminaria japonica, Undaria pinnatifida ve Sargassum fusiforme	Microwave assisted extraction, classical extraction (Xiao et al., 2012), pressurized liquid extraction

The phylogeny also supports the monophyly of cyanobacteria, which distinguishes akinetes and heterocysts based on sequence analysis of the 16S ribosomal-ribonucleic acid (rRNA), hetR, and rbcL genes and a large sampling of the multicellular taxon (Tomitani et al., 2006).

Compared to other prokaryotes, cyanobacteria generally have large cells and complex morphologies and are likely to be found in the fossil record. It is therefore thought that the group of bacteria most centrally associated with the world's past is likely to leave a trace in the sediments. A wide network of cyanobacterial species is given in Figure 12.1.

The direct conversion of CO_2 by solar energy to biofuel by photosynthetic microorganisms such as microalgae and cyanobacteria also has many advantages over conventional biofuel production from plant biomass. Also, in a recent study, cyanobacterial species were identified at the molecular level by polymerase chain reaction

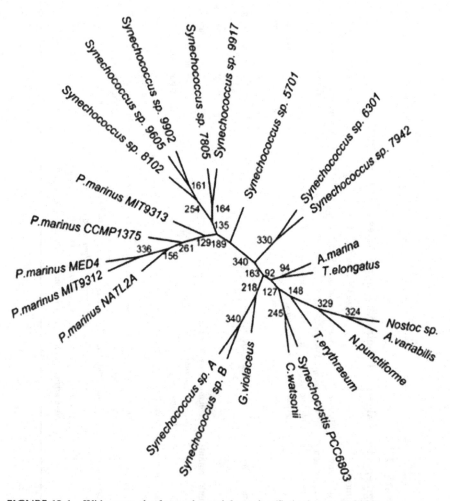

FIGURE 12.1   Wide network of cyanobacterial species (Swingley et al., 2008).

(PCR) and sequencing processes, and according to 16S rRNA sequence analyses, to reach the highest biomass, the newly isolated cyanobacterial species was identified with a 99% similarity rate as *Leptolyngbya* sp. (Taştan, 2016).

12.2 FACTORS AFFECTING MICROALGAE GROWTH

12.2.1 LIGHT

Light is the most important element in the development of microalgae in the process of photosynthesis. Three different light variables affect photosynthesis (Marsh, 2008): light intensity, light period, and light restriction. Light intensity is the amount of light energy coming into the culture medium. Although light is a desirable input, too much is harmful to microalgae culture. Under natural conditions, microalgae develop better in shaded areas that do not receive direct and continuous sunlight. Because these areas are places where light intensity decreases and humidity and temperature changes are buffered. Excessive light intensity damages the cells and becomes an obstacle to microalgae growth (Carvalho et al., 2011). Along with the intensity of the light, the spectral quality of the light is also important for microalgae growth and metabolism (Park and Lee, 2001). Active radiations for photosynthesis are radiations of wavelengths between 400 and 700 nm, which form the visible region of the solar radiation spectrum. At the same time, the rays of this wavelength that the human eye can see correspond to 43% of the total energy of the sunlight spectrum. As the amount of light increases up to a certain point, microalgae growth also increases. This point is called the saturation point. Light intensities above the saturation point do not increase the rate of microalgae growth. Even after a certain intensity, light slows down microalgae growth. This light intensity is called the inhibition intensity. The saturation and inhibition intensities depend on environmental factors such as temperature, CO_2 level, and nutritional supplementation (Sorokin and Krauss, 1958).

The light period is a concept that must be taken into account in large-scale production and is related to how long the cells are exposed to light. In artificial lighting, control of the light source is possible. Turning the light on and off is called a flash. In high-density microalgae cultures, flash lighting is more successful than continuous lighting, considering the specific oxygen production rates (Marsh, 2008). Especially for high-intensity lights, dark periods between short flashes increase the efficiency of photosynthesis (Park and Lee, 2001). There is no flash lighting in cultivation systems where daylight is used. In these cases, the light period can be considered a day and night period. In a study on the effect of exposure time on microalgae growth, a linear relationship was observed between the decrease in exposure time and the decrease in BP, except for the 12:12 lighting period. Again, in the same study, there is no significant difference in BP between lighting 24 h a day and lighting 22 h a day (Jacob-Lopes et al., 2009).

Light restriction is a concept related to whether the rays can reach the cells sufficiently or not. Microalgae on the upper surface of the culture medium see light. However, as we go deeper into the culture medium, the cells begin to see no light. This is because the cells shade each other. As the light penetrates the algae culture, the light intensity decreases dramatically due to the cells shading each other. Optical depth, which means the rate of diffused radiation, is taken into account in microalgae-growing medium designs (Kumar et al., 2010). The easiest way to overcome the light restriction is to reduce the depth of the culture medium (Larsdotter, 2006). If microalgae are to be grown in photobioreactors, the permeability of the materials used and the surface-to-volume ratio of the system are effective in capturing light (Wang et al., 2012). Light restriction can be prevented to some extent by choosing shallow ponds in open pond cultivation systems (Sayaner, 2013).

The growth rate of photosynthetic organisms increases with an increase in the amount of light. The increase in the amount of light causes the saturation level to be reached after a certain point. If high light levels continue, inhibition occurs due to the high amount of energy produced by disturbing the balance of the organism, and irreversible damage may occur in the case of photoinhibition. Light inhibition begins to take effect within a few minutes, and in some cultures, it can cause more than 50% damage within 15–20 min (Ogbonna et al., 1996; Lee, 2001).

Light is the most important element in the development of microalgae in the process of photosynthesis. Photoautotrophic microalgae perform photosynthesis using carbon dioxide and light. Depending on the light source, they convert the absorbed energy into chemical energy such as adenosine triphosphate and nicotinamide adenine dinucleotide phosphate. Generally, they use light with a wavelength of 400–700 nm. The absorbed wavelength may vary depending on the microalgae species. For example, green microalgae absorb the best light energy between 450–475 nm and 630–675 nm with chlorophyll (green) pigments during photosynthesis. Light directly affects the photosynthetic activity of microalgae (Zeng et al., 2011). Photosynthetic efficiency is expressed as the conversion of light energy into chemical energy during photoautotrophic development (Brennan and Owende, 2010). Photosynthetic activity differs according to the light spectrum. Light is a limiting environmental condition for the life of microalgae.

Excessive light intensity affects the growth of microalgae by damaging the cells, and low light levels affect growth. Growth rate and lipid production (LP) may vary according to the lighting level. Light intensity and efficient access to light for microalgae are important factors. In general, it has been reported that the illumination level at which many photosynthetic microorganisms reach saturation is ~200 µmol/m²/s (Teo et al., 2014).

In a study on the effect of illumination level on microalgae growth, he stated that the specific growth rate of microalgae increased threefold when they increased the illumination level from 20 to 50 µmol/m²/s (Sharma et al., 2011). One of the factors affecting the development of microalgae is photoperiod time (George et al., 2014). Photoperiod refers to the light/dark hours to which microalgae are exposed during the day (Wahidin et al., 2013). *Nannochloropsis sp.* investigated the effect of photoperiod time on microalgae growth and LP and found the best results (maximum cell concentration: 6.5×10^7 cells/mL, lipid: 31.3% dry weight) at 18 h light and 6 h dark cycle. Lee et al. (2015) stated that during the photoperiod for carbon removal, the extension of the dark cycle is both an advantage and a disadvantage for nitrogen and phosphorus removal. In addition, it was stated that it would be advantageous to keep the light period longer for microalgal BP (Lee et al., 2015). Photoperiod times and light intensities used in the cultivation of some microalgae species are given in Table 12.6 (Elcik and Çakmakcı, 2017).

TABLE 12.6
Photoperiod Times and Light Intensities Used in the Cultivation of Some Microalgae (Elcik and Çakmakcı, 2017)

Microalgae	Light/Dark (h)	Lighting Level (μmol/m²/s)
Chlorella vulgaris	14/10	250
Tetraselmis suecica	12/12	125
Phaeodactylum tricornutum	12/12	125
Chlorococcum sp. RAP-13	13/11	40
Scenedesmus obliquus[a]	14/10	250
Desmodesmus sp. EJ9-6	15/9	120
Nannochloropsis oculata	24/0	150
Cheatoceros sp.	16/8	40
Spirulina platensis	12/12	27
Botryococcus braunii-TN101	16/8	30

Growth is heterotrophic in the species marked ([a]) and autotrophic in other species.

12.2.2 TEMPERATURE

Temperature is a measure of heat intensity. The effect of temperature on photosynthesis varies depending on the type of microalgae, the CO_2 content of the environment, the light intensity, and the exposure time. If the light is a limiting factor, the temperature has little effect on the rate and amount of photosynthesis. If the light intensity is not limiting, if CO_2 is limiting, photosynthesis increases to a certain extent with temperature increase. Respiration, which is another life activity of chlorophyll-containing organisms, also increases with an increase in temperature up to a certain level. As the temperature increases, the respiratory rate increases, and increasing respiration increases biomass losses. Temperature also affects the chemical composition of microalgae. The temperature value at which photosynthesis and respiration decrease varies according to the microalgae species.

Temperature is an important parameter for microalgae, like every living thing. In particular, high temperatures, which affect cell morphology and physiology, increase the metabolic rate, while low temperatures restrict growth. Microalgae can develop at wide temperature values, such as 5°C–35°C (Rashid et al., 2014). Although the optimum temperature varies for each microalgae species, it is generally in the range of 20°C–30°C for most industrially attractive microalgae species. Temperature for microalgae is a stress factor that affects the cell's lipid composition, nutrient-carbon uptake, and growth rate. It reduces the fluidity of the cell membrane at low temperatures. To cope with this situation, the cell begins to produce unsaturated fatty acids (Juneja and Murthy, 2018). On the other hand, at high temperatures, the cell begins to increase its saturated fatty acid content.

Microalgae and cyanobacteria can live in different temperature ranges. However, they react immediately to temperature changes that directly affect their metabolism and physiological activities. For this reason, the species to be produced should be

TABLE 12.7

Optimum Growth Temperatures of Some Microalgae (Elcik and Çakmakçı, 2017)

Microalgae Species	Cultivation Temperature (°C)
Chlorella sp. ADE5	30 ± 2
Spirulina platensis	25 ± 1
Botryococcus braunii	26 ± 1
Desmodesmus sp. EJ9-6	24 ± 1
Desmodesmus communis	18–25
Chlorella zofingiensis	25 ± 1
Coccomyxa actinabiotis	24 ± 2
Platymonas subcordiformis	25
Chlorella vulgaris	22 ± 2
Anabaena sp. PCC 7120	30
Chodatella sp.	25
Scenedesmus obliquus	30 ± 3
Chaetoceros sp.	35
Monoraphidium sp. SB2	25–35

produced in the appropriate system, taking into account the optimal temperature range. The optimum growth temperatures of some microalgae are given in Table 12.7 (Elcik and Çakmakcı, 2017).

The effects of temperature and light on microalgae culture cannot be studied independently of each other. The optimum light value varies with temperature, and for good annual productivity, the light-temperature relationship of the grown species should be known exactly. In a study by Abu-Rezq et al. (2010), they performed some tests for the best growth performance of *D. salina* (as seen in Figure 12.2), based on samples taken from a regionally isolated location. The results of the experiments showed that the sample of *D. salina* preferred high salinities (45 psu), low temperatures (20°C), phosphate concentrations of 30 g/m³/day³, high light intensities (18×103 lux²), and high pH levels of up to 9.18 (without using CO_2 gas) for their optimum growth (Abu-Rezq et al., 2010).

12.2.3 pH

Another parameter that affects the production of microalgae and cyanobacteria is pH. pH is the negative logarithm of the H^+ ion concentration in any solution. Although it has a chemical value of 0–14, the pH range in which living organisms live is usually 3–10. The pH value is important for the nutrient uptake and metabolic activities of microalgae. The pH can change in the environment depending on the nutrients taken up by the microalgae. The pH increases with carbon dioxide and nitrate intake and decreases with ammonia intake. Each microalgae species can grow specifically in a certain pH range. pH is in the range of 7–9, suitable for good growth in the culture of

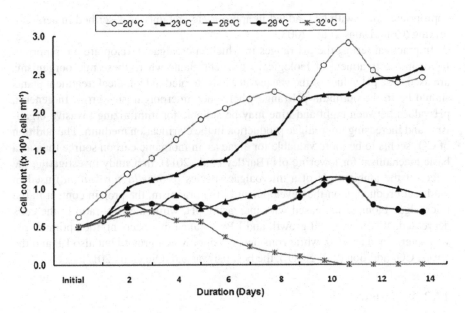

FIGURE 12.2   Cell number of *D. salina* depending on temperature (Abu-Rezq et al., 2010).

many algae species (Lavens and Sorgeloos, 1996; Katarzyna et al., 2015). The optimum range is between 8.2 and 8.7. In general, above pH 10 and 11, the growth of microalgae is inhibited. However, some species can also show optimum growth in acidic or alkaline conditions. For example, optimum pH values for *Spirulina platensis* and *C. littorale* species were determined to be 9 and 4, respectively (Zeng et al., 2011). The growing system itself also affects the pH value. The pH values of the pool and photobioreactor environments, fed from the same source and producing almost the same amount of biomass, were on average 8 and 10, respectively (Tsai et al., 2012).

In the microalgal cultivation process, the pH is mostly controlled with carbon dioxide, carbonate, and bicarbonate (Elcik and Çakmakcı, 2017). These carbon species have a pH-buffering effect. At pH levels below 5, the majority of dissolved inorganic carbons are in the form of CO_2. At pH 6.6, CO_2 and HCO_3^- are in equal amounts; at pH 8.3, almost all are in the form of HCO_3^-. Therefore, pH should be controlled to increase the carbon dioxide utilization of microalgae in the microalgae cultivation process. Microalgae use CO_2 by photosynthesizing, and the pH rises. To control the rising pH, hydrochloric acid or acetic acid is used. Acetic acid is more advantageous than hydrochloric acid as it also contributes to the growth of microalgae as a carbon source (Rashid et al., 2014).

The pH value changes the metabolic activities of microalgae. Failure to provide the appropriate pH for the culture causes the cells to lyse, the contents to pass into the medium, and the culture to die. This problem is overcome by the aeration of the culture. In very dense cultures, an increase in pH occurs over time. In extreme pH values, the most appropriate pH value of the culture changes the metabolic activities of microalgae due to the danger of complete collapse of the culture due to disruption of cellular processes (Rashid et al., 2014). In this case, the pH balance is kept in the appropriate range with CO_2 added in appropriate amounts as described in aeration/mixing (Conk Dalay et al., 2008).

In practical terms, the pH ranges in which microalgae develop are an important factor in the treatment of biological treatment plants where these microorganisms are used. The pH values of the wastewater to be treated in biological treatment plants should be in the optimum pH range where such microorganisms grow. In general, pH values between eight and nine may be suitable for minimizing invasive organisms and increasing microalgae production in the clarification medium. The addition of CO_2 seems to be more valuable for algae as an inorganic carbon source than as a basic mechanism for lowering pH (Bartley et al., 2014). In a study investigating the effect of the ambient pH of a microalgae species (*Chlorella sorokiniana*) used in biodiesel production on the development of the organism, the protein content in the microalgae biomass increased with increasing pH, while the C/N ratio in the cells decreased. When only cell growth and LP are taken into account, the optimum pH was determined as ~6.0, while considering not only cell growth but also LP and the lowest CO_2 addition (2.01 g CO_2), the best pH was 8.0 (Qiu et al., 2017).

12.2.4 SALINITY

Salinity refers to the concentration of water-soluble salt in the living environment or culture media of microorganisms. Salinity is one of the most important ecological factors affecting microalgae growth, metabolism, and distribution. It has been reported that the photosynthesis and protein synthesis capacities of cells decrease due to the increase in salinity and that the growth rate decreases with a further increase in salinity, thus a significant loss in production (Zhang et al., 2010). Depending on the salinity of the living environment, some microalgae species can grow in fresh and saltwater environments, just like fish. Some examples of these are given in Table 12.8.

TABLE 12.8

Microalgae Species Growing in Fresh and Salt Waters (Rashid et al., 2014; Heimann and Huerlimann, 2015; Rani-Borges et al., 2021)

Freshwater	Saltwater (Sea)
Amphora sp.,[a] *Chlorella* sp., *Chlorella vulgaris, Chlorella sorokiniana, Chlamydomonas reinhardtii, Chlamydomonas pyrenoidosa Desmodesmusarmatus, Desmodesmus asymmetricus, Dolichospermum flosaquae, Euglena gracilis, Isochrysis aff. galbana (T-ISO)*	*Botryococcus braunii, Chaetoceros muelleri Chlorococcum littorale, Chrysotila carterae, Dunaliella salina, Nannochloropsis oculata, Nannochloropsis* sp., *Nannochloropsis salina, Phaeodactylum tricornutum, Tetraselmis chuii, Tetraselmis suecica, Tisochrysis lutea, Synechococcus* sp.
Mesotaenium sp., *Microcystis aeruginosa Navicula* sp.,[a] *Pavlova salina, Raphidocelis subcapitata, Scenedesmus quadricauda, Scenedesmus obliquus, Spirulina platensis, Tetraedron* sp.	

[a] These microalgae live in fresh water but are also known to be salt tolerant.

Salinity is one of the important factors affecting the BP of microalgae (Rashid et al., 2014). The optimum salinity value required for the cultivation of each microalgae species varies (Mata et al., 2010). Salinity is a complex stress factor that affects net lipid productivity in microalgae cells. Microalgae species can tolerate salinity stress to some extent. Marine microalgae that thrive in a saline environment can thrive better in a high-salt growth environment and can tolerate salinity stress to some extent. Freshwater microalgae can thrive in less salty nutrient media. However, in cultivation, excessive salinity stress inhibits photosynthesis, which reduces biomass and net lipid productivity (Martínez-Roldán et al., 2014). For example, the marine microalgae *nannochloropsis salina* usually grows at a salinity of 34 psu (practical salinity unit), while *nannochloropsis salina* is between 68 and 8 psu. It does not grow under Barley et al. (2013). For the cultivation of *chlorella vulgaris* and *chlorella sp.*, which are classified as freshwater microalgae and produced industrially, the best salinity value was determined as 30 psu (Shaleh, 2004; Fathi and Asem, 2013).

12.2.5 CARBON

Microalgae are single-celled photosynthetic organisms with relatively simple requirements for growth that need sunlight to convert water and carbon dioxide into proteins, amino acids, lipids, polysaccharides, carotenoids, and other bioactive compounds.

Since microalgae are composed of ~50% carbon, 10% nitrogen, and 2% phosphorus, the most essential nutrients for their growth are carbon, nitrogen, and phosphorus. In addition, minerals such as calcium, potassium, magnesium, iron, zinc, molybdenum, cobalt, selenium, and nickel are essential nutrients in trace amounts. Since they are photosynthetic creatures with CO_2/O_2 converters, they can take carbon dioxide from the atmosphere or from special production environments that come out as a result of combustion and which are planned and prepared for microalgae production. The use of microalgae is being investigated to reduce the carbon dioxide gas emitted from the chimney as a result of the combustion of coal, natural gas, biomass, and other fuels, and these studies will be beneficial in terms of both preventing environmental pollution by reducing greenhouse gases and producing biodiesel.

About half of the microalgae biomass consists of carbon by dry weight. Microalgae can use both organic and inorganic carbon sources. Autotrophic microalgae use inorganic carbon (CO_2, salts, and light as energy sources for growth), while heterotrophic microalgae use organic (an external organic compound, although not photosynthetic, as an energy source) carbon. Some algae are mixotrophic (Lee, 1980). Autotrophic microalgae use carbon dioxide (CO_2) > bicarbonate (HCO_3^{-1}) > carbonate (CO_3^{-2}) as the inorganic carbon source in order of priority (Christenson and Sims 2011). Carbon dioxide given to the microalgae culture is usually supplied from the atmosphere. Since the solubility of CO_2 in water is low, continuous feeding is required with the help of an air pump. CO_2 given to the system can react with water and lower the pH. It can prevent microalgae growth at low pH. To prevent this, high-density inoculation can be used to ensure that all of the CO_2 is used by the microalgal culture, or alkaline chemicals can be applied to increase the pH (Na, Ca, Mg, and K hydroxides). He stated that microalgae growth rate and BP increased at a 1%–15% CO_2 level

(Rashid et al., 2014). Some scientists have suggested that the key to the design of microalgae production systems is the CO_2 supply (Lundquist et al., 2010). Carbon, which is the most important nutrient for microalgae, should be fed to the ponds in the form of CO_2 in general. However, the storage of dissolved CO_2 in the culture medium is limited due to alkalinity. If an excessive amount of CO_2 is fed, it will be released into the atmosphere from the pool surface. Efficient use of CO_2 is necessary, as pumping CO_2 into pools creates parasitic energy losses. For this reason, CO_2 needs to be added frequently in controlled amounts.

12.2.6 NITROGEN

Nitrogen is the most critical nutrient in microalgae growth, as in all living things. It is necessary for growth, reproduction, and other physiological activities. Although the use of nitrogen in microalgae varies according to the species, they mostly use nitrogen in the form of nitrate, nitrite, ammonium, and urea. They generally use ammonium, urea, nitrate, and nitrite in order of priority. Because some species of cyanobacteria are diazotrophic, they are capable of fixing atmospheric nitrogen (N_2) in the gaseous state. They can also thrive in natural environments lacking nitrogen. Nitrogenous substances are not added to the nutrient media used for their reproduction. For example, Anabeana, Nostoc, and Oscillatoria species are used for agricultural purposes by multiplying in BG11 nutrient medium. Agricultural, domestic, and industrial wastewater can be used as an inexpensive nitrogen source, containing nitrogen sources in different forms, for the reproduction of nitrogen-demanding microalgae. A gradual purification and microalgae production can be carried out by using the wastewater in the next stage for the reproduction of microalgae with nitrogen fixation ability.

The use of nitrogen changes the pH of the microalgal culture. When microalgae use ammonium, H^+ ions are released, and pH decreases. This reaction is shown in equation 2.1.

$$NH_4^+ + 7CO_2 + 17.72H_2O \Rightarrow C_{7.6}H_{8.1}O_{2.5}N + 7.6O_2 + 15.2H_2O + H^+ \quad (2.1.)$$

In cases of excessive use of ammonium, the pH may decrease, stopping microalgae growth. He stated that oil production can increase 2–3 times under limiting conditions (Breuer et al., 2012). If nitrate nitrogen is found in excess in the growth medium and microalgae use nitrate, the pH of the environment increases. When one mole of nitrate is used, this time 1 mole of OH, not hydrogen, is produced. This reaction is shown in equation 2.2.

$$NO_3^- + 5.7CO_2 + 5.4H_2O \Rightarrow C_{5.7}H_{9.8}O_{2.3}N + 8.25O_2 + OH^- \quad (2.2.)$$

In cases of excessive use of nitrate, the pH may increase above ten and affect microalgae growth. For this reason, the pH needs to be adjusted if ammonium and nitrate are used as nitrogen sources. Nitrogen deficiency is a chemical stress factor for microalgae other than diazotrophic cyanobacteria (Pancha et al., 2014). Nitrogen limitation is the most effective method for improving microalgae lipid accumulation, which not only causes lipid accumulation but also causes a gradual change in lipid composition from free fatty acids to triacylglycerol (Widjaja et al., 2009). Schenk et al. (2008) stated that many algae species contain approximately 10%–30% oil on a dry weight basis under normal conditions, and oil production can increase two to three times under nitrogen-limiting conditions. It has been determined that *Chlorella vulgaris, Chlorella zofingiensis, Neochloris oleoabundans,* and *Scenedesmus obliquus* species can accumulate more than 35% lipid by dry weight under nitrogen-limiting growing conditions (Schenk et al., 2008).

12.2.7 PHOSPHORUS

Phosphorus has a significant effect on the growth and metabolism of microalgae. Phosphorus accounts for <1% of the total algal biomass (~0.03%–0.06%) and is an important component for the sustainable growth of microalgae (Hanon et al., 2010). Phosphorus is an essential nutrient for cell development and microalgae metabolism and is present in basic cell components such as phospholipids, nucleic acids, or nucleotides (Ruiz-Martinez et al., 2014). It plays an important role in processes such as energy transfer, cellular signal transmission, photosynthesis, and respiration within the cell (Courchesne et al., 2009; Singh et al., 2015). The absence of phosphorus in an environment results in the suppression of photosynthesis and affects microalgae growth.

Phosphorus is also effective in LP, and studies have shown that lipid contents and LP of microalgae increase at low phosphorus concentrations (Singh et al., 2015). Therefore, when designing strategies to obtain higher lipid and biomass under environmental stresses, the combination of low nitrates and optimum phosphate concentrations can increase lipid and BP in one growth cycle (Belotti et al., 2014). Liang et al. (2013), in their study with *Chlorella* sp., determined that the lipid content and LP of microalgae increased at low phosphorus concentrations (Liang et al., 2013).

The ratio of nitrogen and phosphorus, the macronutrients required in most culture media, should be 16N:1P. Optimization of nutrients required for algal culture development provides economic benefits by reducing production costs (Gross, 2013; Blair et al., 2014).

Some phosphorus may remain in the tail waters resulting from the treatment of domestic and industrial wastewater and the irrigation of pan-grown rice in agriculture. By growing microalgae in these waters, waters lower in phosphorus can be obtained. This may be beneficial in reducing the risk of eutrophication in receptive environments such as streams, lakes, and seas. In a study conducted on this subject, 87.9% chemical oxygen demand (COD), 84% phosphate, and 100% total nitrogen (TN) were used in anaerobic digester wastewater (ADWWT) by using bacteria and microalgae cultures together and diluting ADWWT using treatment plant discharge water at a level suitable for mixotrophic microalgal growth. I expense has been provided. In addition, WWT provided by microalgal biomass was found to be more effective than bacterial treatment. Under the same conditions, 64.6% COD and 54.8% phosphate removal were achieved with microalgae culture, while 45.7% COD and 46.6% phosphate removal efficiency were obtained as a result of bacterial activity. When the results obtained are evaluated, it has been determined that bacteria and microalgae compete to consume the nutrients in the environment, and thus a more effective treatment is provided (Keriş Şen, 2019).

12.2.8   HYDRAULIC AND BIOMASS HOLDING TIME

Voltolina et al. (1999) achieved the removal of 79.4% of NH_4-N in wastewater by producing algae in wastewater in their study. They revealed the growth potential of algae in a controlled environment in wastewater, depending on the temperature and hydraulic retention time (Voltolina et al., 1999). The main factor that needs to be known in the decision phase regarding which nitrogen and phosphate elements are the limiting factors is the stoichiometry of the phytoplankton species that causes eutrophication. If an assumption is made that 1 μg P and 10 μg N are required for the formation of 1 μg Chlorophyll-a, it can be said that the system is limited by phosphorus in the case of N/P10. In the case of N/P = 10, the system is not limited by either. If these ratios are extended for all phytoplankton, it is stated that it is a safer approach to consider phosphorus limiting in the case of N/P > 20 and nitrogen limiting in the case of N/P < 5 (Muslu, 2001).

All organic materials of vegetable and animal origin, whose main components are carbohydrate compounds, can be used as a biomass energy source. Among these energy sources, microalgae seem promising for biofuel production with their advantages such as high photosynthetic activities, high BP, and rapid proliferation.

Electricity, ethanol, hydrogen, methane, and biodiesel can be produced from microalgae by biochemical methods, as well as syngas, biological coal, biodiesel, and electricity by using thermochemical methods.

Microalgae generally live autotrophically and perform photosynthesis using their pigments. By performing photosynthesis, they convert carbon dioxide, water, and sunlight into biomass.

Biofuel production technology from microalgae basically consists of four stages: isolation and characterization of microalgae, microalgal BP, harvesting, and product processing processes. While many factors are important in the development of biofuel production from microalgae, the microalgae cultivation process plays a key role among them (Junying et al., 2013). Also, Lee et al. (2015) stated that it would be advantageous to keep the light period longer for microalgal BP (Lee et al., 2015).

CO_2 given to the system can react with water and lower the pH. To prevent this, a high-density vaccine should be used to ensure that all of the CO_2 is used by the microalgal culture (Rashid et al., 2014). Also, they stated that microalgae growth rate and BP increased at a 1%–15% CO_2 level.

In cases of increased biomass concentration in the bioreactor, increasing the mixing speed to ensure proper mixing may damage the microalgal cells. Good mixing and gas transfer efficiency can be achieved with mechanical mixing used in open systems, but hydrodynamic stress is likely to occur. Therefore, this problem can be avoided by using a sufficient number of partitions (curtains) and controlling turbulence. BP can be increased by up to 75% if the appropriate mixing system is selected (Kumar et al., 2010).

Salinity is one of the factors affecting microalgal BP (Rashid et al., 2014). The optimum salinity value required for the cultivation of each microalgae species varies

(Mata et al., 2010). Marine microalgae such as *Synechococcus* sp., *Nannochloropsis salina, Chlorococcum littorale,* and *Botryococcus braunii* can grow better in a high-salt growth environment. Freshwater microalgae such as *chlorella vulgaris and microcystisa rugosa* can grow in less salty nutrient media (Rashid et al., 2014). For example, the marine microalgae *Nannochloropsis salina* usually grows at a salinity of 34 psu. *Nannochloropsis salina* does not grow above 68 psu and below 8 psu (Bartley et al., 2013).

12.3 MICROALGAE GROWING SYSTEMS

12.3.1 TRADITIONAL SYSTEMS

Microalgae cultivation is generally carried out in open and closed systems. Open systems can be called open pools, and closed systems can be called photobioreactors.

12.3.1.1 Open Systems

Under phototrophic growth conditions, microalgae absorb solar energy and absorb carbon dioxide from nutrients in the air and aquatic life. Reactor selection is the main factor affecting microalgae production and productivity. Open pond production is the cheapest method for large-scale BP because it is the most common, oldest, and simplest. Poor light intensity, temperature fluctuations, and the effects of pH and dissolved oxygen concentration can limit the growth parameters of microalgae. Low biomass productivity (0.1–1.5 g/L) causes operating weakness, and the formation of unwanted species makes system control difficult (Eleren and Burak, 2019). It is performed in various pools, such as circular, unmixed, and raceways. It is low-cost, but climatic conditions play an important role in the outdoor pool. More than 98% of commercial microalgae biomass is produced in open ponds. Due to the limiting effect of light, the depth should be between 20 and 30 cm. In open systems, a process similar to the natural life of microalgae is involved. *Chlorella* sp., *Spirulina platensis, Hematococcus* sp., and *D. salina* are the most common species commercially produced in racetrack-type pools (Figure 12.3) (Amotz, 2009). Racetrack pools are shallow (20–50 cm deep) and are the most preferred pool type for microalgae production. By operating the Kantali wheel continuously, the culture

(a)   (b)

FIGURE 12.3 Examples of the circular stirred ponds (a) and racing open ponds (b) from Japan (Sayaner, 2013).

medium is mixed, and the precipitation of microalgae is tried to be prevented. In this production method, the decrease in the CO_2 transfer rate with the increase in the culture density decreases the production efficiency (Elcik and Çakmakcı, 2017). Harvesting is required to prevent this situation.

Open systems have several advantages and disadvantages. The most important advantages are low investment and operating costs, reducing the production cost by using wastewater as a food source, not needing cooling, and using areas that are not suitable for agriculture. System control is weak in open-pond cultivation. Evaporation losses, nutrient and light limitations, the inability to keep the temperature at an optimum level, and insufficient homogenization reduce efficiency, and it is difficult to achieve the desired results for this reason. Insufficient control of the system may result in the inability to produce microalgae in the desired amount and the proliferation of undesirable species (Oktor, 2018).

Algae ponds are the most common method used for microalgae production. These systems are built in the form of closed loop and oval-shaped circulation channels, in the range of 0.2–0.5 m, to be mixed with a wheel to ensure homogeneity. Algae pools can be made of concrete, fiber glass, or membranes. These pools are made of loop channels in the form of closed channels. Mixing and circulation are provided by an impeller. Channels are made of concrete or compacted earth. The channels can also be covered with white plastic material. The impeller is always operated to prevent the sedimentation of the culture in the channel. Pools with water channels are used in the production of *Spirulina* (*Arthrospira platensis* and *A. maxima*), *D. salina*, *Chlorella vulgaris*, and *Haematococcus pluvialis* (*for astaxanthin*) species and in some WWT plants (Darzins et al., 2010). The most widely used open-system water channel pools. Pools with water channels can be operated individually or in series, as several of them are connected. Figure 12.3 shows the pond facility with water channels, each of which is ~0.4 ha in size, where microalgae production for additional food is made. Also, in Figure 12.3, the production facility of the Cyanotech company, consisting of pools with water channels, is where Spirulina is produced (Sayaner, 2013).

12.3.1.2 Closed Systems

This technology was mainly applied to overcome some of the disadvantages of outdoor pool systems. Closed-culture systems for microalgae have some advantages over open systems. Such systems allow sunlight to be distributed over a larger surface area. In this way, higher cell density is achieved due to the efficient use of lighting (Van der Hulst, 2012). According to the geometry of the reactors used in growing microalgae in closed systems, takes place in horizontal tubular, flat plate, helical tubular photobioreactors (Figure 12.4) (Oktor, 2018). Photobioreactors provide a protected environment for microalgae cultures. Contrary to open ponds, it has a lower risk of contamination and is suitable for the cultivation of one type of microalgae (Brennan and Owende, 2010). It has advantages such as better pH and temperature control, better mixing, fewer evaporation losses, high BP, and a high surface volume ratio compared to open pools. In addition, there are some disadvantages, such as excessive dissolved oxygen levels inhibiting photosynthesis, overheating, cell damage, and deterioration of the materials used for lighting.

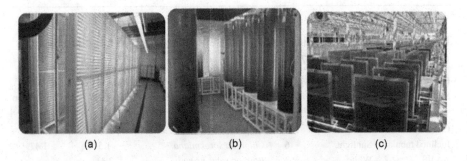

FIGURE 12.4 Microalgae production in a closed horizontal tubular (a), flat plate (b), and helical tubular (c) photobioreactors (Oktor, 2018).

Tubular photobioreactors are preferred by production farms for commercial-scale microalgae production because they minimize contamination, are suitable for continuous culture, require a smaller area compared to open systems, are easy to control and sterilize parameters, and ultimately reach high cell densities and high productivity.

In the tubular photobioreactor, microalgal culture circulates between the feeding chamber and the reactor with the help of a pump. Thus, the mixing of nutrients in the culture medium is ensured, gas transfer is increased, and the collapse of microalgae is minimized (Demirbas, 2010).

12.3.2 INNOVATIVE SYSTEMS

12.3.2.1 Process Processing Methods in Microalgae Systems

Harvesting is the preparation of the grown microalgal biomass for subsequent drying, oil extraction, etc. One or two harvesting applications can be performed to reach the desired microbial concentration. Harvesting also has some difficulties. Difficulties such as the low density of the culture medium, the small size of the microalgae, and the application of large volumes of water increase the cost of production.

We can list the harvesting methods applied as follows:

- **Chemical-based**: Flocculation
- **Electric-based**: Electrophoresis
- **Biological-based**: Natural Flocculation
- **Mechanical-based**: Centrifuge, Filtration, Weight Settlement, DAF (Dissolved Air Flotation)
- Membrane Filtration

Microalgae harvesting is done by physical, chemical, and biological methods. Current technologies used in solid-liquid separation, such as centrifugation, flocculation, gravity settling, air flotation, and electrophoresis, are both costly and energy-consuming to harvest low-cell density microalgal cultures. It is common to use more of them together. The selection of the right harvesting method depends on the

TABLE 12.9

Biomass Production Efficiencies of Different Microalgae Production Systems (Elcik and Çakmakcı, 2017)

Production System	Light Source	Volume (L)	Microalgae Species	Biomass Con. (g/L)	Biomass Production
Tubular reactor	Sunlight	70	*Chaetoceros calcitrans*	2.5	37.3[a]
Circular column	Sunlight	120	*Tetraselmis suecica*	1.7	38.2[a]
Inclined tubular	Sunlight	6	*Chlorella sorokiniana*	1.5	1.47[b]
Air lift	White fluorescent	3	*Botryococcus braunii*	2.31	–
Air-lift tubular	Sunlight	200	*Phaeodactylum tricornutum*	2.38	20[a]
Bubble colon	Sunlight	55	*Haematococcus pluvialis*	1.4	0.55[b]
Tubular	Sunlight	65	*Spirulina platensis*	2.2	32.5[a]
Race track	Sunlight	375	*Neochloris oleoabundans*	2.8	20[a]
Race track	Sunlight	150	*Chlorella variabilis*	1.6	8.1[a]

[a] g/m2 day
[b] g/L day

product characteristics to be obtained. The microalgal harvesting process should not be harmful to post-harvest processing and should not cause contamination. Among the listed methods, membrane filtration seems promising due to its advantages in reducing cost and energy consumption in the harvesting process of microalgae in recent years (Oswald, 1988; Torzillo and Vonshak, 2004; Chisti, 2007; Sen et al., 2013; Elcik and Çakmakcı, 2017; Faried et al., 2017).

In experimental studies, production conditions, microalgae species, and BP purposes are different from each other. Some studies were carried out in open spaces and some in closed spaces. Some systems were operated throughout the year and some during the summer period. The volumes of the systems used are also very important. The BP efficiencies of different microalgae production systems are listed in Table 12.9 (Elcik and Çakmakcı, 2017).

12.4 REDUCING CO_2 EMISSIONS WITH MICROALGAE

Control and reduction of CO_2, which is the biggest cause of global warming among greenhouse gases, can be achieved by taking some precautions. These

- Reducing energy consumption
- Increasing efficiency in energy use
- Turning to fuels with lower carbon content (using natural gas instead of coal)
- Using energy sources that do not emit CO_2, such as renewable and nuclear energy
- Capturing and using CO_2 from fossil fuel use and storage.

The photosynthesis of terrestrial and aquatic plants, from trees to microscopic phytoplankton, has a crucial impact on the global carbon cycle. A better evaluation of this biological process is one of the solutions to be developed to overcome the increase in atmospheric CO_2 concentration and global warming (Kurano et al., 1998). The ability of plants to capture and use CO_2 as part of photosynthesis makes biological CO_2 reduction an attractive option. Terrestrial plants capture enormous amounts of CO_2 from the atmosphere. The use of terrestrial plants for biological CO_2 sequestration is not an economically viable option due to the relatively low concentration of CO_2 in the atmosphere. Aquatic microalgae stand out with their more efficient photosynthesis and rapid reproduction when compared to C_3 and C_4 (carbon cycle species in photosynthesis) plants. Therefore, microalgae can be considered an important candidate for the biological CO_2 reduction method. Another aspect that makes microalgae superior to terrestrial plants as a biological CO_2 reduction tool is that they can be integrated with point sources that emit high concentrations (5%–15%) of CO_2, such as thermal power plants. Since terrestrial plants keep CO_2 in gaseous form, no significant increase in CO_2 sequestration is expected by feeding point CO_2 emissions to terrestrial plants. Because microalgae use CO_2 dissolved in water, they can provide further CO_2 reduction per unit area by evaluating their point CO_2 emissions. CO_2 reduction with microalgae is a biological method to be applied to combat global warming. Microalgae maintain their life cycle by using CO_2 in the atmosphere. Mycoalgae provide CO_2 reduction by binding CO_2 with photosynthesis. However, to combat global warming, massive CO_2 reductions are required. To achieve this goal, possibilities for microalgae to bind more CO_2 in a shorter time are being explored. It has been proven by many studies that more CO_2 binds per unit area in microalgae culture media fed with high concentrations of CO_2 (Kurano et al., 1998; Packer, 2009). Photosynthetic organisms form biomass by binding CO_2. The obtained biomass can be converted into different biofuels by different methods. When biofuels are burned, the CO_2 bound by the biomass is released back into the atmosphere. However, since this is a closed CO_2 cycle, biofuels serve to reduce CO_2 emissions. Although biofuels are a source of CO_2 emissions when used, they contribute to reducing CO_2 emissions since they reduce the use of fossil fuels.

Large-scale microalgae cultivation, which is necessary for CO_2 reduction, is a subject that has been studied for the purpose of biofuel production. However, biofuels such as microalgae-derived biodiesel and bioethanol are not yet economically competitive with fossil-based alternatives. To reduce the cost of microalgae-based biofuels. In this way, unit biofuel costs will be reduced a little more by taking advantage of economies of scale. To replace fossil fuels with biofuels, a significant amount of biofuel must be produced. For this reason, researche on microalgae cultivation on a large scale for the purpose of biofuel sources continues. Although it is possible to obtain different products other than biofuels, the most important product that requires large-scale microalgae cultivation is biofuels. Therefore, microalgae-based biofuel studies are of great importance in projects aiming to reduce CO_2 originating from thermal power plants with microalgae. The main purpose of microalgae cultivation for CO_2 capture is to realize production with the lowest possible costs.

The natural CO_2 concentration (0.03%) in the air is not sufficient for optimum growth and high productivity for microalgae. In algae cultures made at low salinity and near neutral pH, the air must be enriched with CO_2 for adequate growth. An exception to this is Spirulina, which can use carbonate or bicarbonate added to the environment in the form of salt as a carbon source. CO_2 exists in water in the following forms, depending on pH, temperature, and concentrations of nutrients. In many culture media, the buffering system is weak. Rapid assimilation of CO_2 or bicarbonate in the environment by fast-growing algae causes an increase in pH as a result of the above equation, due to the secretion of OH^- ions by the algae. In dense cultures, the pH must be balanced to keep the cultured algae between their optimum values and to prevent the depletion of carbon in the medium (Becker, 1995).

Even a small amount of CO_2 addition accelerates microalgae growth compared to the CO_2 concentration in the atmosphere. In this regard, 2–3 fold increases in microalgae growth were obtained at 15% CO_2 and 20% CO_2 levels (Hauck et al., 1996; Lee et al., 1996). In Figure 12.5, the photo of the air-fed culture medium containing 0.038% CO_2 (left) and the air-fed culture medium (right) containing 1% CO_2 at the same flow rate is given after 36 h of the growth period. The biomass amounts of Chlamydomonas reinhardtii, freshwater microalgae, were determined by measuring chlorophyll α in the tubes. It has been determined that the amount of biomass in the air-fed culture containing 1% CO_2 is twice the amount of biomass in the other culture (Packer, 2009).

Hsueh et al. (2009) observed the growth values of *Nannochloropsis oculta* species using gas mixtures with different CO_2 concentrations in a 1-L bubble column photobioreactor. The CO_2 concentrations of the gas mixtures examined in the study were adjusted to 0.036% (air), 5%, 8%, 10%, 20%, and 40%. According to these concentrations, the carbon load of the culture medium is 1.4×10^{-2}, respectively; 1.9, 3.0, 3.8, 7.6, and 15.1 g/min-L. In addition to CO_2, gas mixtures contain 6% O_2 and varying amounts of N_2. Figure 12.6 shows the development of Nannochlopsis oculta species growing with gas mixtures at different CO_2 concentrations and the change in the pH value of the culture. It has been observed that the microalgae growth rate

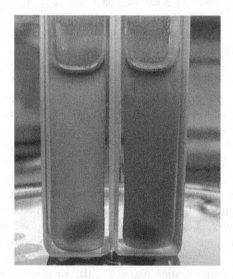

FIGURE 12.5 Microalgae growth status with two different CO_2 concentrations (Packer, 2009).

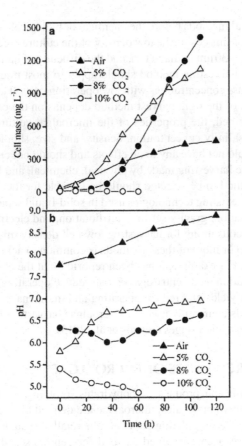

FIGURE 12.6 Development of *Nannochloropsis oculata* species and change of culture pH value over time at different CO_2 concentrations (Hsueh et al., 2009).

increases with increasing CO_2 concentration from air to 8%. However, at a 10% CO_2 concentration, microalgae could not grow. It has been stated that the low pH value of the culture medium may be the reason why no improvement can be observed (Hsueh et al., 2009).

12.5 HARVESTING MICROALGAE

Microalgal harvesting arises from a concentration and separation process to generate an algal cake, paste, or sludge of 15%–25% or more dry solids from a dilute biomass of 0.02%–0.06% dry solids. It is possible to obtain the desired microalgae concentration by harvesting applications consisting of one or two stages (Barros et al., 2015). In the two-stage harvesting process, firstly, algal mud is obtained at a concentration of 2%–7% suspended solids (SS), and in the second step, the algal cake is obtained at a concentration of 15%–25% SS. Harvesting of microalgae is about 20%–30% of the total cost of microalgae production (Grima et al., 2003; Uduman et al., 2010). Pretreatments can be applied to increase microalgae harvest yields and reduce operating costs. When harvesting, it is possible to reduce operating and maintenance costs by using mechanical methods before the chemical or biological concentration stages. The main reason for this cost is the low density of the culture medium, the small size of the microalgae (3–30 μm in diameter), and the difficulty of handling large volumes of water. An ideal harvesting method should apply to most microalgal species and achieve high biomass concentrations with low operating costs (Danquah et al., 2009). Since the selection of the right harvest method depends on the characteristics of the product to be obtained, the properties of the microalgal culture medium, such as moisture content, salinity concentration, density, and size, should be within acceptable ranges. It should not have any toxic effects and should not cause contamination. Also, in microalgae harvesting made by physical, chemical, and biological methods, the harvesting method to be selected should be suitable for the reuse of the culture medium. However, existing technologies used in solid-liquid separation, such as centrifugation, flocculation, gravity settling, air flotation, and electrophoresis, are both costly and energy-consuming for harvesting low-cell density microalgal cultures. A combination of two or more of these methods is common to achieve high separation rates at a low cost. For example, it has been reported that the combined use of flocculation, sedimentation, and centrifugation can reduce operating costs. It is possible to increase harvest yields and reduce operating and maintenance costs by using the mechanical methods commonly used for microalgae harvesting before the chemical or biological concentration stages (Barros et al., 2015).

12.6 COMMERCIAL USES OF MICROALGAE

While algae contain and produce carbohydrates, proteins, essential amino acids, vitamins, and bioactive molecules, microalgae are defined as microbial sources that have the property of accumulating some commercially valuable metabolites. Most forms of algae have been consumed as food for centuries and are used for their health effects. Algae are known as green algae, brown algae, and red algae according to their colors. They produce important pigments such as chlorophyll a, b, and c, β-carotene, astaxanthin, phycocyanin, xanthophyll, and phycoerythrosine. These pigments are frequently used in foods, pharmaceuticals, textiles, and the cosmetics industry (Borowitzka, 2013). In contrast to macroalgae (seaweeds), where agar, alginic acid, and carrageenan have been commercially produced for many years, microscopic microalgae have gained newer commercial use in recent years and are collected for their polysaccharide content. Therefore, microalgae containing at least 30,000 species are natural resources that need to be explored (Priyadarshani and Rath, 2012; Mobin and Alam, 2017).

Microalgae can be used as a raw material source for many types of biofuels. For example, anaerobic degradation of microalgal biomass can produce methane, biodiesel from microalgal oils, and biohydrogen through photobiological reactions.

12.6.1 BIODIESEL

Biodiesel, which has an important share in biofuels, is a chemically non-toxic, renewable, and biodegradable fuel type. Biodiesel, a mixture of fatty acid methyl esters, is usually produced by the transesterification of oil from animal fats or oily crops such as rapeseed, corn, soybean, palm, and castor beans. However, these raw material sources have low oil yields and require high production costs for water, soil, and fertilizers. Moreover, if all the transport fuel consumed in the world is replaced with biodiesel, biodiesel obtained from oil crops, waste edible oils, and animal fats will not be able to meet this demand. For example, meeting only half of current US transportation fuel needs with biodiesel would require unsustainably large plantations for petroleum products (Priyadarshani and Rath, 2012). Using the average oil yield per hectare from various crops, the area required to meet 50% of US transport fuel needs is given in Table 12.10.

According to Table 12.10, microalgae appear to be the only source of biodiesel with the potential to completely displace fossil diesel. Microalgae generally have the potential to double their biomass within 24 h. Biomass doubling times during exponential growth are generally about 3.5 h and are short (Chisti, 2007). Moreover, unlike other oil crops, microalgae grow extremely fast, and many are extremely rich in oil. Although the oil content in microalgae can exceed 80% by weight of dry biomass, oil levels between 20% and 50% are quite common in practice (Table 12.11).

The oil yield or mass of oil produced per unit volume in algal biomass depends on the algal growth rate and the oil content of the biomass. Microalgae with high oil yields are preferred for biodiesel production.

Therefore, it is important to seek alternative sources for biodiesel production in pursuit of sustainable development. Plant-based biofuels such as microalgae have been successfully commercialized over a 10-year period to replace petroleum-based diesel in the transport and industrial sectors. Algae, which can produce oil with biological carbon capture potential, is very important for biodiesel production, and its sustainability has been proven by regulations (Zhu et al., 2017). One of the main differences between petroleum diesel and algae-/microalgae-derived biodiesel is that the fatty acid chains are different. This situation provides an extra advantage to algal biodiesel in terms of engine performance.

TABLE 12.10

Comparison of Some Biodiesel Sources (Priyadarshani and Rath, 2012)

Biomass	Oil Yield (L/ha)	Required Agricultural Land (M ha)	Percentage of Existing U.S. Agricultural Land[a]
Sweetcorn	172	1.54	846
Soya	446	594	326
Canola	1.190	223	122
Jatropha	1.892	140	77
Coconut oil	2.689	99	54
Palm oil	5.950	45	24
Microalgae[b]	136.90	2	1.1
Microalgae[c]	58.700	4.5	2.5

[a] To meet 50% of all transportation fuel needs of the United States.

[b] The 70% oil (by weight) in biomass.

[c] The 30% oil in biomass (by weight).

TABLE 12.11

The Fat Content of Some Microalgae Species (Chisti, 2007)

Microalgae	Fat Content (% Dry Mass)
Botryococcus braunii	25–75
Chlorella sp.	28–32
Crypthecodinium cohnii	20
Cylindrotheca sp.	16–37
Dunaliella primolecta	23
Isochrysis sp.	25–33
Monallanthus salina	20
Nannochloris sp.	20–35
Nannochloropsis sp.	31–68
Neochloris oleoabundans	35–54
Nitzschia sp.	45–47
Phaeodactylum tricornutum	20–30
Schizochytrium sp.	50–77
Tetraselmis sueica	15–23

Microalgae produce many different types of lipids, hydrocarbons, and other complex oils, depending on the species. Not all algae oils are sufficient or suitable for making biodiesel, but suitable oils are widely available. Using microalgae to produce biodiesel does not compromise the production of food, feed, and other products from plant crops, as does plant biomass. Potentially, oil-producing heterotrophic microorganisms grown on a natural organic carbon source such as sugar, rather than microalgae, could be used to make biodiesel. However, heterotrophic production is not as efficient as the use of photosynthetic microalgae. This is because renewable sources of organic carbon necessary for the growth of heterotrophic microorganisms are ultimately produced by photosynthesis, often in crop plants.

Lipids often accumulate intracellularly, so lipid recovery processes are complex, both on a laboratory scale and on a large scale. Lipid extraction with organic solvents generally facilitates cell lysis. The dewatering or drying processes of microbial biomass make cellular degradation methods more energy-intensive and expensive. Autoclaving, bead beating, high-speed homogenization, high-pressure homogenization, ultrasonication for extraction, microwave irradiation, and thermolysis are the methods applied to break the cell wall. It can be converted into renewable energy by many chemical methods, such as microbial conversion, liquefaction, gasification, incineration, pyrolysis, and transesterification. The major obstacle to biomass conversion is the complex cellular structure of cell wall components. Organic compounds bound to the cell wall consist of hemicellulose and cellulose, which resist biodegradability. Cell wall stiffness, based on cellulose and hemicellulose, ensures the biodegradability of the polymer.

Therefore, biomass pretreatment is necessary to facilitate cell wall digestibility and the production of bioproducts such as biofuels. Pretreatments such as mechanical, thermal, biological, and chemical are widely used in algal cell wall disruption (Ananthi et al., 2021).

The lipid content and fatty acid methyl ester profile can also be variable for the same algae species. Therefore, algal strain selection is one of the most important steps to reduce cost and time in large-scale cultivation for biodiesel production. Algal strain selection, cultivation method, culture conditions, and chemical composition of algae strongly affect production costs as well as engine performance and exhaust emissions. A comparison of the basic properties of biodiesel obtained from microalgae with ASTM standards (American standards) is given in Table 12.12 (Piloto-Rodríguez et al., 2017).

Saturated fatty acids in algal biomass provide more efficient cetane numbers, oxidative stability, combustion temperature, and viscosity in engine performance. A higher content of saturated fatty acids worsens cold flow but improves oxidative stability and combustion. Higher unsaturated fatty acid content, especially polyunsaturated fatty acids, improves cold flow but leads to poorer oxidative stability and combustion. The total ratio of saturated and unsaturated fatty acids in fatty acid methyl ester is the most important factor determining biodiesel quality. Therefore, examining the fatty acid profile of algae biodiesel is the first step in uncovering a commercial product. The effects of different algae species on engine performance are indicated in Table 12.13 (Murad and Al-Dawody, 2020).

Microalgae-based biodiesel production generally consists of the stages of extraction of oils from microalgae, removal of excess solvent, and production of biodiesel catalyzed by a homogeneous or heterogeneous catalyst (Hossain et al., 2008). The conventional transesterification method has many disadvantages that limit the commercial-scale production of biodiesel from microalgae. For example, this method is difficult to implement, has high energy consumption, and is expensive. However, many waste liquids are formed during the purification of biodiesel, and the disposal of these wastes poses a problem (Pragya et al., 2013). In addition to the conventional method, there are many methods developed for biodiesel production from microalgae in the literature (Patil et al., 2012; Cheng et al., 2014).

TABLE 12.12
Comparison of Fuel Properties of Microalgal Biodiesel (Piloto-Rodríguez et al., 2017)

Fuel Properties	Diesel Fuel (ASTM D7675)	Biodiesel (ASTM D6751)	Algal Biodiesel
Density (g/cm^3)	0.84–0.90	0.86–0.90	0.85–0.87
Kinematic viscosity (mm^2/s)	1.9–4.1	1.9–6.0	2.0–5.2
Heating value (MJ/kg)	40–45	39	37–41
Acid value (mg KOH/g)	0.7–1.0	0.5	0.37–0.42
Iodine value (g I/100 g)	120	25	19.0
Flash point (°C)	62	100	115
Melting point (°C)	−16	−11.6	−6
Cetane number	60	47	37–72

TABLE 12.13

Effects of Different Algae Types on Engine Performance (Murad and Al-Dawody, 2020)

Microalgal	Blend Ratio (%)	Engine Features	Results
Spirulina sp.	B100	Single cylinder, four stroke, 17.5 compression ratio, 3.7 kW	Decrease in brake thermal efficiency, torque, exhaust gas temperature, carbon dioxide nitrogen oxide, increase in SFC when using B100
Chlorella	B0, B100, raw alg oil	Kirloskar diesel engine, single cylinder, water cooled, capacity 5.2 kW	Reducing emissions of CO, Unburned Hydrocarbons, NO_x and smoke opacity
Scenedesmus obliquus	B10, B20	Kirloskar, single-cylinder four-stroke, water-cooled, diesel engine	Lower BSFC, EGT for B20, higher brake thermal efficiency, lower emissions
Chlorella vulgaris	B10, B20	Kirloskar TV-I DI, four stroke, singlef cylinder, capacity 3.6 kW, water cooled	Reduction in NO_x, CO and hydrocarbon emissions for B20
Schizochytrium sp.	B20, B40 ve B60	Kirloskar, AV1, single cylinder, vertical, four stroke, capacity 3.7 kW	Poor atomization and rising exhaust gas temperatures gave lower thermal efficiency, lower HC, higher BSFC, NO_x and CO, decreased ignition delay
Ankistrodesmus braunii ve *Nannochloropsis. brauni*	B0, B50, crude algae oil	Ricardo E6 single-cylinder indirect injection diesel engine	Reducing the output torque of the engine and increasing the noise caused by combustion, decreasing the compression ratio of the engine caused the combustion noise to decrease

12.6.1.1 Biodiesel Production Process from Microalgae

The growth cycle of algae, unlike vegetative biomass, is very short (1–10 days), thus allowing several harvests per month. The harvesting technology applied depends on the type of algae grown; It includes two main processes, namely collection and algae biomass concentration. Separation of water content is relevant to this biomass harvesting for the subsequent biodiesel production process. The process includes two specific steps:

1. Separation and densification of microalgae from bulk suspension by microseparators, electrophoresis-precipitation, flotation, and flocculation
2. Dewatering, filtration, or centrifugation of microalgae slurry

The first step in algal harvesting is sieving or filtering. Microseparators and vibrating screens are the standard screening devices used. High harvest rates and a removal efficiency of 95% can be achieved using vibrating screen filtration. Various centrifuge systems, such as solid spray discs, solid bowl decanters, and hydrocyclones, are used to improve the concentration. Centrifugation represents the fastest method, with up to 95% biomass recovery under optimal conditions, and is therefore preferred over other methods. For large algae systems (closed photobioreactor), 73 kg of water is required to process 1 kg of algae biomass, which requires high energy consumption.

Therefore, applying flocculation reduces energy consumption. In flocculation, firstly, there are unstable impurities in the colloid and aggregate formation during neutralization; secondly, the adsorption of organic matter by the aggregate through the bridging process; and finally, the removal of particulate aggregates by precipitation, filtration, and scavenging.

In the flotation process, air bubbles adhere to the algae mass and lift them to the surface of the liquid, where they are easier to collect. After the algae suspension is concentrated on 5%–15% dry matter, further slurry dehydration is required to proceed to the next step of biodiesel production. A significant amount of energy (11.22 MJ/kg) is required for drying algal biomass. Therefore, drying is a major economic concern because the energy required to dry algae can account for the largest share of processing costs. The extraction of moisture content from algae biomass is necessary because moisture inhibits further processing steps such as lipid extraction and transesterification. Many different dehydration techniques can be used to dry algae suspensions, such as sun drying, freeze drying, flash drying, rotary drying, spray drying, toroidal drying, and fluid bed drying. All special technologies for thickening and dewatering algal slurry have different characters. The microalgae type, size, morphology, and composition of the medium used for cultivation affect the effectiveness of these technologies. It is important to analyze all available technologies before choosing which one to implement (Bošnjaković and Sinaga, 2020).

Dehydration of the biomass is followed by lipid extraction from the algae biomass. The process is much more demanding than oil extraction from terrestrial cultures. The extraction of lipids from algae mass is usually done by a mechanical process (expeller press) or by chemicals and solvents. Applying an extruder is a simple and efficient process of mechanically crushing algae cells to obtain oil. The applied pressure should not be too high, as it may cause a decrease in lipid quality. Other mechanical techniques used include bead beating. The application of chemical solvents for the extraction of algal lipids, such as n-hexane, methanol, and ethanol, represents the most suitable method as it is valid and widely used in laboratory research (the Folch method is the most popular). However, the effectiveness of this method largely depends on microalgae strains. The two-step process can reduce saponification and increase esterification efficiency (with the homogeneous catalyst method, the efficiency of the conversion of lipid to biodiesel can be increased to 90%). Other technologies aimed at improving algal lipid extraction, such as the autoclaving supercritical technique and the osmotic technique, are still under development. Oil extraction technology from wet algae biomass technique is also applied to eliminate energy consumption for replaceable solvent extraction, vacuum shelf drying, electroporation techniques, and osmotic dehydration of algae biomass from slurry (Lundquist et al., 2010).

The third step in biodiesel production after oil extraction is the transesterification process. For more than 50 years, the triglyceride (TG) transesterification process has been applied in biodiesel synthesis. During this process, TG reacts with alcohol in the presence of a catalyst to form fatty acid esters. During biodiesel synthesis, glycerol and fatty acid esters (TG) are used to synthesize shorter-chain alcohol esters, and the resulting by-product is glycerol. Methanol and ethanol are most often used as the acyl receptors during biodiesel synthesis.

The transesterification reaction consists of several steps;

- **Step 1**: the reaction of TG with alcohol produces an ester molecule and a diglyceride molecule from a TG molecule.
- **Step 2**: another ester molecule and monoglyceride are formed from one diglyceride molecule.
- **Step 3**: MG reacts with a third molecule of ester and an alcohol molecule to make glycerol. Transesterification of a TG molecule requires three alcohol molecules, while the reaction results in the formation of three shorter-chain compounds (fatty acid esters (biodiesel) and glycerol) per TG molecule.

In addition to the main process of ester synthesis, other reactions may occur during biodiesel production, depending on the conditions. Moisture in the reaction medium can cause oil hydrolysis. The resulting free fatty acids can react with an alkali catalyst to produce soap.

Considering that the chemical reactions of biodiesel production are quite slow, catalysts are used to accelerate them. Depending on the type of catalyst used in the process, biodiesel production methods can be chemical or biotechnological. According to the aggregation state of the reactants, the process is classified as homogeneous or heterogeneous catalysis. In the chemical method, alkali and acid catalysts are used. The biotechnological method differs from the chemical one in that instead of chemical reagents, enzymes are used to accelerate the reaction. Developments in biodiesel production technologies have also led to the emergence of a biodiesel production method called supercritical conditions, that is; when no catalyst is used during the reaction. When the transesterification process is complete, the produced glycerol and remaining alcohol are separated, while the produced biodiesel is purified from glycerol, soaps, glycerides, and residual glycerol (Makareviciene and Skorupskaite, 2019).

As an alternative to the traditional transesterification process, *in-situ* transesterification provides a positive advantage in industrial production. It is to eliminate the need for certain steps (i.e., lipid extraction) from the algae harvested in the *in-situ* transesterification process to make biodiesel. Therefore, *in-situ* transesterification helps reduce equipment installation and maintenance costs and energy consumption. However, *in-situ* transesterification requires dried microalgae to improve lipid extraction and achieve a high conversion rate to biodiesel. Although the percentage of energy consumption for dried BP varies slightly from study to study, it is generally accepted that the microalgae drying step (up to <5% moisture) consumes more than 70% of the total cost of biodiesel production. Wet *in-situ* transesterification can further eliminate the drying step of microalgae, significantly reducing capital and operating costs.

The wet *in-situ* transesterification process uses partially dehydrated wet microalgae containing 60%–80% moisture content. Wet algae are often undesirable for the biodiesel production process in terms of high energy consumption for heating and additional chemical input to compensate for the water barrier to catalysts. The wet *in-situ* transesterification process generally requires the use of an acid catalyst, as water causes soap formation when an alkaline catalyst is used. Another parameter that may be important for the wet *in-situ* transesterification process is the requirement to break down the cell walls of untreated wet algae directly from harvest to extract lipid oil from the interior of the microalgae cells. The properties of the cells and their walls vary according to the microalgae species. Therefore, physical solutions such as microwave, sonication, and mixing, used with a combination of additional chemical inputs, such as solvents or ionic liquids, are more commonly preferred for use in wet *in-situ* transesterification, especially in low-temperature regions. For wet *in-situ* transesterification carried out above 200°C, the researchers took advantage of the water content in microalgae. Most studies have focused on the supercritical state of water or alcohol, both as a reagent and as a solvent. A large amount of energy input is required to heat the reactants to the desired temperature. This is limiting in terms of the economic feasibility of wet *in-situ* transesterification. Wet *in-situ* transesterification conducted in a mid-temperature region tends to take advantage of both low- and high-temperature regions. Increasing the reaction temperature to the range of 100°C and 200°C can also still achieve a high conversion rate to biodiesel without the need for additional chemicals compared to wet *in-situ* transesterification in a temperature region below 100°C (Kim et al., 2019).

The EU has strengthened the EU sustainability criteria for bioenergy by expanding its scope to cover all fuels produced from biomass, regardless of their final energy use. The Directive has been in effect since December 24, 2018 and should be enacted in EU countries by June 30, 2021 (European Parliament, 2018). In the USA, the Environmental Protection Agency (EPA) has published Renewable Fuel Standards, and in this context, fuel standards until 2030 are (1) 0.34% for cellulosic biofuel, (2) 2.1% for biomass-based diesel, (3) 2% for advanced biofuels (algal biofuels), and 93, (4) 11.56% for conventional renewable fuels (Environmental Protection Agency, 2020).

12.6.2 Bio-Gase

Methane fermentation applications using algae have attracted significant attention due to the production of beneficial byproducts such as biogas. Biogas consists of organic matter decomposed by anaerobic microorganisms and is predominantly composed of methane (55%–75%) and carbon dioxide (25%–45%) (Harun and Danquah, 2011).

The methane produced can be used as a fuel gas or for electricity generation. The residual biomass as a result of anaerobic decomposition can be used as fertilizer. Thus, sustainability can be achieved and production costs can be reduced. Microalgae do not have lignin in their structure, and their cellulose content is low. For this reason, the decomposition of organic materials for methane production is easier, and process stability is ensured (Vergara-Fernández et al., 2008). Biomethane yields produced from different algae species are given in Table 12.14.

TABLE 12.14
Biomethane Yields from Different Algal Species (Elcik and Çakmakcı, 2017)

Biomass	Methane Yield (m³/kg)	References
Laminaria sp.	0.26–0.28	Chynoweth (2005)
Gracilaria sp.	0.28–0.4	Bird et al. (1990)
Macrocystis	0.39–0.41	Chynoweth (2005)
L. Digitata	0.5	Morand and Briand (1999)
Ulva sp.	0.2	Briand and Morand (1997)

Organic loading, temperature, pH, and holding times are the main parameters affecting methane yield in biogas production with anaerobic decomposition processes. In general, high residence times and organic loading rates result in high methane yields. In addition, anaerobic degradation can be carried out under mesophilic (35°C) and thermophilic (55°C) conditions (Harun et al., 2010). (Otsuka and Yoshino, 2004) reported that they obtained 180 mL/g (volatile solid) methane production as a result of anaerobic decomposition of the *algae Ulva* sp.

After the biodiesel production process from microalgae, algal biomass comes out as waste. These wastes can be converted into biogas by fermentation and provide electrical energy input. Algal biomass consists of a mixture of organic and inorganic matter. The organic part consists of complex polymeric macromolecules such as proteins, polysaccharides, lipids, and nucleic acids. Polymers appear in particulate or colloidal form. The anaerobic digestion process (ADP) converts organic matter into end products (methane and carbon dioxide), new biomass, and inorganic residue. Several groups of microorganisms are involved in substrate transformation, and the overall process consists of multiple steps with many intermediates. In general, the process can be simplified into four successive steps: (1) hydrolysis, (2) fermentation or acidogenesis, (3) acetogenesis, (4) methanogenesis, and (5) electrical energy.

The steps can be explained as follows (Bohutskyi and Bouwer, 2013):

1. Hydrolysis of colloid and particulate biopolymers to monomers.
2. Fermentation or acidogenesis of amino acids and sugars to intermediates (propionate, butyrate, lactate, ethanol, etc.), acetate, hydrogen, and formate.
3. Oxidation of long-chain fatty acids and fermentation of alcohol to volatile fatty acids, and hydrogenation. Anaerobic oxidation or acetogenesis of intermediates such as volatile fatty acids to acetate, carbon dioxide, and hydrogen. This reaction is carried out by obligate and facultative hydrogen-producing species.
4. Conversion of acetate to methane by acetoclastic methanogens. The conversion of molecular hydrogen and carbon dioxide to methane by hydrogenophilic methanogens.
5. Conversion of methane into electrical energy with a gas türbine

The same group of microorganisms that are primary fermenters perform the first three steps. These biological processes are sometimes referred to as acidogenesis or the acid phase.

Other important biological processes in ADP include:

- Conversion of various monocarbon compounds (e.g., formate, methanol) to acetic acid. This reaction is carried out by homoacetogenic bacteria.
- Reduction of sulfur compounds to hydrogen sulfide by sulfur-reducing bacteria.

Maintaining the environmental and operational parameters applicable to micro-algae is one of the key factors for effective methane production. The main environmental factors are temperature, pH, alkalinity, and redox potential. Operational parameters such as C:N:P ratio, presence of essential micronutrients, organic loading rate, hydraulics , solids retention time, and incoming salts and toxicant concentrations are subject to strict control and regulation. Accumulation of certain intermediates or byproducts, such as volatile fatty acids, ammonia and hydrogen sulfide can lead to inhibition of methane production (Chen et al., 2008).

Biogas is formed during Anaerobic Digestion (AD) and has two main components: methane (about 55%–70% by volume) and carbon dioxide (30%–40%). Depending on the source of the biogas, other minor components include nitrogen (<2%), hydrogen, oxygen (<1%), hydrogen sulfide (0–50 ppm), and other sulfide compounds and volatile organic compounds (VOC).

Large amounts of harmful VOCs can be produced during the digestion of household waste. Carbon dioxide is not a harmful inert gas, but the presence of carbon dioxide in biogas reduces its calorific value. Removing carbon dioxide is an expensive process, and power generation equipment often operates with carbon dioxide concentrations of up to 40%–50%. The most abundant sulfur compound in biogas is hydrogen sulfide, but other reduced sulfur chemicals (e.g., sulfites and thiols) are also present. The main source of sulfur in biogas is the degradation of sulfur-containing amino acids (cysteine and methionine). Hydrogen sulfide in concentrations higher than 300–500 ppm can form unhealthy and dangerous sulfur dioxide (SO_2) and sulfuric acid (H_2SO_4), which corrode pipeline metal parts, storage tanks, compressors, and engines. Before biogas application, desulfurization plants need to be installed. Another corrosive pollutant is ammonia (NH_3). The combustion of biogas with a high concentration of ammonia increases the emission of nitrogen oxides into the atmosphere. Both hydrogen sulfide and ammonia are pollutants that pose health risks. Other compounds of concern in biogas are siloxanes. They are organic polymers of silicone from a wide variety of industrial, personal care, pharmaceutical, and other products. These organic compounds can be oxidized to silicon dioxide and accumulate in valves, gas turbines, and engines, causing erosion and reducing operating efficiency (Buswell and Mueller, 1952).

It has the lowest oxidation state and the largest theoretical methane yield, more than twice the methane yield from proteins, glycerol, and carbohydrates. The theoretical methane yield is related to the average carbon oxidation state of the substrate. Macroalgae with high carbohydrate content and cyanobacteria with high protein content are theoretically weaker feedstocks for methane production, while microalgae with high lipid content have a higher potential methane yield.

12.6.3 BIOHYDROGEN

Hydrogen is a clean energy carrier with great potential that can be used as an alternative fuel. There are various technologies in hydrogen production. Biological hydrogen production processes are more environmentally friendly and less energy-intensive when compared to conventional processes. Hydrogen can be produced biologically by biophotolysis, photofermentation, dark fermentation, or a combination of these processes. The most important obstacle to the commercialization of these processes is the low hydrogen production efficiency and speed. The acceleration of commercialization will be provided by the discovery of genetically modified microorganisms, processes with improved operating conditions, and combined processes. Green algae and cyanobacteria break down water molecules into H+ ions and O_2 by direct and indirect biophotolysis pathways (Oncel, 2013). In direct photolysis, solar energy directly splits water into hydrogen ions and oxygen through photosynthetic reactions. The H^+ ions produced are converted into H_2 gas by the hydrogenase enzyme (Bahadar and Khan, 2013). Since the iron-hydrogenase activity used in the biophotolysis process is sensitive to oxygen, special conditions are required. Since there is no hydrogenase enzyme in green plants, only CO_2 reduction occurs. Cyanobacteria have hydrogenase enzymes and can produce hydrogen when suitable conditions are provided. Chlamydomonas reinhardtii, one of the green algae, can produce hydrogen under anaerobic conditions or use it as an electron donor. The hydrogen ions produced are converted into hydrogen gas by the hydrogenase enzyme.

In the indirect photolysis mechanism, fermentable organic nutrients produced by microalgae are used by photosynthetic bacteria to produce H_2 (Lakaniemi, 2012).

Indirect biophotolysis consists of the following steps (Genç, 2009):

a. Biomass formation in the photosystem
b. Production of two moles of acetate and four moles of hydrogen per one mole of glucose in cells by aerobic dark fermentation
c. Conversion of two moles of acetate to hydrogen

Zhang and Melis (2002) reported that Chlamydomonas reinhardtii species produced 10%–20% H_2 in the presence of 2–7 mM methylamine hydrochloride (Bahadar and Khan, 2013). Also, for the same species, Kruse et al. (2005) reported a maximum H_2 production rate of 4 mL/h and 540 mL H_2/L culture; Torzillo et al. (2009) reported 5.8 mL/h and 504 mL H_2/L culture (Zhang and Melis, 2002; Kruse et al., 2005).

Kawaguchi et al.'s (2001) H_2 production was performed using the biomass of *Chlamydomonas reinhardtii* and *Dunaliella tertiolecta*, respectively, in a mixed culture of *Rhodobium marinum* and *Lactobacillus amylovorus* species (Lakaniemi, 2012). In their study, they reached 0.85 (*C. reinhardtii*) and 1.55 (*D. tertiolecta*) mmol H_2/h L culture H_2 production rate (Kawaguchi et al., 2001; Lakaniemi, 2012).

12.6.4 BIOETHANOL

Bioethanol is a fuel produced by the fermentation of starch or cellulose-based products from woody plants such as wheat, corn, and sugar beet, used by blending with gasoline with certain mixing ratios. Sugarcane, sugar beet, corn, potato, sweet sorghum, agricultural wastes, wheat, woody, and algae can be used as raw materials in the production of bioethanol.

Ethanol has become an important alternative fuel or fuel additive in recent years. Bioethanol fuel reduces lead, sulfur, carbon monoxide, and particulate emissions. Ethanol, which is used as a fuel in Brazil, has contributed to the reduction of carbon emissions by reducing the total emissions by 15%. Bioethanol production can be realized by using the rich carbohydrate content in the structure of microalgae in the fermentation of bacteria as a food source (Bahadar and Khan, 2013).

Ethanol production takes place in three steps.

- First, it is hydrolyzed by enzymatic condensation to produce large carbohydrates and cell wall starch.
- Next, yeasts such as Saccharomyces cerevisiae are added to convert the sugars to ethanol.
- Finally, the produced ethanol is purified by distillation.

C. vulgaris and *chlorococcum* sp. are commonly used in bioethanol production due to their high starch content (Bahadar and Khan, 2013). Wu et al. (2014), in their study of *Gracilaria* sp. used the sequential acid and enzyme hydrolysis method for confectionery in the bioethanol production study that they carried out using the type. In this method, they developed, and they converted 1 g of dry Gracilaria species into 0.236 g of ethanol (Wu et al., 2014).

Harun and Danquah (2011) investigated the effects of time, temperature, microalgae loading, and acid concentration variables on acid pretreatment applied in bioethanol production from algae. In their study, the highest bioethanol concentration (7.20 g/L) was obtained at a microalgae concentration of 15 g/L at 140°C in a reaction time of 30 min using 1% sulfuric acid (Harun and Danquah, 2011). Borines et al. (2013) used *Sargassum* spp. for bioethanol production and obtained 89% ethanol conversion as a result of fermentation of Saccharomyces cerevisiae species at 40°C at pH 4.5 for 48 h. Ye Lee et al. (2013) obtained 6.65 g/L bioethanol from *Saccharina japonica* by pretreatment with low concentration (0.06%) sulfuric acid. In their study, they stated that the glucan content and enzymatic digestibility of *Saccharina japonica* increased by removing non-cellulosic components from the algal biomass after low-intensity acid pretreatment (Ye Lee et al., 2013). Bioethanol fermentation from microalgae requires less energy than biodiesel production. In addition, the unwanted CO_2 by-product can be used in microalgae cultivation. However, commercial production of bioethanol from microalgae is still under investigation (Bahadar and Khan, 2013).

12.6.5 BIOFERTILIZER

Cyanobacteria, which are in the microalgae group, use light to fix CO_2 and N_2 and thus contribute to plant development. These creatures produce hormones that promote plant growth, control soil-borne pathogens, and improve soil quality. Cyanobacteria are found in far eastern countries such as India and China, especially in production conditions where plant nutrient loss and exploitation are high in rice agriculture. It has been widely used for years as a biofertilizer because it fixes nitrogen, facilitates soil stabilization, adds organic matter to the soil, secretes plant growth-promoting substances, and improves the physical-chemical properties of the soil. Biofertilizers used for agricultural purposes include blue-green algae, azolla, mycorrhiza, and various bacteria that promote plant growth (Hegde et al., 1999). More than 70 species of cyanobacteria (Cyanophyta) are known to nitrogen-fixing cyanobacteria and are divided into two main groups: unicellular forms (aerobic-anaerobic) and filamentous forms (with heterocysts-non-heterocysts) (Bergman et al., 1997; Madigan et al., 1997; Burlage et al., 1998). Cyanobacteria are the first colony-forming terrestrial ecosystem and the oldest oxygenic phototrophic microorganisms in the world. They play a leading role in the disintegration of rocks and the formation of a biological soil crust by producing carbon-nitrogen, producing oxygen, and stabilizing aggregates and soil water levels in desert environments. Many of these functions are possible with excessive production of exopolysaccharides (Chamizo et al., 2019). Cyanobacterial cells are usually surrounded by a polysaccharide sheath or capsule, which can retain water and confer a great tolerance to desiccation (Whitton, 1992). Beneficial effects of cyanobacterial inoculation have also been reported on several other crops grown in different field conditions, such as rice, barley, oats, tomatoes, radishes, cotton, sugarcane, maize, peppers, and lettuce (Ananya and Ahmad, 2014). The effect of cyanobacteria on meeting the nitrogen needs of plants occurs in different amounts depending on the plant variety, application method, cyanobacterial species, and environmental conditions. As seen in Table 12.15, while the rate of meeting the nitrogen requirement in rice plants is between 30% and 50%, this rate has reached 100% with soil and foliar application in sunflowers.

The most abundant microbial components of microbiotic shells are filamentous cyanobacteria, which exert a mechanical effect on soil particles as they form a sticky

---

TABLE 12.15

The Rate of Meeting the Nitrogen Requirement of Different Cultivated Plants by Cyanobacteria Application

Herb	Cyanobacteria Species	The Rate of Meeting the Total Nitrogen Requirement of the Plant from the Nitrogen Produced by the Cyanobacteria (%)	Literature
Sweetcorn	*Nostoc sp.*	17–40	Maqubela et al. (2009)
Wheat	*Anabaena*	23–28	Prasanna et al. (2013)
Paddy	*Nostoc commune*	50	Pereira et al. (2009)
Paddy	*Azolla-Anabaena*	50	Mian (2002)
Paddy	*Azolla*	30–50	Choudhury and Kennedy (2004)
Sunflower	*Oscillatoria annae*	100[a]	Bhuvaneshwari et al. (2011)

[a] Application as soil cyano pit and foliar cyano spray.

network and bind soil particles on the surface of polysaccharide sheath materials (Belnap and Gardner, 1993; Malam Issa et al., 2007).

The effects of blue-green algae on soil and plant growth are summarized below.

1. As they have a filamentous structure and produce adhesives, they improve soil structure and increase porosity. Thanks to their jelly structure, they increase the water-holding capacity.

   They produce substances that promote growth, such as hormones (auxin, gibberellin), vitamins, and amino acids (Taher et al., 2011).

2. After their death and decay, they increase the organic matter in the soil (Balasubramanian et al., 2013).

3. They reduce the effect of soil salinity on plants (Vargas et al., 1998).

   Kaushik and Krishna Murti (1981) determined that the application of cyanobacteria to the physical and chemical properties of alkaline soils resulted in a significant improvement in the aggregation state of the soil, significantly reduced pH, electrical conductivity, and exchangeable sodium, and also significantly increased the hydraulic conductivity of the soil. Similar positive effects are expected in acid soils with the studies to be carried out in this study.

4. They prevent the growth of weeds (Abo-Hashesh and Hallenbeck, 2012).

5. They increase the availability of soil phosphorus by the plant by producing organic acid (Roger and Reynaud, 1982; Rodríguez et al., 2006).

6. Cyanobacterial fertilizers are a promising alternative to prevent soil pollution caused by agrochemicals (Ibraheem, 2007).

7. In addition, some researchers have stated that soil-insoluble phosphorus is converted into a soluble form by these bacteria. (Roger and Kulasooriya, 1980).

It has been determined that pH, which is one of the soil properties affecting the growth of cyanobacteria, generally prefers neutral to slightly alkaline pH for optimum growth. pH is a very important factor in the formation and diversity of cyanobacteria (Singh, 1961). Acid soils are one of the most stressful environments for cyanobacteria, and these microorganisms are not normally found at pH values below 4 or 5. It is known that soil pH has a selective effect on the native algae flora, especially cyanobacteria, and their amount in the soil (Nayak and Prasana, 2007). The diversity and abundance of cyanobacteria living in fresh water and soil are seen at alkaline pH values. However, it has been determined that some planktonic picocyanobacteria are also found in environments where the pH drops to 4.5 and a few filamentous forms with true branching decrease to ~4.0 (Whitton and Potts, 2012).

Cyanobacteria can also secrete plant growth hormones as secondary metabolites, promote the transport of nutrients from the soil to plants, and improve the chemical properties of the soil. Their different morphologies and physiological characteristics allow for wide distribution in the ecosystem and tolerance to environmental stresses. The compounds produced by cyanobacteria and their benefits in terms of the pellet-plant system are shown in Figure 12.7 (Singh et al., 2018).

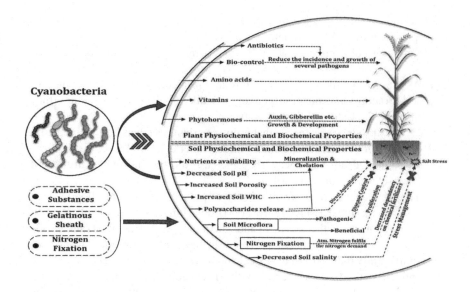

FIGURE 12.7   An overview of the benefits of cyanobacteria on plants and soil (Singh et al., 2018).

It has been reported that cyanobacteria can improve the nutrient content of the soil with their C and N fixative properties, especially in arid environments (Mayland and Mcintosh, 1966; Jeffries et al., 1992; Lange et al., 1994). The researchers applied cyanobacteria to organic material-poor soil in a pot experiment and observed that the microbial activity in the soil increased according to the results of the experiment. In addition, they stated that the application of cyanobacteria improves the physical properties of the soil, its nutritional status, and the yield of plants grown in this soil (Nisha et al., 2007).

Being photosynthetic organisms, they regulate biological activity by increasing the organic matter in the soil. Thanks to their ability to bind N_2, cyanobacteria can survive harsh environmental conditions (Paerl et al., 1995; Bergman et al., 1997, Irisarri et al., 2007). In addition to carbon compounds, a significant amount of organic nitrogen is also present in the formed organic matter (5%). This nitrogen is a potential source of nitrogen that could be useful for next season's crop production (Rogers and Burns, 1994; Zaady et al., 1998). Cyanobacteria are used as biofertilizers in modern agriculture due to their ability to add atmospheric nitrogen to the soil (Mandal et al., 1999; Ladha and Reddy, 2003). However, it has been determined that the application of nitrogen fertilizer at 60–90 kg N/ha and higher levels to the fields where cyanobacteria will be applied, especially in rice agriculture, prevents the growth of Azolla and cyanobacteria (Singh et al., 1988). It has been known for a long time (El-Zeky et al., 2005). In a field experiment, it was observed that soil biological activities such as the number of cyanobacteria and bacteria, CO_2 output, dehydrogenase, and nitrogenase enzymes increased when blue-green algae were applied. According to the results, it has been shown that chemical fertilizers can be reduced by applying ¼ and ½ of the recommended dose of cyanobacteria (Hegazi et al., 2010). Cyanobacteria that fix

TABLE 12.16

Classification of Nitrogen-Fixing Cyanobacteria According to Their Morphological Structures (Kalyanasundaram et al., 2020)

Form	Cyanobacteria Sample
Filamentous	Anabaena, Aulosira, Calothrix, Cylindrospermum, Fischerella
Heterocystous	Hapalosiphon, Mastigocladus, Nostoc, Scytonema, Stigonema
Form	Tolypothrix, Westiella and Westiellopsis
Unicellular	*Aphanothece, Chrococcidiopsis, Cyanothece, Dermocarpa*
Colonical form	*Gloeothece, Myxosarcina, Synechococcus, Synechocystis and Xenococcus*
Non-Heterocystous	*Oscillatoria, Trichodesmium, Microcoleus, Pseudanabaena*
Filamentous form	*Plectonema and Lyngbya*

nitrogen can have different morphological structures. Their nitrogen fixation forms, according to their morphological structures, are given in Table 12.16.

As the use of cyanobacteria as biofertilizer, especially in rice agriculture, increases, the need for chemical fertilizers in modern agriculture will decrease. This will contribute to the reduction of greenhouse gas emissions by reducing the use of fossil fuels, especially natural gas, which is used in the production of chemical fertilizers. Algae also help reduce greenhouse gases by fixing nitrogen from the atmosphere and using carbon dioxide in respiration. Phytohormone production of cyanobacteria was measured and applied to wheat in axenic and field conditions. Some of the cyanobacteria produced cytokinin and indole acetic acid hormones, and this application caused an increase in germination, shoot length, tillering, number of lateral roots, spike length, and grain weight in wheat (Hussain and Hasnain, 2011).

12.6.5.1 Biofertilizer Effect of Other Microalgae

Algae biomass contains a high percentage of macronutrients, significant amounts of micronutrients, and amino acids (El-Fouly et al., 1992; Mahmood, 2016). As a new biofertilizer, algal biomass contains macronutrients as well as micronutrients, some growth regulators, polyamines, natural enzymes, carbohydrates, proteins, and vitamins applied to increase plant growth and yield (Shaaban, 2001a,b; Abd El-Moniem and Abd-Allah, 2008). In addition, algae biomass applied to the soil improves soil properties, which has a positive effect on the nutritional status of plants (Al-Gosaibi, 1994) (Table 12.17).

In addition to the direct application of microalgae to plants or soil, applications can be made to increase the fertility of the soil by applying them to compost and organic fertilizers. In one of them, compost prepared from fruit and tree residues, *Chlorella* sp. and *Scenedesmus* sp. inoculum, was inoculated as a mixed culture, and its effect on cauliflower growth was determined. It has been shown that the application of a compost and microalgae mixture provides the nitrogen requirement and product yield of the cauliflower plant in a way that is equivalent to chemical fertilizer applications (Díaz-Pérez et al., 2023). Such microalgae and organic fertilizer applications are considered valuable in terms of reducing the negative environmental effects of chemical fertilizers, reducing carbon emissions, and ensuring the sustainability of

TABLE 12.17

Functions of Cyanobacteria and Algae in Different Plants (Kalyanasundaram et al., 2020)

Algae/Cyanobacteria	Plants	Function (Contribution)
Cyanobacteria	Rice	Biological nitrogen fixation and increased nitrogen efficiency
Spirulina platensis, Ascophyllum nodosum and Baltic green macroalgae	Wheat	Increase of 100 grain weight and dry mass of shoot
Jania rubens, Laurencia obtuse, Corallina elongate	Maize	Improvement in root and polyphenolic and anti-oxidant contents
Ulva lactuca, Caulerpa sertularioides, Padina gymnospora, and Sargassum liebmannii	Tomato	Biostimulants-germination and growth
Spirulina platensis, Amphora cofeaeformis	Cucumber	Significant increase in yield and fruit quality Nematode control
Chlorella vulgaris, Scenedesmus quadricauda	Sugar beet	Increased root length, upregulation of genes involved in nutrient acquisition
Ascophyllum nodosum	Wine grape	Increased anthocyanin content
Chlorella vulgaris	Banana	Improved yield, bunch and hand weight

agricultural production. Although the propagation of microalgae and mixing it into compost and organic fertilizers are currently considered expensive applications, it is expected that research studies will continue to grow microalgae in wastewater and sell them at cheap prices.

12.6.6 ANIMAL FEED STOCK

The protein content of algae varies from species to species. For example, in *Spirulina*, this ratio was 70% in dry matter, while it was 30%–40% in red algae, 20% in green algae, and 10%–11% in brown algae (Givens, 1997). The oil content of algae is low, varying between 1% and 5%. However, the essential fatty acids it contains are much higher than those of other land plants. *Spirulina, chlorella*, and *scenesdesmus* are the main microalgae grown for animal feed. Microalgae are used to a great extent in aquaculture (Parmar et al., 2011). The level of unsaturated fatty acids and protein content determines the nutritional value of microalgal aquaculture feeds. Algae are also a storehouse of minerals and vitamins. The amount of alpha and beta-carotene in brown and red algae is 2–7 mg/100 g DM. The nutrient composition of dried *Spirulina platensis* used in animal nutrition research (Varga et al., 2002) is given in Table 12.18.

It has been stated that the most striking feature of Spirulina compared to synthetic pigments is that it is a natural colorant. Although synthetic pigments are carcinogenic, Spirulina is a natural colorant. Due to this feature, it has been reported that it is used because it provides natural coloration in chicken eggs and fish feeds in aquaculture (Challem, 1981). Microalgae can also be considered a food source for cats, dogs, cows, horses, and poultry (Spolaore et al., 2006), which was determined

TABLE 12.18
Nutrient Composition of Spirulina Platensis
(Varga et al., 2002)

Nutrient Composition	% Dry Matter
Dry matter	93–97
Crude protein	55–60
Oil	6–8
Carbohydrate	12–20
Ash	6–8
Cellulose	8–10
Chlorophyll	1–1.5

to increase the rate (Becker and Ventkateraman, 1981). Two products are obtained during the cultivation and production of microalgae. One is biomass, and the other is oil extract obtained from biomass. The biomass is a whole consisting of long-chain ω-3 fatty acids naturally trapped inside the algae cells and the algal cell and contains natural antioxidants such as beta-carotene and vitamins A, E, and C produced by algae (Grinstead et al., 2000). Biomass is suitable to be added to animal feeds. Today, cell metabolites or dried biomass obtained from many other species, especially *Chlorella, Dunaliella, Haematococcus,* and *Spirulina,* are commercially evaluated. These metabolites find a market in pharmaceutical and nutraceutical health products and cosmetics. In addition, they are used in the nutrition of terrestrial and aquatic creatures due to the proteins, vitamins, minerals, and pigments they contain (Duru and Yılmaz, 2013). Microalgae in WWT can be used for purposes such as coliform inhibition, COD and BOD (biochemical oxygen demand) removal, N and/or P removal, as well as heavy metal removal. However, while microalgae treat wastewater, they also produce important pigments within their bodies, the main ones being chlorophyll a, b, and c, β-carotene, Astaxanthin, Phycocyanin, Xanthophyll, and Phycoerythrosine. These pigments are frequently used in the foods, pharmaceuticals, textiles, and cosmetics industries. It can be recovered as biofuel, biofertilizer, and animal feed. It has a great contribution to "zero carbon" by capturing CO_2. Studies on using microalgae as a raw material source for biodiesel production are increasing, especially in European countries. Determining microalgae species with high oil content and growth rate and using them in purification will be a completely innovative environmental technology. Microalgae and cyanobacteria form phytoplankton. Because they are used as nutrients for fish and other aquatic organisms, they form the starting point of the food chain in nature. In many sources on microalgal biotechnology, cyanobacteria are evaluated together with microalgae (Borowitzka, 1988). Among the blue-green algae, rapidly growing cyanobacterial genera such as *Spirulina, Oscillatoria,* and *Anabaena* are frequently used in food and nutrition studies because they produce pigments, fatty acids, carbohydrates, proteins, and many other nutritional compounds (Yıldız, 2001). *Spirulina platensis* (Gom.) Geitler (=Arthrospira) is a filamentous, spiral-shaped prokaryotic organism from the class

Cyanophyceae (Blue-green algae), which is the most cultivated and widely used in cosmetics, medicine, and various industrial fields as human and animal food (Koru and Cirik 2002). The aquaculture sector is growing by 10% every year to balance the diminishing natural resources. Adding *Spirulina* to fish feeds helps solve the two biggest problems for breeders. The first is to increase the resistance against infections and diseases in farmed fish, and the second is that it contributes to enhancing flavor, texture, and skin color. As *Spirulina* increases the flavor of the feed, therefore, the growth rates of fish larvae increase and there is less feed loss. Fish is flavorful and contains less belly fat. It grows faster, tastes better, and is more resistant to diseases. The use of *Spirulina* improves the cost/performance ratio of fish feeds (Vonshak and Richmond, 1988). Red-colored phycobiliproteins, phycoerythrin, blue phycobiliproteins too, it is called phycocyanin. *Porphyridium* sp. is known as a source of fluorescent pink pigment, and important phycobiliproteins are identified as b-phycoerythrin and phycoerythrin. Lutein is a natural colorant of the carotenoid group. This xanthophyll is found in the retina. Lutein is found in green leafy vegetables (spinach, kale, etc.), egg yolks, and some flowers (marigolds). Lutein, as a color pigment, gives foods a bright yellow color. It is used in the feed industry. Lutein is added to chicken feeds to increase yolk yellowness.

12.7   MICROALGAE AND WASTEWATER TREATMENT

12.7.1   Microalgae Species Used in Wastewater Treatment

Microalgae known for their wide diversity in gene content, morphology, and physiology, are categorized under a variety of phylogenetic lineages. Microalgae are also photoautotrophs that make photosynthesis in membrane-bound organelles (chloroplasts). Microalgae are distributed in four main eukaryotic groups: chromalveolata, archaeplastida, excavate, and rhizaria subgroups. However, most microalgae show a close relationship with Chromalveolata compared to Archaeplastida (El-Sheekh et al., 2021). Microalgae-based wastewater treatment process, one of the most promising technologies for the advanced treatment and nutrient recovery of wastewater, has gained ever-increasing popularity in recent years (Abdelfattah et al., 2022). The pollution of agricultural, industrial, and municipal wastewater with a wide range of organic and inorganic contaminants, such as microplastics, heavy metals, xenobiotics, and high concentrations of nitrates, phosphates, and carbon composites, makes it difficult on the food chain and thus the basis of human life (Wollmann et al., 2019). Conventional WWT plants, by focusing on the removal of SS mostly mechanically and the reduction of activated sludge and biological oxygen demand, break down organic and inorganic materials (Kuśnierz et al., 2022). It is very important to limit nitrogen and phosphorus from these compounds in order to prevent eutrophication in surface water resources such as lakes and rivers (Qin et al., 2020). However, since the degradation potential of these compounds is limited by conventional treatment methods, they increase and accumulate the organic content of groundwater (Chan et al., 2022). Microalgae have metabolic flexibility due to their ability to perform photoautotrophic, mixotrophic, or heterotrophic metabolisms and thus represent promising biological systems for the treatment of wastewater sources.

Microalgae biomass generated from wastewater streams presents a big potential for sustainable bioproducts such as proteins, fatty acids, pigments, biofertilizers/biochar, and animal feed (Wollmann et al., 2019). It also allows bacteria to reduce the purification performance of the bacteria and the total energy cost of the oxygen source by providing additional oxygen through direct absorption or conversion of water pollutants into a harmless form or through photosynthesis (Anwar et al., 2019). In other words, microalgae assimilate organic carbon, nitrogen, and phosphorus and use CO_2 through photosynthesis to produce O_2, which is used by heterotrophic bacteria to decompose these nutrients. In addition, CO_2, inorganic nitrogen and phosphorus released as a result of aerobic metabolic activities can be used by microalgae as nutrients in the next photosynthesis steps. This symbiotic relationship between microalgae and bacteria was demonstrated to be able to preserve algae from toxic constituents in wastewater while increasing the removal of hazardous contaminants (Li et al., 2019). Therefore, microalgae appear to provide significant benefits over traditional cultivation and treatment processes: (1) microalgae assimilates many pollutants and wastewater is treated at a lower cost; (2) microalgae can reduce nutrient concentration in wastewater and meet discharge standards; (3) microalgae can be converted to carbon-neutral fuel with inorganic carbon sources; (4) microalgae have the potential to reduce production costs and greenhouse gas emissions associated with fossil-based fertilizers; and (5) microalgae biogas can be converted into valuable products such as biofuel, fertilizer, and animal feed (Li et al., 2019). The four main components of microalgae are proteins, carbohydrates, lipids, and nucleic acids, and their proportions strongly depend on the kind of algae. Table 12.19 summarizes the composition of commonly used microalgae species in the literature. The rate of different constituents among species varies notably. However, proteins in general create the largest percentage, and later come lipids and carbohydrates (Li et al., 2019; Gan et al., 2021).

It is well known that algae are photosynthetic and are able to attack any environment. Their significant role in nutrient cycling and oxygen generation is most notable in many ecosystems. Biomass acquired from microalgae can be used as an adsorbent.

12.7.2 Cyanobacteria and Their Species Used in Wastewater Treatment

As long as the biomass produced is reused, the use of cyanobacteria in WWT is an environmentally friendly process with no secondary pollution and provides efficient nutrient recycling. Cyanobacteria need high N and P requirements to survive and thrive, so this cycle further strengthens their ability to use nutrient-rich wastewater. In addition, assimilated nitrogen and phosphorus can be converted into algal biomass as biofertilizer after the elimination of heavy metals, and treated water can be discharged into water bodies or used in hydroelectric water plants (Sood et al., 2015).

Cyanobacteria are monocellular, colonial, and filamentous bacteria, divided by binary fission or multiple fission. Cyanobacteria taxonomically, the monocellular forms are divided by dual fission or multiple fission (El-Sheekh et al., 2021). In other words, cyanobacteria, whose production systems are the same as those of microalgae because they are photosynthetic, are the only prokaryotic group of organisms that contain 16S rRNA and chlorophyll-a, can fix carbon dioxide and atmospheric

TABLE 12.19

Composition of Microalgae Species Commonly Used in Wastewater Treatment (Li et al., 2019)

Representative Species	Composition (% Dry w/w)			
	Carbohydrate	Protein	Lipid	Nucleic Acid
Anabaena cylindrica	25–30	43–56	4–7	NM
Ankistrodesmus sp.	10.8	31.1	24–31	NM
Aphanizomenon flos-aquae	23	62	3	NM
Arthrospira maxima	13–16/20	60–71	6–7	NM
Botryococcus braunii	14.1	22	14.5/25–75	NM
Chlamydomonas reinhardtii	17	48	16.6–25.3	NM
Chlorella sp.	12–17	51–58	10–48	NM
Chlorella pyrenoidas	26	57–60.4	2–37	NM
Chlorella vulgaris	9–17	51–58	5–58	4–5
Chlorogloeopsis fritschii	44	50	7	NM
Dunaliella salma	16.3–32	25.7–57	6–25.3	NM
Dunaliella tertiolecta	12.2	20.0	15.0/16.7–71	NM
Euglena gracilis	14–18	39–61	14–38	NM
Haematococcus pluvialis	27	48	15/25	NM
Isochrysis galbana	17	27	7–40	NM
Nannochloropsis sp.	35.9	28.8	12–53	NM
Nitzschia sp.	9.2	16.8	12.1/16–47	NM
Prymnesium parvum	25–33	28–45	22–38	1–2
Porphyridium cruentum	32.1/40–57	28–39	6.5/9–18.8/	NM
Spirogyra sp.	33–64	6–20	11–21	NM
Synechococcus sp.	15	63	11	5
Scenedesmus obliquus	10–17	11.8–56	11–55	3–6
Scenedesmus quadricauda	NM	47	1.9–18.4	NM
Scenedesmus dimorphus	16–52	8–43	16–40	NM
Spirulina (Arthrospira) platensis	8–24.9	31.4–68.2	4–16.6	2–5
Spirulina maxima	13–16	60–71	4–9	3–4.5

nitrogen, have the ability to perform photosynthesis, and produce organic carbon and oxygen. They are also called blue-green algae and mostly live in aquatic or terrestrial habitats. Cyanobacteria are photosynthetic organisms of ecological, evolutionary, and economic importance (Mejean et al., 2010).

Toxic algae are found in both freshwater and marine systems. The cyanobacteria in these organisms can change the colors of lake and pool waters to red and green when they multiply suddenly. The toxins formed by these algae are called cyanotoxins for short. The toxins generally produced by cyanobacteria are secondary metabolites and are known to be produced by a large number of organisms (Jeong et al., 2020).

Microalgae and cyanobacteria are utilized in phycoremediation to eliminate nutrients and toxins from wastewater and carbon dioxide from the air. Also, algae and cyanobacteria are efficiently degrading pollutants by cleaning out some enzymes that reduce the toxic compounds into basic non-toxic compounds (Touliabah et al., 2022).

Traditionally, cyanobacteria are described not only microscopically but also as toxins. Morphological identifications make it difficult to distinguish between toxic and non-toxic species, due to some genera including both toxic and non-toxic species. The overgrowth of cyanobacteria in freshwater has increased worldwide, and the toxins produced by many species have been found to be harmful to humans and other organisms. At least 30 of more than 1,500 species of cyanobacteria have been shown to have the potential to produce toxic compounds. However, among the potentially toxic species, different compounds with some chemical and toxicological effects have been isolated, purified, and characterized (Saker et al., 2005). Cyanobacterial toxins have groups of compounds with very different chemical structures. These are known in two groups: cytotoxins and biotoxins, which are toxic to humans. Although cytotoxins are not lethal to humans and animals, they are relatively more toxic to algae and mammalian cells. Toxin-producing cyanobacteria can be categorized according to their toxicological properties. In these categories, neurotoxins (anatoxin-a, anatoxin-a (S), saxitoxin, and neosaxitoxin); tumor promoters (from microcystins and lipopolysaccharides); dermatotoxins/irritant toxin (lyngbiatoxin A, apysiatoxins, and lipopolysaccharides); hepatotoxins (microsistin, nodularins, and cylindrospermopsin); when based on their chemical structure, cyclic peptides (microsistin and nodularin); neurotoxic alkaloids (neurotoxins and cylindrospermopsin); and lipopolysaccharides, cyanotoxins are collected in three main groups (Msagati et al., 2006). Recently, it has been reported that cyclic peptide hepatotoxins, which are of interest to most scientists, especially considering their wide distribution in water resources, are more effective than neurotoxic alkaloids or lipopolysaccharides.

In Table 12.20, it is seen that cyanobacteria have a wide range of distribution areas. Thermal areas are suitable places for extremophile creatures, and these areas also host cyanobacteria. Research keeps going on for all of these toxin groups. Ultimate findings have included: (1) chromosome loss and DNA strand fracture by cylindrospermopsin; (2) inhibition of nuclear protein phosphatase by microcystin with alterations in phosphorylation statute of the p53 tumor-suppressor protein; and (3) conjugation of nodularin and microcystins to glutathione by glutathione S-transferases as a first step in cyanobacterial hepatotoxin detoxication in the brine shrimp Artemia salina (Codd et al., 2005). Proven to exist in aquatic and terrestrial environments, freshwater, marine, moist soil, bare soil, and deserts, cyanobacteria form a large group of prokaryotic photosynthetic organisms, some of which have unique nitrogen fixation abilities (Latysheva et al., 2012). They occur as plankton and sometimes in the form of phototropic biofilms.

They can keep alive in heavy environmental terms owing to their eccentric features, such as Sood et al. (2015).

a. existence of an external mucilaginous sheath outside the cell wall assists them to survive in tough conditions,
b. nitrogen-fixing capability helps them to survive in low N terms (as seen in Table 12.21),
c. generation of allelochemicals that assist in contesting with other organisms in the vicinity.

TABLE 12.20
Principle Groups of Cyanobacterial Toxins and Their Sources[a]

Toxin	Number of Structural Variants	Structure and Activity	Toxigenic Genera[b]
Hepatotoxins			
Microcystins	71	Cyclic heptapeptides; hepatoxic, protein phosphatase- inhibition, membrane integrity and conductance disruption, tumour promoters	*Microcystis, Anabaena, Nostoc, Anabaenopsis, Planktothrix, Oscillatoria, Hapalosiphon*
Nodularins	9	Cyclic pentapeptides; hepatotoxins, protein phosphatase- inhibition, membrane integrity and conductance disruption, tumour promoters, carcinogenic	*Nodularia, Theonella* (sponge-containing cyanobacterial symbionts)
Cylindro-spermopsins	3	Guanidine alkaloids; necrotic injury to liver (also to kidneys, spleen, lungs, intestine), protein synthesis- inhibitor, genotoxic	*Cylindrospermopsis, Aphanizomenon, Umezakia, Anabaena,[c] Raphidiopsis[d]*
Neurotoxins			
Anatoxin-a (including homoanatoxin-a)	5	Alkaloids; postsynaptic, depolarising neuromuscular blockers	*Anabaena, Oscillatoria, Phormidium, Aphanizomenon, Rhaphidiopsis[e]*
Anatoxin-a (s)	1	Guanidine methyl phosphate ester; inhibits acetylcholinesterase	*Anabaena*
Saxitoxins	20	Carbamate alkaloids, sodium channel-blockers	*Aphanizomenon, Anabaena, Lyngbya, Cylindrospermopsis,[f] Planktothrix*
Dermatotoxins and Cytotoxins			
Lyngbyatoxin-a	1	Alkaloids; inflammatory agents, protein kinase C activators	*Lyngbya, Schizothrix, Oscillatoria*
Aplysiatoxins	2	Alkaloids; inflammatory agents, protein kinase C activators	*Lyngbya, Schizothrix, Oscillatoria*
Endotoxins			
Lipopoly-saccharide	Many	Lipopolysaccharides; inflammatory agents, gastrointestinal irritants	All?

[a] Updated from Cohen (1997), Falconer (1998), Codd et al. (1999), and Sivonen and Jones (1999).

[b] Not all species within a genus, or strains within species, produce the particular toxin, except for lipopolysaccharide which is a marker from Gram negative prokaryotes including cyanobacteria.

[c] Schembri et al. (2001).

[d] Li et al. (2001).

[e] Namikoshi et al. (2003).

[f] Lagos et al. (1999) and Molica et al. (2002).

The dominance and variety of cyanobacterial genera in distinct aquatic bodies exemplify the tolerance of this group to a broad array of pollutants. This has generated an interest in the scientific community to detect their potential in WWT. As seen in Table 12.22, the enormous diversity of cyanobacteria in various ecological habitats has been studied by scientists around the world. However, most of the available studies reveal the predominance of cyanobacteria in different freshwater and polluted habitats and/or reservoirs.

Reports on WWT potentials reflect that cyanobacteria are effective in treating a wide variety of wastewater in terms of nutrient removal, water quality healing, and heavy metal removal.

12.7.2.1 Wastewater Pollutants Removed by Cyanobacteria

Cyanobacteria have thrived in a wide variety of conditions and are known as prokaryotic microorganisms that increase water productivity by fixing atmospheric nitrogen and carbon. Most cyanobacterial species are considered to be one of the most useful bioaccumulators because of their biomass, widespread distribution, availability of low-cost reproductive technology, adaptive metabolisms, and high absorption capacities. Heavy metal removal of microalgae and cyanobacteria by physicochemical methods is shown in Table 12.23. Cyanobacterial species can reduce aromatic hydrocarbons and xenobiotics to less toxic or non-toxic components and use them as nutrients, limiting the toxic effects of the existing environment, as seen in Table 12.24 (Touliabah et al., 2022).

Depending on the conclusions reported by a variety of studies, the dual practice of cyanobacteria and microalgae to treat wastewater appears to be a possible solution to nutrient recovery. In addition, as shown in Table 12.25, it was understood that the concentration of N and P in wastewater had major effects on the uptake of microalgal nutrients, their binding to surfaces, and the yield of the biochemical composition.

The microalgae and cyanobacteria group has become a suitable candidate for WWT and BP because this group has good growth rates as well as high photosynthetic

---

TABLE 12.21

Types of Nitrogen-Fixing and Non-Fixing Aerobic Bacteria (Mcewan et al., 1998)

Aquaspirillum	N fixing:	A fasciculus, A. itersonii, A. magnetotacticum, A. Peregrinum
	Non-N fixing:	A. anulus, A. aquaticum, A. autotropicum, A. bengal, A. delicatum, A. dispar, A, giesbergeri, A. gracile, A. metamorphum, A. polymorphum, A, psychrophilum, A. putridiconchylium, A. serpens, A. sinosum
Chrornatium	N fixing:	C. gracile, C. minus, C. minutissimun, C. vinosum, C. violascens, C. warmingii, C. weissei
	Non-N fixing:	C. buderi, C. okenii, C. purpuratum, C. salexigens, C. tepidum
Clostridium	N fixing:	C. aceticum, C. acetobutylicum, C. arcticum, C. beijerinckii, C. butyricum, C. cellobioparum, C. formicoaceticum, C. kluyveri, C. papyrosolvens, C. pasteurianum, C. thermosaccarolyticum

(Continued)

TABLE 12.21 (*Continued*)
Types of Nitrogen-Fixing and Non-Fixing Aerobic Bacteria
(Mcewan et al., 1998)

	Non-N fixing:	*C. acidiurici, C. aerotolerans, C. aminovalericum,*
		C. argentinense, C. aurantibutyricum, C. baratii, C. barkeri,
		C. bifermentans, C. botulinum, C. bryantii, C. butyricum,
		C. cadaveris, C. carnis, C. celerecrescens, C. cellulolyticum,
		C. cellulovorans, C. chartatabidum, C. chauvoei,
		C. clostridiiforme, C. coccoides, C. cochlearium,
		C. cocleatum, C. colinum, C. collagenovorans,
		C. cylindrosporum, C, difjcile, C. disporicum, C. durum,
		C. fallax, C. felsineum, C. fervidum, C. ghoni, C. glycolicum,
		C. haemolyticum, C. indolis, C. innocuum, C. intestinale,
		C. josui, C. lentocellum, C. leptum, C. limosum,
		C. lituseburense, C. madisonii, C. magnum,
		C. malenomanatum, C. methylpentosum, C. nexile, C. novyi,
		C. oceanicum, C. oroticum, C. oxalicum, C. paraputrijicum,
		C. perfringens, C. pfennigii, C. polysaccharolyticum,
		C. populeti, C. propionicum, C. proteolyticum, C. puniceum,
		C. purineinolyticum, C. putrefaciens, C. putrijicum,
		C. quercicolum, C. ramosum, C. rectum, C. saccharolyticum,
		C. sartagoforme, C. scatolgenes, C. scindens, C. septicum,
		C. sordellii, C. sphenoides, C. spiroforme, C. sporogenes,
		C. sporosphaeroides, C. stercorarium, C. sticklandii,
		C. subterminale, C. symbiosum, C. tertium, C. tetani,
		C. tetanomorphwn, C. thermoaceticum,
		C. thermoautotrophicum, C. thermobutyricum,
		C. thermocellum, C. thermocopriae,
		C. thermohydrosulfuricum, C. thermolacticum,
		C. thermosulfirogenes, C. tyrobutyricum
Desulfotomaculum	N fixing:	*D. nigrificans, D. orientisi, D. Ruminis*
	Non-N fixing:	*D. acetoxidans, D. antarcticum, D. geothermicum,*
		D. guttoideum, D. kuznetsovii, D. sapomandens,
		D. thermoacetoxidans
Desulfovibrio	N fixing:	*D. africanus, D. desulfiricans, D. gigas, D. salexigans,*
		D. Vulgaris
	Non-N fixing:	*D. carbinplicus, D. fructosouorans, D. furfuralis,*
		D. giganteus, D. piger, D. simplex, D. sulfodisrnutans
Methanobacterium	N fixing:	*M. formicium, M. ivanovii, M. thermautrophicum*
	Non-N fixing:	*M. alcafiphium, M. bryantii, M. espanolae, M. palustre,*
		M. thermoaggregans, M. thermoalcaliphilum, M.
		thermoformicicium, M. uliginosum, M. wolfei
Rhodospirillum	N fixing:	*R. fulvum, R. molischianum, R. photometricum, R. rubrum,*
		R, salexigens
	Non-N fixing:	*R. centenum, R. mediosalinum, R. salinarum*

(*Continued*)

TABLE 12.21 *(Continued)*
Types of Nitrogen-Fixing and Non-Fixing Aerobic Bacteria
(Mcewan et al., 1998)

Vibrio	N fixing:	*V. diazotrophicus, V. natriegens, V, pelagius*
	Non-N fixing:	*V. aestuarianus, V. albensis, V. alginolyticus, V. anguillarum, V. campbellii, V. cholerea, V. costicola, V. fischeri, V. gazogenes, V. harveyi, V. logei, V. marinus, V. mediterranei, V. metschnikovii, V. nereis, V. nigripulchritudo, V. ordalii, V. orientalis, V. parahaemolyticus, V. proteolyticus, V. psychroerythrus, V. salmonicidia, V. splendidus, V. tubiashii*

TABLE 12.22
Cyanobacterial Variety of some Water Bodies of the World (Sood et al., 2015)

Location	Type of Contamination	Commonly Occurring Cyanobacterial Genera
Shallow lake, Santa Olalla, Southwestern Spain	NA	*Anabaena* spp., *Anabaenopsis* spp., *Aphanocapsa delicatissima, Aphanothece clathrata, Chroococcus disperses, Leptolyngbya* sp., *Limnothrix amphigranulata, Merismopedia tenuissima, Microcystis aeruginosa, Oscillatoria* sp., *Pseudanabaena limnetica, Raphidiopsis mediterranea*
Lake Bogoria, Kenya	NA	*Leptolyngbya, Spirulina, Oscillatoria*-like, *Planktothricoides, Synechocystis, Arthrospira* and *Anabaenopsis*
Lake Taihu and Lake Chaohu, China	NA	*Microcystis*
Lake Ulungur, Xinjiang, China	NA	*Microcystis* spp.
Lake Taihu, China	Agricultural intensification and industrial pollution	*Microcystis* sp.
Lake Gregory, Nuwara Eliya, Sri Lanka	Agricultural and industrial activities	*Synechococcus* sp., *Microcystis aeruginosa, Calothrix* sp., *Leptolyngbya* sp., *Limnothrix* sp.
Mendota, Kegoonsa, Wingactra and Monona Lakes	Agricultural and urban run off	*Microcystis, Aphanizomenon, Chroococcus, Anabaena* and *Cylindrospermopsis*
Lake Fryxell, McMurdo Dry Valleys, Antarctica	NA	*Nostoc* sp., *Hydrocoryne* cf. *spongiosa, Nodularia* cf. *harveyana, Leptolyngbya* spp., *Phormidium* cf. *autumnale*
Lake Pamvotis (suburban Mediterranean Lake), Greece	NA	*Microcystis* sp., *Anabaena* sp./ *Aphanizomenon* sp.

(Continued)

TABLE 12.22 (*Continued*)
Cyanobacterial Variety of some Water Bodies of the World (Sood et al., 2015)

Location	Type of Contamination	Commonly Occurring Cyanobacterial Genera
Lake Loosdrecht, the Netherlands	NA	*Aphanizomenon, Planktothrix, Microcystis, Synechococcus, Prochlorothrix hollandica, "Oscillatoria limnetica* like", *Limnothrix/Pseudanabaena* group
Western basin of Lake Erie	NA	*Microcystis* and *Planktothrix*
Lake Mendota, Wisconsin, USA	Agricultural and urban run off	*Aphanizomenon* and *Microcystis*

NA: not available.

productivity and a high tolerance to toxic effects in wastewater. It has been proven by studies that growing microalgae in different wastewaters, such as domestic, agricultural, and industrial wastewater, can effectively remove N, P, and metal ions. It is known as bioaccumulation and bio-adsorption, which is environmentally friendly and economically possible, especially when compared to classical methods.

12.7.2.2 Pollutant Removal Mechanism of Cyanobacteria

Important advances have been made in the detection and analysis of cyanobacterial toxins (Du et al., 2019). For example, as they can provide an indication of the potential for toxin production at the single colony or filament level, it has been reported that PCR and DNA microarray tests are required for microcystin, nodular and cylindrospermopsin biosynthesis, and peptide and polyketide synthesis (Kumari and Rai, 2020). Timely detection of real toxin production by single colonies and filaments is also probable before bloom development by using immunoassays. Quantitative analysis can likely be done with an enzyme-linked immunosorbent assay with a minimum limit of detection of nearly 0.015 ng per colony or filament. In this way, toxins in single-washed colonies can be easily measured (Roy-Lachapelle et al., 2021). Analysis of only a few colonies before inflorescence formation can not only determine whether an inflorescence will be toxic but can also be used to map the extent of toxin production and quantify the contribution of cyanobacterial species or morphotypes to the overall microcystin pool at a given time. The practice of immunodetection methods to quantify the microcystin ingredients of colonies and filaments can also be used for the formulation and correction of monitoring beginnings, threshold values, and guideline levels (Codd et al., 2005).

In addition, microscopy techniques are one of the most commonly used techniques for morphologically determining the diversity of microorganisms, their ecology, and their physiological adaptations. For example, a scanning electron microscope, transmission electron microscope, and confocal laser scanning microscope can be used in addition to the classical light microscope (Strunecký et al., 2019).

TABLE 12.23
Heavy Metals (HMs) Removal by Conventional Physio-Chemical Methods, Microalgae, and Cyanobacteria (El-Sheekh et al., 2021)

HMs	Approaches	Chemical/Biological Agents	HMs Removal
Chromium (Cr)	Conventional	Adsorbed by nanoscale zerovalent iron (nZVI)	99.9%
	Algae	*S. quadricauda*	23.98–25.19 mg/g
	Cyanobacteria	*Limnococcus sp.* and *Leptolyngbya* sp.	50%
Cadmium (Cd)	Conventional	Polyelectrolyte-Coated Industrial Waste Fly Ash	99%
	Algae	Microalgae-endophyte symbiotic system (*C. vulgaris* and water hyacinth*)*	99.2%
	Cyanobacteria	*N. calcicole*	98% of 2.5 pm
Arsenic (Ar)	Conventional	Combined pre-oxidation, coagulation, adsorption, and co-precipitation	Decrease from 423 to 6.8 mg/L
	Algae	*Cladophora sp.*	100% bioabsorption of 80 g/L
	Cyanobacteria	*Synechocystis sp.* PCC6803, *Microcystis sp.* PCC7806, and *Nostoc sp.* PCC7120	80% of 100 pm sodium arsenite
Lead (Pb)	Conventional	Adsorption using acid-activated clay	92.4%
	Algae	*P. typicum* and *S. quadricauda* var *quadrispina*	5–20 mg/L
	Cyanobacteria	*P. ambiguum*	5–20 mg/L
Mercury (Hg)	Conventional	Inorganic salts (NaCl and $MgCl_2$)	5%–10%
	Algae	*S. obtusus* XJ-15	95 mg/g
	Cyanobacteria	*N. paludosum*	96%

12.7.2.3 Cyanotoxins, Their Formation, and Their Effects on Treated Water Quality

The overgrowth of cyanobacteria is an important problem that is increasing in frequency and threatening all water resources (Choi et al., 2022). However, eutrophication, which occurs as a result of organic pollution in water bodies, prepares the environment for the deterioration of the water quality of lakes and dam lakes and the formation of cyanobacterial species (Xu et al., 2021). Many different types of cyanobacteria are known, which are toxin producers that cause various diseases in humans and death in birds and aquatic organisms. Many different species of cyanobacteria are known, which are toxin producers that cause various diseases in humans, death in birds, and aquatic organisms, and are divided into hepatotoxins, neurotoxins, and dermatotoxins according to their effects (Nowruzi and Porzani, 2021).

In fact, the features that distinguish cyanobacteria from other algae groups and bring them closer to bacteria are their prokaryotic nature, the murein structure of their cell walls, and the presence of specialized cells (heterocysts) that have the ability to fix nitrogen. They have the ability to synthesize chlorophyll-a and at least one

TABLE 12.24

Potential for Laboratory-Scale Removal of Cyanobacteria and Organic Pollutants (Touliabah et al., 2022)

Pollutants	Algae Species	Organic Pollutants	Degradation Rate (%)
Dyes	*Nostoc linckia* HA 46	Toxic reactive red 198 dye	94
	N. muscorum	Tartrazine	70
	Nostoc linckia	Azo dye	81.97
	Oscillatoria rubescens	Basic Fuchsin (5 ppm)	94
	Phormidium ceylanicum	Acid Red 97	89
	Ph. animale	Remazol Black B (RBB)	99.66
	Chroococcus minutus	Amido Black 10B (100 mg/L)	55
	Gloeocapsa pleurocapsoides	FF Sky Blue (100 mg/L)	90
Hydrocarbon	*Prototheca zopfii*	Saturated aliphatic hydrocarbons	49 ± 11
	Prototheca zopfii	Aromatic compounds	26.5 ± 14.5
	Oscillatoria sp.	Pyrene	95
Phenol	*Anabeana variabilis*	O-nitrophenol (ONP)	100
Pesticides and herbicides	*Oscillatoria quadripunctulata*	Biocides	40

phycobilin as pigments (Irankhahi et al., 2023). The cell wall is composed of cellulose and pectin. The reserve nutrient is glycogen, as are the proteins cyanophycin and volitin.

Cyanobacteria are phototrophic and need microorganisms with anaerobic metabolism for water, carbon dioxide, inorganic compounds, and light (Demoulin et al., 2019). Because they are the primary producers, they have very important duties in the biosphere. Some species can grow even in nutrient-poor environments through nitrogen fixation with a small number of specialized cells called "heterocysts" and thus increase the productivity of the environment (Zheng et al., 2021). In agricultural practices, some nitrogen-fixing cyanobacteria species are used as fertilizer in order to increase the nitrogen level of the soil. In addition, they have the ability to develop spore cells called "akinet" from vegetative cells among the cells forming the thread (trichome) in filamentous forms. Akinets are one of the mechanisms responsible for the survival of cyanobacterial species in unfavorable conditions (Kimura et al., 2020).

Many cyanobacteria species can move vertically in water thanks to their cytoplasmic character and cylindrical gas-filled gas sacs. Their reproduction is vegetative or spore-forming; sexual reproduction is not observed. Since algae are distributed in competitive environments, they have developed some adaptation and defense mechanisms. The most important adaptation observed in cyanobacteria is that they produce extremely bioactive secondary metabolites (cyanotoxins) that are toxic to eukaryotes and mammals. Cyanotoxins, which have high activity, affect the environment and disrupt the ecosystem. Poisoning caused by cyanotoxins occurs when toxin-producing cyanobacteria are taken directly into the body or by consuming living

TABLE 12.25
Removal of Nutrients from Various Wastewaters and Biomass Increase in Microalgae Species under Variable Environmental Conditions (El-Sheekh et al., 2021)

Algae	Wastewater Type	Total Nitrogen (TN)	Total Phosphorus (TP)	Biomass Production
Cladophora	Secondary effluent	TN (48%)	TP (13%)	26 g/m²/day
Algal consortium	Raw domestic	TN (2.52 g/m²/day)	TP (1.25 g/m²/day)	29.7–48.9 g/m²/day
C. vulgaris	Synthetic domestic wastewater	TN (98.69%)	TP (86.07%)	17.94×10⁶ cell/L
Sceuedesnuus sp. ZTY1	Domestic wastewater	TN (90%)	TP (90%)	3.6×10⁶ cells m/L
C. sorokimana AK-1	Swine wastewater	TN (90%)	TP (90%)	6.5 g/L
C. pyreuoidosa	Domestic wastewater	NH/-N (95%) COD (78%)	TP (81%)	1.71 g/L
S. plateusis	Swine wastewater	N/A	TP (41.6%)	N/A
C. sorokiinamu	Municipal and piggery wastewater	NH₃ (100%)	Orthophosphate (60%)	1 g/L
Consortia of algae and bacteria	Municipal wastewater	TN (83%)	TP (100%)	0.018 mg/L/h
S. obliquus and C. vulgaris	Urban wastewater sample	TN (98%)	TP (100%)	900 mg/L/day
C. vulgaris	Anaerobic sludge	TN (1.6–9.8 mg/L/day)	TP (3.0 mg/L/day)	39–195 mg/L/day
Cyauobacteria	Agricultural runoff	TN (95%)	TP (96%)	86 mg TSS/L/day
Chlorella sp.	Municipal wastewater	TN (75%)	TP (93%)	5.6 g/m²/day
H. nbescceus	Wastewater	TN (83%)	TP (85%)	6.3 g/m²/day

"N/A" refers to "not available".

things such as fish and mussels that feed on these creatures and have cyanotoxin accumulation in their bodies (Pei et al., 2020).

Increased cyanotoxin levels due to the increase in cyanobacteria are a significant threat, especially to fish populations. Cyanotoxins, which pose a hazard to drinking water supplies, pose a serious potential health threat to humans, such as the liver, nerve synapses, nerve axons, skin, and intestinal tract, particularly through eating and drinking by mouth or consuming food that has accumulated cyanotoxins (Zi et al., 2018). Mainly nitrogen, phosphorus, and inorganic carbon input of nutrients coming to water resources with wastewater causes excessive algae growth, so water resources and sensitive water areas can become eutrophic and potential cyanobacteria breeding areas and threaten the environment and ecosystem. As a result, the most effective method to deal with cyanobacteria is to control water resources and prevent their formation before overgrowth. It is of great importance to create emergency action plans in order to respond quickly to negative situations that may occur due to the increase of harmful cyanobacteria and their toxins, to minimize the damage to humans and the environment, and to intervene at the source.

12.8 CASE STUDIES FOR ALGAEL-BASED PHYCOREMEDIATION APPLICATIONS

Despite the benefits offered by algae-based biofuels, full-scale applications of algae cultivation for biofuel production in artificial environments are very expensive, especially with current technologies. It has been understood that the production of biofuels with microalgae grown in artificial systems cannot economically obtain the energy of the desired quality and amount. It is an inevitable fact that large energy inputs are required for the cultivation and harvesting of algae in the desired quality and quantity in artificial environments. Therefore, the integration of algae production from wastewater and industrial emissions (CO_2) is economically beneficial for cost reduction and optimization. The simultaneous implementation of WWT, flue gas use, and microalgae cultivation for biofuel production, while reducing the energy cost of biofuel production processes from microalgae, is considered as an extremely attractive option for the environmental sustainability of freshwater and air resources. Microalgae can reproduce very easily in domestic and industrial wastewater, which is rich in carbon and nutrients.

An important input for algal biomass and biofuel production is the wastewater required for algal growth. With the help of microalgae, biological treatment of wastewater can be carried out. In wastewater recovery applications, the treatment method varies depending on the reuse purpose. In conventional treatment applications used in the treatment of urban and domestic wastewater, the treatment stages are as follows: Primary treatment is the treatment of wastewater by physical, mechanical, and/or chemical processes. Secondary treatment is the purification of SS and biodegradable organic matter in wastewater with biological treatment. Tertiary Treatment/ Advanced Treatment: with biological treatment, in addition to the treatment of SS and biodegradable organic matter in wastewater, it is also the treatment of nutrients in wastewater.

The concentrations of N and P in wastewater, which algae use as nutrients and thus purify, vary considerably depending on the type and stage of the treatment process. Urban wastewater (also called sewage) contains total concentrations of N and P in amounts in the industrial scale range of 10–100 mg/L. After secondary treatment, total N and P fall into the ranges of 20–40 mg/L and 1–10 mg/L, respectively. These concentrations are quite suitable for microalgal cultivation. The N:P molar ratio of sewage wastewater is between 11 and 13 (calculated based on NO_3^- and PO_4^{3-}). Typical algal biomass contains 6.6% N and 1.3% P by dry weight, with an 11.2 molar N:P ratio, a composition similar to that found in sewage wastewater (Lage et al., 2018).

Ensuring the high amount and quality of biomass efficiency of microalgae grown in wastewater makes it a suitable approach for high-quality and quantity microalgae cultivation as a renewable and sustainable source for energy generation. The efficient growth of microalgae in wastewater depends on a variety of variables. As in every growth medium, parameters such as pH and temperature of the growth medium of microalgae, N, P, and organic carbon concentrations, light input, O_2 and CO_2 presence are critical variables (Ip et al., 1982).

Although nutrient concentrations such as N and P are high in domestic wastewater and some industrial wastewater, they can inhibit algae growth because they are usually found in high ammonia form (Ip et al., 1982). Other factors that have a negative effect on algae growth in wastewater, especially industrial wastewater, include toxins such as cadmium, mercury, and/or organic chemicals. Pathogenic bacteria or predatory zooplankton are biotic factors that have a negative effect on algae growth. These variables vary from one WWT plant to another, depending on the wastewater characteristics (Pittman et al., 2011).

At the stage of microalgae production from wastewater, contamination formation in the culture medium, alkalinity deficiency in the nitrogen removal stage, nitrate accumulation, etc., biological process-based problems and problems arising from long-term operating processes can be encountered. In the literature, there are many studies on microalgae production from wastewater. In these studies, subjects such as biomass growth potential, nutrient removal efficiency, oil production capacity, and reproductive kinetics of microalgae have been investigated.

12.8.1 DOMESTIC WASTEWATERS

Microalgae remove N and P from domestic wastewater quite effectively. In studies conducted on domestic wastewater in the literature, different types of algae (*Chlamydomonas reinhardtii*, *Scenedesmus obliquus*, *Botryococcus braunii*, *Chlorella* sp., *Micractinium* sp., and *actinastrum* sp.) have different BP potentials (25–2,000 mg/L/day) according to wastewater characteristics and WWT plant type) and different lipid efficiencies (8–505 mg/L/day) (Ip et al., 1982; Martınez et al., 2000; Órpez et al., 2009; Woertz et al., 2009; Kong et al., 2010).

Li et al. (2011) conducted studies with *Chlorella* sp. during a 14-day trial period in which they investigated the nutrient removal, biomass, and biodiesel production efficiency in urban wastewater. At the end of these studies using *Chlorella* sp.

microalgae, NH_4^+, TN, total phosphorus (TP), and COD removal efficiencies were found to be 93.9%, 89.1%, 80.9%, and 90.8%, respectively, while the biomass yield was 0.92 g/L/day and the biodiesel yield was determined to be 0.12 g/L/day (Li et al., 2011).

In a study conducted with *chlorella minutissima* microalgae, which can grow well in high concentrations in a domestic WWT plant in India and was determined to be the dominant species of the next steps of the oxidation pond system, this species can grow heterotrophically in the dark, live in a wide pH range and in the presence of salt, and grow mixotrophically in the light. Furthermore, it has been determined that it can use various organic carbon substrates by acting as a nitrogen source, as well as ammonia or nitrate as a nitrogen source. After 10 days of growth, it was observed that, with the growth of these algae, biomass productivity could reach 379 mg/L under photoheterotrophic conditions and 73.03 mg/L under photoautotrophic conditions. At the end of the study, it was concluded that *chlorella minutissima* could be a good candidate to provide high biomass efficiency in high-speed pool systems that treat domestic wastewater and can be preferred as an alternative energy source (Bhatnagar et al., 2010).

In another study, the biomass yield and nitrogen and phosphorus of *Chlorella* sp. microalgae species in wastewater with different levels of nutrient content were taken from three different points of the St. Paul Metro domestic WWT plant inlet, outlet, and concentrated top water by using a biocoil photobioreactor. The effects of CO_2 and pH on growth were also investigated in their study. The results showed that the optimal pH for the *C. reinhardtii* strain was around 7.5, with a maximum of 2.0 g/L in the biocoil. The algal dry biomass conversion rate showed that the day has passed. It was determined that the oil content of the strain of *Chlamydomonas reinhardtii* was 25.25% (w/w) by dry biomass weight, and according to the results of the analysis made in the centrifuged overhead wastewater of the Biocoil reactor, 55.8 mg of nitrogen and 17.4 mg of phosphorus were effectively removed per liter per day (Kong et al., 2010).

Zhou et al. (2012) have investigated, wastewater nutrient removal efficiency and lipid accumulation of *auxenochlorella* protothecoides UMN280 microalgae in domestic wastewater culture growth media in batch and semi-continuous sowing at different hydraulic holding times. The results of the 6-day batch production of *auxenochlorella* protothecoides UMN280 showed that with a high growth rate (0.490/day), high biomass productivity (269 mg/L/day), and high lipid productivity (78 mg/L/day), the maximum removal efficiencies from wastewaters for TN, TP, COD, and total organic carbon were 59%, 81%, 88%, and 96%, respectively (Zhou et al., 2012).

In the study conducted by Welch (2013), the cultivation of microalgae in the effluent of the domestic WWT plant, the optimization of the culture medium, and the nutrient removal efficiency were investigated. In the study in which *Chlorella* and *Chlamydomonas* species were used, different amounts of carbon dioxide were supplied to two algae cultures under the same ambient conditions by mixing air. When the chlorophyll-a and cell count results were evaluated in the study, it was seen that adding carbon dioxide to the culture medium increased the growth rate of microalgae.

In the second stage of the study, two separate microalgae culture environments were created by providing only carbon dioxide and air to the effluent of the domestic WWT plant. Similar to the results of the previous experiment, carbon dioxide given to the culture medium increased the growth rate. In addition, the results of the study carried out with the effluent of the domestic WWT plant have been obtained to support the idea that the lack of nutrients will increase oil production (Welch, 2013).

In another study in which *chlorella kessleri* microalgae used municipal wastewater with/without CO_2 supplementation as culture medium, the lipid content corresponding to 4.5–57.4 mg/L/day LP was determined as 8%–25.9% on a dry weight basis (Li et al., 2012).

Li et al. (2011) studied on *Chlorella kessleri* microalgae, which were used in the treatment of concentrated municipal wastewater, and they determined the lipid content corresponding to 29.6–91.0 mg/L/day LP as 14.4%–24.2% on a dry weight basis (Li et al., 2011).

12.8.2 DISTILLERY WASTEWATERS

The type of organic acid present in the wastewater was also found to affect the growth of microalgae. As per a study recently conducted by Ren et al. (2018) on different types of dark fermented molasses wastewater (ethanol type, propionate type, and butyrate type), it was found that butyrate-type wastewater was the most suitable for the growth of *Scenedesmus* species that gave 10%–40% higher BP and LP of 166.8 and 64.8 mg/L/day, respectively, as compared to propionate type and ethanol type wastewater (Ren et al., 2018).

On the other hand, Gaurav et al. (2016) applied a pretreatment strategy for cane molasses, wherein the authors passed it via a strong acid-cation exchange resin to reduce the hindrance in the heterotrophic cultivation of *chlorella pyrenoidosa* due to metal ions. The applied strategy enhanced biomass growth (1,220 mg/L) by 57% as compared to when molasses was untreated. Furthermore, the study reported a lipid yield of 0.66 g/g with 10 g/L treated cane molasses (Gaurav et al., 2016).

Likewise, in another study, Dos Santos et al. (2016) conducted preliminary studies to evaluate the cultivation of Spirulina maxima in a media containing 0.1% and 1% (v/v) vinasse under autotrophic, heterotrophic, and mixotrophic conditions in batch mode. The preliminary results with 0.1% and 1% (v/v) vinasse showed higher specific growth rates (0.54–1.02 day^{-1}, respectively) in the heterotrophic mode during the exponential growth phase, while the mixotrophic mode was found effective in enhancing the growth of microalgae during the stationary growth phase. Therefore, the authors applied the cyclic two-stage cultivation strategy to enhance the biomass growth of Spirulina maxima in vinasse. For cyclic two-stage cultivation, the autotrophic mode was maintained for 12 h during the course of the light phase of the photoperiod (light intensity: 70–200 μmol photons/m^2/s), followed by the heterotrophic mode for 12 h during the course of the dark phase of the photoperiod (vinasse concentration: 3.0% v/v). The adopted strategy of cyclic two-stage cultivation consisted of three cycles with 75% removal of suspension and reinsertion of the medium having a vinasse concentration of 3% (v/v), which was separated through autotrophic rest periods of a few days between cycles. The study reported an increase in BP between

70 and 87 mg/L/day on the seventh day of each cycle and enhanced the tolerance of microalgae to vinasse (from 1% to 3%). The study also reported a lipid content of 7.2%–8.7% w/w and a higher protein content of 74.3%–77.3% w/w of microalgae dry weight, which is suitable for nutrition in both animals and humans (Dos Santos et al., 2016).

In another study, Souza and Costa (2007) evaluated the suitability of molasses for mixotrophic cultivation of microalgae *Spirulina platensis* by analyzing the effect of light intensity (32.5, 45.5, and 58.5 μmol/m²/s) and molasses concentration (0.25, 0.5, and 0.75 g/L) on BP. They found that until the 11th day, growth was obtained mainly in heterotrophic mode, and the concentration of molasses was a significant factor that influenced maximum BP. However, after the 11th day, light intensity also played a key role in enhancing biomass growth (maximum biomass concentration: 2.94 g/L; BP: 290 mg/L/day at 0.75 g/L molasses concentration and 45.5 μmol/m²/s light intensity) (Souza and Costa, 2007).

Given the toxic nature of distillery wastewater, several studies have assessed the feasibility of microalgae cultivation using it. The growth response of each microalgae species varies depending on the concentration and source of wastewater, which resulted in different BP and LP, as shown in Table 12.26.

12.8.3 Dairy Wastewater

Gentili (2014) investigated that dairy industry wastewater is used for algae production, but the biggest problem encountered is reported to be the low biomass yield of grown algae. A literature search has shown that the highest biomass yield of algae grown in dairy (dairy) wastewater is below 0.7 g/L (Blier et al., 1995; El-Sikaily et al., 2007; Christenson and Sims, 2011).

Another study (Lu et al., 2016) showed that the limiting factor for algal growth in dairy wastewaters is ammonia nitrogen deficiency. In this study, dairy industry wastewater was mixed with slaughterhouse wastewater with a higher ammonia and nitrogen content. The results showed that mixing wastewater at a low cost improved the nutrient profile and biomass conversion. Algae grown in mixed wastewater have high protein (55.98%–66.91%) and oil content (19.10%–20.81%). Thus, the main problem of the low biomass yield of algae grown in dairy industry wastewater has been solved. The grown algae can be used to produce animal feed and biofuels as well as significantly reduce the nutrient content of wastewater (Lu et al., 2016).

In another study, a maximum of 1.23 g/L dry biomass was obtained in 7 days with *chlorella vulgaris* species grown in dairy industry wastewater, while organic and inorganic matter removal efficiencies were determined at the end of 10 days, such as BOD, COD, SS, TN, and TP, which were found to be 85.61%, 80.62%, 29.10%, 85.47%, and 65.96%, respectively. In the study, it was determined that the BP efficiency was greatly affected by the nutrient reduction in the dairy industry wastewater, but the biodiesel produced by *C. vulgaris* in the dairy industry wastewater was in good agreement with the American Society for Testing and Materials (ASTM)-D6751 and European Standards (EN) 14214 standards. Therefore, the use of dairy effluent for microalgae cultures has been evaluated as a useful and practical strategy as an advanced, environmentally friendly treatment process (Choi, 2016).

TABLE 12.26
Biomass and Lipid Productivity of Different Microalgae Species Cultivated in Distillery Wastewater

Microalgae sp.	Wastewater Concentration	Culture	Biomass (mg/L/day)	Lipid (mg/L/day)	References
Chlorella pyrenoidosa – Rhodosporidium toruloides (Microalgae-yeast)	Distillery wastewater + domestic wastewater (1:1)	Mixotrophic	7,250 mg/L	4,600 mg/L	Ling et al. (2014)
Arthrospira maxima sp.	30% (v/v) sugarcane vinasse	Mixotrophic	201	–	Montalvo et al. (2019)
Desmodesmus sp.	Wastewater sugarcane vinasse	Heterotrophic	101.1 mg/L/h	–	de Mattos and Bastos (2016)
Micractinium sp. ME05	10% (v/v) sugarcane vinasse	Mixotrophic	320	3,400	Engin et al. (2018)
Botryococcus sp.	15 g/L molasses	Mixotrophic	3,005 mg/L	67.7	Yeesang and Cheirsilp (2014)
Scenedesmus sp.	Butyrate-type dark fermented molasses	Heterotrophic	166.8	64.8	Ren et al. (2018)
Spirulina platensis	50% (v/v) anaerobic digested molasses	Mixotrophic	1,230	25.03%	Kaushik et al. (2006)
Spirulina maxima sp.	5 g/L beet vinasse	Mixotrophic (batch mode)	150	–	Barrocal et al. (2010)
Neochloris oleoabundans sp.	6% (v/v) anaerobic digested vinasse +1 g/L NaHCO$_3$	Mixotrophic	1.2 mg/L	17.7%	Olguín et al. (2015)
Braunii Chlorella vulgaris sp.	2 g/L anaerobic digested vinasse	Mixotrophic	70	17	Marques et al. (2013)

COD removal: 95.34%, TN removal: 51.18%, TP removal: 89.29%.

Tsolcha et al. (2016) evaluated the BP potential despite the removal of pollutants from wastewater by diluting aerobically treated whey wastewater with water in the range of 0.005–0.35 as a substrate for the production of Choricystis-like algae under photoheterotrophic conditions. According to the results of the study, COD, TN, and PO_4^{3-} removal efficiencies from wastewater were found to be 92.3%, 97.3%, and 99.7%, respectively, while the lipid content of the produced algae biomass ranged from 9.2% to 13.4%, corresponding to 60.8–119.5 mg/L LP. They concluded that the ratio of saturated and monounsaturated fatty acids in the produced lipids reached 79%, and therefore, the investigated treated whey wastewater could be a very suitable source for biodiesel production with an algae-based system (Tsolcha et al., 2016).

In another study, *Chlorella vulgaris* was tried to be grown in four different growing media (Blue Green Medium (BG11), Blue Green Medium (BG11) + Whey, Bold's Basal Medium (BMM) + Whey, and tap water). As a result of the trials lasting 21 days, the highest increase in cell number, biomass, and proportional lipid content was obtained in *Chlorella vulgaris* grown in Bold's Basal Medium + Whey. Biomass and proportional oil amounts were determined as 10.14 g/L and 20.7%, respectively. In the experiments, it was concluded that whey can be used successfully as a food source in the production of *chlorella vulgaris* (Koç and Duran, 2017). The results of studies carried out on dairy wastewater in the literature are summarized in Table 12.27.

12.8.4 SWINE WASTEWATERS

It has been reported in the literature that due to the possibility of antibiotic residues in the microalgae biomass being produced by phycoremediation of swine production facility wastewater, there may be problems related to the safety of its use (Vickers, 2017).

On the other hand, some researchers have suggested that the microalgae biomass to be obtained due to its rich carbohydrate content can be an interesting raw material source for bioethanol production at low costs (Ho et al., 2013) and therefore used with a developed application that will eliminate its disadvantages for biomethane production in the context of circular economy (Perazzoli et al., 2016).

For example, the treatability and bioethanol production efficiency of pig wastewater plant wastewater in a microalgae-based treatment system were investigated by Michelon et al. (2016). At the end of this study, it was revealed in the literature (Silva and Bertucco, 2019) that the biomass that can be obtained with an average of 50% carbon content from swine production wastewater can result in an estimated annual production of 15.051 L bioethanol/ha/year by performing 324 L bioethanol conversion per one-ton biomass. In addition, in this study, the potential to convert the produced biomass to biomethane is considered to be 44 ± 2.5 L-CH$_4$, which is estimated to be produced from one ton of biomass. Perazzoli et al. (2016) calculated 2,044,000 LN biomethane/ha/year. In general, it has been determined that the inhibition caused by antibiotics present in the wastewater at a concentration of about 1 mg/L in the phycoremediation of pig production wastewater with a circular economic approach using a microalgae consortium including *Chlorella* spp. can be prevented by increasing the hydraulic retention time (~11 days). Thus, it has been demonstrated that BP potential can be improved by considering the integration of substrates (i.e., microalgae and pig manure) and microalgae cellular composition changes (Michelon et al., 2016; Perazzoli et al., 2016).

TABLE 12.27

Microalgae Treatment and Biomass Conversion Efficiency of Dairy Industry Wastewater

Microalgae sp.	Wastewater Type	Dilution	Influent (mg/L)	COD Removal (%)	TN Removal (%)	TP Removal (%)	BP (mg/L/day)	LP (mg/L/day)	References
Scenedesmus quadricauda	Dairy wastewater	No dilution	TOC: 170.1; TN: 85.9; PO_4^{3-}: 8.7	64.47[a]	86.21	89.83[b]	58.75	–	Daneshvar et al. (2018)
Acutodesmus dimorphus	Dairy wastewater	No dilution	COD: 2593.3; NH_4-N: 277.4; PO_4^{3-}: 5.96	90	100	100[b]	210	50	Chokshi et al. (2016a)
Chlorella vulgaris	Dairy wastewater	NaClO pretreatment	COD: 1,338; TN: 65; TP: 60	63.1	77.8	99.5	450	51	Qin et al. (2014)
Chlorella pyrenoidosa	Untreated dairy wastewater	75% DW (dilution with tap water)	TP: 60; NO_3^-: 66.4; PO_4^{3-}: 31	–	60[c]	87[b]	453	3.5[d]	Kothari et al. (2012)
Chlorella pyrenoidosa	Treated dairy wastewater	75% DW (dilution with tap water)	NO_3^-: 14.3; PO_4^{3-}: 3.4	–	49[c]	83[b]	326.6	2[d]	Kothari et al. (2012)
Chlorella vulgaris	Dairy wastewater + slaughter	1:1	COD: 6,000; TN: 260	77.05	60.5	20.23	432	20.25[e]	Lu et al. (2016)
Chlorella vulgaris	Dairy ADW	No dilution	COD: 356; TN: 28.9	80.62	85.47	65.96	123	–	Choi (2016)

[a] TOC.
[b] Phosphate.
[c] Nitrate.
[d] Lipid(g).
[e] Lipid %.
DW, dairy wastewater; BP, biomass production; LP, lipid production.

Moreover, Michelon et al. (2022) reported that it is possible to remove antibiotics such as tetracycline, chlortetracycline, oxytetracycline, and doxycycline detected in swine wastewater with microalgae-based technologies, but it can be a much more effective tool, especially for the tetracyclines. The microalgae biomass harvested after the removal of tetracyclines has been found to have a carbohydrate-rich content ($\geq$50%). The study results proved that treatment of swine wastewater containing these antibiotic residues using microalgae-based processes may be an environmentally friendly alternative to providing raw materials for bioethanol and biomethane production (Michelon et al., 2022).

12.8.5 FOOD WASTEWATERS

Anaerobic digestion is the most widely used method for the treatment of high-strength food wastewater as it generates energy in the form of biogas (Choi, 2016). However, the anaerobic digested food industrial wastewater still contains high amounts of organic and inorganic nutrients, which require additional treatment before their discharge into water bodies (Gupta et al., 2019). Nutrients present in anaerobic digested food industrial wastewater can be used in agriculture. Nevertheless, direct use of Anaerobic Digestion Waste (ADW) in agriculture is not suggested since its carbon-to-nitrogen ratio might not be tolerated by the plants. The use of microalgae for nutrient removal from anaerobically digested food processing industrial wastewater is a good option, as the obtained biomass can be used for biodiesel production (Choi, 2016). Predominantly, most anaerobic digested food industrial wastewater contains high ammonium concentrations, which inhibit microalgae growth, although the inhibitory concentration of ammonium differs from species to species (Ho et al., 2013).

One way to lower the inhibitory effect of ammonium is through the dilution of wastewater (Shin et al., 2015). Dilution of anaerobically digested food wastewater with municipal wastewater was found to be effective in reducing the concentration of ammonium to such an extent that it could be tolerated by microalgae. Shin et al. (2015) tested different dilution ratios of food wastewater (1/10, 1/20, and 1/30) on a flask scale. The authors found that the dilution ratio of 1/20 (COD: 336.5 mg/L, TN: 136.36 mg/L, and TP: 4.1 mg/L) was the best to achieve higher BP and LP (50.75 and 15.59 mg/L/day, respectively), as well as COD (66.4%) and nutrient removal (90%). The authors also found that the dilution ratio of 1/20 enhanced the BP and LP of Scenedesmus bijuga by 1.2 and 1.9 times, respectively, compared to the control medium.

In another study in which *Chlorella* sp. mixotrophic microalgae used acidogenically digested swine wastewater culture medium, biomass productivity/maximum biomass was found to be 0.276 g/L/day, with the NH_3-N, Total-N, PO_4^{3-}-P, and COD pollutants' values reduced to 60.39, 38.5, 20.21, and 751.33 mg/L, respectively (Hu et al., 2013).

12.9 CONCLUSION

Depletion of fossil fuels, increasing demand for energy, and global climate change concerns have led humanity to search for an alternative sustainable fuel source. In this context, vegetable oils, animal fats, and domestic/industrial waste oils have been seen as raw materials for biofuel production. However, biofuels produced, especially from vegetable oils, were limited to a maximum of 1.7% in the renewable energy category until 2030 in 2018 with the Renewable Energy Directive II, as they needed agricultural lands to compete with food crops such as soy, sunflower, and canola and consume clean water resources. In contrast, microalgae have emerged as a potential raw material for biofuel production. Algae have an oil content of 20%–80%, which can be converted into different types of fuel, such as kerosene oil and biodiesel. Diesel production from algae is economical and easy. In addition, microalgae can be cultured without the need for agricultural lands or ecological landscapes, and they can provide an opportunity to use treatment technologies as integrated systems with WWT and carbon dioxide capture technologies to prevent global climate change. In recent years, with the development of technologies in which algal biomass and biofuel production are integrated with wastewater and flue gas treatment, the competitiveness of microalgae-based phycoremediation technologies with conventional systems has increased. Especially in wastewater recovery applications, the treatment method with microalgae varies depending on the purpose of reuse. Most of the domestic and industrial wastewater provides a suitable environment for most photosynthetic microalgae. When microalgae are exposed to such suitable environments, they are biotechnological products that have the potential to grow rapidly and convert solar energy into cellular material and heat. So, microalgae-based phycoremediation technologies are seen as a good ecological investment for countries with a sustainable green economy and environmental understanding.

During the daytime, algae absorb carbon dioxide in the atmosphere like a great generator and produce oxygen, which is vital for the lives of humans and other living things. In addition to the biomass obtained from algae, the high commercial value of the valuable metabolites they accumulate in their cellular structures and the benefits such as the use of some species in environmental applications, especially in the treatment of wastewater, increase the current interest in microalgae and make it an area where intensive research is carried out in biotechnology. Because, while achieving high biomass efficiency, appropriate conditions for each wastewater type still need to be investigated in order to prevent some refractory pollutants found in the wastewater structure from inhibiting algae growth.

AUTHOR CONTRIBUTION

Conceive: G.Y.T, M.A.G.; Design: G.Y.T., M.A.G.; Supervision: G.Y.T., M.A.G.; Literature Review: G.Y.T., M.A.G., E.Ö.; Writer: G.Y.T., M.A.G., E.Ö.; Critical Reviews: G.Y.T., M.A.G.

CONFLICT OF INTEREST

The authors have declared no conflicts of interest. A statement concerning the conflicts of interest of all authors is mandatory.

REFERENCES

Abalde, J., Betancourt, L., Torres, E., Cid, A., & Barwell, C. (1998). Purification and characterization of phycocyanin from the marine cyanobacterium Synechococcus sp. IO9201. *Plant Science, 136*(1), 109–120.

Abdelfattah, A., Ali, S. S., Ramadan, H., El-Aswar, E. I., Eltawab, R., Ho, S.-H., Elsamahy, T., Li, S., El-Sheekh, M. M., & Schagerl, M. (2022). Microalgae-based wastewater treatment: Mechanisms, challenges, recent advances, and future prospects. *Environmental Science and Ecotechnology, 13*, 100205.

Abd El-Moniem, E. A., & Abd-Allah, A. S. E. (2008). Effect of green alga cells extract as foliar spray on vegetative growth, yield and berries quality of superior grapevines. *American-Eurasian Journal of Agricultural & Environmental Sciences, 4*(4), 427–433.

Abiusi, F., Wijffels, R. H., & Janssen, M. (2020). Oxygen balanced mixotrophy under day-night cycles. *ACS Sustainable Chemistry & Engineering, 8*(31), 11682–11691.

Abo-Hashesh, M., & Hallenbeck, P. C. (2012). Fermentative hydrogen production. In: Hallenbeck, P. C. (eds.), *Microbial Technologies in Advanced Biofuels Production* (pp. 77–92). Berlin, Germany: Springer.

Abu-Rezq, T. S., Al-Hooti, S., & Jacob, D. A. (2010). Optimum culture conditions required for the locally isolated Dunaliella salina. *Journal of Algal Biomass Utilization, 1*(2), 12–19.

Al-Gosaibi, A. M. (1994). Use of algae as a soil conditioner for improvement of sandy soils in Al-Ahsa, Saudi Arabia. *Journal of Agricultural Sciences, Mansoura Univiversiy (Egypt), 1*(6), 475–479.

Ananthi, V., Brindhadevi, K., Pugazhendhi, A., & Arun, A. (2021). Impact of abiotic factors on biodiesel production by microalgae. *Fuel, 284*, 118962.

Ananya, A. K., & Ahmad, I. Z. (2014). Cyanobacteria "the blue green algae" and its novel applications: A brief review. *International Journal of Innovation and Applied Studies, 7*(1), 251.

Antony, A. R., & Ramanan, R. (2021). Genetic engineering of microalgae for the production of high-value metabolites: Status and prospects. In: Sarada, R., Schenk, P., & Shekh, A. (eds.), *Microalgal Biotechnology* (pp. 209–253). Cambridge: Royal Society of Chemistry.

Anwar, M., Lou, S., Chen, L., Li, H., & Hu, Z. (2019). Recent advancement and strategy on bio-hydrogen production from photosynthetic microalgae. *Bioresource Technology, 292*, 121972.

Asha P., Singh, N. K., Ashok P., Gnansounou, E., & Datta M. (2011). Cyanobacteria and microalgae: a positive prospect for biofuels.

Aydin-Sisman, G. (2019). Mikroalg Teknolojisi ve Çevresel Kullanımı, *Harran Üniversitesi Mühendislik Dergisi, 4*(1), 81–92.

Bahadar, A., & Khan, M. B. (2013). Progress in energy from microalgae: A review. *Renewable and Sustainable Energy Reviews, 27*, 128–148.

Balasubramanian, D., Arunachalam, K., Arunachalam, A., & Das, A. K. (2013). Effect of water hyacinth (Eichhornia crassipes) mulch on soil microbial properties in lowland rainfed rice-based agricultural system in Northeast India. *Agricultural Research, 2*, 246–257.

Barrocal, V. M., García-Cubero, M. T., González-Benito, G., & Coca, M. (2010). Production of biomass by Spirulina maxima using sugar beet vinasse in growth media. *New biotechnology, 27*(6), 851–856.

Barros, A. I., Gonçalves, A. L., Simões, M., & Pires, J. C. (2015). Harvesting techniques applied to microalgae: A review. *Renewable and Sustainable Energy Reviews, 41*, 1489–1500.

Bartley, M. L., Boeing, W. J., Corcoran, A. A., Holguin, F. O., & Schaub, T. (2013). Effects of salinity on growth and lipid accumulation of biofuel microalga Nannochloropsis salina and invading organisms. *Biomass and Bioenergy, 54*, 83–88.

Bartley, M. L., Boeing, W. J., Dungan, B. N., Holguin, F. O., & Schaub, T. (2014). pH effects on growth and lipid accumulation of the biofuel microalgae Nannochloropsis salina and invading organisms. *Journal of Applied Phycology, 26*, 1431–1437.

Becker, E. W. (1995). *Microalgae: Biotechnology and Microbiology* (vol. 10). Cambridge: Cambridge University Press.

Becker, E., & Ventkateraman, L. V. (1981). Biotechnology and exploitation of algae-the Indian approach-edited by RD fox. All Indian Coordinated Project on Algae Department of Science and Technolgy, India.

Beijerinck, M. W. (1890). Culturversuche mit Zoochlorellen, Lichenengonidien und anderen niederen Algen. *Botanische Zeitung, 48*, 781–785.

Belnap, J., & Gardner, J. S. (1993). Soil microstructure in soils of the Colorado Plateau: The role of the cyanobacterium *Microcoleus vaginatus*. The Great Basin Naturalist, *53*, 40–47.

Belotti, G., de Caprariis, B., De Filippis, P., Scarsella, M., & Verdone, N. (2014). Effect of *Chlorella vulgaris* growing conditions on bio-oil production via fast pyrolysis. *Biomass and Bioenergy, 61*, 187–195.

Ben-Amotz, J.E.W. Polle, D.V. Subba Rao., (2009). (Eds.), The Alga Dunaliella: Biodiversity, Physiology, Genomics and Biotechnology, Science Publishers, Enfield, NH, USA.

Bergman, B., Gallon, J. R., Rai, A. N., & Stal, L. J. (1997). N2 fixation by non-heterocystous cyanobacteria. *FEMS Microbiology Reviews, 19*(3), 139–185.

Bhatnagar, A., Bhatnagar, M., Chinnasamy, S., & Das, K. C. (2010). Chlorella minutissima: A promising fuel alga for cultivation in municipal wastewaters. *Applied Biochemistry and Biotechnology, 161*, 523–536.

Bhuvaneshwari, B., Subramaniyan, V., & Malliga, P. (2011). Comparative studies of cyanopith and cyanospray biofertilizers with chemical fertilizers on sunflower (*Helianthus annuus* L.). *International Journal of Environmental Sciences, 1*(7), 1515–1525.

Bird, C. J., Nelson, W. A., Rice, E. L., Ryan, K. G., & Villemur, R. (1990). A critical comparison of Gracilaria chilensis and G. sordida (Rhodophyta, Gracilariales). *Journal of applied phycology, 2*, 375–382.

Bird K., Chynoweth D., Jerger D., Effects of marine algal proximate composition on methane yields, *Journal of applied phycology*, 2 (3), 207–213, 1990. 225.

Blair, M. F., Kokabian, B., & Gude, V. G. (2014). Light and growth medium effect on *Chlorella vulgaris* biomass production. *Journal of Environmental Chemical Engineering, 2*(1), 665–674.

Blier, R., Laliberte, G., & De la Noüe, J. (1995). Tertiary treatment of cheese factory anaerobic effluent with Phormidium bohneri and Micractinum pusillum. *Bioresource Technology, 52*(2), 151–155.

Bohutskyi, P., & Bouwer, E. (2013). Biogas production from algae and cyanobacteria through anaerobic digestion: A review, analysis, and research needs. In: Lee, J. W. (ed.), *Advanced Biofuels and Bioproducts* (pp. 873–975). New York: Springer.

Borines, M. G., de Leon, R. L., & Cuello, J. L. (2013). Bioethanol production from the macroalgae *Sargassum* spp. *Bioresource Technology, 138*, 22–29.

Borowitzka, M. A. (1988). Vitamins and fine chemicals from microalgae. In: Borowitzka, M. A., & Borowitzka, L. J. (Eds), *Micro-Algal Biotechnology* (pp. 173–196). Cambridge, UK: Cambridge University Press.

Borowitzka, M. A. (2013). High-value products from microalgae-their development and commercialisation. *Journal of Applied Phycology, 25*, 743–756.

Borowitzka, M. A., & Moheimani, N. R. (eds.). (2013). *Algae for Biofuels and Energy* (Vol. 5, pp. 133–152). Dordrecht: Springer.

Barrocal, V. M., García-Cubero, M. T., González-Benito, G., & Coca, M. (2010). Production of biomass by Spirulina maxima using sugar beet vinasse in growth media. *New biotechnology, 27*(6), 851–856.

Bošnjaković, M., & Sinaga, N. (2020). The perspective of large-scale production of algae biodiesel. *Applied Sciences, 10*(22), 8181.

Boussiba S, Richmond AE (1979) Isolation and characterization of phycocyanins from the Blue-Green Alga Spirulina platensis. Arch. Microbiol. 120: 155–159.

Brennan, L., & Owende, P. (2010). Biofuels from microalgae: A review of technologies for production, processing, and extractions of biofuels and co-products. *Renewable and Sustainable Energy Reviews, 14*(2), 557–577.

Breuer, G., Lamers, P. P., Martens, D. E., Draaisma, R. B., & Wijffels, R. H. (2012). The impact of nitrogen starvation on the dynamics of triacylglycerol accumulation in nine microalgae strains. *Bioresource Technology, 124,* 217–226.

Briand, X., & Morand, P. (1997). Anaerobic digestion of Ulva sp. 1. Relationship between Ulva composition and methanisation. *Journal of applied phycology,* 9, 511–524.

Burlage, R. S., Atlas, R., Stahl, D., Sayler, G., & Geesey, G. (eds.). (1998). *Techniques in Microbial Ecology* (pp. 8–14). Demand: Oxford University Press.

Burlew, J. S. (1953). Algal culture from laboratory to pilot plant. *BioScience,* 3(5), 11.

Buswell, A. M., & Mueller, H. F. (1952). Mechanism of methane fermentation. *Industrial & Engineering Chemistry, 44*(3), 550–552.

Cadoret, J. P., Garnier, M., & Saint-Jean, B. (2012). Microalgae, functional genomics and biotechnology. *Advances in Botanical Research, 64,* 285–341.

Carvalho, A. P., Silva, S. O., Baptista, J. M., & Malcata, F. X. (2011). Light requirements in microalgal photobioreactors: An overview of biophotonic aspects. *Applied Microbiology and Biotechnology, 89,* 1275–1288.

Challem, J. J. (1981). Spirulina: What It Is, the Health Benefits It Can Give You. New Canaan, CT: Keats Publishing Inc.

Chamizo, S., Adessi, A., Mugnai, G., Simiani, A., & De Philippis, R. (2019). Soil type and cyanobacteria species influence the macromolecular and chemical characteristics of the polysaccharidic matrix in induced biocrusts. *Microbial Ecology, 78,* 482–493.

Chan, S. S., Khoo, K. S., Chew, K. W., Ling, T. C., & Show, P. L. (2022). Recent advances biodegradation and biosorption of organic compounds from wastewater: Microalgae-bacteria consortium-A review. *Bioresource Technology, 344,* 126159.

Charles, C. N., Msagati, T., Swai, H., & Chacha, M. (2019). Microalgae: An alternative natural source of bioavailable omega-3 DHA for promotion of mental health in East Africa. *Scientific African, 6,* e00187.

Chaudhry, F. N., & Malik, M. F. (2017). Factors affecting water pollution: A review. *Journal of Ecosystem & Ecography, 7*(1), 225–231.

Chen, Y., Cheng, J. J., & Creamer, K. S. (2008). Inhibition of anaerobic digestion process: A review. *Bioresource Technology, 99*(10), 4044–4064.

Cheng, J., Huang, R., Li, T., Zhou, J., & Cen, K. (2014). Biodiesel from wet microalgae: Extraction with hexane after the microwave-assisted transesterification of lipids. *Bioresource Technology, 170,* 69–75.

Cheng, P., Huang, J., Song, X., Yao, T., Jiang, J., Zhou, C., … Ruan, R. (2022). Heterotrophic and mixotrophic cultivation of microalgae to simultaneously achieve furfural wastewater treatment and lipid production. *Bioresource Technology, 349,* 126888.

Chisti, Y. (2007). Biodiesel from microalgae. *Biotechnology Advances, 25*(3), 294–306.

Choi, H. J. (2016). Dairy wastewater treatment using microalgae for potential biodiesel application. *Environmental Engineering Research, 21*(4), 393–400.

Choi, J. S., Park, Y. H., Kim, S., Son, J., Park, J., & Choi, Y. E. (2022). Strategies to control the growth of cyanobacteria and recovery using adsorption and desorption. *Bioresource Technology, 365,* 128133.

Chokshi, K., Pancha, I., Ghosh, A., & Mishra, S. (2016). Microalgal biomass generation by phycoremediation of dairy industry wastewater: an integrated approach towards sustainable biofuel production. *Bioresource technology, 221,* 455–460.

Choudhury, A. T. M. A., & Kennedy, I. R. (2004). Prospects and potentials for systems of biological nitrogen fixation in sustainable rice production. *Biology and Fertility of Soils, 39,* 219–227.

Christenson, L., & Sims, R. (2011). Production and harvesting of microalgae for wastewater treatment, biofuels, and bioproducts. *Biotechnology Advances*, *29*(6), 686–702.

Chynoweth, D. P. (2005). Renewable biomethane from land and ocean energy crops and organic wastes. *Hortscience*, *40*(2), 283–286.

Ciferri, O. (1983). Spirulina, the edible microorganism. *Microbiological Reviews*, *47*(4), 551–578.

Codd, G. A., Lindsay, J., Young, F. M., Morrison, L. F., & Metcalf, J. S. (2005). Harmful cyanobacteria: From mass mortalities to management measures. In: Huisman, J., Matthijs, H. C. P., & Visser, P. M. (eds.), *Harmful Cyanobacteria* (pp. 1–23). Dordrecht: Springer.

Cohen, Z. (1997). The chemicals of Spirulina. In: Vonshak, A. (ed.), *Spirulina Platensis Arthrospira* (pp. 193–222). Boca Raton, FL: CRC Press.

Conk Dalay, M., İmamoplu, E., & Öncel, S. (2008). Mikroalgal Biyokütle Üretimi Gçin DüĞük Maliyetli Fotobiyoreaktör Tasarımı. TÜBĞTAK MAG, Proje (104M354).

Courchesne, N. M. D., Parisien, A., Wang, B., & Lan, C. Q. (2009). Enhancement of lipid production using biochemical, genetic and transcription factor engineering approaches. *Journal of Biotechnology*, *141*(1–2), 31–41.

Crini, G., & Lichtfouse, E. (2019). Advantages and disadvantages of techniques used for wastewater treatment. *Environmental Chemistry Letters*, *17*, 145–155.

Daneshvar, E., Santhosh, C., Antikainen, E., & Bhatnagar, A. (2018). Microalgal growth and nitrate removal efficiency in different cultivation conditions: Effect of macro and micronutrients and salinity. *Journal of Environmental Chemical Engineering*, *6*(2), 1848–1854.

Danquah, M. K., Gladman, B., Moheimani, N., & Forde, G. M. (2009). Microalgal growth characteristics and subsequent influence on dewatering efficiency. *Chemical Engineering Journal*, *151*(1–3), 73–78.

Darzins, A., Pienkos, P., & Edye, L. (2010). Current status and potential for algal biofuels production. *A Report to IEA Bioenergy Task*, *39*(13), 403–412.

Del Mondo, A., Sansone, C., & Brunet, C. (2022). Insights into the biosynthesis pathway of phenolic compounds in microalgae. *Computational and Structural Biotechnology Journal*, *20*, 1901–1913.

Demirbas, A. (2010). Use of algae as biofuel sources. *Energy Conversion and Management*, *51*(12), 2738–2749.

Demoulin, C. F., Lara, Y. J., Cornet, L., François, C., Baurain, D., Wilmotte, A., & Javaux, E. J. (2019). Cyanobacteria evolution: Insight from the fossil record. *Free Radical Biology and Medicine*, *140*, 206–223.

Díaz-Pérez, F. J., Díaz, R., Valdés, G., Valdebenito-Rolack, E., & Hansen, F. (2023). Effects of microalgae and compost on the yield of cauliflower grown in low nutrient soil. *Chilean Journal of Agricultural Research*, *83*(2), 181–194.

Dos Santos, R. R., Araújo, O. D. Q. F., de Medeiros, J. L., & Chaloub, R. M. (2016). Cultivation of Spirulina maxima in medium supplemented with sugarcane vinasse. *Bioresource Technology*, *204*, 38–48.

Du, X., Liu, H., Yuan, L., Wang, Y., Ma, Y., Wang, R., ... Zhang, H. (2019). The diversity of cyanobacterial toxins on structural characterization, distribution and identification: A systematic review. *Toxins*, *11*(9), 530.

Duru, M. D., & Yılmaz, H. K. (2013). Mikroalglerin pigment kaynağı olarak balık yemlerinde kullanımı. *Türk Bilimsel Derlemeler Dergisi*, *2*, 112–118.

Ebenezer, V., Medlin, L. K., & Ki, J. S. (2012). Molecular detection, quantification, and diversity evaluation of microalgae. *Marine Biotechnology*, *14*, 129–142.

El-Zeky, M. M., Metwaly, G. S., & Aref, E. M. (2005). Using of cyanobacteria or azolla as alternative sources of nitrogen for rice production. *Journal of Agricultural Chemistry and Biotechnology*, *30*(9), 5567–5577.

Elcik, H., & Çakmakcı, M. (2017). Mikroalg üretimi ve mikroalglerden biyoyakıt eldesi. *Journal of the Faculty of Engineering and Architecture of Gazi University*, *32*(3), 795–820.

Eleren, S. Ç., & Burak, Ö. (2019). Sürdürülebilir ve çevre dostu biyoyakıt hammaddesi: Mikroalgler. *Pamukkale Üniversitesi Mühendislik Bilimleri Dergisi, 25*(3), 304–319.

El-Fouly, M. M., Abdalla, F. E., & Shaaban, M. M. (1992). Multipurpose large scale production of microalgae biomass in Egypt. In *Proceedings of the 1st Egyptian Etalian Symposium on Biotechnology*, Assuit, Egypt, Nov. 21–23, 305–314.

El-Sheekh, M., El-Dalatony, M. M., Thakur, N., Zheng, Y., & Salama, E. S. (2021). Role of microalgae and cyanobacteria in wastewater treatment: Genetic engineering and omics approaches. *International Journal of Environmental Science and Technology, 19*, 2173–2194.

El-Sikaily, A., El Nemr, A., Khaled, A., & Abdelwehab, O. (2007). Removal of toxic chromium from wastewater using green alga Ulva lactuca and its activated carbon. *Journal of Hazardous Materials, 148*(1–2), 216–228.

Engin, I. K., Cekmecelioglu, D., Yücel, A. M., & Oktem, H. A. (2018). Heterotrophic growth and oil production from Micractinium sp. ME05 using molasses. *Journal of applied phycology, 30*, 3483–3492.

Environmental Protection Agency (2020). Renewable Fuel Standard Program: Standards for 2020 and Biomass-Based Diesel Volume for 2021 and Other Changes: https://www.govinfo.gov/content/pkg/FR-2020-02-06/pdf/2020-00431.pdf.

European Parliament (2018). 15. Directive (EU) 2018/2001 of the European Parliament and of the Council of 11 December 2018 on the Promotion of the Use of Energy from Renewable Sources (Text with EEA relevance.), PE/48/2018/REV/1. https://eur-lex.europa.eu/eli/dir/2018/2001/oj.

Faried, M., Samer, M., Abdelsalam, E., Yousef, R.S., Attia, Y.A., Ali, A.S. (2017) Biodiesel production from microalgae: Processes, technologies and recent advancements. Renewable and Sustainable Energy Reviews, 79, 893–913.

Farrar, W. V. (1966). Tecuitlatl; a glimpse of Aztec food technology. *Nature, 211*, 341–342.

Fathi, M., & Asem, A. (2013). Investigating the impact of NaCl salinity on growth, β-carotene, and chlorophyll a in the content life of halophytes of algae *Chlorella* sp. *Aquaculture, Aquarium, Conservation & Legislation, 6*(3), 241–245.

Galès, A., Bonnafous, A., Carré, C., Jauzein, V., Lanouguère, E., Le Floc'h, E., … Fouilland, E. (2019). Importance of ecological interactions during wastewater treatment using high rate algal ponds under different temperate climates. *Algal Research, 40*, 101508.

Gan, Y. Y., Chen, W. H., Ong, H. C., Lin, Y. Y., Sheen, H. K., Chang, J. S., & Ling, T. C. (2021). Effect of wet torrefaction on pyrolysis kinetics and conversion of microalgae carbohydrates, proteins, and lipids. *Energy Conversion and Management, 227*, 113609.

Gaurav, K., Srivastava, R., Sharma, J. G., Singh, R., & Singh, V. (2016). Molasses-based growth and lipid production by Chlorella pyrenoidosa: A potential feedstock for biodiesel. *International Journal of Green Energy, 13*(3), 320–327.

Genç, N. (2009). Biyolojik hidrojen üretim prosesleri. *Balıkesir Üniversitesi Fen Bilimleri Enstitüsü Dergisi, 11*(2), 17–36.

Gentili, F. G. (2014). Microalgal biomass and lipid production in mixed municipal, dairy, pulp and paper wastewater together with added flue gases. *Bioresource Technology, 169*, 27–32.

Givens, D.I. (1997). Sources of N-3 Polyunsaturated Fatty Acids Additional to Fish Oil for livestock diets, New Meats Congress, Bristol- England.

George, B., Pancha, I., Desai, C., Chokshi, K., Paliwal, C., Ghosh, T., & Mishra, S. (2014). Effects of different media composition, light intensity and photoperiod on morphology and physiology of freshwater microalgae Ankistrodesmus falcatus: A potential strain for bio-fuel production. *Bioresource Technology, 171*, 367–374.

Ghosh, A., Sarkar, S., Gayen, K., & Bhowmick, T. K. (2020). Effects of carbon, nitrogen, and phosphorus supplements on growth and biochemical composition of *Podohedriella* sp. (MCC44) isolated from northeast India. *Environmental Progress & Sustainable Energy, 39*(4), e13378.

Goswami, R. K., Agrawal, K., & Verma, P. (2022). Phycoremediation of nitrogen and phosphate from wastewater using *Picochlorum* sp.: A tenable approach. *Journal of Basic Microbiology*, *62*(3–4), 279–295.

Grima, E. M., Belarbi, E. H., Fernández, F. A., Medina, A. R., & Chisti, Y. (2003). Recovery of microalgal biomass and metabolites: Process options and economics. *Biotechnology Advances*, *20*(7–8), 491–515.

Grinstead, G. S., Tokach, M. D., Dritz, S. S., Goodband, R. D., & Nelssen, J. L. (2000). Effects of Spirulina platensis on growth performance of weanling pigs. *Animal Feed Science and Technology*, *83*(3–4), 237–247.

Gross, M. A. (2013). Development and optimization of algal cultivation systems (Doctoral dissertation, Iowa State University).

Gupta, S., Pawar, S. B., & Pandey, R. A. (2019). Current practices and challenges in using microalgae for treatment of nutrient rich wastewater from agro-based industries. *Science of the Total Environment*, *687*, 1107–1126.

Hanon, M., Gimbel, J., Tlan, M., Rasala, B., & Mayfield, S. (2010). Biofuels from algae, challenge and potential. *Biofuels*, *1*(5), 763–784.

Harun, R., & Danquah, M. K. (2011). Influence of acid pre-treatment on microalgal biomass for bioethanol production. *Process Biochemistry*, *46*(1), 304–309.

Harun, R., Singh, M., Forde, G. M., & Danquah, M. K. (2010). Bioprocess engineering of microalgae to produce a variety of consumer products. *Renewable and Sustainable Energy Reviews*, *14*(3), 1037–1047.

Hauck, J. T., Scierka, S. J., & Perry, M. B. (1996). Effects of simulated flue gas on growth of microalgae. *Preprints of Papers, American Chemical Society, Division of Fuel Chemistry, Orlando, FL*, *41(CONF-960807)*.

Hegazi, A. Z., Mostafa, S. S., & Ahmed, H. M. (2010). Influence of different cyanobacterial application methods on growth and seed production of common bean under various levels of mineral nitrogen fertilization. *Nature and Science*, *8*(11), 183–194.

Hegde, D. M., Dwivedi, B. S., & Sudhakara Babu, S. N. (1999). Biofertilizers for cereal production in India: A review. *Indian Journal of Agricultural Science*, *69*(2), 73–83.

Heimann, K., & Huerlimann, R. (2015). Microalgal classification: Major classes and genera of commercial microalgal species. In: Kim, S. K. (ed.), *Handbook of Marine Microalgae* (pp. 25–41). Cambridge, MA: Academic Press.

Ho, S. H., Huang, S. W., Chen, C. Y., Hasunuma, T., Kondo, A., & Chang, J. S. (2013). Characterization and optimization of carbohydrate production from an indigenous microalga Chlorella vulgaris FSP-E. *Bioresource Technology*, *135*, 157–165.

Hossain, A. S., Salleh, A., Boyce, A. N., Chowdhury, P., & Naqiuddin, M. (2008). Biodiesel fuel production from algae as renewable energy. *American Journal of Biochemistry and Biotechnology*, *4*(3), 250–254.

Hsueh, H. T., Li, W. J., Chen, H. H., & Chu, H. (2009). Carbon bio-fixation by photosynthesis of *Thermosynechococcus* sp. CL-1 and *Nannochloropsis oculta*. *Journal of Photochemistry and Photobiology B: Biology*, *95*(1), 33–39.

Hu, B., Zhou, W., Min, M., Du, Z., Chen, P., Ma, X., ... Ruan, R. (2013). Development of an effective acidogenically digested swine manure-based algal system for improved wastewater treatment and biofuel and feed production. *Applied Energy*, *107*, 255–263.

Hu, J., Nagarajan, D., Zhang, Q., Chang, J. S., & Lee, D. J. (2018). Heterotrophic cultivation of microalgae for pigment production: A review. *Biotechnology Advances*, *36*(1), 54–67.

Hussain, A., & Hasnain, S. (2011). Phytostimulation and biofertilization in wheat by cyanobacteria. *Journal of Industrial Microbiology and Biotechnology*, *38*(1), 85–92.

Ibraheem, I. (2007). Cyanobacteria as alternative biological conditioners for bioremediation of barren soil. *Egyptian Journal of Phycology*, *8*(1), 99–117.

İlter, I., Akyıl, S., Koç, M., & Kaymak-Ertekin, F. (2017). Alglerden elde edilen ve gıdalarda doğal renklendirici olarak kullanılan pigmenter ve fonksiyonel özellikleri. *Türk Tarım-Gıda Bilim ve Teknoloji Dergisi*, *5*(12), 1508–1515.

Ip, S. Y., Bridger, J. S., Chin, C. T., Martin, W. R. B., & Raper, W. G. C. (1982). Algal growth in primary settled sewage: The effects of five key variables. *Water Research*, *16*(5), 621–632.

Irankhahi, P., Riahi, H., Shariatmadari, Z., & Aghashariatmadari, Z. (2023). Diversity and distribution of heterocystous cyanobacteria across solar radiation gradient in terrestrial habitats of Iran. *Rostaniha*, *23*(2), 264–281.

Irisarri, P., Gonnet, S., Deambrosi, E., & Monza, J. (2007). Cyanobacterial inoculation and nitrogen fertilization in rice. *World Journal of Microbiology and Biotechnology*, *23*, 237–242.

Jacob-Lopes, E., Scoparo, C. H. G., Lacerda, L. M. C. F., & Franco, T. T. (2009). Effect of light cycles (night/day) on CO2 fixation and biomass production by microalgae in photobioreactors. *Chemical Engineering and Processing: Process Intensification*, *48*(1), 306–310.

Jalilian, N., Najafpour, G. D., & Khajouei, M. (2020). Macro and micro algae in pollution control and biofuel production: A review. *ChemBioEng Reviews*, *7*(1), 18–33.

Jeffries, D. L., Klopatek, J. M., Link, S. O., & Bolton Jr, H. (1992). Acetylene reduction by cryptogamic crusts from a blackbrush community as related to resaturation and dehydration. *Soil Biology and Biochemistry*, *24*(11), 1101–1105.

Jeong, Y., Cho, S. H., Lee, H., Choi, H. K., Kim, D. M., Lee, C. G., … Cho, B. K. (2020). Current status and future strategies to increase secondary metabolite production from cyanobacteria. *Microorganisms*, *8*(12), 1849.

Johnston, H. W. (1970). The biological and economic importance of algae, Part 3. *Edible algae of fresh and brackish waters. Tuatara*, *18*(1), 19–35.

Jones, C. S., & Mayfield, S. P. (2012). Algae biofuels: Versatility for the future of bioenergy. *Current Opinion in Biotechnology*, *23*(3), 346–351.

Juneja, A., & Murthy, G. S. (2018). Model predictive control coupled with economic and environmental constraints for optimum algal production. *Bioresource Technology*, *250*, 556–563.

Junying, Z. H. U., Junfeng, R. O. N. G., & Baoning, Z. O. N. G. (2013). Factors in mass cultivation of microalgae for biodiesel. *Chinese Journal of Catalysis*, *34*(1), 80–100.

Kafle, A., Timilsina, A., Gautam, A., Adhikari, K., Bhattarai, A., & Aryal, N. (2022). Phytoremediation: Mechanisms, plant selection and enhancement by natural and synthetic agents. *Environmental Advances*, *8*, 100203.

Kaloudas, D., Pavlova, N., & Penchovsky, R. (2021). Phycoremediation of wastewater by microalgae: A review. *Environmental Chemistry Letters*, *19*, 2905–2920.

Kalyanasundaram, G. T., Ramasamy, A., Rakesh, S., & Subburamu, K. (2020). Microalgae and cyanobacteria: Role and applications in agriculture. In: Arumugam, M., Kathiresan, S., & Nagaraj, S. (eds.), *Applied Algal Biotechnology*. Hauppauge, NY: Nova Science Publishers.

Katarzyna, L., Sai, G., & Singh, O. A. (2015). Non-enclosure methods for non-suspended microalgae cultivation: Literature review and research needs. *Renewable and Sustainable Energy Reviews*, *42*, 1418–1427.

Kaushik, B. D. (1994). Algalization of rice in salt-affected soils. *Annals of Agricultural Research*, *14*, 105–106.

Kaushik, R., Prasanna, R., & Joshi, H. C. (2006). Utilization of anaerobically digested distillery effluent for the production of Spirulina platensis (ARM 730).

Kawaguchi, H., Hashimoto, K., Hirata, K., & Miyamoto, K. (2001). H2 production from algal biomass by a mixed culture of Rhodobium marinum A-501 and Lactobacillus amylovorus. *Journal of Bioscience and Bioengineering*, *91*(3), 277–282.

Keris-Sen, U. D., Sen, U., Soydemir, G., & Gurol, M. D. (2014). An investigation of ultrasound effect on microalgal cell integrity and lipid extraction efficiency. *Bioresource technology*, *152*, 407–413.

Keriş Şen, Ü. D. (2019). Anaerobik çürütücü atıksularının mikroalg reaktörlerinde arıtılması. Uludağ Üniversitesi Mühendislik Fakültesi Dergisi, *24*(3), 89–108. doi: 10.17482/uumfd.530127.

Kharecha, P., Kasting, J., & Siefert, J. (2005). A coupled atmosphere-ecosystem model of the early Archean Earth. *Geobiology*, *3*(2), 53–76.

Kim, B., Heo, H. Y., Son, J., Yang, J., Chang, Y. K., Lee, J. H., & Lee, J. W. (2019). Simplifying biodiesel production from microalgae via wet in situ transesterification: A review in current research and future prospects. *Algal Research, 41*, 101557.

Kimura, S., Nakajima, M., Yumoto, E., Miyamoto, K., Yamane, H., Ong, M., … Asami, T. (2020). Cytokinins affect the akinete-germination stage of a terrestrial filamentous cyanobacterium, Nostoc sp. HK-01. *Plant Growth Regulation, 92*, 273–282.

Kiran, B., Pathak, K., Kumar, R., & Deshmukh, D. (2017). Phycoremediation: An eco-friendly approach to solve water pollution problems. In: Kalia, V. C., & Kumar, P. (eds.), *Microbial Applications Vol. 1: Bioremediation and Bioenergy* (pp. 3–28). New York: Springer International Publishing.

Koberg, M., Cohen, M., Ben-Amotz, A., & Gedanken, A. (2011). Bio-diesel production directly from the microalgae biomass of Nannochloropsis by microwave and ultrasound radiation. *Bioresource Technology, 102*(5), 4265–4269.

Kobayashi, M., Kakizono, T., Nishio, N., Nagai, S., Kurimura, Y., & Tsuji, Y. (1997). Antioxidant role of astaxanthin in the green alga Haematococcus pluvialis. *Applied Microbiology and Biotechnology, 48*, 351–356.

Koç, C., & Duran, H. (2017). Determination of the effect of whey as a nutritional supplement in different growth medium regarding to its potential to biodiesel feedstock production. *Anadolu Tarım Bilimleri Dergisi, 32*(3), 309–315.

Kong, Q. X., Li, L., Martinez, B., Chen, P., & Ruan, R. (2010). Culture of microalgae *Chlamydomonas reinhardtii* in wastewater for biomass feedstock production. *Applied biochemistry and Biotechnology, 160*, 9–18.

Koru, E., & Cirik, S. (2002). Biochemical composition of Spirulina biomass in open-air system. Proceedings of ICNP, Trabzon, pp. 97–100.

Kosourov, S., Nagy, V., Shevela, D., Jokel, M., Messinger, J., & Allahverdiyeva, Y. (2020). Water oxidation by photosystem II is the primary source of electrons for sustained H2 photoproduction in nutrient-replete green algae. *Proceedings of the National Academy of Sciences, 117*(47), 29629–29636.

Kothari, R., Pathak, V. V., Kumar, V., & Singh, D. P. (2012). Experimental study for growth potential of unicellular alga Chlorella pyrenoidosa on dairy waste water: an integrated approach for treatment and biofuel production. *Bioresource technology, 116*, 466–470.

Kobayashi, M., Kakizono, T., Nishio, N., Nagai, S., Kurimura, Y., & Tsuji, Y. (1997). Antioxidant role of astaxanthin in the green alga Haematococcus pluvialis. *Applied Microbiology and Biotechnology, 48*, 351–356.

Kruse, O., Rupprecht, J., Bader, K. P., Thomas-Hall, S., Schenk, P. M., Finazzi, G., & Hankamer, B. (2005). Improved photobiological H2 production in engineered green algal cells. *Journal of Biological Chemistry, 280*(40), 34170–34177.

Kumar, A., Ergas, S., Yuan, X., Sahu, A., Zhang, Q., Dewulf, J., … Van Langenhove, H. (2010). Enhanced CO2 fixation and biofuel production via microalgae: Recent developments and future directions. *Trends in Biotechnology, 28*(7), 371–380.

Kumari, N., & Rai, L. C. (2020). Cyanobacterial diversity: Molecular insights under multifarious environmental conditions. In: Singh, P. K., Kumar, A., Singh, V. K., & Shrivistava, A. K. (eds.), *Advances in Cyanobacterial Biology* (pp. 17–33). London: Academic Press.

Kurano, N., Sasaki, T., & Miyachi, S. (1998). Carbon dioxide and microalgae. In Inui, T., Anpo, M., Izui, K., Yanagida, S., & Yamaguchi, T. (eds.), *Studies in Surface Science and Catalysis* (Vol. 114, pp. 55–63). Amsterdam: Elsevier.

Kuśnierz, M., Domańska, M., Hamal, K., & Pera, A. (2022). Application of integrated fixed-film activated sludge in a conventional wastewater treatment plant. *International Journal of Environmental Research and Public Health, 19*(10), 5985.

Lacroux, J., Jouannais, P., Atteia, A., Bonnafous, A., Trably, E., Steyer, J. P., & van Lis, R. (2022). Microalgae screening for heterotrophic and mixotrophic growth on butyrate. *Algal Research, 67*, 102843.

Ladha, J. K., & Reddy, P. M. (2003). Nitrogen fixation in rice systems: State of knowledge and future prospects. *Plant and Soil, 252*, 151–167.

Lage, S., Gojkovic, Z., Funk, C., & Gentili, F. G. (2018). Algal biomass from wastewater and flue gases as a source of bioenergy. *Energies, 11*(3), 664.

Lakaniemi, A.-M. (2012). Microalgal cultivation and utilization in sustainable energy production.

Lange, O. L., Meyer, A., Zellner, H., & Heber, U. (1994). Photosynthesis and water relations of lichen soil crusts: Field measurements in the coastal fog zone of the Namib Desert. *Functional Ecology, 8*(2), 253–264.

Larsdotter, K. (2006). Wastewater treatment with microalgae: A literature review. *Vatten, 62*(1), 31.

Latysheva, N., Junker, V. L., Palmer, W. J., Codd, G. A., & Barker, D. (2012). The evolution of nitrogen fixation in cyanobacteria. *Bioinformatics, 28*(5), 603–606.

Lavens, P., & Sorgeloos, P. (1996). Manual on the production and use of live food for aquaculture (No. 361). Food and Agriculture Organization (FAO).

Lee, C. S., Lee, S. A., Ko, S. R., Oh, H. M., & Ahn, C. Y. (2015). Effects of photoperiod on nutrient removal, biomass production, and algal-bacterial population dynamics in lab-scale photobioreactors treating municipal wastewater. *Water Research, 68*, 680–691.

Lee, J. S., Sung, K. D., & Kim, M. S. (1996). Current aspects of carbon dioxide fixation by microalgae in Korea. *Preprints of Papers, American Chemical Society, Division of Fuel Chemistry, 41*(CONF-960807).

Lee, R. E. (1980) *Phycology.* New York: Cambridge University Press.

Lee, Y. K. (2001). Microalgal mass culture systems and methods: Their limitation and potential. *Journal of Applied Phycology, 13*, 307–315.

Li, K., Liu, Q., Fang, F., Luo, R., Lu, Q., Zhou, W., ... Ruan, R. (2019). Microalgae-based wastewater treatment for nutrients recovery: A review. *Bioresource Technology, 291*, 121934.

Li, Y., Chen, Y. F., Chen, P., Min, M., Zhou, W., Martinez, B., ... Ruan, R. (2011). Characterization of a microalga Chlorella sp. well adapted to highly concentrated municipal wastewater for nutrient removal and biodiesel production. *Bioresource Technology, 102*(8), 5138–5144.

Li, Y., Zhou, W., Hu, B., Min, M., Chen, P., & Ruan, R. R. (2012). Effect of light intensity on algal biomass accumulation and biodiesel production for mixotrophic strains *Chlorella kessleri* and *Chlorella protothecoide* cultivated in highly concentrated municipal wastewater. *Biotechnology and Bioengineering, 109*(9), 2222–2229.

Liang, K., Zhang, Q., Gu, M., & Cong, W. (2013). Effect of phosphorus on lipid accumulation in freshwater microalga Chlorella sp. *Journal of Applied Phycology, 25*, 311–318.

Ling, J., Nip, S., Cheok, W. L., de Toledo, R. A., & Shim, H. (2014). Lipid production by a mixed culture of oleaginous yeast and microalga from distillery and domestic mixed wastewater. *Bioresource Technology, 173*, 132–139.

Little, S. M., Senhorinho, G. N., Saleh, M., Basiliko, N., & Scott, J. A. (2021). Antibacterial compounds in green microalgae from extreme environments: A review. *Algae, 36*(1), 61–72.

Lu, Q., Zhou, W., Min, M., Ma, X., Ma, Y., Chen, P., Zheng, H., Doan, Y. T., Liu, H., & Chen, C. (2016). Mitigating ammonia nitrogen deficiency in dairy wastewaters for algae cultivation. *Bioresource Technology, 201*, 33–40.

Lundquist, T. J., Woertz, I. C., Quinn, N. W. T., & Benemann, J. R. (2010). A realistic technology and engineering assessment of algae biofuel production. *Energy Biosciences Institute, 1–152*.

Macías-Sánchez, MD; Mantell, C; Rodríguez, M; Martínez de la Ossa, E; Lubián, LM; Montero, O. (2009). Comparison of supercritical fluid and ultrasound-assisted extraction of carotenoids and chlorophyll a from *Dunaliella salina. Talanta, 77*, 948–952.

Macías-Sánchez, M. D., Serrano, C. M., Rodríguez, M. R., & de la Ossa, E. M. (2009). Kinetics of the supercritical fluid extraction of carotenoids from microalgae with CO2 and ethanol as cosolvent. *Chemical Engineering Journal, 150*(1), 104–113.

Madigan, M. T., Martinko, J. M., & Parker, J. (1997). *Brock Biology of Microorganisms* (Vol. 11). Upper Saddle River, NJ: Prentice hall.

Mahmood, A. (2016). Buğday yetiştirilen koşullarda farklı mikroalg noküslasyonlarının toprağın mikrobiyal karbon, azot ve fosfor bağlanmasına katkısıt (Master's thesis, Fen Bilimleri Enstitüsü).

Makareviciene, V., & Skorupskaite, V. (2019). Transesterification of microalgae for bio-diesel production. In: Basile, A., & Dalena, F. (eds.), *Second and Third Generation of Feedstocks* (pp. 469–510). Amsterdam: Elsevier.

Malam Issa, O., Défarge, C., Le Bissonnais, Y., Marin, B., Duval, O., Bruand, A., … Annerman, M. (2007). Effects of the inoculation of cyanobacteria on the microstructure and the structural stability of a tropical soil. *Plant and Soil, 290*, 209–219.

Mandal, B., Vlek, P. L. G., & Mandal, L. N. (1999). Beneficial effects of blue-green algae and Azolla, excluding supplying nitrogen, on wetland rice fields: a review. *Biology and Fertility of Soils, 28*, 329–342.

Maqubela, M. P., Mnkeni, P. N. S., Issa, O. M., Pardo, M. T., & D'acqui, L. P. (2009). Nostoc cyanobacterial inoculation in South African agricultural soils enhances soil structure, fertility, and maize growth. *Plant and Soil, 315*, 79–92.

Marques, S. S. I., Nascimento, I. A., de Almeida, P. F., & Chinalia, F. A. (2013). Growth of Chlorella vulgaris on sugarcane vinasse: the effect of anaerobic digestion pretreatment. *Applied biochemistry and biotechnology, 171*, 1933–1943.

Marsh, A. A. (2008). A study into the cultivation of algae for carbon dioxidesequestration from a power plant and it's use as a bio fuel. Cardiff School of Engineering Cardiff University.

Martınez, M. E., Sánchez, S., Jimenez, J. M., El Yousfi, F., & Munoz, L. (2000). Nitrogen and phosphorus removal from urban wastewater by the microalga Scenedesmus obliquus. *Bioresource Technology, 73*(3), 263–272.

Martínez-Roldán, A. J., Perales-Vela, H. V., Cañizares-Villanueva, R. O., & Torzillo, G. (2014). Physiological response of *Nannochloropsis* sp. to saline stress in laboratory batch cultures. *Journal of Applied Phycology, 26*, 115–121.

de Mattos, L. F. A., & Bastos, R. G. (2016). COD and nitrogen removal from sugarcane vinasse by heterotrophic green algae Desmodesmus sp. *Desalination and Water Treatment, 57*(20), 9465–9473.

Mata, T. M., Martins, A. A., & Caetano, N. S. (2010). Microalgae for biodiesel production and other applications: A review. *Renewable and Sustainable Energy Reviews, 14*(1), 217–232.

Mayland, H. F., & McIntosh, T. H. (1966). Availability of biologically fixed atmospheric nitrogen-15 to higher plants. *Nature, 209*, 421–422.

Mcewan, C. E., Gatherer, D., & Mcewan, N. R. (1998). Nitrogen-fixing aerobic bacteria have higher genomic GC content than non-fixing species within the same genus. *Hereditas, 128*(2), 173–178.

Mejean, A., Mann, S., Vassiliadis, G., Lombard, B., Loew, D., & Ploux, O. (2010). In vitro reconstitution of the first steps of anatoxin-a biosynthesis in Oscillatoria PCC 6506: From free L-proline to acyl carrier protein bound dehydroproline. *Biochemistry*, *49*(1), 103–113.

Mendes, R. L., Nobre, B. P., Cardoso, M. T., Pereira, A. P., & Palavra, A. F. (2003). Supercritical carbon dioxide extraction of compounds with pharmaceutical importance from microalgae. *Inorganica Chimica Acta*, *356*, 328–334.

Mian, M. H. (2002). Azobiofer: A technology of production and use of Azolla as biofertiliser for irrigated rice and fish cultivation. In: Kennedy, I. R., & Choudhury, A. T. M. A. (eds.), *Biofertilisers in Action: Rural Industries Research and Development Corporation* (pp. 45–54). Canberra: RIRDC.

Michelon, W., Da Silva, M. L. B., Mezzari, M. P., Pirolli, M., Prandini, J. M., & Soares, H. M. (2016). Effects of nitrogen and phosphorus on biochemical composition of microalgae polyculture harvested from phycoremediation of piggery wastewater digestate. *Applied Biochemistry and Biotechnology*, *178*, 1407–1419.

Michelon, W., Matthiensen, A., Viancelli, A., Fongaro, G., Gressler, V., & Soares, H. M. (2022). Removal of veterinary antibiotics in swine wastewater using microalgae-based process. *Environmental Research*, *207*, 112192.

Mobin, S., & Alam, F. (2017). Some promising microalgal species for commercial applications: A review. *Energy Procedia*, *110*, 510–517.

Montalvo, G. E. B., Thomaz-Soccol, V., Vandenberghe, L. P., Carvalho, J. C., Faulds, C. B., Bertrand, E., ... & Soccol, C. R. (2019). Arthrospira maxima OF15 biomass cultivation at laboratory and pilot scale from sugarcane vinasse for potential biological new peptides production. *Bioresource technology*, *273*, 103–113.

Morand, P., & Briand, X. (1999). Anaerobic digestion of Ulva sp. 2. Study of Ulva degradation and methanisation of liquefaction juices. *Journal of applied phycology*, *11*, 164–177.

Morocho-Jácome, A. L., Ruscinc, N., Martinez, R. M., de Carvalho, J. C. M., Santos de Almeida, T., Rosado, C., ... Baby, A. R. (2020). (Bio) Technological aspects of microalgae pigments for cosmetics. *Applied Microbiology and Biotechnology*, *104*, 9513–9522.

Mousavi, S. A., & Khodadoost, F. (2019). Effects of detergents on natural ecosystems and wastewater treatment processes: A review. *Environmental Science and Pollution Research*, *26*, 26439–26448.

Msagati, T. A., Siame, B. A., & Shushu, D. D. (2006). Evaluation of methods for the isolation, detection and quantification of cyanobacterial hepatotoxins. *Aquatic Toxicology*, *78*(4), 382–397.

Mu, N., Mehar, J. G., Mudliar, S. N., & Shekh, A. Y. (2019). Recent advances in microalgal bioactives for food, feed, and healthcare products: Commercial potential, market space, and sustainability. *Comprehensive Reviews in Food Science and Food Safety*, *18*(6), 1882–1897.

Murad, M. E., & Al-Dawody, M. F. (2020). Biodiesel production from spirulina microalgae and its impact on diesel engine characteristics: Review. *Al-Qadisiyah Journal for Engineering Sciences*, *13*, 158–166.

Muslu, Y. (2001). Göl ve Haznelerde Su Kalitesi Yönetimi ve Alg Kontrolü. İSKİ, İstanbul.

Nair, A., & Chakraborty, S. (2020). Synergistic effects between autotrophy and heterotrophy in optimization of mixotrophic cultivation of *Chlorella sorokiniana* in bubble-column photobioreactors. *Algal Research*, *46*, 101799.

Naranjo-Ortiz, M. A., & Gabaldón, T. (2019). Fungal evolution: Major ecological adaptations and evolutionary transitions. *Biological Reviews*, *94*(4), 1443–1476.

Nayak, S., Prasanna, R., Prasanna, B. M., & Sahoo, D. B. (2007). Analysing diversity among Indian isolates of Anabaena (Nostocales, Cyanophyta) using morphological, physiological and biochemical characters. *World Journal of Microbiology and Biotechnology*, *23*, 1575–1584.

Ni, H., Chen, Q. H., He, G. Q., Wu, G. B., & Yang, Y. F. (2008). Optimization of acidic extraction of astaxanthin from Phaffia rhodozyma. *Journal of Zhejiang University Science B*, *9*(1), 51–59.

Nisha, R., Kaushik, A., & Kaushik, C. P. (2007). Effect of indigenous cyanobacterial application on structural stability and productivity of an organically poor semi-arid soil. *Geoderma*, *138*(1–2), 49–56.

Nowicka-Krawczyk, P., Komar, M., & Gutarowska, B. (2022). Towards understanding the link between the deterioration of building materials and the nature of aerophytic green algae. *Science of the Total Environment*, *802*, 149856.

Nowruzi, B., & Porzani, S. J. (2021). Toxic compounds produced by cyanobacteria belonging to several species of the order nostocales: A review. *Journal of Applied Toxicology*, *41*(4), 510–548.

Ogbonna, J. C., Yada, H., Masui, H., & Tanaka, H. (1996). A novel internally illuminated stirred tank photobioreactor for large-scale cultivation of photosynthetic cells. *Journal of Fermentation and Bioengineering*, *82*(1), 61–67.

Oktor, K. (2018). Mikroalglerin Çevre Teknolojilerindeki Yeri. *Çevre Bilim Ve Teknoloji*, pp. 337–338.

Olguín, E. J., Dorantes, E., Castillo, O. S., & Hernández-Landa, V. J. (2015). Anaerobic digestates from vinasse promote growth and lipid enrichment in Neochloris oleoabundans cultures. *Journal of applied phycology*, *27*, 1813–1822.

Oncel, S. S. (2013). Microalgae for a macroenergy world. *Renewable and Sustainable Energy Reviews*, *26*, 241–264.

Órpez, R., Martínez, M. E., Hodaifa, G., El Yousfi, F., Jbari, N., & Sánchez, S. (2009). Growth of the microalga *Botryococcus braunii* in secondarily treated sewage. *Desalination*, *246*(1–3), 625–630.

Oswald, W. J., & Golueke, C. G. (1960). Biological transformation of solar energy. *Advances in Applied Microbiology*, *2*, 223–262.

Oswald, W. J., Gotaas, H. B., Golueke, C. G., Kellen, W. R., Gloyna, E. F., & Hermann, E. R. (1957). Algae in waste treatment [with discussion]. *Sewage and Industrial Wastes*, *29*(4), 437–457.

Oswald, W.J., (1988). The role of microalgae in liquid waste treatment and reclamation. In: Lemhi, C.A., Waaland. J.R. (Eds.). Algae and Human Affairs. Cambridge University Press, Cambridge, pp. 255–281

Oswald, W. J. 1988a. Large-scale algal culture systems (engineering aspects). In Borowitzka, M. A. & Borowitzka, L. J. [Eds.] Micro-algal Biotechnology. Cambridge University Press, Cambridge, pp. 357–94.

Otsuka, K., & Yoshino, A. (2004, November). A fundamental study on anaerobic digestion of sea lettuce. In *Oceans' 04 MTS/IEEE Techno-Ocean'04 (IEEE Cat. No. 04CH37600)* (Vol. 3, pp. 1770–1773). IEEE.

Packer, M. (2009). Algal capture of carbon dioxide; biomass generation as a tool for greenhouse gas mitigation with reference to New Zealand energy strategy and policy. *Energy Policy*, *37*(9), 3428–3437.

Paerl, H. W., Pinckney, J. L., & Kucera, S. A. (1995). Clarification of the structural and functional roles of heterocysts and anoxic microzones in the control of pelagic nitrogen fixation. *Limnology and Oceanography*, *40*(3), 634–638.

Pancha, I., Chokshi, K., George, B., Ghosh, T., Paliwal, C., Maurya, R., & Mishra, S. (2014). Nitrogen stress triggered biochemical and morphological changes in the microalgae Scenedesmus sp. CCNM 1077. *Bioresource Technology*, *156*, 146–154.

Pandey, S. (2006). Water pollution and health. *Kathmandu University Medical Journal (KUMJ)*, *4*(1), 128–134.

Park, K. H., & Lee, C. G. (2001). Effectiveness of flashing light for increasing photosynthetic efficiency of microalgal cultures over a critical cell density. *Biotechnology and Bioprocess Engineering*, *6*, 189–193.

Parmar, A., Singh, N. K., Pandey, A., Gnansounou, E., & Madamwar, D. (2011). Cyanobacteria and microalgae: A positive prospect for biofuels. *Bioresource Technology*, *102*(22), 10163–10172.

Pasquet, V., Chérouvrier, J. R., Farhat, F., Thiéry, V., Piot, J. M., Bérard, J. B., ... & Picot, L. (2011). Study on the microalgal pigments extraction process: Performance of microwave assisted extraction. *Process Biochemistry*, *46*(1), 59–67.

Pasquet, V., Morisset, P., Ihammouine, S., Chepied, A., Aumailley, L., Berard, J. B., ... & Picot, L. (2011). Antiproliferative activity of violaxanthin isolated from bioguided fractionation of Dunaliella tertiolecta extracts. *Marine drugs*, 9(5), 819–831.

Patil V, Källqvist T, Olsen E, Vogt G, Gislerød HR (2007) Fatty acid composition of 12 microalgae for possible use in aquaculture feed. Aquacul Int 15:1–9.

Patil, G., Chethana, S., Sridevi, A. S., & Raghavarao, K. S. M. S. (2006). Method to obtain C-phycocyanin of high purity. *Journal of chromatography A*, *1127*(1-2), 76–81.

Patil, G., & Raghavarao, K. S. M. S. (2007). Aqueous two phase extraction for purification of C-phycocyanin. *Biochemical Engineering Journal*, *34*(2), 156–164.

Patil, G., & Raghavarao, K. S. M. S. (2007). Integrated membrane process for the concentration of anthocyanin. *Journal of food engineering*, *78*(4), 1233–1239.

Patil, P. D., Gude, V. G., Mannarswamy, A., Cooke, P., Nirmalakhandan, N., Lammers, P., & Deng, S. (2012). Comparison of direct transesterification of algal biomass under supercritical methanol and microwave irradiation conditions. *Fuel*, *97*, 822–831.

Pei, Y., Xu, R., Hilt, S., & Chang, X. (2020). Effects of cyanobacterial secondary metabolites on phytoplankton community succession. In: Mérillon, J.-M., & Ramawat, K. G. (eds.), Co-Evolution of Secondary Metabolites (pp. 323–344). Cham: Springer.

Perazzoli, S., Bruchez, B. M., Michelon, W., Steinmetz, R. L., Mezzari, M. P., Nunes, E. O., & da Silva, M. L. (2016). Optimizing biomethane production from anaerobic degradation of *Scenedesmus* spp. biomass harvested from algae-based swine digestate treatment. *International Biodeterioration & Biodegradation*, *109*, 23–28.

Pereira, I., Ortega, R., Barrientos, L., Moya, M., Reyes, G., & Kramm, V. (2009). Development of a biofertilizer based on filamentous nitrogen-fixing cyanobacteria for rice crops in Chile. *Journal of Applied Phycology*, *21*, 135–144.

Piloto-Rodríguez, R., Sánchez-Borroto, Y., Melo-Espinosa, E. A., & Verhelst, S. (2017). Assessment of diesel engine performance when fueled with biodiesel from algae and microalgae: An overview. *Renewable and Sustainable Energy Reviews*, *69*, 833–842.

Pittman, J. K., Dean, A. P., & Osundeko, O. (2011). The potential of sustainable algal biofuel production using wastewater resources. *Bioresource Technology*, *102*(1), 17–25.

Pomoni-Papaioannou, F., & Kostopoulou, V. (2008). Microfacies and cycle stacking pattern in Liassic peritidal carbonate platform strata, Gavrovo-Tripolitza platform, Peloponnesus, Greece. *Facies*, *54*, 417–431.

Pragya, N., Pandey, K. K., & Sahoo, P. K. (2013). A review on harvesting, oil extraction and biofuels production technologies from microalgae. *Renewable and Sustainable Energy Reviews*, 24, 159–171.

Prasad, R., Gupta, S. K., Shabnam, N., Oliveira, C. Y. B., Nema, A. K., Ansari, F. A., & Bux, F. (2021). Role of microalgae in global CO2 sequestration: Physiological mechanism, recent development, challenges, and future prospective. *Sustainability*, *13*(23), 13061.

Prasanna, R., Chaudhary, V., Gupta, V., Babu, S., Kumar, A., Singh, R., ... Nain, L. (2013). Cyanobacteria mediated plant growth promotion and bioprotection against Fusarium wilt in tomato. *European Journal of Plant Pathology*, *136*, 337–353.

Priyadarshani, I., & Rath, B. (2012). Commercial and industrial applications of micro algae: A review. *Journal of Algal Biomass Utilization*, *3*(4), 89–100.

Priyadharshini, S. D., Babu, P. S., Manikandan, S., Subbaiya, R., Govarthanan, M., & Karmegam, N. (2021). Phycoremediation of wastewater for pollutant removal: A green approach to environmental protection and long-term remediation. *Environmental Pollution*, 290, 117989.

Qin, B., Zhou, J., Elser, J. J., Gardner, W. S., Deng, J., & Brookes, J. D. (2020). Water depth underpins the relative roles and fates of nitrogen and phosphorus in lakes. *Environmental Science & Technology*, 54(6), 3191–3198.

Qin, L., Shu, Q., Wang, Z., Shang, C., Zhu, S., Xu, J., ... & Yuan, Z. (2014). Cultivation of Chlorella vulgaris in dairy wastewater pretreated by UV irradiation and sodium hypochlorite. *Applied biochemistry and biotechnology*, 172, 1121–1130.

Qin, L., Yu, Q., Ai, W., Tang, Y., Ren, J., & Guo, S. (2014). Response of cyanobacteria to low atmospheric pressure. *Life Sciences in Space Research*, 3, 55–62.

Qiu, R., Gao, S., Lopez, P. A., & Ogden, K. L. (2017). Effects of pH on cell growth, lipid production and CO2 addition of microalgae Chlorella sorokiniana. *Algal Research*, 28, 192–199.

Rani-Borges, B., Moschini-Carlos, V., & Pompêo, M. (2021). Microplastics and freshwater microalgae: What do we know so far? *Aquatic Ecology*, 55, 363–377.

Raposo, A. M. B. (2017). Exploitation of bioactive molecules in the processing of microalgal biomass into biodiesel (Doctoral dissertation, Universidade do Algarve (Portugal)).

Rashid, N., Rehman, M. S. U., Sadiq, M., Mahmood, T., & Han, J. I. (2014). Current status, issues and developments in microalgae derived biodiesel production. *Renewable and Sustainable Energy Reviews*, 40, 760–778.

Ren, H. Y., Kong, F., Ma, J., Zhao, L., Xie, G. J., Xing, D., ... Ren, N. Q. (2018). Continuous energy recovery and nutrients removal from molasses wastewater by synergistic system of dark fermentation and algal culture under various fermentation types. *Bioresource Technology*, 252, 110–117.

Rodríguez, H., Fraga, R., Gonzalez, T., & Bashan, Y. (2006). Genetics of phosphate solubilization and its potential applications for improving plant growth-promoting bacteria. *Plant and Soil*, 287, 15–21.

Roger, P. A., & Kulasooriya, S. A. (1980). *Blue-Green Algae and Rice*. Los Baños, Philippines: International Rice Research Institute.

Roger, P. A., & Reynaud, P. A. (1982). Free-living blue-green algae in tropical soils. In: Dommergues, R., & Diem, G. H. (eds.), *Microbiology of Tropical Soils and Plant Productivity* (pp. 147–168). Berlin/Heidelberg, Germany: Springer Science & Business Media.

Rogers, S. L., & Burns, R. G. (1994). Changes in aggregate stability, nutrient status, indigenous microbial populations, and seedling emergence, following inoculation of soil with Nostoc muscorum. *Biology and Fertility of Soils*, 18, 209–215.

Roy-Lachapelle, A., Solliec, M., Sauvé, S., & Gagnon, C. (2021). Evaluation of ELISA-based method for total anabaenopeptins determination and comparative analysis with on-line SPE-UHPLC-HRMS in freshwater cyanobacterial blooms. *Talanta*, 223, 121802.

Ruen-ngam D., Shotipruk A., Pavasant P., (2011). Comparison of Extraction Methods for Recovery of Astaxanthin from Haematococcus pluvialis. Sep. Sci. Technol. 46, 64–70.

Ruen-ngam, D., Shotipruk, A., Pavasant, P., Machmudah, S., & Goto, M. (2012). Selective extraction of lutein from alcohol treated Chlorella vulgaris by supercritical CO_2. *Chemical engineering & technology*, 35(2), 255–260.

Ruiz-Martinez, A., Serralta, J., Pachés, M., Seco, A., & Ferrer, J. (2014). Mixed microalgae culture for ammonium removal in the absence of phosphorus: Effect of phosphorus supplementation and process modeling. *Process Biochemistry*, 49(12), 2249–2257.

Saadatnia, H., & Riahi, H. (2009). Cyanobacteria from paddy fields in Iran as a biofertilizer in rice plants. *Plant, Soil and Environment*, *55*(5), 207–212.

Sahu, D., Priyadarshani, I., & Rath, B. (2012). Cyanobacteria-as potential biofertilizer. *CIBTech Journal of Microbiology*, *1*(2–3), 20–26.

Saker, M. L., Jungblut, A. D., Neilan, B. A., Rawn, D. F., & Vasconcelos, V. M. (2005). Detection of microcystin synthetase genes in health food supplements containing the freshwater cyanobacterium Aphanizomenon flos-aquae. *Toxicon*, *46*(5), 555–562.

Salla, A. C. V., Margarites, A. C., Seibel, F. I., Holz, L. C., Brião, V. B., Bertolin, T. E., ... Costa, J. A. V. (2016). Increase in the carbohydrate content of the microalgae Spirulina in culture by nutrient starvation and the addition of residues of whey protein concentrate. *Bioresource Technology*, *209*, 133–141.

Sarada, R., Vidhyavathi, R., Usha, D., & Ravishankar, G. A. (2006). An efficient method for extraction of astaxanthin from green alga Haematococcus pluvialis. *Journal of agricultural and food chemistry*, *54*(20), 7585–7588.

Sarada, R. M. G. P., Pillai, M. G., & Ravishankar, G. A. (1999). Phycocyanin from Spirulina sp: influence of processing of biomass on phycocyanin yield, analysis of efficacy of extraction methods and stability studies on phycocyanin. *Process biochemistry*, *34*(8), 795–801.

Saravanan, A., Kumar, P. S., Jeevanantham, S., Karishma, S., Tajsabreen, B., Yaashikaa, P. R., & Reshma, B. (2021). Effective water/wastewater treatment methodologies for toxic pollutants removal: Processes and applications towards sustainable development. *Chemosphere*, *280*, 130595.

Sathinathan, P., Parab, H. M., Yusoff, R., Ibrahim, S., Vello, V., & Ngoh, G. C. (2023). Photobioreactor design and parameters essential for algal cultivation using industrial wastewater: A review. *Renewable and Sustainable Energy Reviews*, *173*, 113096.

Sayaner, O. (2013). Mikroalg Yetiştiriciliği Yoluyla Baca Gazı Kaynaklı Karbondioksit Azaltımı (Doctoral dissertation, Enerji Enstitüsü).

Schagerl, M., Ludwig, I., El-Sheekh, M., Kornaros, M., & Ali, S. S. (2022). The efficiency of microalgae-based remediation as a green process for industrial wastewater treatment. *Algal Research*, *66*, 102775.

Schenk, P. M., Thomas-Hall, S. R., Stephens, E., Marx, U. C., Mussgnug, J. H., Posten, C., & Hankamer, B. (2008). Second generation biofuels: high-efficiency microalgae for biodiesel production. *Bioenergy research*, *1*, 20–43.

Schleyer, G., & Vardi, A. (2020). Algal blooms. *Current Biology*, *30*(19), R1116–R1118.

Schmidt, R. A., Wiebe, M. G., & Eriksen, N. T. (2005). Heterotrophic high cell-density fed-batch cultures of the phycocyanin-producing red alga Galdieria sulphuraria. *Biotechnology and bioengineering*, *90*(1), 77–84.

Sen, B., Alp, M. T., Sonmez, F., Kocer, M. A. T., & Canpolat, O. (2013). Relationship of algae to water pollution and waste water treatment. *Water treatment*, *14*, 335–354.

Shaaban, M. M. (2001a). Green microalgae water extract as foliar feeding to wheat plants. *Pakistan Journal of Biological Sciences* 4(6): 628–632.

Shaaban, M. M. (2001b). Nutritional status and growth of maize plants as affected by green microalgae as soil additives.

Shaleh, S. R. M. (2004). Optimum growth parameters for both indoor and outdoor propagation of microalgae, Chlorella vulgaris and Isochrysis galbana (Doctoral dissertation, Ph. D., Universiti Putra Malaysia, Department of Science, Serdang).

Sharma, Y. C., Singh, B., & Korstad, J. (2011). A critical review on recent methods used for economically viable and eco-friendly development of microalgae as a potential feed-stock for synthesis of biodiesel. *Green Chemistry*, *13*(11), 2993–3006.

Shin, D. Y., Cho, H. U., Utomo, J. C., Choi, Y. N., Xu, X., & Park, J. M. (2015). Biodiesel production from Scenedesmus bijuga grown in anaerobically digested food wastewater effluent. *Bioresource Technology*, *184*, 215–221.

Shuba, E. S., & Kifle, D. (2018). Microalgae to biofuels: 'Promising' alternative and renewable energy: Review. *Renewable and Sustainable Energy Reviews, 81*, 743–755.

Silva, C. E. D. F., & Bertucco, A. (2019). Bioethanol from microalgal biomass: A promising approach in biorefinery. *Brazilian Archives of Biology and Technology, 62*, e19160816.

Singh, A. K., Singh, P. P., Tripathi, V., Verma, H., Singh, S. K., Srivastava, A. K., & Kumar, A. (2018). Distribution of cyanobacteria and their interactions with pesticides in paddy field: A comprehensive review. *Journal of Environmental Management, 224*, 361–375.

Singh, A. L., Singh, P. K., & Lata, P. (1988). Effects of different levels of chemical nitrogen (urea) on azolla and blue-green algae intercropping with rice. *Fertilizer Research, 17*, 47–59.

Singh, A., Pal, D. B., Kumar, S., Srivastva, N., Syed, A., Elgorban, A. M., ... Gupta, V. K. (2021). Studies on zero-cost algae based phytoremediation of dye and heavy metal from simulated wastewater. *Bioresource Technology, 342*, 125971.

Singh, J., & Saxena, R. C. (2015). An introduction to microalgae: Diversity and significance. In: Kim, S. W. (ed.), *Handbook of Marine Microalgae* (pp. 11–24). Cambridge, MA: Academic Press.

Singh, P., Guldhe, A., Kumari, S., Rawat, I., & Bux, F. (2015). Investigation of combined effect of nitrogen, phosphorus and iron on lipid productivity of microalgae Ankistrodesmus falcatus KJ671624 using response surface methodology. *Biochemical Engineering Journal, 94*, 22–29.

Singh, R. N. (1961). *Role of Blue-Green Algae in Nitrogen Economy of Indian Agriculture.* New Delhi: Indian Council of Agricultural Research.

Soni, B., Kalavadia, B., Trivedi, U., & Madamwar, D. (2006). Extraction, purification and characterization of phycocyanin from Oscillatoria quadripunctulata—Isolated from the rocky shores of Bet-Dwarka, Gujarat, India. *Process Biochemistry, 41*(9), 2017–2023.

Sood, A., Renuka, N., Prasanna, R., & Ahluwalia, A. S. (2015). Cyanobacteria as potential options for wastewater treatment. *Phytoremediation: Management of Environmental Contaminants, 2*, 83–93.

Sorokin, C., & Krauss, R. W. (1958). The effects of light intensity on the growth rates of green algae. *Plant Physiology, 33*(2), 109–113.

Sosa-Hernández, J. E., Romero-Castillo, K. D., Parra-Arroyo, L., Aguilar-Aguila-Isaías, M. A., García-Reyes, I. E., Ahmed, I., ... & Iqbal, H. M. (2019). Mexican microalgae biodiversity and state-of-the-art extraction strategies to meet sustainable circular economy challenges: High-value compounds and their applied perspectives. *Marine Drugs, 17*(3), 174.

Souza, M. D. R. A. Z. D., & Costa, J. A. V. (2007). Mixotrophic cultivation of microalga Spirulina platensis using molasses as organic substrate.

Spolaore, P., Joannis-Cassan, C., Duran, E., & Isambert, A. (2006). Commercial applications of microalgae. *Journal of Bioscience and Bioengineering, 101*(2), 87–96.

Srivastava, S., Bhargava, A., Srivastava, S., & Bhargava, A. (2022). Biological synthesis of nanoparticles: Algae. In: Srivastava, S., & Bhargava, A. (eds.), *Green Nanoparticles: The Future of Nanobiotechnology* (pp. 139–171). New York: Springer Nature.

Strunecký, O., Kopejtka, K., Goecke, F., Tomasch, J., Lukavský, J., Neori, A., ... Koblížek, M. (2019). High diversity of thermophilic cyanobacteria in Rupite hot spring identified by microscopy, cultivation, single-cell PCR and amplicon sequencing. *Extremophiles, 23*, 35–48.

Swingley, W. D., Blankenship, R. E., & Raymond, J. (2008). Insights into cyanobacterial evolution from comparative genomics. In: Herrero, A. (ed.), *The Cyanobacteria: Molecular Biology, Genomics and Evolution* (pp. 21–43). Norfolk, United Kingdom: Caister Academic Press.

Taher, H., Al-Zuhair, S., Al-Marzouqi, A. H., Haik, Y., & Farid, M. M. (2011). A review of enzymatic transesterification of microalgal oil-based biodiesel using supercritical technology. *Enzyme Research*, 2011, 25.

Taştan, B. E. (2016). Termik Santral Kömür Emisyonlarına Dirençli Leptolyngbya sp. Biyokütle Optimizasyonu ve Kömür Emisyonlarından CO2 Fiksasyonu. *KSÜ Doğa Bilimleri Dergisi, 19*(4), 362–372.

Teo, C. L., Atta, M., Bukhari, A., Taisir, M., Yusuf, A. M., & Idris, A. (2014). Enhancing growth and lipid production of marine microalgae for biodiesel production via the use of different LED wavelengths. *Bioresource Technology, 162*, 38–44.

Tomitani, A., Knoll, A. H., Cavanaugh, C. M., & Ohno, T. (2006). The evolutionary diversification of cyanobacteria: Molecular-phylogenetic and paleontological perspectives. *Proceedings of the National Academy of Sciences, 103*(14), 5442–5447.

Tomitani, A., Okada, K., Miyashita, H., Matthijs, H. C., Ohno, T., & Tanaka, A. (1999). Chlorophyll b and phycobilins in the common ancestor of cyanobacteria and chloroplasts. *Nature, 400*(6740), 159–162.

Torzillo, G., & Vonshak, A. (2004). Applied course on production and monitoring of microalgal growth. In: Richmond, A. (ed.,) *Handbook of Microalgal Mass Cultures* (vol. 1, pp. 57–82). Oxford: Blackwell Science.

Torzillo, G., Scoma, A., Faraloni, C., Ena, A., & Johanningmeier, U. (2009). Increased hydrogen photoproduction by means of a sulfur-deprived Chlamydomonas reinhardtii D1 protein mutant. *International Journal of Hydrogen Energy, 34*(10), 4529–4536.

Touliabah, H. E. S., El-Sheekh, M. M., Ismail, M. M., & El-Kassas, H. (2022). A review of microalgae-and cyanobacteria-based biodegradation of organic pollutants. *Molecules, 27*(3), 1141.

Tsai, D. D. W., Ramaraj, R., & Chen, P. H. (2012). Growth condition study of algae function in ecosystem for CO2 bio-fixation. *Journal of Photochemistry and Photobiology B: Biology, 107*, 27–34.

Tsolcha, O. N., Tekerlekopoulou, A. G., Akratos, C. S., Bellou, S., Aggelis, G., Katsiapi, M., … Vayenas, D. V. (2016). Treatment of second cheese whey effluents using a Choricystis-based system with simultaneous lipid production. *Journal of Chemical Technology & Biotechnology, 91*(8), 2349–2359.

Uduman, N., Qi, Y., Danquah, M. K., Forde, G. M., & Hoadley, A. (2010). Dewatering of microalgal cultures: A major bottleneck to algae-based fuels. *Journal of Renewable and Sustainable Energy, 2*(1), 012701.

Ullmann, J., & Grimm, D. (2021). Algae and their potential for a future bioeconomy, landless food production, and the socio-economic impact of an algae industry. *Organic Agriculture, 11*(2), 261–267.

van der Hulst C. (2012). Microalgae Cultivation Systems: Analysis of Microalgae Cultivation Systems and LCA for Biodiesel Production. (MSc Thesis, Utrecht University, Heidelberglaan, Utrecht, Netherlands).

Varga, L., Szigeti, J., Kovács, R., Földes, T., & Buti, S. (2002). Influence of a Spirulina platensis biomass on the microflora of fermented ABT milks during storage (R1). *Journal of Dairy Science, 85*(5), 1031–1038.

Vargas, M. A., Rodriguez, H., Moreno, J., Olivares, H., Campo, J. D., Rivas, J., & Guerrero, M. G. (1998). Biochemical composition and fatty acid content of filamentous nitrogen-fixing cyanobacteria. *Journal of Phycology, 34*(5), 812–817.

Varshney, P., Mikulic, P., Vonshak, A., Beardall, J., & Wangikar, P. P. (2015). Extremophilic micro-algae and their potential contribution in biotechnology. *Bioresource Technology, 184*, 363–372.

Vergara-Fernández, A., Vargas, G., Alarcón, N., & Velasco, A. (2008). Evaluation of marine algae as a source of biogas in a two-stage anaerobic reactor system. *Biomass and Bioenergy, 32*(4), 338–344.

Vickers, N. J. (2017). Animal communication: When I'm calling you, will you answer too? *Current Biology*, *27*(14), R713–R715.

Viskari, P. J., & Colyer, C. L. (2003). Rapid extraction of phycobiliproteins from cultured cyanobacteria samples. *Analytical biochemistry*, *319*(2), 263–271.

Voltolina, D., Cordero, B., Nieves, M., & Soto, L. P. (1999). Growth of Scenedesmus sp. in artificial wastewater. *Bioresource Technology*, *68*(3), 265–268.

Vonshak, A., & Richmond, A. (1988). Mass production of the blue-green alga Spirulina: An overview. *Biomass*, *15*(4), 233–247.

Wahidin, S., Idris, A., & Shaleh, S. R. M. (2013). The influence of light intensity and photo-period on the growth and lipid content of microalgae Nannochloropsis sp. *Bioresource Technology*, *129*, 7–11.

Wang, B., Lan, C. Q., & Horsman, M. (2012). Closed photobioreactors for production of microalgal biomasses. *Biotechnology Advances*, *30*(4), 904–912.

Welch, K. (2013). Growth optimization of microalgae on treated wastewater effluent (Doctoral dissertation, The Florida State University).

Whitton, B. A. (1992). Diversity, ecology, and taxonomy of the cyanobacteria. In: Mann, N. H., & Carr, N. G. (eds.), *Photosynthetic Prokaryotes* (pp. 1–51). New York: Springer.

Whitton, B. A., & Potts, M. (2012). Introduction to the cyanobacteria. In *Ecology of cyanobacteria II: their diversity in space and time* (pp. 1–13). Dordrecht: Springer Netherlands.

Widjaja, A., Chien, C. C., & Ju, Y. H. (2009). Study of increasing lipid production from fresh water microalgae *Chlorella vulgaris*. *Journal of the Taiwan Institute of Chemical Engineers*, *40*(1), 13–20.

Woertz, I., Feffer, A., Lundquist, T., & Nelson, Y. (2009). Algae grown on dairy and municipal wastewater for simultaneous nutrient removal and lipid production for biofuel feedstock. *Journal of Environmental Engineering*, *135*(11), 1115–1122.

Wollmann, F., Dietze, S., Ackermann, J. U., Bley, T., Walther, T., Steingroewer, J., & Krujatz, F. (2019). Microalgae wastewater treatment: Biological and technological approaches. *Engineering in Life Sciences*, *19*(12), 860–871.

Wu, F. C., Wu, J. Y., Liao, Y. J., Wang, M. Y., & Shih, L. (2014). Sequential acid and enzymatic hydrolysis in situ and bioethanol production from Gracilaria biomass. *Bioresource Technology*, *156*, 123–131.

Xiao, X., Si, X., Yuan, Z., Xu, X., & Li, G. (2012). Isolation of fucoxanthin from edible brown algae by microwave-assisted extraction coupled with high-speed countercurrent chromatography. *Journal of separation science*, *35*(17), 2313–2317.

Xiao, Y., Gan, N., Liu, J., Zheng, L., & Song, L. (2012). Heterogeneity of buoyancy in response to light between two buoyant types of cyanobacterium *Microcystis*. *Hydrobiologia*, *679*, 297–311.

Xu, S., Jiang, Y., Liu, Y., & Zhang, J. (2021). Antibiotic-accelerated cyanobacterial growth and aquatic community succession towards the formation of cyanobacterial bloom in eutrophic lake water. *Environmental Pollution*, *290*, 118057.

Ye Lee, J., Li, P., Lee, J., Ryu, H. J., & Oh, K. K. (2013). Ethanol production from Saccharina japonica using an optimized extremely low acid pretreatment followed by simultaneous saccharification and fermentation. *Bioresource Technology*, *127*, 119–125.

Yeesang, C., & Cheirsilp, B. (2014). Low-cost production of green microalga Botryococcus braunii biomass with high lipid content through mixotrophic and photoautotrophic cultivation. *Applied biochemistry and biotechnology*, *174*, 116–129.

Yen, H. W., Hu, I. C., Chen, C. Y., Ho, S. H., Lee, D. J., & Chang, J. S. (2013). Microalgae-based biorefinery-from biofuels to natural products. *Bioresource Technology*, *135*, 166–174.

Yıldız Töre, G. (2001). Spirulina sp.(cyanophyceae) kültürü üzerine araştırmalar (Master's thesis, Ege Üniversitesi).

Yousuf, A. (2020). Fundamentals of microalgae cultivation. In: Yousuf, A. (ed.), *Microalgae Cultivation for Biofuels Production* (pp. 1–9). Cambridge, MA: Academic Press.

Yuliasni, R., Kurniawan, S. B., Marlena, B., Hidayat, M. R., Kadier, A., Ma, P. C., & Imron, M. F. (2023). Recent progress of phytoremediation-based technologies for industrial wastewater treatment. *Journal of Ecological Engineering, 24*(2), 208–220.

Zaady, E., Groffman, P., & Shachak, M. (1998). Nitrogen fixation in macro-and microphytic patches in the Negev desert. *Soil Biology and Biochemistry, 30*(4), 449–454.

Zeng, X., Danquah, M. K., Chen, X. D., & Lu, Y. (2011). Microalgae bioengineering: From CO2 fixation to biofuel production. *Renewable and Sustainable Energy Reviews, 15*(6), 3252–3260.

Zhang, L., & Melis, A. (2002). Probing green algal hydrogen production. *Philosophical Transactions of the Royal Society of London. Series B: Biological Sciences, 357*(1426), 1499–1509.

Zhang, T., Gong, H., Wen, X., & Lu, C. (2010). Salt stress induces a decrease in excitation energy transfer from phycobilisomes to photosystem II but an increase to photosystem I in the cyanobacterium Spirulina platensis. *Journal of Plant Physiology, 167*(12), 951–958.

Zheng, H., Wang, Y., Li, S., Nagarajan, D., Varjani, S., Lee, D. J., & Chang, J. S. (2022). Recent advances in lutein production from microalgae. *Renewable and Sustainable Energy Reviews, 153*, 111795.

Zheng, Y., Chiang, T. Y., Huang, C. L., Feng, X. Y., Yrjälä, K., & Gong, X. (2021). The predominance of proteobacteria and cyanobacteria in the cycas dolichophylla coralloid roots revealed by 16S rRNA metabarcoding. *Microbiology, 90*, 805–815.

Zhou, W., Li, Y., Min, M., Hu, B., Zhang, H., Ma, X., … Ruan, R. (2012). Growing wastewater-born microalga Auxenochlorella prototecoides UMN280 on concentrated municipal wastewater for simultaneous nutrient removal and energy feedstock production. *Applied Energy, 98*, 433–440.

Zhu, L. (2015). Biorefinery as a promising approach to promote microalgae industry: An innovative framework. *Renewable and Sustainable Energy Reviews, 41*, 1376–1384.

Zhu, L., Nugroho, Y. K., Shakeel, S. R., Li, Z., Martinkauppi, B., & Hiltunen, E. (2017). Using microalgae to produce liquid transportation biodiesel: What is next? *Renewable and Sustainable Energy Reviews, 78*, 391–400.

Zi, J., Pan, X., MacIsaac, H. J., Yang, J., Xu, R., Chen, S., & Chang, X. (2018). Cyanobacteria blooms induce embryonic heart failure in an endangered fish species. *Aquatic Toxicology, 194*, 78–85.

Index

Note: **Bold** page numbers refer to tables and *italic* page numbers refer to figures.

Printed in the United States
by Baker & Taylor Publisher Services